Genomes

COLD SPRING HARBOR MONOGRAPH SERIES

The Lactose Operon
The Bacteriophage Lambda
The Molecular Biology of Tumour Viruses
Ribosomes
RNA Phages
RNA Polymerase
The Operon
The Single-Stranded DNA Phages
Transfer RNA:
    Structure, Properties, and Recognition
    Biological Aspects
Molecular Biology of Tumor Viruses, Second Edition:
    DNA Tumor Viruses
    RNA Tumor Viruses
The Molecular Biology of the Yeast *Saccharomyces*:
    Life Cycle and Inheritance
    Metabolism and Gene Expression
Mitochondrial Genes
Lambda II
Nucleases
Gene Function in Prokaryotes
Microbial Development
The Nematode *Caenorhabditis elegans*
Oncogenes and the Molecular Origins of Cancer
Stress Proteins in Biology and Medicine
DNA Topology and Its Biological Effects
The Molecular and Cellular Biology of the Yeast *Saccharomyces*:
    Genome Dynamics, Protein Synthesis, and Energetics
    Gene Expression
    Cell Cycle and Cell Biology
Transcriptional Regulation
Reverse Transcriptase
The RNA World
Nucleases, Second Edition
The Biology of Heat Shock Proteins and Molecular Chaperones
Arabidopsis
Cellular Receptors for Animal Viruses
Telomeres
Translational Control
DNA Replication in Eukaryotic Cells
Epigenetic Mechanisms of Gene Regulation
C. elegans II
Oxidative Stress and the Molecular Biology of Antioxidant Defenses
RNA Structure and Function
The Development of Human Gene Therapy
The RNA World, Second Edition
Prion Biology and Diseases
Translational Control of Gene Expression
Stem Cell Biology
Prion Biology and Diseases, Second Edition
Cell Growth: Control of Cell Size
The RNA World, Third Edition
The Dog and Its Genome
Telomeres, Second Edition
Genomes

Genomes

Perspectives from the 10th Anniversary Issue of
Genome Research

EDITED BY

Hillary E. Sussman
Maria A. Smit
Cold Spring Harbor Laboratory

COLD SPRING HARBOR LABORATORY PRESS
Cold Spring Harbor, New York

Genomes

Monograph 46
All rights reserved
©2006 by Cold Spring Harbor Laboratory Press,
 Cold Spring Harbor, New York
Printed in the United States of America

Publisher	John Inglis
Acquisition Editor	Alexander Gann
Development Director	Jan Argentine
Project Coordinator	Inez Sialiano
Production Editor	Kathleen Bubbeo
Compositor	Compset Inc.
Production Manager	Denise Weiss
Desktop Editor	Susan Schaefer
Cover Designer	Denise Weiss

Front Cover Artwork: A conceptual representation of genomes, comprised of spiraling chemical bases of DNA, as a selection of organisms for which genome sequence data are available. Illustration by Bang Wong, ClearScience/www.clearscience.info.

Library of Congress Cataloging-in-Publication Data
Genomes / edited by Hillary E. Sussman and Maria Smit.
   p. cm. -- (Cold Spring Harbor monograph series ; 46)
   Includes bibliographical references and index.
   ISBN 0-87969-806-3 (hardcover : alk. paper) -- ISBN 0-87969-807-1 (pbk. : alk. paper)
   1. Genomics. 2. Genomes.   I. Sussman, Hillary E. II. Smit, Maria.
III. Genome research. IV. Series.
   [DNLM: 1. Genomics. 2. Genome.   QU 58.5 G335 2006]
QH447.G46515 2006
572.8'6--dc22
                                                                                    2006001768

10 9 8 7 6 5 4 3 2 1

All World Wide Web addresses are accurate to the best of our knowledge at the time of printing.

Authorization to photocopy items for internal or personal use, or the internal or personal use of specific clients, is granted by Cold Spring Harbor Laboratory Press, provided that the appropriate fee is paid directly to the Copyright Clearance Center (CCC). Write or call CCC at 222 Rosewood Drive, Danvers, MA 01923 (508-750-8400) for information about fees and regulations. Prior to photocopying items for educational classroom use, contact CCC at the above address. Additional information on CCC can be obtained at CCC Online at http://www.copyright.com/.

All Cold Spring Harbor Laboratory Press publications may be ordered directly from Cold Spring Harbor Laboratory Press, 500 Sunnyside Blvd., Woodbury, New York 11797-2924. Phone: 1-800-843-4388 in Continental U.S. and Canada. All other locations: (516) 422-4100. FAX: (516) 422-4097. E-mail: cshpress@cshl.edu. For a complete catalog of all Cold Spring Harbor Laboratory Press publications, visit our World Wide Web Site http://www.cshlpress.com/.

Contents

Preface, vii

Foreword by James D. Watson, ix

1 Insights on Biology and Evolution from Microbial Genome Sequencing, *1*
   C.M. Fraser-Liggett

2 Changing Perspectives in Yeast Research Nearly a Decade after the Genome Sequence, *19*
   K. Dolinski and D. Botstein

3 Genomics of the Fungal Kingdom: Insights into Eukaryotic Biology, *41*
   J.E. Galagan, M.R. Henn, L.-J. Ma, C.A. Cuomo, and B. Birren

4 The *Arabidopsis* Genome: A Foundation for Plant Research, *71*
   M. Bevan and S. Walsh

5 Grains of Knowledge: Genomics of Model Cereals, *97*
   A.H. Paterson, M. Freeling, and T. Sasaki

6 Genomics in *Caenorhabditis elegans*: So Many Genes, Such a Little Worm, *117*
   L.W. Hillier, A. Coulson, J.I. Murray, Z. Bao, J.E. Sulston, and R.H. Waterston

7 *Drosophila melanogaster:* A Case Study of a Model Genomic Sequence and Its Consequences, *143*
   M. Ashburner and C.M. Bergman

8 Unraveling Genomic Regulatory Networks in the Simple Chordate, *Ciona intestinalis*, 159
   W. Shi, M. Levine, and B. Davidson

9 Fish Genomics and Biology, 177
   H. Roest Crollius and J. Weissenbach

10 Xenomics, 199
   E. Amaya

11 Chicken Genome: Current Status and Future Opportunities, 221
   D.W. Burt

12 Advances in Livestock Genomics: Opening the Barn Door, 237
   J.E. Womack

13 The Canine Genome, 255
   E.A. Ostrander and R.K. Wayne

14 Impact of Genomics on Research in the Rat, 281
   J. Lazar, C. Moreno, H.J. Jacob, and A.E. Kwitek

15 The Mouse Genome, 313
   J.L. Guénet

16 Genomics of the Future: Identification of Quantitative Trait Loci in the Mouse, 345
   L. Flaherty, B. Herron, and D. Symula

17 Comparing the Human and Chimpanzee Genomes: Searching for Needles in a Haystack, 357
   A. Varki and T.K. Altheide

18 Structure and Function of the Human Genome, 395
   P.F.R. Little

19 Emerging Technologies in DNA Sequencing, 413
   M.L. Metzker

20 Genome Annotation Past, Present, and Future: How to Define an Open Reading Frame at Each Locus, 439
   M.R. Brent

Index, 465

Preface

THE ERA OF GENOME SEQUENCING BEGAN TO UNFOLD about ten years ago with the publication of the complete genome sequence of *Haemophilus influenzae* (1995). Since then, the genomes of hundreds of bacteria and model organisms, such as the yeast *Saccharomyces cerevisiae* (1996), the nematode *Caenorhabditis elegans* (1998), and the fruit fly *Drosophila melanogaster* (2000), as well as the human (2001, 2004), have been completely sequenced and many more species boast a range of resources that have in turn revealed the innate complexity of genomes and engendered even more intense inquiry.

The impact of these genomic resources is neither species—nor discipline—specific. Molecular understanding of model species has proven to be translatable, with insights gained in each holding promise for unraveling perplexities in all. Studies in other species ameliorate the human condition, if not by ultimately enabling better prevention, diagnoses, and treatment of human diseases, then, for example, economically, via the improvement of crops that feed the world, per bioremediation, or through the development of effective insecticides to combat infectious parasites. Furthermore, the usefulness of genomic technology is far-reaching, fueling an increasing number of studies using genomic tools and data in other fields and building bridges, most notably, to evolutionary biology and genetics.

In these pages you will find a collection of articles reprinted from the Tenth Anniversary Issue of the journal *Genome Research*, which report on a selection of model species' genomes—from microbes to human—and which strive collectively to illustrate how genomic resources have made a difference in our understanding of genome biology and evolution. They serve as a summation of data and analyses, reflect the rapid progress in the development of novel technologies that have conceptually advanced our understanding of genome structure and function, and address biological conundrums, such as the genotype–phenotype connection, that motivate current research. They ask, "What biological questions were posed with the advent of large-scale genomic resources, how were they elucidated, and what remains enigmatic?"

These articles also serve as a reminder of the historical intent behind the movements to create such genomic resources and pay homage to the inspired minds fundamental to undertaking such endeavors. Most importantly, they speak to the future of genomic research. There is still promise in the sequence and in other genomic data that is yet to be fulfilled. As we look forward to further insights gleaned from increasingly sophisticated mapping resources, newly sequenced genomes, multispecies genome comparisons, the identification of quantitative trait loci, and the construction of functional and regulatory networks, for example, we acknowledge the achievements along the path that led us here. In addition to these articles on the genomes of model organisms, you will also find updates on emerging technologies in DNA sequencing and in gene prediction, two areas of deep interest and timely relevance.

We thank the authors of these commissioned articles for their vigorous efforts in writing thorough and enjoyable accounts of where we have been, where we are, and where we hope to be. We also offer fond gratitude to the other Editors of *Genome Research*, Aravinda Chakravarti, Evan Eichler, Richard Gibbs, Eric Green, Rick Myers, and Bill Pavan, who conceived of the idea of publishing the celebratory issue from which these articles are reprinted and for their interminable insight, expertise, and sound guidance.

HILLARY E. SUSSMAN
MARIA A. SMIT

Foreword

WHEN THE HUMAN GENOME PROJECT WAS BEING STARTED, I don't recall any serious discussions as to what would happen after we knew the three billion letters of the human genome. Fifteen years, the time interval we thought the project would require, was too far ahead for serious planning. Three years into the project, I unexpectedly bumped into Senator Tom Harkin of Iowa at Reagan National Airport who expressed keen disappointment that we had not yet started sequencing human DNA. Two of his sisters had already died prematurely of breast cancer, and he wanted soon to announce on the Senate floor that the Human Genome Project had helped identify the breast cancer genes at the root of his family's cancer. Happily only three more years passed before the discoveries of the genes *BRCA1* and *BRCA2* opened up the possibility of identifying potential victims of hereditary breast cancer.

By then manual DNA sequencing was in its last throes. Lee Hood's vision of machines replacing mistake-prone technicians was the rush of the moment. Central to all early planning was the belief that before serious sequencing began, physical maps of overlapping 100,000–500,000 base pair cloned DNA fragments must be obtained. Our first sequencing efforts tackled the genomes of already well-known model organisms like *E. coli*, *S. cerevisiae*, *Caenorhabditis*, and *Drosophila*. Not only are these genomes much smaller than that of the human, they also contain much less of the largely functionless selfish DNA generated by gone amok transposable elements inserting themselves throughout their respective genomes.

From the start, a key imperative was to reduce the cost of sequencing to much lower than the 1998 figure of 10 dollars per base pair. Our projected three billion dollars final cost anticipated at least a 20-fold improvement before the project ended. Soon after the ABI 377 sequencer began filling up the big genome centers, we knew that goal would be reached.

Sequencing at even lower costs only became practical in 2000 when capillary sequencing machines came into widespread deployment. They allowed almost mistake-free DNA sequences to be obtained in the 10 cent cost range, letting NIHGR move on to support the sequencing of the mouse genome before the last ambiguous regions of the human genome became sorted out. As I write, the complete sequence of a boxer dog has just been described. Physical maps of its genome were not required as improved whole genome shotgun sequencing methods provide by themselves highly accurate vertebrate-sized genomes. The dog genome, with its 2.4 billion base pairs and so slightly smaller than its human equivalent, took only 24 months to complete. It cost some 30 million dollars and so largely accurate base pair callouts cost less than 2 cents.

Another 100-fold reduction in sequencing costs is already the goal of DNA technology hotshots. Until their new dawn arrives, possibly as soon as 2007, this book's 20 chapters and some 335 genomes are our current best approximation as to how DNA and its genes have changed since their first appearance on the Earth more than three billion years ago.

JAMES D. WATSON
13 December 2005

1

Insights on Biology and Evolution from Microbial Genome Sequencing

Claire M. Fraser-Liggett
The Institute for Genomic Research
Rockville, Maryland 20850

D URING THE PAST 10 YEARS, GENOMICS-BASED APPROACHES have had a profound impact on the field of microbiology and our understanding of microbial species. Since the first report on the complete genome sequence of *Haemophilus influenzae* in 1995 (Fleischmann et al. 1995), nearly 300 other prokaryotic genome sequences have been completed (http://www.genomesonline.org/; http://cmr.tigr.org/tigr-scripts/CMR/CmrHomePage.cgi), with another 750 projects underway. In the early days of microbial genomics, our ignorance about the extent of species diversity was reflected in the assumption that the complete sequence of 20–30 carefully chosen representatives of the bacterial and archaeal domains of life would provide a sufficient amount of information for follow-up investigations. As sequence data began to accumulate, it quickly became clear that we had underestimated the wealth of genetic and biochemical diversity in the prokaryotic world. Indeed, the completion of the sequence of *Escherichia coli* O157:H7 by Perna and colleagues in 2001 revealed that this new isolate contained more than 1300 strain-specific genes as compared with *E. coli* K-12. These genes encode proteins involved in virulence and expanded metabolic capabilities, as well as several prophages. This was a striking example of the fact that two members of the same species could differ in gene content by almost 30%. Today, many genomics efforts are focused on sequencing multiple isolates and strains and providing new insights into species diversity and the dynamic nature of the prokaryotic genome.

Because of their larger genome sizes, genome sequencing efforts on fungi and unicellular eukaryotes were slower to get started than projects

Genomes, ©2006 Cold Spring Harbor Laboratory Press 0-87969-806-3

focused on prokaryotes; however, today there are a number of genome sequences available from both of these groups of organisms that have led to significant improvements in overall sequence annotation and also shed considerable light on novel aspects of their biology (see Chapters 2 and 3).

A CHANGING VIEW OF THE MICROBIAL WORLD

The microbial world can be classified into four groups that differ in many aspects of their biology: Bacteria and Archaea, which represent the prokaryotes, single-celled Eukarya, and viruses. During the past 10 years, a large and phylogenetically diverse number of microbial species has been targeted for genome analysis. Extremes in genome size (<500 kb to almost 10 Mbp) and gene content have also been revealed by these studies, with no absolute boundaries between viral, bacterial, archaeal, fungal, and protist genomes (Fig. 1).

When one considers the more than 20-fold difference in bacterial genome size (Fig. 1), a question that emerged is whether or not one can define a minimal set of genes essential for life. The notion of a minimal genome has been explored through a number of both experimental (Hutchison III et al. 1999; Sassetti et al. 2003; Krause and Balish 2004) and theoretical (Koonin 2003; Klasson and Andersson 2004) approaches. One of the first studies employed transposon mutagenesis in the minimal organisms *Mycoplasma genitalium* and *Mycoplasma pneumoniae*, based on the assumption that there would be a limited number of genes encoding proteins with redundant functions, thereby making it easier to identify essential versus nonessential genes. The results of this study suggested that as many as 130 of the 480 predicted coding sequences may not be essential in vitro (Hutchison III et al. 1999). However, one limitation of the transposon approach is that the mutants that are generated contain disruptions only in single genes. To date, there has been no experimental validation that a minimal genome containing 350 functional genes would support life. Indeed, successive rounds of *M. genitalium* mutagenesis revealed that fitness of the cultures is gradually reduced (Peterson and Fraser 2001). While it might be possible to compensate for reduced fitness through gene loss by supplementation of the growth medium, this notion points out the intricate relationship between the definition of a minimal genome and the cellular environment. More recently, new methods have been described for the deletion of large, non-essential regions of the *E. coli* genome (Kolisnychenko et al. 2002; Goryshin et al. 2003; Hashimoto et al. 2005) and are allowing for correlations between genotype and phenotype.

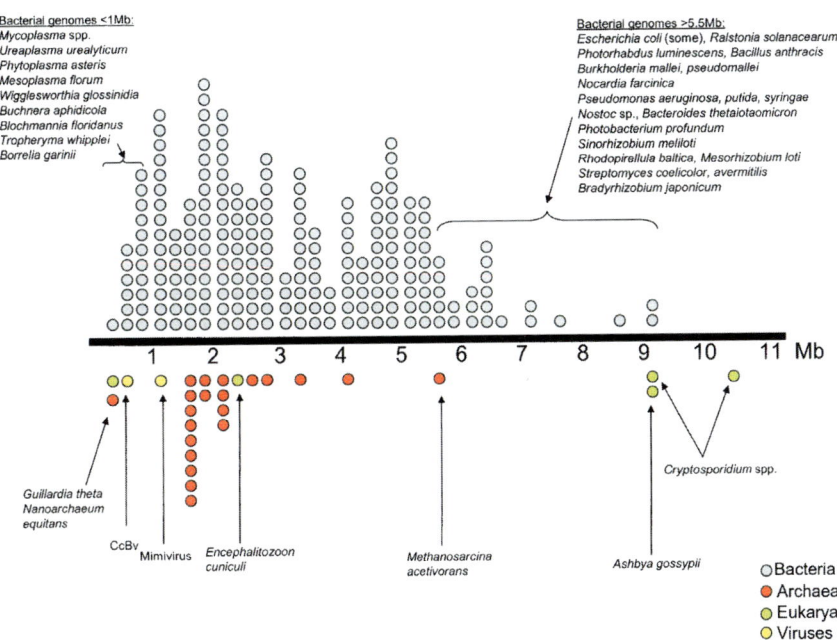

Figure 1. Depiction of overlapping genome size in members of the Bacteria (blue), Archaea (red), Eukarya (green), and viruses (yellow), in the size range (~0.5–10.5 Mb) in which this overlap has been found to occur. Number of circles at a given point on the scale indicates the number of completed genomes of a specific size. Circles representing unusually small (<1 Mb) or large (>5.5 Mb) bacterial genomes are labeled with the species name. Reprinted with permission from Elsevier © 2005, from Ward and Fraser (2005).

While these techniques show great promise for engineering reduced bacterial genomes, other new approaches for generating 5- to 6-kb segments of DNA from oligonucleotides (Smith et al. 2003) and for microchip-based multiplex gene synthesis (Tian et al. 2004) represent a significant step toward an era of synthetic biology that will also enable some of the current hypotheses about minimal gene sets to be tested experimentally.

MULTIPLE FORCES ARE SHAPING MICROBIAL GENOMES

Comparative genomics approaches have revealed that the prokaryotic genome is a dynamic entity, different in many respects from more stable multicellular eukaryotic genomes. Multiple forces have shaped the prokaryotic genome during its evolution; these include gene loss/genome reduction, genome rearrangement, expansion of functional capabilities

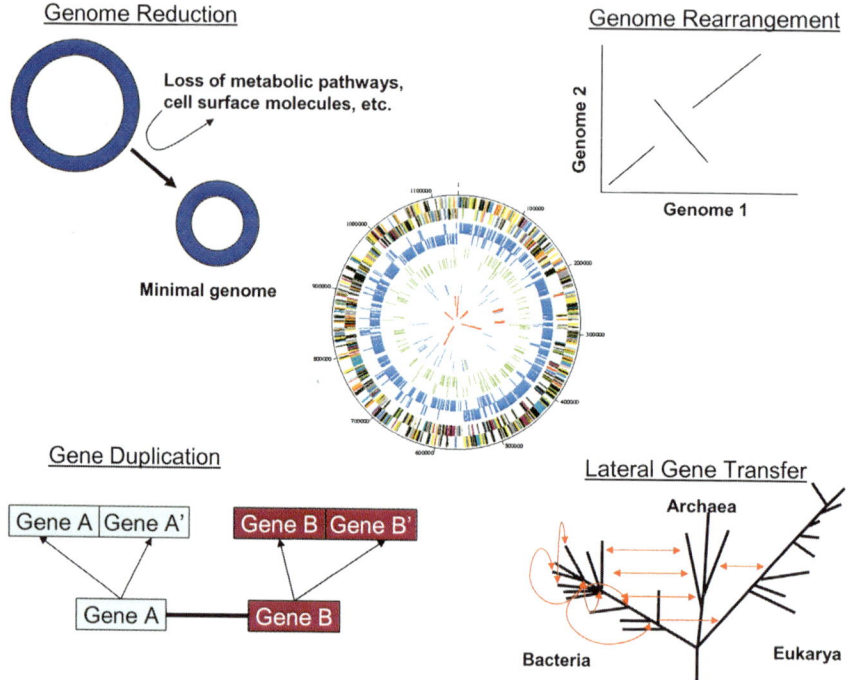

Figure 2. Multiple forces, including genome reduction, genome rearrangement, gene duplication, and acquisition of new genes via lateral gene transfer, are shaping microbial genomes. Details of each of these processes can be found in the text.

through gene duplication, and acquisition of functional capabilities through lateral gene transfer (Fig. 2).

Genome projects on various obligate intracellular pathogens and endosymbionts have provided several windows on the process of reductive evolution (Andersson and Kurland 1998; Moran and Plague 2004). Although these organisms are similar in that they contain significantly reduced genomes (≤1 Mbp) that are missing one or more key metabolic pathways, thereby making them dependent on their hosts for survival, it is clear that there are multiple solutions to a minimal genome. Evidence has suggested that the transition to obligate intracellular species often involves deletions of large segments of DNA early on in the process, likely catalyzed by genome instability mediated by insertion sequence (IS) elements or other mobile DNA that are eventually lost from the genomes (Moran and Mira 2001; Moran and Plague 2004; Belda et al. 2005; see also the example below from comparative analysis of *Burkholderia* spp). While it was initially thought that genes targeted for loss during this process were ones that were no longer necessary for survival in a host environment, there are increasing numbers of examples from comparative genomics projects that suggest some gene loss may be beneficial to the microbe. This is true in the case of certain pathogens, including *Mycobacterium tuberculosis* (Tsolaki et al. 2004), which has lost metabolic pathways and become more virulent, and *Shigella* (Nakata et al. 1993) and *Bordetella pertussis* (Parkhill et al. 2003), which have lost genes encoding cell surface antigens, a situation that may allow for enhanced ability to evade the host immune system.

Genome rearrangements mediated by IS elements can also play a major role in genome plasticity. The extent of such rearrangements often reflects the lifestyle of the organism. In general, obligate intracellular organisms that exist in relative isolation contain few mobile elements, and their genomes tend to be stable over long periods of time. At the other extreme, free-living bacteria often contain large numbers of IS elements and repetitive DNA sequences that may mediate homologous recombination. One example of the role of IS elements in mediating genome-wide rearrangements comes from the comparative analysis of *Burkholderia mallei* (Nierman et al. 2004), an obligate mammalian pathogen that causes glanders, and *Burkholderia pseudomallei* (Holden et al. 2004), an environmental soil-dwelling organism. The *B. mallei* genome contains 171 complete or partial IS elements that collectively represent >3% of the sequence (Nierman et al. 2004). The *B. mallei* genome is 1.4 Mb smaller than that of its closest relative, *B. pseudomallei*, and most of the synteny break points between the two genomes are bounded by IS elements. In addition, two syntenic portions of chromosome 1 in *B. pseudomallei* that are found on chromosome 2 in *B. mallei* are flanked by IS elements, lending further support to the idea that genome rearrangement can play a large role in genome structural alteration in certain species (Nierman et al. 2004).

Gene duplication and functional diversification is yet another mechanism for generating diversity within microbial genomes. Gene paralogs (genes related by duplication) can represent as much as 50% of the larger microbial genomes, and an interesting subset of such genes are lineage-specific duplications that presumably are responsible, in part, for species-specific biology. A recent analysis of 115 completed prokaryotic genome sequences by Konstantinidis and Tiedje (2004) using the Clusters of Orthologous Groups (COGs) database (Tatusov et al. 2003) revealed that larger genomes are disproportionately enriched in genes encoding proteins involved in regulation, secondary metabolism, and transport. The inverse correlation was observed with proteins involved in translation and DNA processing. This analysis provides a possible explanation for why species with larger genomes are more apt to dominate environments where nutrients are scare, because they are more versatile in terms of their ability to sense and respond to changing environmental conditions.

Another source of genome variability that plays an important role in prokaryotic genome evolution is lateral gene transfer (LGT), which brings new genes into a genome and provides a means for rapid adaptations to changing demands on an organism (Boucher et al. 2003). For example, acquisition of virulence determinants on pathogenicity islands appears to play a major role in pathogen evolution. Phylogenetic analysis of multiple strains of *Staphylococcus aureus* indicated that the diversification of the highly variable RD13 region, encoding putative pathogenesis-related proteins, has likely occurred by LGT and recombination (Fitzgerald et al. 2003). Comparative analysis between *S. aureus* and its close relative, *Staphylococcus epidermidis*, demonstrated that genome islands in non-syntenic regions of the genome, likely acquired by LGT, are the primary source of variations in pathogenicity and virulence between these species (Gill et al. 2005). The acquisition of a circular plasmid with 99.6% similarity with the *Bacillus anthracis* toxin-encoding plasmid, pXO1, by *Bacillus cereus* G9241 was likely responsible for the emergence of a strain of *B. cereus* able to cause a disease resembling inhalation anthrax (Hoffmaster et al. 2004). Although homologs of the *B. anthracis* pXO2-encoded capsule genes were not found in this strain of *B. cereus*, a polysaccharide capsule cluster encoded on a second, previously unidentified plasmid, pBC218, was identified. The virulence of this stain was confirmed in an A/J mouse model of anthrax. LGT probably also plays an important role in generating diversity among environmental bacteria. Several lines of evidence have suggested extensive LGT from archaeal species to the hyperthermophilic bacteria, *Thermotoga maritima* (Nelson et al. 1999; Nesbo et al. 2002; Mongodin et al. 2005).

As prokaryotic genome sequences began to accumulate, there were a number of attempts to generate whole-genome phylogenies; however, these often resulted in trees that were incongruent with phylogenies based on rRNA and suggested that it may be difficult, if not impossible, to reconstruct the Tree of Life given the extent of LGT (Doolittle 1999). While there has been considerable debate about the frequency of LGT (Gogarten et al. 2002; Lawrence and Hendrickson 2003), the source of laterally transferred genes (Daubin et al. 2003a), the most robust methods for detecting LGT (Koonin et al. 2001; Ragan 2001; Snel et al. 2002), and its impact on phylogeny (Bapteste et al. 2004), more recent analyses have suggested that it is possible to extract a coherent phylogenetic pattern from analysis of a "core" set of genes (Daubin et al. 2003b; Phillipe and Douady 2003). While a more detailed discussion of LGT is beyond the scope of this review, there appears to be a consensus that it is perhaps most appropriate that the evolution of prokaryotic species is best depicted as a network of vertically and laterally transferred genes, rather than as a single tree. A recent report by Kunin and colleagues (2005) has suggested that certain organisms may serve as hubs for rapid LGT across species. One implication of this hypothesis is that a gene(s) conferring a selective advantage may traverse across many species with a small number of LGT events.

Given the extent of LGT that has been described in numerous studies, a question that can be posed is whether or not it is possible to define the pan-genome for any given bacterial species, that is, the total number of genes associated with all strains of an organism. A recent study by Tettelin and colleagues (2005) that examined diversity in eight isolates of *Streptococcus agalactiae* has suggested that the number of genes associated with this species may be theoretically unlimited. The pan-genome can be divided into three parts: a core genome shared by all strains, a set of dispensable genes shared by some but not all isolates, and a set of strain-specific genes associated with each isolate examined. The core genome encodes basic aspects of *S. agalactiae* biology, while the dispensable and strain-specific genes contribute to the genetic diversity of the species and the ability to colonize certain niches. This contrasts with *B. anthracis*, for which the pan-genome can adequately be described by four genome sequences. This difference in the type of pan-genome may reflect several factors, including the different lifestyles of the two organisms (i.e., exposure of *S. agalactiae* to diverse environments vs. occupation of a more isolated biological niche by *B. anthracis*), the ability of each species to acquire and stably incorporate foreign DNA, and an advantage in niche adaptation by acquisition of laterally transferred DNA.

Comparative genomics of unicellular eukaryotes has also come of age with the completion of genome sequencing projects on a range of organisms, including a number of apicomplexa (*Plasmodium* spp., *Theileria* spp., *Cryptosporidium* spp.), trypanosomes (*Trypanosoma brucei*, *T. cruzi*, and *Leishmania major*), amoebae (*Dictyostelium discoideum* and *Entamoeba histolytica*), microsporidia (*Encephalitozoon cuniculi*), and nucleomorphs (*Guillarida theta*). It is of interest that there are a number of parallels between unicellular eukayotes and prokaryotes. As observed with bacterial species, these unicellular eukaryotes differ tremendously in genome size, genome organization (chromosome number, gene density, and the presence and size of introns), and gene number. Genome reduction is also a force in the unicellular eukaryotic world, as evidenced by the minimal genomes of *G. theta* (0.55 Mb) (Douglas et al. 2001) and *E. cuniculi* (2.51 Mb) (Katinka et al. 2001). In these species, gene density, at approximately one gene per 1.2 Mb, approaches that observed in the prokaryotic world, and many metabolic genes have been lost, making these species dependent on their hosts for energy and small molecules. Other unicellular eukaryotes, such as two species of *Theileria* (Gardner et al. 2005; Pain et al. 2005) and *Cryptosporidium* (Abrahamsen et al. 2004; Xu et al. 2004) protists, represent organisms with moderately compact genomes with a gene density of approximately one gene per 2 Mb. In these cases, genome reduction also results from gene loss, particularly with regard to metabolic genes and plastid-related genes (Keeling 2004), together with reduced intron content (*Cryptosporidium*) or reduced intergenic space (*Theileria*). Despite the fact that a number of unicellular genomes have undergone reductive evolution, recent studies on two *Cryptosporidium* spp. (Abrahamsen et al. 2004; Xu et al. 2004) and the protist *Entameoeba histolytica* (Loftus et al. 2005) have lent support to the idea of unicellular eukaryote acquisition of bacterial genes involved in cellular metabolism through LGT.

APPLICATIONS OF MICROBIAL GENOMICS

Approximately 40% of the bacterial species that have been targeted for genome analysis represent important human pathogens. Comparative in silico methods are allowing for correlations to be made between genotype and phenotype in many instances. For example, the Chlamydiaceae represent a group of closely related obligate intracellular pathogens that cause a range of diseases in mammalian and avian hosts. Genome analysis of several members of this clade has revealed a limited number of variable metabolic and cell surface genes, clustered in the replication termination region, that account for much of the differences in tissue and host tropism between species (Read et al. 2003; Thomson et al. 2005). An unexpected finding from Chlamydia genome sequencing projects was the discovery of a number of genes with similarity to enterobacterial virulence factors (Read et al. 2003), suggesting that the Chlamydiae may have been reservoirs for virulence genes at some distant point in the evolution of the enterobacteria.

Transcriptomic and proteomic approaches have also provided insights into genes involved in virulence, the molecular basis of host specificity, and host–pathogen interactions. One advantage of such large-scale approaches is the ability to monitor global changes in gene and protein expression in both the pathogen and the host during the infectious process. Another is that they can be used to study genes and proteins whose function is unknown. Recent transcriptome studies of *Neisseria meningitidis*—a causative agent of septicemia and meningococcal meningitis—provide an excellent example of how transcriptome analysis can be exploited (Grifantini et al. 2002; Dietrich et al. 2003). These studies showed that there were distinct sets of genes that were differentially regulated during two key steps in the meningococcal infection of human cells—the initial interaction with epithelial cells in the respiratory tract, and the later interaction with endothelial cells in the blood–brain barrier. These differentially regulated genes—which encode membrane transporters, transcription factors, general metabolic pathways proteins, and a number of hypothetical proteins—are obvious candidates for further studies that in turn could lead to novel approaches to preventing diseases caused by *N. meningitidis*.

One of the goals of genome-enabled research on human pathogens is the development of novel diagnostics, antimicrobial compounds, and vaccines. While progress is being made on all fronts, there have been a number of successes in the application of genome sequence data to the identification of novel vaccine candidates. A new method, reverse vaccinology, has been described that allows for identification of potential vaccine candidates based on genomic information, rather than the more traditional approach toward vaccine development pioneered by Pasteur more than two centuries ago, which requires growing the infectious agent as a first step (Rappuoli 2000). This approach has been successfully applied to rapid vaccine development against a number of human pathogens, including *Neisseria meningitidis* (Pizza et al. 2000; Tettelin et al. 2000), *Streptococcus pneumoniae* (Ross et al. 2001; Wizemann et al. 2001), *Chlamydia* spp. (Grandi 2003), as well as the viral pathogen that causes severe acute respiratory disease (SARS) (Bukreyev et al. 2004; Yang et al. 2004). While reverse vaccinology has proven to be very promising in generating a protective immune response in various animal models of disease, the results of ongoing clinical trials of these novel vaccine candidates will provide the ultimate test of the effectiveness of this approach to find new vaccines of benefit to humans.

Because of their unique metabolic properties, a variety of environmental organisms with potential utility in catabolic degradation of toxic compounds or other bioremediation processes have also been targeted for sequencing and functional analysis. As one example, *Geobacter sulfurreducens* is a member of the δ-proteobacteria and has the ability to precipitate soluble metals such as iron and uranium as a by-product of electron transport. Genome analysis of *G. sulfurreducens* (Methe et al. 2003) revealed the presence of a large number of c-type cytochromes and suggested the existence of a large number of redundant electron transport networks. A subsequent study demonstrated that electron transport in *G. sulfurreducens* occurs through direct contact of the pili with iron oxides and suggests that it may be possible to engineer biologically based conductive materials (Reguera et al. 2005). Continued exploration of the metabolic capabilities of microbes with potential bioremediation and biotransformation capabilities will be facilitated by the increasing availability of genome sequence data, together with the development of tools and databases for reconstruction of metabolic pathways such as EcoCyc (Keseler et al. 2005), the Biocatalysis/Biodegradation Database (Hou et al. 2004), and OptStrain (Pharkya et al. 2004), and new approaches for gene and genome synthesis (Smith et al. 2003; Tian et al. 2004).

Metabolic engineering of microbes is an area of longstanding industrial interest, especially for the production of small molecules. Genomics-based methodologies, including comparative DNA sequencing, transcriptome, proteome, and metabolome profiling, together with in silico modeling and simulation, have become important tools in bioengineering strategies (for review, see Lee et al. 2005). The potential of this approach to generate predictive models of cellular behavior was recently demonstrated by Fong et al. (2005), who engineered *E. coli* strains with improved production of lactic acid based on the genome-scale metabolic analysis together with adapted evolution of the new strains. The use of "omics" data to generate predictive models is still in its infancy and is limited by several factors, including a comprehensive lack of information on regulatory networks and the challenges of integrating data across multiple scales and times. However, one of the most exciting areas of future genome-enabled research will be in systems biology.

METAGENOMICS: ANOTHER NEW FRONTIER

Despite all of the progress in microbial genomics in the past 10 years, it is important to remember that essentially all of the projects completed to date have been focused on species that can be grown in culture. Given that >99% of the prokaryotes in the environment cannot be cultured in the laboratory, we are still greatly limited in our knowledge about the physiology and ecology of microbial communities (Schloss and Handelsman 2005). While small-subunit ribosomal RNA (rRNA) genes have been used in surveys of diversity in the uncultured prokaryotic world for some time, this information cannot provide any insights into the biology or species interactions in communities. The more recent application of high-throughput sequencing technology together with newer algorithms for sequence assembly have given rise to the field of metagenomics (Riesenfeld et al. 2004), which has provided unprecedented access to and information about uncultured microbial communities. Two studies published in 2004 demonstrated the power of this approach, particularly with simple microbial communities.

Using a whole-genome shotgun approach, Tyson et al. (2004) were able to reconstruct two almost complete genome sequences of *Leptospirillum* group II and *Ferroplasma* type II and the partial sequence of three other species from a low-complexity acid mine drainage biofilm growing underground within a pyrite ore body. This was possible because the community was dominated by a small number of distinct species. Genome analysis for each organism revealed specific pathways for carbon and nitrogen fixation and energy generation. A larger project published shortly thereafter by Venter and colleagues (2004) used a whole-genome shotgun strategy to explore the diversity in the Sargasso Sea. A total of almost 1.5 billion base pairs of DNA was generated, and it was estimated that this single sample of sea water contained ≥1800 species based on sequence relatedness. While it was not possible in this study to reconstruct nearly complete genome sequences because of the diversity of the sample, it was possible to identify >1.2 million new genes, including >700 new rhodopsin-like photoreceptors that may be involved in a new form of phototrophy in the oceans.

Because of the current limitations in assembling nearly complete genome sequence data from complex communities, there has been a renewed interest in developing methods for culturing recalcitrant microbial species (Rappe et al. 2002; Sait et al. 2002; Tyson and Banfield 2005). Several recent successes in these efforts have been reported that will most certainly facilitate follow-up functional genomics studies.

CONCLUSIONS

Although we have made tremendous progress in the past decade in the field of microbial genomics, work to date represents just the tip of the iceberg given the estimated number of microbial species on Earth. With the accumulation of more sequence data from cultivated isolates and expansion of metagenomics efforts, it is likely that the coming decade will be filled with new insights into the strange and often unpredictable microbial world. Systems-based approaches that integrate DNA sequence with data from transcriptome, proteome, and metabolome studies will begin to reveal the intricate workings of a microbial cell (Fig. 3).

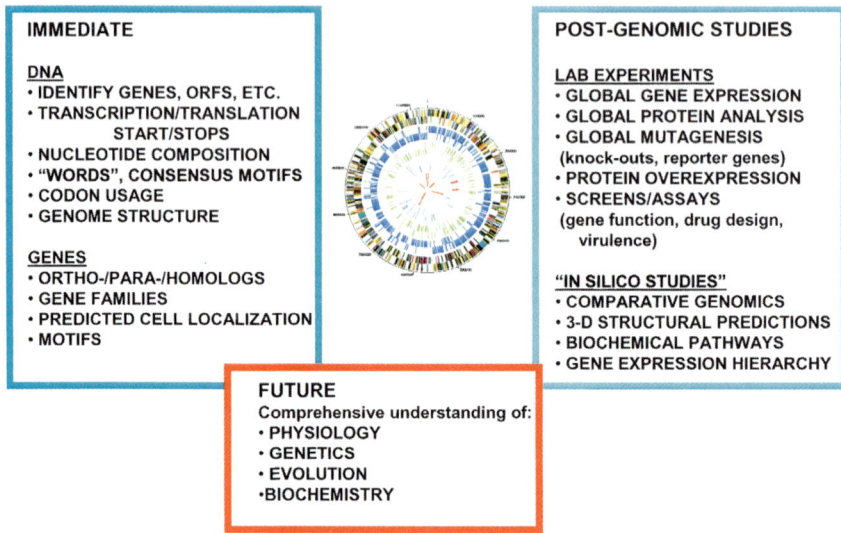

Figure 3. Applications and future directions in microbial genomics. The availability of microbial DNA sequence has provided a new foundation for follow-up studies, both in vitro and in silico. The ultimate goal is to integrate data from these multiple approaches to achieve a new systems-level understanding of the microbial cell.

ACKNOWLEDGMENTS

I thank all of my colleagues at The Institute for Genomic Research who have contributed to TIGR's efforts in microbial genomics during the past decade, and TIGR's outside collaborators who have contributed their expertise to these projects.

REFERENCES

Abrahamsen, M.S., Templeton, T.J., Enomoto, S., Abrahante, J.E., Zhu, G., Lancto, C.A., Deng, M., Liu, C., Widmer, G., Tzipori, S. et al. 2004. Complete genome sequence of the apicomplexan, *Cryptosporidium parvum*. *Science* **304:** 441–445.

Andersson, S.G. and Kurland, C.G. 1998. Reductive evolution of resident genomes. *Trends Microbiol.* **6:** 263–268.

Bapteste, E., Boucher, Y., Leigh, J., and Doolittle, W.F. 2004. Phylogenetic reconstruction and lateral gene transfer. *Trends Microbiol.* **12:** 406–411.

Belda, E., Moya, A., and Silva, F.J. 2005. Genome rearrangement distances and gene order phylogeny in γ-Proteobacteria. *Mol. Biol. Evol.* **22:** 1456–1467.

Boucher, Y., Douady, C.J., Papke, R.T., Walsh, D.A., Boudreau, M.E., Nesbo, C.L., Case, R.J., and Doolittle, W.F. 2003. Lateral gene transfer and the origins of prokaryotic groups. *Annu. Rev. Genet.* **37:** 283–328.

Bukreyev, A., Lamarinde, E.W., Buchholz, U.J., Vogel, L.N., Elkins, W.R., St. Claire, M., Murphy, B.R., Subbarao, K., and Collins, P.L. 2004. Mucosal immunization of African green monkeys with an attenuated parainfluenza virus expressing the SARA coronavirus spike protein for the prevention of SARS. *Lancet* **363:** 2122–2127.

Daubin, V., Lerat, E., and Perriere, G. 2003a. The source of laterally transferred genes in bacterial genomes. *Genome Biol.* **4:** R57.

Daubin, V., Moran, N.A., and Ochman, H. 2003b. Phylogenetics and the cohesion of bacterial genomes. *Science* **301:** 829–832.

Dietrich, G., Kurz, S., Hubner, C., Aepinus, C., Theiss, S., Gickenberger, M., Panzner, U., Weber, J., and Frosch, M. 2003. Transcriptome analysis of *Neisseria meningitidis* during infection. *J. Bacteriol.* **185:** 155–164.

Doolittle, W.F. 1999. Phylogenetic classification and the universal tree. *Science* **284:** 2124–2129.

Douglas, S., Zauner, S., Fraunholz, M., Beaton, M., Penny, S., Deng, L.T., Wu, X., Reith, M., Cavalier-Smith, T., and Maier, U.G. 2001. The highly reduced genome of an enslaved algal nucleus. *Nature* **410:** 1040–1041.

Fitzgerald, J.R., Reid, S.D., Ruotsalainen, E., Tripp, T.J., Liu, M., Cole, R., Kuusela, P., Schlievert, P.M., Jarvinen, A., and Musser, J.M. 2003. Genome diversification in *Staphylococcus aureus*: Molecular evolution of a highly variable chromosomal region encoding the Staphylococcal exotoxin-like family of proteins. *Infect. Immun.* **71:** 2827–2838.

Fleischmann, R.D., Adams, M.D., White, O., Clayton, R.A., Kirkness, E.F., Kerlavage, A.R., Bult, C.J., Tomb, J.F., Dougherty, B.A., Merrick, J.M., et al. 1995. Whole-genome random sequencing and assembly of *Haemophilus influenzae* Rd. *Science* **269:** 496–512.

Fong, S.S., Burgard, A.P., Herring, C.D., Knight, E.M., Blattner, F.R., Maranas, C.D., and Palsson, B.Ø. 2005. In silico design and adaptive evolution of *Escherichia coli* for production of lactic acid. *Biotechnol. Bioeng.* **91:** 643–648.

Gardner, M.J., Bishop, R., Shah, T., de Villiers, E.P., Carlton, J.M., Hall, N., Ren, Q., Paulsen, I.T., Pain, A., Berriman, M., et al. 2005. Genome sequence of *Theileria parva*, a bovine pathogen that transforms lymphocytes. *Science* **309:** 134–137.

Gill, S.R., Fouts, D.E., Archer, G.L., Mongodin, E.F., Deboy, R.T., Ravel, J., Paulsen, I.T., Kolonay, J.F., Brinkac, L., Beanan, M., et al. 2005. Insights on evolution of virulence and resistance from the complete genome analysis of an early methicillin-resistant *Staphylococcus aureus* strain and a biofilm-producing methicillin-resistant *Staphylococcus epidermidis* strain. *J. Bacteriol.* **187:** 2426–2438.

Gogarten, J.P., Doolittle, W.F., and Lawrence, J.G. 2002. Prokaryotic evolution in light of gene transfer. *Mol. Biol. Evol.* **19:** 2226–2238.

Goryshin, I.Y., Naumann, T.A., Apodaca, J., and Reznikoff, W.S. 2003. Chromosomal deletion formation system based on Tn5 double transposition: Use for making minimal genomes and essential gene analysis. *Genome Res.* **13:** 644–653.

Grandi, G. 2003. Rational antibacterial vaccine design through genomic technologies. *Int. J. Parasitol.* **33:** 615–620.

Grifantini, R., Bartolini, E., Muzzi, A., Draghi, M., Frigimelica, E., Berger, J., Ratti, G., Petracca, R., Galli, G., Agnusdei, M., et al. 2002. Previously unrecognized vaccine candidates against group B meningococcus identified by DNA microarrays. *Nature Biotechnol.* **20:** 914–921.

Hashimoto, M., Ichimura, T., Mizoguchi, H., Tanaka, K., Fujimitsu, K., Keyamura, K., Ote, T., Yamakawa, T., Yamazaki, Y., Mori, H., et al. 2005. Cell size and nucleoid organization of engineered *Escherichia coli* cells with a reduced genome. *Mol. Microbiol.* **55:** 137–149.

Hoffmaster, A.R., Ravel, J., Rasko, D.A., Chapman, G.D., Chute, M.D., Marston, C.K., De, B.K., Sacchi, C.T., Fitzgerald, C., Mayer, L.W., et al. 2004. Identification of anthrax toxin genes in a *Bacillus cereus* associated with an illness resembling inhalation anthrax. *Proc. Natl. Acad. Sci.* **101**: 8449–8454.

Holden, M.T., Titball, R.W., Peacock, S.J., Cerdeno-Tarraga, A.M., Atkins, T., Crossman, L.C., Pitt, T., Churcher, C., Mungall, K., Bentley, S.D., et al. 2004. Genomic plasticity of the causative agent of melioidosis, *Burkholderia pseudomallei*. *Proc. Natl. Acad. Sci.* **101**: 14220–14245.

Hou, B.K., Ellis, L.B., and Wackett, L.P. 2004. Encoding microbial metabolic logic: Predicting biodegradation. *J. Ind. Microbiol. Biotechnol.* **31**: 261–272.

Hutchison III, C.A., Peterson, S.N., Gill, S.R., Cline, R.T., White, O., Fraser, C.M., Smith, H.O., and Venter, J.C. 1999. Global transposon mutagenesis and a minimal *Mycoplasma* genome. *Science* **286**: 2165–2169.

Katinka, M.D., Duprat, S., Cornillot, E., Metenier, G., Thomarat, F., Prensier, G., Barbe, V., Peyretaillade, E., Brottier, P., Wincker, P., et al. 2001. Genome sequence and gene compaction of the eukaryotic parasite *Encephalitozoon cuniculi*. *Nature* **414**: 450–453.

Keeling, P.J. 2004. Reduction and compaction in the genome of the apicomplexan parasite *Cryptosporidium parvum*. *Dev. Cell* **6**: 614–616.

Keseler, I.M., Collado-Vides, J., Gama-Castro, S., Ingraham, J., Plaey, S., Paulsen, I.T., Peralta-Gil, M., and Kapr, P.J. 2005. EcoCyc: A comprehensive database resource for *Escherichia coli*. *Nucleic Acids. Res.* **33**: D334.

Klasson, L. and Andersson, S.G. 2004. Evolution of minimal-gene-sets in host-dependent bacteria. *Trends Microbiol.* **12**: 37–43.

Kolisnychenko, V., Plunkett III, G., Herring, C.D., Feher, T., Posfai, J., Blattner, F.B., and Posfai, G. 2002. Engineering a reduced *Escherichia coli* genome. *Genome Res.* **12**: 640–647.

Konstantinidis, K.T. and Tiedje, J.M. 2004. Trends between gene content and genome size in prokaryotic species with larger genomes. *Proc. Natl. Acad. Sci.* **101**: 3160–3165.

Koonin, E.V. 2003. Comparative genomics, minimal gene-sets and the last universal common ancestor. *Nat. Rev. Microbiol.* **1**: 127–136.

Koonin, E.V., Makarova, K.S., and Aravind, L. 2001. Horizontal gene transfer in prokaryotes: Quantification and classification. *Annu. Rev. Microbiol.* **55**: 709–742.

Krause, D.C. and Balish, M.F. 2004. Cellular engineering in a minimal microbe: Structure and assembly of the terminal organelle of *Mycoplasma pneumoniae*. *Mol. Microbiol.* **51**: 917–924.

Kunin, V., Goldovsky, L., Darzentas, N., and Ouzounis, C.A. 2005. The net of life: Reconstructing the microbial phylogenetic network. *Genome Res.* **15**: 954–959.

Lawrence, J.G. and Hendrickson, H. 2003. Lateral gene transfer: When will adolescence end? *Mol. Microbiol.* **50**: 725–727.

Lee, S.Y., Lee, D.-Y., and Kim, T.Y. 2005. Systems biotechnology for strain improvement. *Trends Biotechnol.* **23**: 349–358.

Loftus, B., Anderson, I., Davies, R., Alsmark, U.C., Samuelson, J., Amedeo, P., Roncaglia, P., Berriman, M., Hirt, R.P., Mann, B.J., et al. 2005. The genome of the protist parasite *Entamoeba histolytica*. *Nature* **433**: 865–868.

Methe, B.A., Nelson, K.E., Eisen, J.A., Paulsen, I.T., Nelson, W., Heidelberg, J.F., Wu, D., Wu, M., Ward, N., Beanan, M.J., et al. 2003. Genome of *Geobacter sulfurreducens*: Metal reduction in subsurface environments. *Science* **302**: 1967–1969.

Mongodin, E.F., Hance, I.R., Deboy, R.T., Gill, S.R., Daugherty, S., Huber, R., Fraser, C.M., Stetter, K., and Nelson, K.E. 2005. Gene transfer and genome plasticity in *Thermotoga maritima*, a model hyperthermophilic species. *J. Bacteriol.* **187:** 4935–4944.

Moran, N.A. and Mira, A. 2001. The process of genome shrinkage in the obligate symbiont *Buchnera aphidocola*. *Genome Biol.* **2:** research0054.

Moran, N.A. and Plague, G.R. 2004. Genomic changes following host restriction in bacteria. *Curr. Opin. Genet. Dev.* **14:** 627–633.

Nakata, N., Tobe, T., Fukuda, I., Suzuki, T., Konatsu, K., Yoshikawa, M., and Sasakawa, C. 1993. The absence of a surface protease, OmpT, determines the intracellular spreading ability of *Shigella*: The relationship between the ompT and kcpA loci. *Mol. Microbiol.* **9:** 459–468.

Nelson, K.E., Clayton, R.A., Gill, S.R., Gwinn, M.L., Dodson, R.J., Haft, D.H., Hickey, E.K., Peterson, J.D., Nelson, W.C., Ketchum, K.A., et al. 1999. Evidence for lateral gene transfer between Archaea and bacteria from genome sequence of *Thermotoga maritima*. *Nature* **399:** 323–329.

Nesbo, C.L., Nelson, K.E., and Doolittle, W.F. 2002. Suppressive subtractive hybridization detects extensive genomic diversity in *Thermotoga maritima*. *J. Bacteriol.* **184:** 4475–4488.

Nierman, W.C., DeShazer, D., Kim, H.S., Tettelin, H., Nelson, K.E., Feldblyum, T., Ulrich, R.L., Ronning, C.M., Brinkac, L.M., Daugherty, S.C., et al. 2004. Structural flexibility in the *Burkholderia mallei* genome. *Proc. Natl. Acad. Sci.* **101:** 14246–14251.

Pain, A., Renauld, H., Berriman, M., Murphy, L., Yeats, C.A., Weir, W., Kerhornou, A., Aslett, M., Bishop, R., Bouchier, C., et al. 2005. Genome of the host-cell transforming parasite *Theileria annulata* compared with *T. parva*. *Science* **309:** 131–133.

Parkhill, J., Sebaihia, M., Preston, A., Murphy, L.D., Thomson, N., Harris, D.E., Holden, M.T., Churcher, C.M., Bentley, S.D., Mungall, K.I., et al. 2003. Comparative analysis of the genome sequences of *Bordetella pertussis*, *Bordetella parapertussis* and *Bordetella bronchispetica*. *Nat. Genet.* **35:** 32–40.

Perna, N., Plunkett III, G., Burland, V., Mau, B., Glasner, J.D., Rose, D.J., Mayhew, G.F., Evans, P.S., Gregor, J., Kirkpatrick, H.A., et al. 2001. Genome sequence of enterohaemorrhagic *Escherichia coli* O157:H7. *Nature* **409:** 529–533.

Peterson, S.N. and Fraser, C.M. 2001. The complexity of simplicity. *Genome Biol.* **2:** 1–8.

Pharkya, P., Burgard, A.P., and Maranas, C.D. 2004. OptStrain: A computational framework for redesign of microbial production systems. *Genome Res.* **14:** 2367–2376.

Phillipe, H. and Douady, C.J. 2003. Horizontal gene transfer and phylogenetics. *Curr. Opin. Microbiol.* **6:** 498–505.

Pizza, M., Scarlato, V., Masignani, V., Giuliani, M.M., Arico, B., Comanducci, M., Jennings, G.T., Baldi, L., Bartolini, E., Capecchi, B., et al. 2000. Identification of vaccine candidates against serogroup B meningococcus by whole-genome sequencing. *Science* **287:** 1816–1820.

Ragan, M.A. 2001. Detection of lateral gene transfer among microbial genomes. *Curr. Opin. Genet. Dev.* **11:** 620–626.

Rappe, M.S., Connon, S.A., Vergin, K.L., and Giovannoni, S.J. 2002. Cultivation of the ubiquitous SAR11 marine bacterioplankton clade. *Nature* **418:** 630–633.

Rappuoli, R. 2000. Reverse vaccinology. *Curr. Opin. Microbiol.* **3:** 445–450.

Read, T.D., Myers, G.S., Brunham, R.C., Nelson, W.C., Paulsen, I.T., Heidelberg, J., Holtzapple, E., Khouri, H., Federova, N.B., Carty, H.A., et al. 2003. Genome sequence of *Chlamydophila caviae*: Examining the role of niche-specific genes in the evolution of the Chlamydiaceae. *Nucleic Acids Res.* **31:** 2134–2147.

Reguera, G., McCarthy, K.D., Mehta, T., Nicoll, J.S., Tuominen, M.T., and Lovely, D.J. 2005. Extracellular electron transfer via microbial nanowires. *Nature* **435**: 1098–1101.

Riesenfeld, C.S., Schloss, P.D., and Handelsman, J. 2004. Metagenomics: Genomic analysis of microbial communities. *Annu. Rev. Genet.* **38**: 525–552.

Ross, B.C., Czajkowski, L., Hocking, D., Margetts, M., Webb, E., Rothel, L., Patterson, M., Agius, C., Camuglia, S., Reynolds, E., et al. 2001. Identification of vaccine candidate antigens from a genomic analysis of *Porphyromonas gingivalis*. *Vaccine* **19**: 4135–4142.

Sait, M., Hugenholtz, P., and Janssen, P.H. 2002. Cultivation of globally distributed soil bacteria from phylogenetic lineages previously only detected in cultivation-independent surveys. *Environ. Microbiol.* **4**: 654–666.

Sassetti, C.M., Boyd, D.H., and Rubin, E.J. 2003. Genes required for mycobacterial growth defined by high density mutagenesis. *Mol. Microbiol.* **48**: 77–84.

Schloss, P.D. and Handelsman, J. 2005. Metagenomics for studying unculturable microorganisms: Cutting the Gordian knot. *Genome Biol.* **6**: 229.1–229.4.

Smith, H.O., Hutchison, C.A., Pfannkoch, C., and Venter, J.C. 2003. Generating a synthetic genome by whole genome assembly: ɸX174 bacteriophage from synthetic oligonucleotides. *Proc. Natl. Acad. Sci.* **100**: 15440–15445.

Snel, B., Bork, P., and Huynen, M.A. 2002. Genomes in flux: The evolution of archaeal and proteobacterial gene content. *Genome Res.* **12**: 17–25.

Tatusov, R.L., Fedorova, N.D., Jackson, J.D., Jacobs, A.R., Kiryutin, B., Koonin, E.V., Krylov, D.M., Mazumder, R., Mekhedov, S.L., Nikolskaya, A.N., et al. 2003. The COG database: An updated version includes eukaryotes. *BMC Bioinformatics* **4**: 41–54.

Tettelin, H., Saunders, N.J., Heidelberg, J., Jeffries, A.C., Nelson, K.E., Eisen, J.A., Ketchum, K.A., Hood, D.W., Peden, J.F., Dodson, R.J., et al. 2000. Complete genome sequence of *Neisseria meningitidis* serogroup B strain MC58. *Science* **287**: 1809–1815.

Tettelin, H., Masignani, V., Cieslewicz, M.J., Donati, C., Medini, D., Ward, N.L., Angiouli, S.V., Crabtree, J., Jones, A.L., Durkin, A.S., et al. 2005. Genome analysis of multiple pathogenic isolates of *Streptococcus agalactiae*: Implications for the microbial "pan-genome". *Proc. Natl. Acad. Sci.* **102**: 13950–13955.

Thomson, N.R., Yeats, C., Bell, K., Holden, M.T., Bentley, S.D., Livingstone, M., Cerdeno-Tarraga, A.M., Harris, B., Doggett, J., Ormond, D., et al. 2005. The *Chlamydophila abortus* genome sequence reveals an array of variable proteins that contribute to interspecies variation. *Genome Res.* **15**: 629–640.

Tian, J., Gong, H., Sheng, N., Zhou, X., Gulari, E., Gao, X., and Church, G. 2004. Accurate multiplex gene synthesis from programmable DNA microchips. *Nature* **432**: 1050–1054.

Tsolaki, A.G., Hirsch, A.E., DeRiemer, K., Enciso, J.A., Wong, M.Z., Hannan, M., Goguet de la Salmoniere, Y.O., Aman, K., Kato-Maeda, M., and Small, P.M. 2004. Functional and evolutionary genomics of *Mycobacterium tuberculosis*: Insights from genomic deletions in 100 strains. *Proc. Natl. Acad. Sci.* **101**: 4865–4870.

Tyson, G.W. and Banfield, J.F. 2005. Cultivating the uncultivated: A community genomics perspective. *Trends Microbiol.* **13**: 411–415.

Tyson, G.W., Chapman, J., Hugenholtz, P., Allen, E.E., Ram, R.J., Richardson, P.M., Solovyev, V.V., Rubin, E.M., Rokhsar, D.S., and Banfield, J.F. 2004. Community structure and metabolism through reconstruction of microbial genomes from the environment. *Nature* **428**: 37–43.

Venter, J.C., Remington, K., Heidelberg, J.F., Halpern, A.L., Rusch, D., Eisen, J.A., Wu, D., Paulsen, I., Nelson, K.E., Nelson, W., et al. 2004. Environmental genome shotgun sequencing of the Sargasso Sea. *Science* **304**: 66–74.

Ward, N. and Fraser, C.M. 2005. How genomics has affected the concept of microbiology. *Curr. Opin. Microbiol.* **8:** 564–571.

Wizemann, T.M., Heinrichs, J.H., Adamou, J.E., Erwin, A.L., Kunsch, C., Choi, G.H., Barash, S.C., Rosen, C.A., Masure, H.R., Tuomanen, E., et al. 2001. Use of a whole genome approach to identify vaccine molecules affording protection against *Streptococcus pneumoniae* infection. *Infect. Immun.* **69:** 1593–1598.

Xu, P., Widme, G., Wang, Y., Ozaki, L.S., Alves, J.M., Serrano, M.G., Puiu, D., Manque, P., Akiyoshi, D., Mackey, A.J., et al. 2004. The genome of *Cryptosporidium hominis*. *Nature* **431:** 1107–1112.

Yang, Z.Y., Kong, W.P., Huang, Y., Roberts, A., Murphy, B.R., Subbarao, K., and Nabel, G.J. 2004. A DNA vaccine induces SARS coronavirus neutralization and protective immunity in mice. *Nature* **428:** 561–564.

2

Changing Perspectives in Yeast Research Nearly a Decade after the Genome Sequence

Kara Dolinski and David Botstein
Lewis-Sigler Institute for Integrative Genomics
Department of Molecular Biology
Princeton University
Princeton, New Jersey 08544

THE FIRST COMPLETE NUCLEOTIDE SEQUENCE OF A EUKARYOTIC genome, that of budding yeast (*Saccharomyces cerevisiae*), was published in 1996 (Goffeau et al. 1996). It was the result of a broad international effort, stimulated by a consensus reached in the United States, nearly a decade earlier, that there should be an extraordinary 15-year effort to sequence the human genome, supported by funding of the order of $3 billion. A particularly significant feature of the National Academy of Sciences report (Alberts et al. 1988) that announced this consensus was the recommendation that the genomic sequences of a few other eukaryotes should be determined first. The eukaryotic genomes chosen were those of the leading "model organisms," because their genomes are significantly smaller than that of the human, and because substantial and successful molecular genetics research communities had already been developed to study them. Largely because of the efforts of these communities, it was already known that many of the proteins carrying out basic cellular functions are highly conserved among all the eukaryotes, suggesting that knowing the sequences of both the model genomes and the human genome would be an important path to understanding them both. Explicitly named were yeast (*Saccharomyces cerevisiae*), a nematode worm (*Caenorhabditis elegans*), and a fruitfly (*Drosophila melanogaster*). The yeast genome, containing ~12 million bp, is only 0.4% the length of the

3-billion-bp human genome, and the worm and fly genomes are ~3.3% and 5.5% the length of the human genome, respectively (numbers from *Saccharomyces* Genome Database [SGD; http://www.yeastgenome.org], UCSC Gold Path [http://genome.brc.mcw.edu/goldenPath/stats.html], GSC at Washington University [http://www.genome.wustl.edu/projects/celegans/], and FlyBase [http://www.flybase.org/annot/release.html#releases], respectively).

Thus, the yeast genome became the pioneer eukaryotic genome, and the yeast research community was the first to profit from knowledge of the complete genome sequence. A dramatic transformation of yeast research ensued that presaged similar transformations of research in the other model organisms, the mouse and the human, as each of these genome sequences became available. The transformation began with technical improvements that greatly accelerated research, especially any research involving identification of pieces of DNA cloned, for example, after a biological selection from clone libraries. Whereas before the sequence, yeast researchers identified clones by mainly genetic and/or physical mapping methods, now a single sequence run sufficed. Technologies unimaginable before (e.g., DNA microarrays containing each and every yeast gene) became commonplace. The same European-led consortium that initiated the sequencing effort undertook to produce the deletion of every yeast open reading frame (ORF) (Winzeler et al. 1999; Giaever et al. 2002), and development of a whole class of genome-scale genetic methods began, a development that is still far from complete.

Today we are seeing the beginning of an even more profound transformation of yeast research, one that is more than technical. The availability of the entire genome sequence has made possible the asking of new kinds of research questions, questions that can be answered only when one has truly comprehensive information about an organism. For example, once the entire genome sequence became known, it became possible, for the first time, to study expression of all the genes at once, where before one could study genes only a few at a time. The very idea of what constitutes "specificity," has been changed by the ability to study expression of all the genes without exception. It is now routine to enumerate, in a single experiment, all the genes in an organism that respond to a specific stimulus or stress.

Comparative analyses of the complete genome sequences of the yeast, worm, fly, mouse, and human genomes forcefully validated the expectation of extreme conservation of sequence and function over evolutionary time (see Chervitz et al. 1998; Rubin et al. 2000; Lander et al. 2001; Venter et al. 2001). There has been, as a result, a "grand unification" in molecular biology, as it became clear that, at least for proteins, sequence similarity more often than not leads to an unambiguous assessment of functional similarity. Thus, through comparative genomics much, if not quite all, of the experimental work elucidating gene or protein function done in one organism illuminates them all, and transfer of annotation from the organism where the experimental data were collected to other organisms has become routine. Since yeast is still, for many basic cellular functions, the most tractable experimental system, much of the annotation of basic cellular functions in all eukaryotes, including especially the human, can be traced back to experiments done in yeast. As described below, the genome databases, especially Gene Ontology (GO), are continuing to facilitate the transfer of annotation so that a newly discovered gene function in yeast soon appears as an annotation for the orthologs in the other eukaryotes.

FROM THE PARTS LIST TO THE SYSTEM LEVEL: GOALS OF POST–GENOME-SEQUENCE YEAST RESEARCH

The most obvious goal of genomic science arose directly from knowing complete genomic sequences: to decipher, annotate, and understand the role of each and every feature of the DNA sequence in the life of the organism. Put another way, we want to understand the reasons that each genomic DNA sequence feature was selected over evolutionary time.

The knowledge of the entire genomic sequences of organisms has motivated a new kind of biological analysis that looks beyond individual genes to the ensemble of all the genes. A second, more ambitious goal of genomic science has thereby emerged: to understand not only what every gene and gene product does for the organism but also how all of these genes, gene products, and functions and their regulation interact together to produce the properties of the organism. The ultimate aim of this new "system-level" biological research (Hartwell et al. 1999; Ideker et al. 2001a), not really practical before genomic sequences, is to account for and to model, in a fully quantitative way, not just how each of the genes participates in the biology of the organism but also how their interactions are controlled to maintain homeostasis over the entire life cycle and environmental range experienced by the organism.

Saccharomyces cerevisiae, by dint of its small genome and its experimental tractability, has become the pioneer in a new era of biology, where all the individual "parts" of the organism (conveniently encoded in the genome) are specified, and where research is aimed at understanding fully and quantitatively how the ensemble of the parts, subassemblies, and regulatory networks work together to produce the robust living organisms we see.

GENES AND THEIR BIOLOGICAL ROLES

In 1995, the total number of yeast genes about whose function something was understood was of the order of two thousand (Hughes et al. 2004); virtually all that was known about these genes derived directly from experiments by inference from mutant phenotypes or direct biochemical assays. The information about these genes began to be collected into the SGD (www.yeastgenome.org) (Balakrishnan et al. 2005) at about this time; as the genomic sequence became available, other databases also came into existence (see Table 1; Mewes et al. 1997; Costanzo et al. 2000; Guldener et al. 2005). SGD remains the primary source for updates to the genome sequence, primary annotation, gene names, and nomenclature, as well as the basic functional information about each gene, which is continually culled from the experimental literature. In what follows we refer the reader to SGD for details about individual genes and global statistics about genes.

Currently, the best estimate for the number of yeast ORFs that actually encode proteins is 5773, of which 1474 (25%) are listed by SGD as "uncharacterized." This means that something biological is known about 4299 yeast genes, approximately a twofold increase since the genome sequence became available. However, for many genes the biological information available is still very limited, and quite a bit of the new information about biological function derives from sequence comparisons; a gene never studied in yeast but apparently orthologous to a well-characterized gene in another organism acquires functional annotation by inference.

Development of diverse genome-scale experimental technologies raised considerable expectations of an acceleration in the discovery rate for the functional roles of individual poorly characterized genes. Surprisingly, this promise still remains largely unfulfilled. Although each of the technologies (i.e., two-hybrid analyses, synthetic lethality methods, and gene coexpression using DNA microarrays) has had many significant successes, most of the new functional annotations of genes continue to derive from research papers describing experiments focused on just a few genes at a time. A major symptom of the problem has been the startlingly limited overlap in the predictions of different genome-scale methods. The reader is referred to an excellent recent summary of these issues (Hughes et al. 2004). Thus the most elementary goal of post-genome research, to annotate the biological role of every gene, remains a challenge.

GENE EXPRESSION TECHNOLOGY AND THE EMERGENCE OF SYSTEM-LEVEL BIOLOGY

In the mid to late 1990s, technology that allowed one for the first time to simultaneously assess gene expression for the entire genome was developed. DNA microarrays, on which each ORF or other sequence feature is represented (DeRisi et al. 1997; Brown and Botstein 1999), has been the dominant approach used in yeast research; at the same time, Serial Analysis of Gene Expression (SAGE), an alternative sequence-based method, was also developed (Velculescu et al. 1997). Although gene expression technology has yet to make a big impact on the rate of biological annotation of yeast genes, it has nevertheless transformed yeast research, by studying gene expression associated with relatively simple biological experiments in a fully comprehensive way. This comprehensiveness has stimulated both experimental and theoretical approaches to understanding regulatory networks and other features of yeast biology at the system level (Hartwell et al. 1999; Ideker et al. 2001a). Rather than attempt a review of all this research (SGD lists, at the time of writing, 299 published genome-scale expression studies), we will limit ourselves to a few paradigm examples, often from our own experience.

Defining Functional or Regulatory Subsystems, or "Modules"

The gene expression technology provided a direct and simple way to enumerate genes whose expression respond to individual stimuli or stresses. Early examples follow: Gasch and colleagues (2000, 2001) studied all the genes responding to a number of diverse stresses, and they defined a generic "environmental stress response" that underlies all stress responses, including temperature change, starvation, oxidative stress, and radiation; Roberts and colleagues (2000) studied all the genes that respond to α-factor; and Ideker et al. (2001b) studied nutritional perturbations to cells growing on galactose. Similarly, genes with characteristic expression changes during sporulation (Chu et al. 1998) or the cell division cycle (Cho et al. 1998; Spellman et al. 1998) were systematically determined. Such studies have associated genes with each other and with particular biological activities, although further work has been required to provide detailed understanding.

However, studies of this kind have provided important new clues to which sets of genes might act together, in concert or in sequence, in a common biological process. For instance, the ribosomal genes comprise one of the most tightly coregulated groups of functionally related genes. In experiments that examine processes as diverse as sporulation (Chu et al. 1998) to response to arsenic (Haugen et al. 2004), the ribosomal genes (both mitochondrial and nuclear) are often the most significantly coregulated genes in the data set. Genome-wide expression technology has confirmed existing and discovered novel components of other major transcriptional networks in yeast, including clusters of genes involved in amino acid metabolism, energy pathways, DNA replication, and the stress response (Eisen et al. 1998; Chua et al. 2004), all of which are, as are the ribosomal proteins, frequently the most significant groups of coregulated genes in a data set. We have also learned that it is quite rare for genes to have unchanging expression levels across different experiments; for example, expression of the yeast actin (*ACT1*) gene, which was traditionally used as a control in Northern blots to ensure that equivalent levels of RNA were loaded in each well, changes significantly in several diverse types of microarray experiments (Fig. 1).

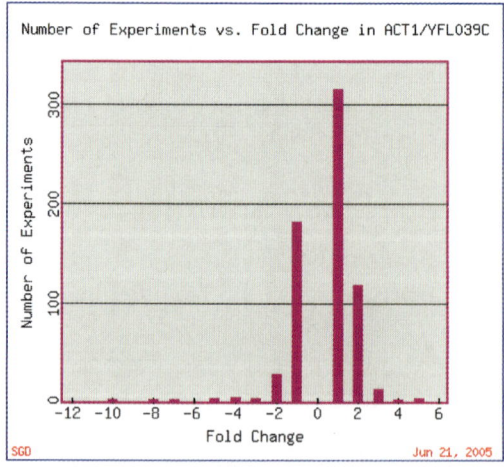

Figure 1. Fold change of *ACT1* gene expression in microarray experiments available at SGD. This figure was generated by the SGD Expression Connection tool at http://db.yeastgenome.org/cgi-bin/expression/expressionConnection.pl. The outlying values (−10, +fivefold) are experiments from the Mnaimneh et al. (2004) data set that profiled expression in strains with essential genes under control of titratable promoters. Other conditions that led to at least fourfold change in *ACT1* expression include expression during sporulation (less than fourfold) (Chu et al. 1998) and prolonged stationary phase (−4.3 fold) (Gasch et al. 2000).

In many cases, later studies have not only confirmed the functional associations among genes based on coexpression but also found the regulators that are responsible; the study by Zhu and colleagues (2000) is one example. These studies not only have resulted in further experimental work, but also, as we summarize below, have stimulated many successful theoretical and analytical efforts in defining and understanding regulatory network behavior in yeast, making yeast once again the leading organism in an emerging field of biology.

Analysis and Display of Genome-Scale Data

The comprehensiveness of these methods led to a new problem: How can such a vast amount of data be analyzed, managed, and presented? Biologists were no longer able to examine these results individually; each of the aforementioned studies was comprised of hundreds of thousands of individual gene expression measurements. A key development in the field was applying clustering algorithms and data visualization tools to allow for analysis and presentation of the large volume of microarray results. An early approach, which is still probably one of the most popular, is the application of hierarchical clustering to group similarly expressed genes together, representing their relative expression levels graphically with colored boxes (Fig. 2A; Weinstein et al. 1997; Eisen et al. 1998; Wen et al. 1998). Several other methods of analyzing gene expression have since been applied, including self-organizing maps (SOM) (Tamayo et al. 1999), simulated annealing clustering (Lukashin and Fuchs 2001), graph-theoretic approaches (Sharan and Shamir 2000), biclustering (Cheng and Church 2000; Tanay et al. 2002; Kluger et al. 2003), and other sophisticated approaches (see Ihmels et al. 2002).

Gene Ontology

As genome-scale data accumulated, it quickly became clear that interpretation would depend critically on high-quality functional annotation. Without reliable underlying annotation, even the best clustering algorithm will not allow one to make sense of the data. Not only was some description needed for as many genes as possible, but that description needed to be written by using a controlled vocabulary, so that it was easy to search and find, for example, all transcription factors known in the genome. For this task, the GO was developed. GO is a structured, controlled vocabulary that describes the biological processes, functions, and

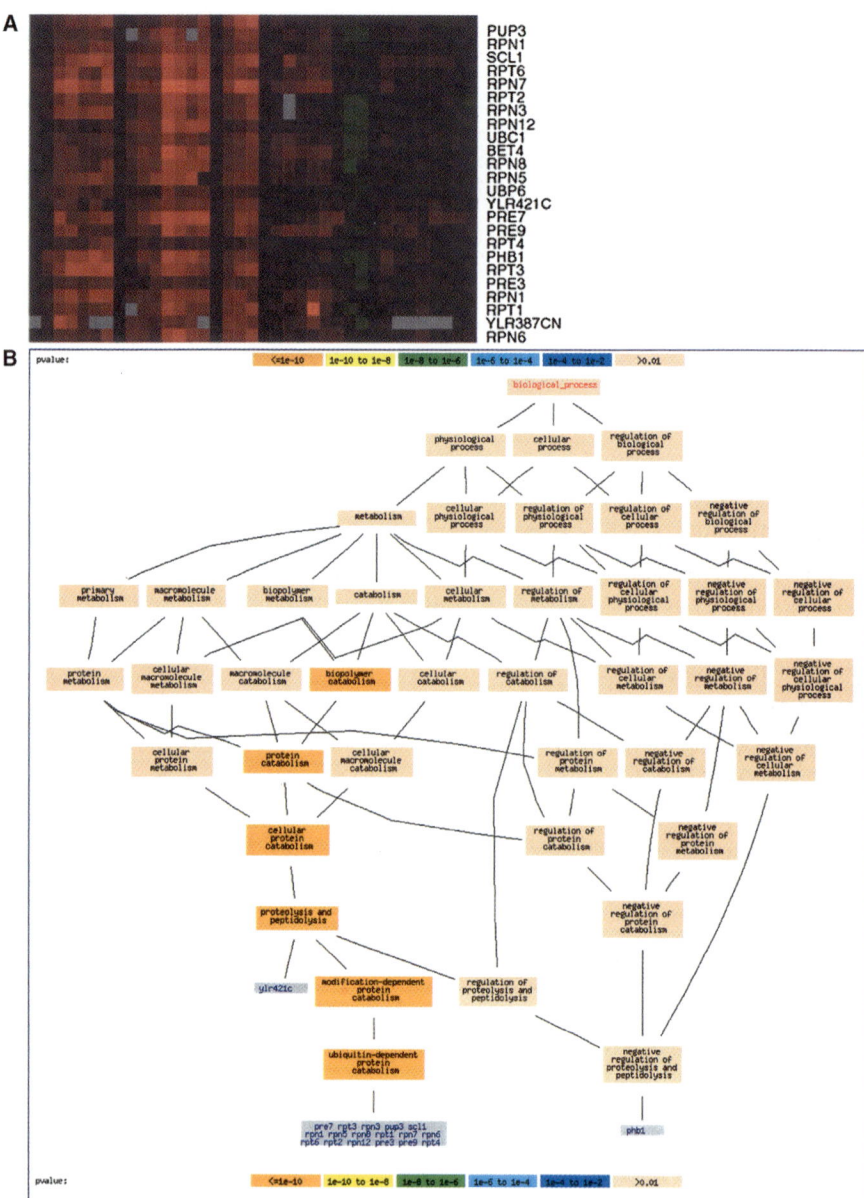

Figure 2. (*A*) Display of a group of genes that exhibit similar expression during the DNA damage response as described in Gasch et al. (2001). Red indicates increased expression, while green indicates decreased expression levels. Each gene's expression is represented by a single row of colored boxes, while each sample is represented by a single column. (*B*) GO Term Finder results with the cluster from *A* as input; the most significant enriched GO Term is "proteolysis and peptidolysis," with a *P*-value of 1.26^{-28}.

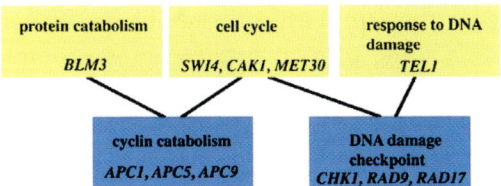

Figure 3. The Gene Ontology: a structured, controlled vocabulary to describe gene products. The diagram above is a small part of the biological process ontology. Parent terms are yellow boxes, child terms are blue boxes, and a sampling of genes associated with each GO term are in italics. Note that child terms can have multiple parents, allowing for more accurate representation of complex relationships among different biological processes.

locations of gene products (Ashburner et al. 2000). The GO is structured such that broad, general terms are parents of more specific terms, child terms can have more than one parent term, and these parent–child relationships are captured in the ontology (Fig. 3).

The initial goal of GO was to improve queries of orthologous genes across different organism databases without having to deal with nomenclature differences. For example, with GO, one would not need to know that "*CDC25*" is the yeast homolog of "*son of sevenless*" to link from the yeast *CDC25* entry in SGD to the *son of sevenless* entry in FlyBase; instead one could bridge the databases by using the GO-controlled vocabulary term "Ras guanyl-nucleotide exchange factor." Soon after its inception, GO was also applied to provide succinct descriptions of genes in expression clusters, facilitating visualization of the biological importance of the data during analysis by, for instance, the Eisen TreeView program (Eisen et al. 1998; Saldanha 2004).

Computing and Validating Inferences from Experiments

The real power of GO emerged when computational biologists began to use it to validate inferences made from analysis of expression data. GO Term Finder is an early tool that facilitates automated analysis of the biological roles of groups of genes. It searches for significant shared GO terms used to describe any group of genes, such as genes that are coexpressed in a microarray experiment (Boyle et al. 2004). For example, a tight cluster of coexpressed genes from the aforementioned study of stress responses (Gasch et al. 2000) is shown in Figure 2A. When these genes are analyzed by using the GO Term Finder, it is clear that this cluster consists of genes involved in proteolysis (Fig. 2B). There are several other tools that similarly utilize GO for analyzing groups of genes (see the GO home page at http://www.geneontology.org). With these tools in hand, yeast researchers began to make several new biological insights.

INSIGHTS INTO THE GLOBAL TRANSCRIPTIONAL NETWORK

These advances stimulated the use of DNA microarray technology to address questions of global gene regulation, using what has turned out to be a powerful combination of computation and experiment (for a recent review, see Chua et al. 2004). Computational methods were developed that go well beyond clustering coregulated genes to identify short DNA sequences that might be *cis*-acting regulatory elements shared among the coregulated genes and the transcription factors that may bind them; an early example is that of Bussemaker et al. (2001). In later studies, probabilistic methods were used to identify regulatory "modules" by using combinations of upstream regulatory sequence and expression data (Wang et al. 2002, 2005; Segal et al. 2003).

Comparative genomics have also been used successfully to examine transcriptional networks. In two independent comparative genomics studies, conserved regulatory elements were identified by aligning the intergenic regions of closely related *Saccharomyces* species and then searching within them for conserved sequence motifs (Cliften et al. 2003; Kellis et al. 2003). Pritsker et al. (2004) identified putative transcription factor binding motifs by using Gibbs sampling to search for significant regulatory elements within promoters of orthologous genes from 13 hemiascomycetous yeasts.

Transcription factor binding sites are now also being globally identified via experimental means by using a combination of chromatin-immunoprecipitation (chIP) and microarray technology. In these experiments, DNA bound to a given transcription factor is isolated by chIP of the epitope-tagged transcription factor; these DNA fragments are then hybridized to a microarray of intergenic regions (chip). These chIP–chip experiments have now identified binding sites for many transcription factors involved in a variety of biological processes (see Khodursky et al. 2000; Iyer et al. 2001; Lieb et al. 2001; Raghuraman et al. 2001; Kurdistani et al. 2002; Lee et al. 2002; Ng et al. 2003).

With traditional microarrays, chIP–chip experiments, and genomic sequence available, several methods have been developed to elucidate transcriptional networks by integrating these different data sources. Lee et al. (2002), Bar Joseph et al. (2003), and Gao et al. (2004) are all prominent examples where chIP–chip and expression data were combined to generate regulatory modules. In the most comprehensive study to date, Harbison and colleagues (2004) used a combination of experimental (chIP–chip), comparative genomics, and motif discovery methods to identify putative DNA binding sites for >200 transcription factors in yeast.

Most impressively, Beer and Tavazoie (2004) recently applied a probabilistic framework to predict gene expression based on sequence information. In their elegant approach, a Bayesian network takes as input different properties of sequence elements upstream of a gene and outputs the likelihood of that gene exhibiting a particular expression pattern. Their combinatorial rules correctly predicted patterns of gene expression for 73% of the yeast genes studied (1898 of 2587 genes in five test sets), with 27% predicted to be in an expression pattern different than their actual expression pattern; the P-value for the prediction of 73% is $<10^{-127}$. They were then able to use their method to predict regulatory elements in the worm.

INTERACTION NETWORKS

With the completed genome also came the opportunity to generate comprehensive genetic and physical interaction maps. Synthetic Genetic Array (SGA) analysis (Tong et al. 2001) uses the comprehensive ORF-deletion collection (Winzeler et al. 1999; Giaever et al. 2002) and a clever genetic selection as the basis for systematically generating double mutants. Assessment of the growth properties of the double mutants generates large-scale genetic interaction maps based on the concept of "synthetic lethality." Synthetic lethality, described first in *Drosophila* by Dobzhansky (1946) and Sturtevant (1956) and in yeast by Novick et al. (1989), occurs when the combination of two mutations causes lethality, while neither mutation by itself is lethal. The SGA technique has provided a means to perform genetic interaction analysis on a large scale, yielding a genetic interaction network containing ~1000 genes and ~4000 interactions (Tong et al. 2004; see also Pan et al. 2004). Large-scale protein–protein interaction networks have also been generated by using the two-hybrid system and mass spectrometry (Ito et al. 2000, 2001; Uetz et al. 2000; Gavin et al. 2002; Ho et al. 2002; Hazbun et al. 2003).

These results have generated increasing theoretical efforts aimed at characterizing regulatory and functional interaction networks. For example, Yeger-Lotem and colleagues (2004) adapted and extended methods that Shen-Orr et al. (2002) applied in *Escherichia coli* to discover significant network motifs in a combined network of regulatory and physical interactions. Shen-Orr et al. (2002) examined interactions between *E. coli* transcription factors and the operons that they regulate to discover "network motifs," or patterns of connections among genes in the network that occur significantly more frequently than in randomized networks. Yeger-Lotem and colleagues (2004) extended this concept to analyze a network that is comprised of both transcription factor–target interactions and protein–protein physical interactions. They found a few two- and three-protein motifs (e.g., two transcription factors interacting to regulate a third gene) and many (63) four-protein motifs, which in almost all cases were combinations of the smaller motifs. These results suggest that smaller motifs serve as building blocks to construct the larger cellular network.

While already having generated much useful biological data, the large-scale methods for profiling genetic and physical interactions for the entire genome (i.e., all 6000 genes by all 6000 genes) are still labor intensive and, as with all high-throughput methods, generate both false-positive and false-negative results. Also, as noted above, there is distressingly poor concordance among results of the several large-scale studies (Hughes et al. 2004). In the studies where authors systematically compared their high-throughput results with those from individual experiments, the amount of overlap between the two is surprisingly low (Ito et al. 2001; Ho et al. 2002). This type of bench marking, while extremely useful, has been rare thus far because of the lack of a comprehensive collection of individual experimental results culled from the literature. Computational methods that integrate multiple types of experimental evidence to verify results, associate interactions with probability scores, and predict novel interactions or gene functions based on these combined interactions can address some of the limitations inherent with these high-throughput methods. Triosh and Barkai (2005) described a method to verify protein–protein interactions by examining whether orthologs of the interaction partners are coexpressed. Similarly, Yu et al. (2004) assessed whether protein–protein or DNA–protein interactions can be confidently transferred from one organism to another by examining "joint" sequence conservation of the interacting proteins. In another approach, Bayesian frameworks have been applied to integrate different types of functional genomics data (e.g., genetic and physical interactions and correlated expression) to generate the probability of a functional link for all possible gene pairs (Jansen et al. 2003; Troyanskaya et al. 2003; Lee et al. 2004). These pairwise correlations can then be used to cluster functionally related genes together and thus can predict functions for previously uncharacterized genes. A different computational approach uses probabilistic decision trees to integrate different types of data in order to predict phenotypes (King et al. 2003) or synthetic lethal interactions (Wong et al. 2004), also meant to lead to functional predictions for uncharacterized genes. For all of these methods, the number of confirmed successful predictions for as-yet-uncharacterized genes is still too small to constitute a robust test of their efficacy.

PUBLIC RESOURCES FOR YEAST GENOMICS

We are confident that the genome-scale experimentation and the integrative analytical approaches sketched above will provide increasing insights into the biology of yeast and, as we have indicated, other eukaryotes. However, if the data are not publicly available in forms that are machine parsable, these studies will not reach their full potential in terms of generating useful biological knowledge. Toward that end, standards for several types of functional genomics data have been created. The Open Biological Ontologies (OBO; http://obo.sourceforge.net/) site is a Web page that provides links to various standards and controlled vocabulary projects, including the Microarray Gene Expression Data (MGED) Society (Spellman et al. 2002; Causton and Game 2003), the GO project (Ashburner et al. 2000), the Proteomics Standards Initiative (standards for protein–protein interactions and mass spectrometry data) (Hermjakob et al. 2004a), and BioPAX (a common exchange format for pathways data).

In addition to data format standards, public databases that provide data in these formats for bulk download are also needed. Table 1 lists some of the public databases that provide genomics data sets for bulk download by various means. The Generic Model Organism Database (GMOD) project is a collaboration among the model organism databases to develop reusable software components suitable for sharing across different database groups; many useful, freely available software components are available at the GMOD Web site (http://www.gmod.org).

CONCLUSION AND SOME THOUGHTS ABOUT THE FUTURE

In the decade since the release of the yeast genome DNA sequence, there has been the expected change in the technology of yeast research as well as a rather surprising change in its goals. Indeed, as we have outlined, most of the new understanding of individual yeast gene functions has

Table 1. Some sources of functional genomics data collections for *Saccharomyces cerevisiae*

Database	Data type	References	URL
SGD	Several	Ball et al. 2001	http://www.yeastgenome.org
CYGD/MIPS	Several	Guldener et al. 2005	http://mips.gsf.de/genre/proj/yeast/
bioGRID	Genetic/physical interaction	Breitkreutz et al. 2003	http://biodata.mshri.on.ca/yeast_grid/
BIND	Genetic/physical interaction	Bader et al. 2003	http://www.blueprint.org/bind/bind.php
DIP	Physical interaction	Xenarios et al. 2002	http://dip.doe-mbi.ucla.edu/dip/Main.cgi
MINT	Physical interaction	Zanzoni et al. 2002	http://160.80.34.4/mint/
IntAct	Physical interaction	Hermjakob et al. 2004b	http://www.ebi.ac.uk/intact/index.html
Deletion Consortium	Phenotype analysis	Giaever et al. 2002; Winzeler et al. 1999	http://www-sequence.stanford.edu/group/yeast_deletion_project/data_sets.html
GEO	MicroArray	Edgar et al. 2002	http://www.ncbi.nlm.nih.gov/geo/
Array Express	MicroArray	Brazma et al. 2003	http://www.ebi.ac.uk/arrayexpress/
YMGV	MicroArray	Marc et al. 2001	http://www.transcriptome.ens.fr/ymgv/
SMD	MicroArray	Gollub et al. 2003	http://smd.stanford.edu/
OPD	Mass Spec/Proteomics	Prince et al. 2004	http://bioinformatics.icmb.utexas.edu/OPD

List of the major sources of yeast functional genomics data; in addition to the main SGD site, yeast genome data are also distributed via SGD Lite (http://sgdlite.princeton.edu), a lightweight yeast genome database, which is built from GMOD components and can be downloaded and installed locally.

come from comparative genomics and relatively little from the high-throughput genomic technologies. The latter have, however, fueled the changes in goals, from a focus on individual genes and their interactions to a focus on the system-level transactions that make the robustly functioning organisms we find in nature.

The future of genome-scale technologies is, nevertheless, very promising. It is not clear whether the slow rate at which new annotations are verified is caused by problems in the data, data analysis and representation, or by a more simple lack of focus on the need for such verification. Some methods, now in early stages of development, will no doubt help: Among these are methods based on natural variation (examples include Brem et al. 2002; Steinmetz et al. 2002; Fay et al. 2004), methods that are not limited to nonessential genes (e.g., synthetic lethality with conditional alleles) (see Tong et al. 2004) or titratable promoter alleles (Mnaimneh et al. 2004), methods that study the locations and movements of intracellular molecules (Ghaemmaghami et al. 2003; Huh et al. 2003), and methods that use more biological information from other species (for example, Harbison et al. 2004).

We seem to be just at the dawn of the ability to construct truly quantitative, let alone comprehensive, models of functional and regulatory network interactions at the system level. The apparently simplest case might well be understanding metabolism at this level (a nascent field already being called "metabolomics"; see Famili et al. 2003; Forster et al. 2003; for a review, see Smedsgaard and Nielsen 2005). To this end, it is clear that we lack most of the required basic measurements, such as the concentrations of metabolites in real time after perturbations in the style of Gasch et al. (2000) and Idecker et al. (2001b). Fortunately, in the post–genome-sequence era, it is much easier to acquire this kind of information on a comprehensive scale, and we believe that this will be the path forward. Another challenge of this nature is to understand the basis on which selection acts on the ensemble of genes, proteins, networks, and systems to produce organisms capable of surviving in new environments.

Finally, there remains the eternal issue of verification. We expect that the need for tests of hypotheses generated by genome scale experiments and quantitative models will persist for a very long time. As has always been the case, every model (and the data used to generate it) must be tested, and to be tested, it must be specified in full and available to the public. The yeast community has an excellent record in this regard, one that we believe is a major reason that yeast continues to be the very model of a model organism.

REFERENCES

Alberts, B.M., Botstein, D., Brenner, S., Cantor, C.R., Doolittle, R.F., Hood, L., McKusick, V.A., Nathans, D., Olson, M.V., Orkin, S. 1988. Mapping and sequencing the human genome. In *National Resource Council*. National Academy Press, Washington, DC.

Ashburner, M., Ball, C.A., Blake, J.A., Botstein, D., Butler, H., Cherry, J.M., Davis, A.P., Dolinski, K., Dwight, S.S., Eppig, J.T., et al. 2000. Gene ontology: Tool for the unification of biology. *Nat. Genet.* **25:** 25–29.

Bader, G.D., Betel, D., and Hogue, C.W. 2003. BIND: The Biomolecular Interaction Network Database. *Nucleic Acids Res.* **31:** 248–250.

Balakrishnan, R., Christie, K.R., Costanzo, M.C., Dolinski, K., Dwight, S.S., Engel, S.R., Fisk, D.G., Hirschman, J.E., Hong, E.L., Nash, R., et al. 2005. Fungal BLAST and Model Organism BLASTP Best Hits: New comparison resources at the *Saccharomyces* Genome Database (SGD). *Nucleic Acids Res.* **33:** D374–D377.

Ball, C.A., Jin, H., Sherlock, G., Weng, S., Matese, J.C., Andrada, R., Binkley, G., Dolinski, K., Dwight, S.S., Harris, M.A., et al. 2001. *Saccharomyces* Genome Database provides tools to survey gene expression and functional analysis data. *Nucleic Acids Res.* **29:** 80–81.

Bar-Joseph, Z., Gerber, G.K., Lee, T.I., Rinaldi, N.J., Yoo, J.Y., Robert, F., Gordon, D.B., Fraenkel, E., Jaakkola, T.S., Young, R.A., and Gifford, D.K. 2003. Computational discovery of gene modules and regulatory networks. *Nat. Biotechnol.* **21:** 1337–1342.

Beer, M.A. and Tavazoie, S. 2004. Predicting gene expression from sequence. *Cell* **117:** 185–198.

Boyle, E.I., Weng, S., Gollub, J., Jin, H., Botstein, D., Cherry, J.M., and Sherlock, G. 2004. GO::TermFinder: Open source software for accessing Gene Ontology information and finding significantly enriched Gene Ontology terms associated with a list of genes. *Bioinformatics* **20:** 3710–3715.

Brazma, A., Parkinson, H., Sarkans, U., Shojatalab, M., Vilo, J., Abeygunawardena, N., Holloway, E., Kapushesky, M., Kemmeren, P., Lara, G.G., et al. 2003. ArrayExpress: A public repository for microarray gene expression data at the EBI. *Nucleic Acids Res.* **31:** 68–71.

Breitkreutz, B.J., Stark, C., and Tyers, M. 2003. The GRID: The General Repository for Interaction Datasets. *Genome Biol.* **4:** R23.

Brem, R.B., Yvert, G., Clinton, R., and Kruglyak, L. 2002. Genetic dissection of transcriptional regulation in budding yeast. *Science* **296:** 752–755.

Brown, P.O. and Botstein, D. 1999. Exploring the new world of the genome with DNA microarrays. *Nat. Genet.* **21:** 33–37.

Bussemaker, H.J., Li, H., and Siggia, E.D. 2001. Regulatory element detection using correlation with expression. *Nat. Genet.* **27:** 167–171.

Causton, H.C. and Game, L. 2003. MGED comes of age. *Genome Biol.* **4:** 351.

Cheng, Y. and Church, G.M. 2000. Biclustering of expression data. *Proc. Int. Conf. Intell. Syst. Mol. Biol.* **8:** 93–103.

Chervitz, S.A., Aravind, L., Sherlock, G., Ball, C.A., Koonin, E.V., Dwight, S.S., Harris, M.A., Dolinski, K., Mohr, S., Smith, T., et al. 1998. Comparison of the complete protein sets of worm and yeast: orthology and divergence. *Science* **282:** 2022–2028.

Cho, R.J., Campbell, M.J., Winzeler, E.A., Steinmetz, L., Conway, A., Wodicka, L., Wolfsberg, T.G., Gabrielian, A.E., Landsman, D., Lockhart, D.J., et al. 1998. A genome-wide transcriptional analysis of the mitotic cell cycle. *Mol. Cell* **2:** 65–73.

Chu, S., DeRisi, J., Eisen, M., Mulholland, J., Botstein, D., Brown, P.O., and Herskowitz, I. 1998. The transcriptional program of sporulation in budding yeast. *Science* **282**: 699–705.

Chua, G., Robinson, M.D., Morris, Q., and Hughes, T.R. 2004. Transcriptional networks: Reverse-engineering gene regulation on a global scale. *Curr. Opin. Microbiol.* **7**: 638–646.

Cliften, P., Sudarsanam, P., Desikan, A., Fulton, L., Fulton, B., Majors, J., Waterston, R., Cohen, B.A., and Johnston, M. 2003. Finding functional features in *Saccharomyces* genomes by phylogenetic footprinting. *Science* **301**: 71–76.

Costanzo, M.C., Hogan, J.D., Cusick, M.E., Davis, B.P., Fancher, A.M., Hodges, P.E., Kondu, P., Lengieza, C., Lew-Smith, J.E., Lingner, C., et al. 2000. The yeast proteome database (YPD) and *Caenorhabditis elegans* proteome database (WormPD): Comprehensive resources for the organization and comparison of model organism protein information. *Nucleic Acids Res.* **28**: 73–76.

DeRisi, J.L., Iyer, V.R., and Brown, P.O. 1997. Exploring the metabolic and genetic control of gene expression on a genomic scale. *Science* **278**: 680–686.

Dobzhansky, T. 1946. Genetics of natural populations, XIII: Recombination and variability in populations of *Drosophila pseudoobscura*. *Genetics* **31**: 269–290.

Edgar, R., Domrachev, M., and Lash, A.E. 2002. Gene Expression Omnibus: NCBI gene expression and hybridization array data repository. *Nucleic Acids Res.* **30**: 207–210.

Eisen, M.B., Spellman, P.T., Brown, P.O., and Botstein, D. 1998. Cluster analysis and display of genome-wide expression patterns. *Proc. Natl. Acad. Sci.* **95**: 14863–14868.

Famili, I., Forster, J., Nielsen, J., and Palsson, B.Ø. 2003. *Saccharomyces cerevisiae* phenotypes can be predicted by using constraint-based analysis of a genome-scale reconstructed metabolic network. *Proc. Natl. Acad. Sci.* **100**: 13134–13139.

Fay, J.C., McCullough, H.L., Sniegowski, P.D., and Eisen, M.B. 2004. Population genetic variation in gene expression is associated with phenotypic variation in *Saccharomyces cerevisiae*. *Genome Biol.* **5**: R26.

Forster, J., Famili, I., Fu, P., Palsson, B.Ø., and Nielsen, J. 2003. Genome-scale reconstruction of the *Saccharomyces cerevisiae* metabolic network. *Genome Res.* **13**: 244–253.

Gao, F., Foat, B.C., and Bussemaker, H.J. 2004. Defining transcriptional networks through integrative modeling of mRNA expression and transcription factor binding data. *BMC Bioinformatics* **5**: 31.

Gasch, A.P., Spellman, P.T., Kao, C.M., Carmel-Harel, O., Eisen, M.B., Storz, G., Botstein, D., and Brown, P.O. 2000. Genomic expression programs in the response of yeast cells to environmental changes. *Mol. Biol. Cell* **11**: 4241–4257.

Gasch, A.P., Huang, M., Metzner, S., Botstein, D., Elledge, S.J., and Brown, P.O. 2001. Genomic expression responses to DNA-damaging agents and the regulatory role of the yeast ATR homolog Mec1p. *Mol. Biol. Cell* **12**: 2987–3003.

Gavin, A.C., Bosche, M., Krause, R., Grandi, P., Marzioch, M., Bauer, A., Schultz, J., Rick, J.M., Michon, A.M., Cruciat, C.M., et al. 2002. Functional organization of the yeast proteome by systematic analysis of protein complexes. *Nature* **415**: 141–147.

Ghaemmaghami, S., Huh, W.K., Bower, K., Howson, R.W., Belle, A., Dephoure, N., O'Shea, E.K., and Weissman, J.S. 2003. Global analysis of protein expression in yeast. *Nature* **425**: 737–741.

Giaever, G., Chu, A.M., Ni, L., Connelly, C., Riles, L., Veronneau, S., Dow, S., Lucau-Danila, A., Anderson, K., Andre, B., et al. 2002. Functional profiling of the *Saccharomyces cerevisiae* genome. *Nature* **418**: 387–391.

Goffeau, A., Barrell, B.G., Bussey, H., Davis, R.W., Dujon, B., Feldmann, H., Galibert, F., Hoheisel, J.D., Jacq, C., Johnston, M., et al. 1996. Life with 6000 genes. *Science* **274:** 546, 563–567.

Gollub, J., Ball, C.A., Binkley, G., Demeter, J., Finkelstein, D.B., Hebert, J.M., Hernandez-Boussard, T., Jin, H., Kaloper, M., Matese, J.C., et al. 2003. The Stanford Microarray Database: Data access and quality assessment tools. *Nucleic Acids Res.* **31:** 94–96.

Guldener, U., Munsterkotter, M., Kastenmuller, G., Strack, N., van Helden, J., Lemer, C., Richelles, J., Wodak, S.J., Garcia-Martinez, J., Perez-Ortin, J.E., et al. 2005. CYGD: The Comprehensive Yeast Genome Database. *Nucleic Acids Res.* **33:** D364–D368.

Harbison, C.T., Gordon, D.B., Lee, T.I., Rinaldi, N.J., Macisaac, K.D., Danford, T.W., Hannett, N.M., Tagne, J.B., Reynolds, D.B., Yoo, J., et al. 2004. Transcriptional regulatory code of a eukaryotic genome. *Nature* **431:** 99–104.

Hartwell, L.H., Hopfield, J.J., Leibler, S., and Murray, A.W. 1999. From molecular to modular cell biology. *Nature* **402:** C47–C52.

Haugen, A.C., Kelley, R., Collins, J.B., Tucker, C.J., Deng, C., Afshari, C.A., Brown, J.M., Ideker, T., and Van Houten, B. 2004. Integrating phenotypic and expression profiles to map arsenic-response networks. *Genome Biol.* **5:** R95.

Hazbun, T.R., Malmstrom, L., Anderson, S., Graczyk, B.J., Fox, B., Riffle, M., Sundin, B.A., Aranda, J.D., McDonald, W.H., Chiu, C.H., et al. 2003. Assigning function to yeast proteins by integration of technologies. *Mol. Cell* **12:** 1353–1365.

Hermjakob, H., Montecchi-Palazzi, L., Bader, G., Wojcik, J., Salwinski, L., Ceol, A., Moore, S., Orchard, S., Sarkans, U., von Mering, C., et al. 2004a. The HUPO PSI's molecular interaction format: A community standard for the representation of protein interaction data. *Nat. Biotechnol.* **22:** 177–183.

Hermjakob, H., Montecchi-Palazzi, L., Lewington, C., Mudali, S., Kerrien, S., Orchard, S., Vingron, M., Roechert, B., Roepstorff, P., Valencia, A., et al. 2004b. IntAct: An open source molecular interaction database. *Nucleic Acids Res.* **32:** D452–D455.

Ho, Y., Gruhler, A., Heilbut, A., Bader, G.D., Moore, L., Adams, S.L., Millar, A., Taylor, P., Bennett, K., Boutilier, K., et al. 2002. Systematic identification of protein complexes in *Saccharomyces cerevisiae* by mass spectrometry. *Nature* **415:** 180–183.

Hughes, T.R., Robinson, M.D., Mitsakakis, N., and Johnston, M. 2004. The promise of functional genomics: Completing the encyclopedia of a cell. *Curr. Opin. Microbiol.* **7:** 546–554.

Huh, W.K., Falvo, J.V., Gerke, L.C., Carroll, A.S., Howson, R.W., Weissman, J.S., and O'Shea, E.K. 2003. Global analysis of protein localization in budding yeast. *Nature* **425:** 686–691.

Ideker, T., Galitski, T., and Hood, L. 2001a. A new approach to decoding life: Systems biology. *Annu. Rev. Genomics Hum. Genet.* **2:** 343–372.

Ideker, T., Thorsson, V., Ranish, J.A., Christmas, R., Buhler, J., Eng, J.K., Bumgarner, R., Goodlett, D.R., Aebersold, R., and Hood, L. 2001b. Integrated genomic and proteomic analyses of a systematically perturbed metabolic network. *Science* **292:** 929–934.

Ihmels, J., Friedlander, G., Bergmann, S., Sarig, O., Ziv, Y., and Barkai, N. 2002. Revealing modular organization in the yeast transcriptional network. *Nat. Genet.* **31:** 370–377.

Ito, T., Tashiro, K., Muta, S., Ozawa, R., Chiba, T., Nishizawa, M., Yamamoto, K., Kuhara, S., and Sakaki, Y. 2000. Toward a protein–protein interaction map of the budding yeast: A comprehensive system to examine two-hybrid interactions in all possible combinations between the yeast proteins. *Proc. Natl. Acad. Sci.* **97:** 1143–1147.

Ito, T., Chiba, T., Ozawa, R., Yoshida, M., Hattori, M., and Sakaki, Y. 2001. A comprehensive two-hybrid analysis to explore the yeast protein interactome. *Proc. Natl. Acad. Sci.* **98:** 4569–4574.

Iyer, V.R., Horak, C.E., Scafe, C.S., Botstein, D., Snyder, M., and Brown, P.O. 2001. Genomic binding sites of the yeast cell-cycle transcription factors SBF and MBF. *Nature* **409:** 533–538.

Jansen, R., Yu, H., Greenbaum, D., Kluger, Y., Krogan, N.J., Chung, S., Emili, A., Snyder, M., Greenblatt, J.F., and Gerstein, M. 2003. A Bayesian networks approach for predicting protein–protein interactions from genomic data. *Science* **302:** 449–453.

Kellis, M., Patterson, N., Endrizzi, M., Birren, B., and Lander, E.S. 2003. Sequencing and comparison of yeast species to identify genes and regulatory elements. *Nature* **423:** 241–254.

Khodursky, A.B., Peter, B.J., Schmid, M.B., DeRisi, J., Botstein, D., Brown, P.O., and Cozzarelli, N.R. 2000. Analysis of topoisomerase function in bacterial replication fork movement: Use of DNA microarrays. *Proc. Natl. Acad. Sci.* **97:** 9419–9424.

King, O.D., Lee, J.C., Dudley, A.M., Janse, D.M., Church, G.M., and Roth, F.P. 2003. Predicting phenotype from patterns of annotation. *Bioinformatics* **19(Suppl 1):** i183–i189.

Kluger, Y., Basri, R., Chang, J.T., and Gerstein, M. 2003. Spectral biclustering of microarray data: Coclustering genes and conditions. *Genome Res.* **13:** 703–716.

Kurdistani, S.K., Robyr, D., Tavazoie, S., and Grunstein, M. 2002. Genome-wide binding map of the histone deacetylase Rpd3 in yeast. *Nat. Genet.* **31:** 248–254.

Lander, E.S., Linton, L.M., Birren, B., Nusbaum, C., Zody, M.C., Baldwin, J., Devon, K., Dewar, K., Doyle, M., FitzHugh, W., et al. 2001. Initial sequencing and analysis of the human genome. *Nature* **409:** 860–921.

Lee, T.I., Rinaldi, N.J., Robert, F., Odom, D.T., Bar-Joseph, Z., Gerber, G.K., Hannett, N.M., Harbison, C.T., Thompson, C.M., Simon, I., et al. 2002. Transcriptional regulatory networks in *Saccharomyces cerevisiae*. *Science* **298:** 799–804.

Lee, I., Date, S.V., Adai, A.T., and Marcotte, E.M. 2004. A probabilistic functional network of yeast genes. *Science* **306:** 1555–1558.

Lieb, J.D., Liu, X., Botstein, D., and Brown, P.O. 2001. Promoter-specific binding of Rap1 revealed by genome-wide maps of protein–DNA association. *Nat. Genet.* **28:** 327–334.

Lukashin, A.V. and Fuchs, R. 2001. Analysis of temporal gene expression profiles: Clustering by simulated annealing and determining the optimal number of clusters. *Bioinformatics* **17:** 405–414.

Marc, P., Devaux, F., and Jacq, C. 2001. yMGV: A database for visualization and data mining of published genome-wide yeast expression data. *Nucleic Acids Res.* **29:** e63.

Mewes, H.W., Albermann, K., Heumann, K., Liebl, S., and Pfeiffer, F. 1997. MIPS: A database for protein sequences, homology data and yeast genome information. *Nucleic Acids Res.* **25:** 28–30.

Mnaimneh, S., Davierwala, A.P., Haynes, J., Moffat, J., Peng, W.T., Zhang, W., Yang, X., Pootoolal, J., Chua, G., Lopez, A., et al. 2004. Exploration of essential gene functions via titratable promoter alleles. *Cell* **118:** 31–44.

Ng, H.H., Robert, F., Young, R.A., and Struhl, K. 2003. Targeted recruitment of Set1 histone methylase by elongating Pol II provides a localized mark and memory of recent transcriptional activity. *Mol. Cell* **11:** 709–719.

Novick, P., Osmond, B.C., and Botstein, D. 1989. Suppressors of yeast actin mutations. *Genetics* **121:** 659–674.

Pan, X., Yuan, D.S., Xiang, D., Wang, X., Sookhai-Mahadeo, S., Bader, J.S., Hieter, P., Spencer, F., and Boeke, J.D. 2004. A robust toolkit for functional profiling of the yeast genome. *Mol. Cell* **16:** 487–496.

Prince, J.T., Carlson, M.W., Wang, R., Lu, P., and Marcotte, E.M. 2004. The need for a public proteomics repository. *Nat. Biotechnol.* **22:** 471–472.

Pritsker, M., Liu, Y.C., Beer, M.A., and Tavazoie, S. 2004. Whole-genome discovery of transcription factor binding sites by network-level conservation. *Genome Res* **14:** 99–108.

Raghuraman, M.K., Winzeler, E.A., Collingwood, D., Hunt, S., Wodicka, L., Conway, A., Lockhart, D.J., Davis, R.W., Brewer, B.J., and Fangman, W.L. 2001. Replication dynamics of the yeast genome. *Science* **294:** 115–121.

Roberts, C.J., Nelson, B., Marton, M.J., Stoughton, R., Meyer, M.R., Bennett, H.A., He, Y.D., Dai, H., Walker, W.L., Hughes, T.R., et al. 2000. Signaling and circuitry of multiple MAPK pathways revealed by a matrix of global gene expression profiles. *Science* **287:** 873–880.

Rubin, G.M., Yandell, M.D., Wortman, J.R., Gabor Miklos, G.L., Nelson, C.R., Hariharan, I.K., Fortini, M.E., Li, P.W., Apweiler, R., Fleischmann, W., et al. 2000. Comparative genomics of the eukaryotes. *Science* **287:** 2204–2215.

Saldanha, A.J. 2004. Java Treeview: Extensible visualization of microarray data. *Bioinformatics* **20:** 3246–3248.

Segal, E., Shapira, M., Regev, A., Pe'er, D., Botstein, D., Koller, D., and Friedman, N. 2003. Module networks: Identifying regulatory modules and their condition-specific regulators from gene expression data. *Nat. Genet.* **34:** 166–176.

Sharan, R. and Shamir, R. 2000. CLICK: A clustering algorithm with applications to gene expression analysis. *Proc. Int. Conf. Intell. Syst. Mol. Biol.* **8:** 307–316.

Shen-Orr, S.S., Milo, R., Mangan, S., and Alon, U. 2002. Network motifs in the transcriptional regulation network of *Escherichia coli*. *Nat. Genet.* **31:** 64–68.

Smedsgaard, J. and Nielsen, J. 2005. Metabolite profiling of fungi and yeast: From phenotype to metabolome by MS and informatics. *J. Exp. Bot.* **56:** 273–286.

Spellman, P.T., Sherlock, G., Zhang, M.Q., Iyer, V.R., Anders, K., Eisen, M.B., Brown, P.O., Botstein, D., and Futcher, B. 1998. Comprehensive identification of cell cycle-regulated genes of the yeast *Saccharomyces cerevisiae* by microarray hybridization. *Mol. Biol. Cell* **9:** 3273–3297.

Spellman, P.T., Miller, M., Stewart, J., Troup, C., Sarkans, U., Chervitz, S., Bernhart, D., Sherlock, G., Ball, C., Lepage, M., et al. 2002. Design and implementation of microarray gene expression markup language (MAGE-ML). *Genome Biol.* **3:** research0046.

Steinmetz, L.M., Sinha, H., Richards, D.R., Spiegelman, J.I., Oefner, P.J., McCusker, J.H., and Davis, R.W. 2002. Dissecting the architecture of a quantitative trait locus in yeast. *Nature* **416:** 326–330.

Sturtevant, A.H. 1956. A highly specific complementary lethal system in *Drosophila melanogaster*. *Genetics* **41:** 118–123.

Tamayo, P., Slonim, D., Mesirov, J., Zhu, Q., Kitareewan, S., Dmitrovsky, E., Lander, E.S., and Golub, T.R. 1999. Interpreting patterns of gene expression with self-organizing maps: Methods and application to hematopoietic differentiation. *Proc. Natl. Acad. Sci.* **96:** 2907–2912.

Tanay, A., Sharan, R., and Shamir, R. 2002. Discovering statistically significant biclusters in gene expression data. *Bioinformatics* **18(Suppl 1):** S136–S144.

Tirosh, I. and Barkai, N. 2005. Computational verification of protein–protein interactions by orthologous co-expression. *BMC Bioinformatics* **6:** 40.

Tong, A.H., Evangelista, M., Parsons, A.B., Xu, H., Bader, G.D., Page, N., Robinson, M., Raghibizadeh, S., Hogue, C.W., Bussey, H., et al. 2001. Systematic genetic analysis with ordered arrays of yeast deletion mutants. *Science* **294:** 2364–2368.

Tong, A.H., Lesage, G., Bader, G.D., Ding, H., Xu, H., Xin, X., Young, J., Berriz, G.F., Brost, R.L., Chang, M., et al. 2004. Global mapping of the yeast genetic interaction network. *Science* **303:** 808–813.

Troyanskaya, O.G., Dolinski, K., Owen, A.B., Altman, R.B., and Botstein, D. 2003. A Bayesian framework for combining heterogeneous data sources for gene function prediction (in *Saccharomyces cerevisiae*). *Proc. Natl. Acad. Sci.* **100:** 8348–8353.

Uetz, P., Giot, L., Cagney, G., Mansfield, T.A., Judson, R.S., Knight, J.R., Lockshon, D., Narayan, V., Srinivasan, M., Pochart, P., et al. 2000. A comprehensive analysis of protein–protein interactions in *Saccharomyces cerevisiae*. *Nature* **403:** 623–627.

Velculescu, V.E., Zhang, L., Zhou, W., Vogelstein, J., Basrai, M.A., Bassett Jr., D.E., Hieter, P., Vogelstein, B., and Kinzler, K.W. 1997. Characterization of the yeast transcriptome. *Cell* **88:** 243–251.

Venter, J.C., Adams, M.D., Myers, E.W., Li, P.W., Mural, R.J., Sutton, G.G., Smith, H.O., Yandell, M., Evans, C.A., Holt, R.A, et al. 2001. The sequence of the human genome. *Science* **291:** 1304–1351.

Wang, W., Cherry, J.M., Botstein, D., and Li, H. 2002. A systematic approach to reconstructing transcription networks in *Saccharomyces cerevisiae*. *Proc. Natl. Acad. Sci.* **99:** 16893–16898.

Wang, W., Cherry, J.M., Nochomovitz, Y., Jolly, E., Botstein, D., and Li, H. 2005. Inference of combinatorial regulation in yeast transcriptional networks: A case study of sporulation. *Proc. Natl. Acad. Sci.* **102:** 1998–2003.

Weinstein, J.N., Myers, T.G., O'Connor, P.M., Friend, S.H., Fornace Jr., A.J., Kohn, K.W., Fojo, T., Bates, S.E., Rubinstein, L.V., Anderson, N.L., et al. 1997. An information-intensive approach to the molecular pharmacology of cancer. *Science* **275:** 343–349.

Wen, X., Fuhrman, S., Michaels, G.S., Carr, D.B., Smith, S., Barker, J.L., and Somogyi, R. 1998. Large-scale temporal gene expression mapping of central nervous system development. *Proc. Natl. Acad. Sci.* **95:** 334–339.

Winzeler, E.A., Shoemaker, D.D., Astromoff, A., Liang, H., Anderson, K., Andre, B., Bangham, R., Benito, R., Boeke, J.D., Bussey, H., et al. 1999. Functional characterization of the *S. cerevisiae* genome by gene deletion and parallel analysis. *Science* **285:** 901–906.

Wong, S.L., Zhang, L.V., Tong, A.H., Li, Z., Goldberg, D.S., King, O.D., Lesage, G., Vidal, M., Andrews, B., Bussey, H., et al. 2004. Combining biological networks to predict genetic interactions. *Proc. Natl. Acad. Sci.* **101:** 15682–15687.

Xenarios, I., Salwinski, L., Duan, X.J., Higney, P., Kim, S.M., and Eisenberg, D. 2002. DIP, the Database of Interacting Proteins: A research tool for studying cellular networks of protein interactions. *Nucleic Acids Res.* **30:** 303–305.

Yeger-Lotem, E., Sattath, S., Kashtan, N., Itzkovitz, S., Milo, R., Pinter, R.Y., Alon, U., and Margalit, H. 2004. Network motifs in integrated cellular networks of transcription-regulation and protein-protein interaction. *Proc. Natl. Acad. Sci.* **101:** 5934–5939.

Yu, H., Luscombe, N.M., Lu, H.X., Zhu, X., Xia, Y., Han, J.D., Bertin, N., Chung, S., Vidal, M., and Gerstein, M. 2004. Annotation transfer between genomes: Protein–protein interologs and protein–DNA regulogs. *Genome Res.* **14:** 1107–1118.

Zanzoni, A., Montecchi-Palazzi, L., Quondam, M., Ausiello, G., Helmer-Citterich, M., and Cesareni, G. 2002. MINT: A Molecular INTeraction database. *FEBS Lett.* **513:** 135–140.

Zhu, G., Spellman, P.T., Volpe, T., Brown, P.O., Botstein, D., Davis, T.N., and Futcher, B. 2000. Two yeast forkhead genes regulate the cell cycle and pseudohyphal growth. *Nature* **406:** 90–94.

3

Genomics of the Fungal Kingdom: Insights into Eukaryotic Biology

James E. Galagan, Matthew R. Henn, Li-Jun Ma,
Christina A. Cuomo, and Bruce Birren
The Broad Institute of Massachusetts Institute of Technology
and
Harvard University
Cambridge, Massachusetts 02141

THE OVER 1.5 MILLION MEMBERS OF THE FUNGAL KINGDOM (Hawksworth 1991) impact nearly all other forms of life as either friend or foe. Fungi play a critical role in the environment through the decomposition of organic material and through symbiotic relationships with prokaryotes, plants (including algae), and animals. In particular, fungi share a long history with human civilization. References in Greek literature, mushroom stones from Mesoamerica dating to 1000–300 BC (Lowy 1971), and dried mushrooms of *Piptoporus betulinus* found in a pouch around a Stone Age man's neck in the Alps (Rensberger 1992) all attest to this long relationship. The relationship can be beneficial, as in the case of biotransformations such as fermentation and the production of antibiotics or extremely detrimental, as demonstrated by the devastating impacts of mycoses, plant diseases, and mycotoxins (Moss 1987).

Found within the 900 million years (Myr) of evolutionary history of the fungi is an enormous biological diversity (Fig. 1). This diversity encompasses four major groups of fungal organism, i.e., ascomycetes, basidiomycetes, zygomycetes, and chytrids. Fungal cellular physiology and genetics share key components with animal and plant cells, including multicellularity, cytoskeletal structures, development and differentiation, sexual reproduction, cell cycle, intercellular signaling, circadian rhythms, DNA methylation, and chromatin modification. The shared origins of the

---

Supplemental material is available online at www.genome.org.

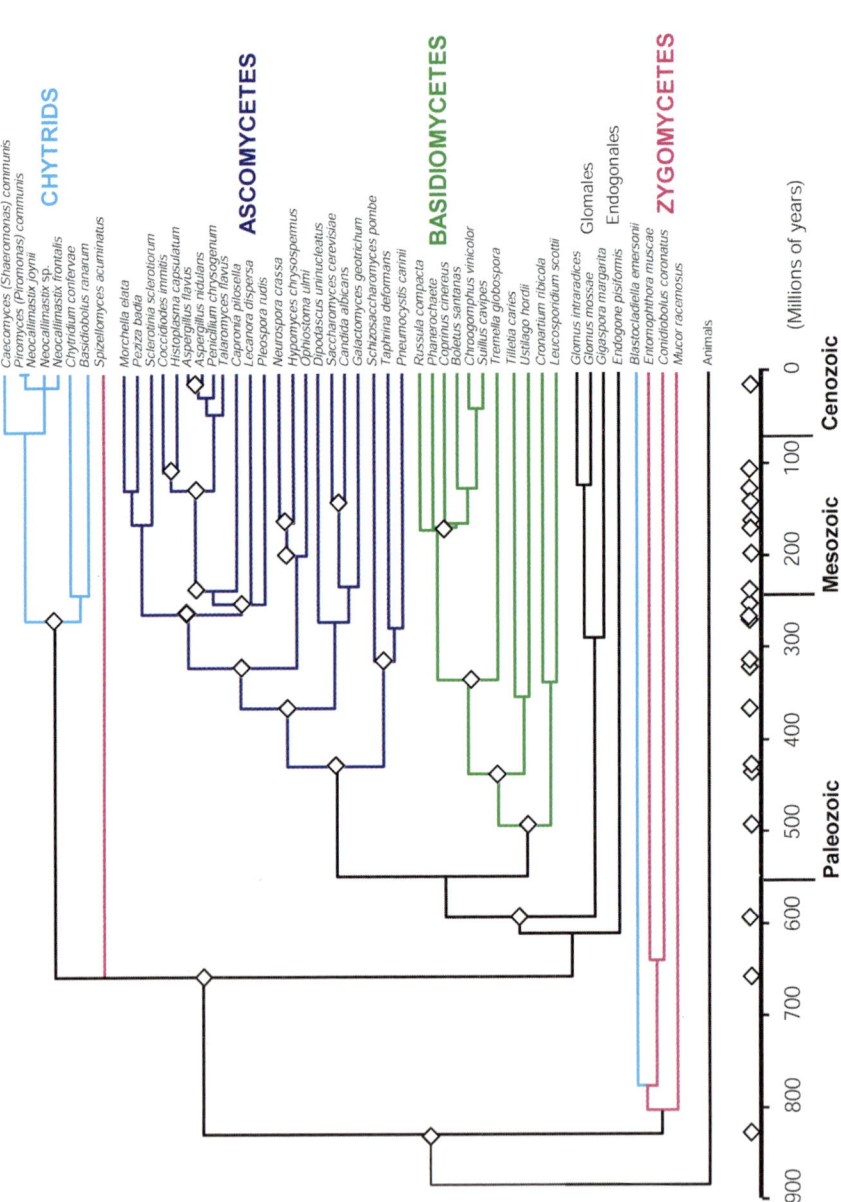

Figure 1. (See facing page for legend.)

genes responsible for these fundamental biological functions between humans and fungi continue to make the understanding of these fungal genes of vital interest to human biology. In addition, their genomes are more easily sequenced and annotated relative to most metazoans and their experimental tractability makes fungi among the most useful model systems in cell biology.

Despite the importance and utility of fungi, until quite recently what was known about their genomes was primarily derived from the sequence of the yeast *Saccharomyces cerevisiae*. But in the last 5 yr, however, there has been an explosion in fungal genomics that has greatly expanded our view of the genetic and physiological diversity of these organisms. We provide here an overview of available fungal genomes and highlight some of the biological insights that have been derived through their analysis. We also discuss insights into the fundamental cellular biology shared between fungi and other eukaryotic organisms. These highlights are not intended to be comprehensive. Specifically, we focus on results derived from whole-genome analysis of fungi other than yeasts, as the genomics of *S. cerevisiae* and related organisms is covered elsewhere in this issue.

FUNGAL GENOMICS' HISTORY AND RESOURCES

The era of fungal genomics—and indeed eukaryotic genomics—was ushered in by the sequencing of the complete genome of the yeast *S. cereviseae*, reported in 1996 (Goffeau et al. 1996). This milestone revolutionized work in yeast and enabled the first global studies of eukaryotic gene function and expression. However, the yeast genome sequence provided only a limited glimpse of the biological diversity of the fungal kingdom. The subsequent completion of *Schizosaccharomyces pombe* (Wood et al. 2002) and *Neurospora crassa* (Galagan et al. 2003) revealed the limits of yeast as a proxy for all other fungi. In particular, the genome of *N. crassa*—the first filamentous fungus to be sequenced—possessed nearly twice as many genes as *S. cerevisiae* and *S. pombe* and lacked homologs to known proteins for over 40% of these genes.

Despite evident need, progress in sequencing fungal genomes was initially slow. To accelerate the pace of fungal genomics, in 2000, a consortium

---

Figure 1. Phylogeny of the fungal kingdom. Major fungal groups colored as indicated by text to *right*. Diamonds indicate evolutionary branch points, and their approximate dating (*bottom*), captured by fungal genomes sequenced in or in progress. Phylogeny based on data from Berbee and Taylor (2001).

of mycologists in collaboration with scientists from the Whitehead Institute/MIT Center for Genome Research—now the Broad Institute—launched the Fungal Genome Initiative (FGI—http://www.broad.mit.edu/annotation/fgi/). The goal of the FGI is to sequence the genomes of fungi from throughout the kingdom. Importantly, the fungi to be sequenced are not selected one at a time without consideration of each other. Rather, they form groups of organisms that maximize their combined value for comparative genomics, evolutionary studies, eukaryotic biology, and medical studies. When the FGI was launched, two fungal genomes were available. Since that time, 23 different fungal genomes have been released (Table 1) through the FGI. These genomes have been matched by a roughly equal number from other centers and projects including the Joint Genome Institute (JGI), the Washington University Genome Sequencing Center, Génolevures, TIGR, the Sanger Institute, the Marine Biological Laboratories (MBL), the Stanford Genome Technology Center, the Duke Center for Genome Research, and the University of British Columbia. These data have been generated through the support of numerous funding agencies, including the National Human Genome Research Institute, the National Science Foundation, the National Institute of Allergy and Infectious Disease, and the US Department of Agriculture and the Department of Energy. Of particular note is the growing partnership between academia and industry, which has resulted in the release of several privately held fungal genome sequences from companies including Monsanto, Syngenta, Biozentrum, Bayer CropScience AG, and Exelixis.

In total, over 40 fungal genomes sequences are currently publicly available with over 40 additional projects underway (Tables 1, 2). These genomes represent important human pathogens, plant pathogens, saprophytes, and model organisms. They also encompass fungi that grow as yeasts, form mycelium or pseudo-hyphae, or are capable of dimorphic (or polymorphic) growth. In addition, they include representatives of all four major fungal groups, i.e., ascomycetes, basidiomycetes, zygomycetes, and chytrids. Importantly, the majority of available fungal genomes fall into clusters of related genomes that enable comparative analysis across a range of evolutionary distances (Fig. 2). These clusters also include related organisms that differ in terms of specific physiological traits (i.e., pathogenicity), thus allowing these traits to be explored through comparison.

Access to these fungal genomic data is available through a growing number of online resources. These resources include the Broad Institute Fungal Genome Initiative Web site (http://www.broad.mit.edu/FGI/), the JGI Integrated Microbial Resource database (http://img.jgi.doe.gov/pub/main.cgi/), the TIGR fungal database (www.tigr.org.tbd/fungal), NCBI

Table 1. Complete fungal and Oomycete genomes[a]

Genus/species	Taxonomy	Sequencing center(s)
Ashbya gossypii[c] (aka *Eremothecium*)	Saccharomycetes	Biozentrum an Syngenta AG
Aspergillus fumigatus[c]	Eurotiomycetes	Sanger Institute & TIGR
Aspergillus nidulans[c]	Eurotiomycetes	Broad Institute & Monsanto
Aspergillus terreus	Eurotiomycetes	Broad Institute
Botrytis cinerea	Leotiomycetes	Broad Institute & Syngenta AG; Genoscope
Candida albicans[c,d]	Saccharomycetes	Stanford Genome Technology Center, Sanger Institute, & Broad Institute
Candida glabrata[c]	Saccharomycetes	Génolevures
Candida guilliermondii (anamorph of *Pichia*)	Saccharomycetes	Broad Institute
Candida lusitaniae (aka *Clavispora*)	Saccharomycetes	Broad Institute
Candida tropicalis	Saccharomycetes	Broad Institute & Génolevures
Chaetomium globosum	Sordariomycetes	Broad Institute
Coprinus cinereus	Homobasidiomycetes	Broad Institute
Coccidiodes immitis[b,c,d]	Eurotiomycetes	Broad Institute
Coccidioides posadasii[c,d]	Eurotiomycetes	TIGR & Broad Institute
Coprinus cinereus	Homobasidiomycetes	Broad Institute
Cryptococcus neoformans (anamorph of *Filobasidiella*)[b,c]	Homobasidiomycetes	Broad Institute, Genome Sciences Center Canada, Duke Center for Genome Research, Stanford Genome Technology Center, & TIGR

(*Continued*)

Table 1. (Continued)

Genus/species	Taxonomy	Sequencing center(s)
Debaryomyces hansenii[c]	Saccharomycetes	Génolevures
Encephalitozoon cuniculi	Microsporidia	Genescope
Fusarium graminearum[c] (aka *Gibberella zeae*)	Sordariomycetes	Broad Institute
Kluyveromyces lactis[c]	Saccharomycetes	Génolevures
Magnaporthe grisea[c]	Sordariomycetes	Broad Institute
Neurospora crassa[c]	Sordariomycetes	Broad Institute
Phanerochaete chrysosporium[c]	Homobasidiomycota	Joint Genome Institute
Phytophthora ramorum	Oomycete	Joint Genome Institute
Phytophthora sojae	Oomycete	Joint Genome Institute
Podospora anserine	Sordariomycete	CNRS & Genoscope
Rhizopus oryzae[c]	Zygomycota	Broad Institute
Saccharomyces bayanus[b,d]	Saccharomycetes	Washington University Genome Sequencing Center, Broad Institute, Génolevures
Saccharomyces castellii	Saccharomycetes	Washington University Genome Sequencing Center
Saccharomyces cerevisiae[b,c]	Saccharomycetes	Stanford Genome Technology Center, Sanger Institute, & Broad Institute

Saccharomyces kluyveri	Saccharomycetes	Washington University Genome Sequencing Center & Génolevures
Saccharomyces kudriazevii	Saccharomycetes	Washington University Genome Sequencing Center
Saccharomyces mikatae[c]	Saccharomycetes	Broad Institute & Washington University Genome Sequencing Center
Saccharomyces paradoxus[c]	Saccharomycetes	Broad Institute
Schizosaccharomyces pombe	Schizosaccharomycetes	Sanger Institute
Sclerotinia sclerotiorum	Leotiomycetes	Broad Institute
Stagonospora nodorum[c] (anamorph of *Phaeosphaeria*)	Dothideomycetes	Broad Institute & International Stagonospora nodorum Genomics Consortium
Uncinocarpus reesei	Eurotiomycetes	Broad Institute
Ustilago maydis[c]	Ustilaginomycota	Broad Institute, Bayer CropScience AG, & Exelixis
Yarrowia lipolytica[c]	Saccharomycetes	Génolevures
Zygosaccharomyces rouxii	Saccharomycetes	Génolevures

[a]Status as of September 10, 2005.
[b]Multiple strain.
[c]Annotated.
[d]Some strains/species still in progress.
A table including URLs is included in Supplemental material.

Table 2. Fungal and Oomycete genome projects in progress[a]

Genus/species	Taxonomy	Sequencing Center(s)
Alternaria brassicicola	Dothideomycetes	Washington University Genome Sequencing Center
Aspergillus clavatus	Eurotiomycetes	TIGR
Aspergillus fischerianus	Eurotiomycetes	TIGR
Aspergillus flavus	Eurotiomycetes	TIGR
Aspergillus niger	Eurotiomycetes	Joint Genome Institute
Aspergillus parasiticus	Eurotiomycetes	University of Oklahoma
Batrachochytrium dendrobatidis[b]	Chytridiomycete	Broad Institute & Joint Genome Institute
Blastomyces dermatitidis (anamorph of Ajellomyces)	Eurotiomycetes	Washington University Genome Sequencing Center
Candida dubliniensis	Saccharomycetes	Sanger Institute
Candida parapsilosis	Saccharomycetes	Sanger Institute
Fusarium oxysporum	Sordariomycetes	Broad Institute
Fusarium verticillioides	Sordariomycetes	Broad Institute & Syngenta AG
Glomus intraradices	Glomeromycetes	Joint Genome Institute
Histoplasma capsulatum (anamorph of Ajellomyces)[b]	Eurotiomycetes	Broad Institute & Washington University Genome Sequencing Center
Kluyveromyces marxianus	Saccharomycetes	Génolevures
Kluyveromyces thermotolarans	Saccharomycetes	Génolevures
Kluyveromyces waltii	Saccharomycetes	Broad Institute
Laccaria bicolor	Homobasidiomycetes	Joint Genome Institute
Lodderomyces elongisporus	Saccharomycetes	Broad Institute
Melampsora larici-populina	Urediniomycetes	Joint Genome Institute
Mycosphaerella fijiensis	Dothideomycetes	Joint Genome Institute
Nectria haematococca	Sordariomycetes	Joint Genome Institute
Nosema locustae[c] (aka Anionospora)	Microsporidia	Marine Biological Laboratory
Peronospora parasitica	Oomycetes	Washington University Genome Sequencing Center
Phakopsora meibomiae	Urediniomycetes	Joint Genome Institute

(Continued)

Table 2. (*Continued*)

Genus/species	Taxonomy	Sequencing Center(s)
Phakopsora pachyrhizi	Urediniomycetes	Joint Genome Institute
Phytophthora capsici	Oomycete	Joint Genome Institute
Phytophthora infestans	Oomycete	Broad Institute & Sanger Institute
Pichia angusta	Saccharomycetes	Génolevures
Pichia farinose	Saccharomycetes	Génolevures
Piromyces sp.	Chytridiomycete	Joint Genome Institute
Pneumocystis carinii[b]	Pneumocystidomycetes	Broad Institute & University of Cincinnati
Puccinia graminis	Urediniomycetes	Broad Institute
Pyrenophora tritici-repentis	Dothidcemycetes	Broad Institute
Saccharomyces exiguus	Saccharomycetes	Génolevures
Saccharomyces servazzii	Saccharomycetes	Génolevures
Schizosaccharomyces japonicus	Schizosaccharomycetes	Broad Institute
Schizosaccharomyces octosporus	Schizosaccharomycetes	Broad Institute
Trichoderma reesei (anamorph of *Hypocrea jecorina*)	Sordariomycetes	Joint Genome Institute
Trichoderma virens	Sordariomycetes	Joint Genome Institute
Xanthoria parietina	Lecanoromycetes	Joint Genome Institute

[a]Status as of September 10, 2005.
[b]Multiple strains.
[c]Annotated.
A table including URLs is included in Supplemental material.

Entrez (http://www.ncbi.nlm.nih.gov/entrez/query.fcgi?db=genomeprj), the Munich Information Center for Protein Sequences (MIPS—http://mips.gsf.de/projects/fungi/), MetaDB (http://www.neurotransmitter.net/metadb/), and the Genomes Online database (http://www.genomesonline.org/). Particularly useful species-specific fungal databases include the Saccharomyces Genome Database (SGD—http://www.yeastgenome.org/) and CandidaDB (http://genolist.pasteur.fr/CandidaDB/). Of note is the Fungal Genetics Stock Center (FGSC http://www.fgsc.net/), which provides access to clones and other experimental resources in conjunction with several fungal genome projects. A more complete list of online resources is presented in the Supplemental material.

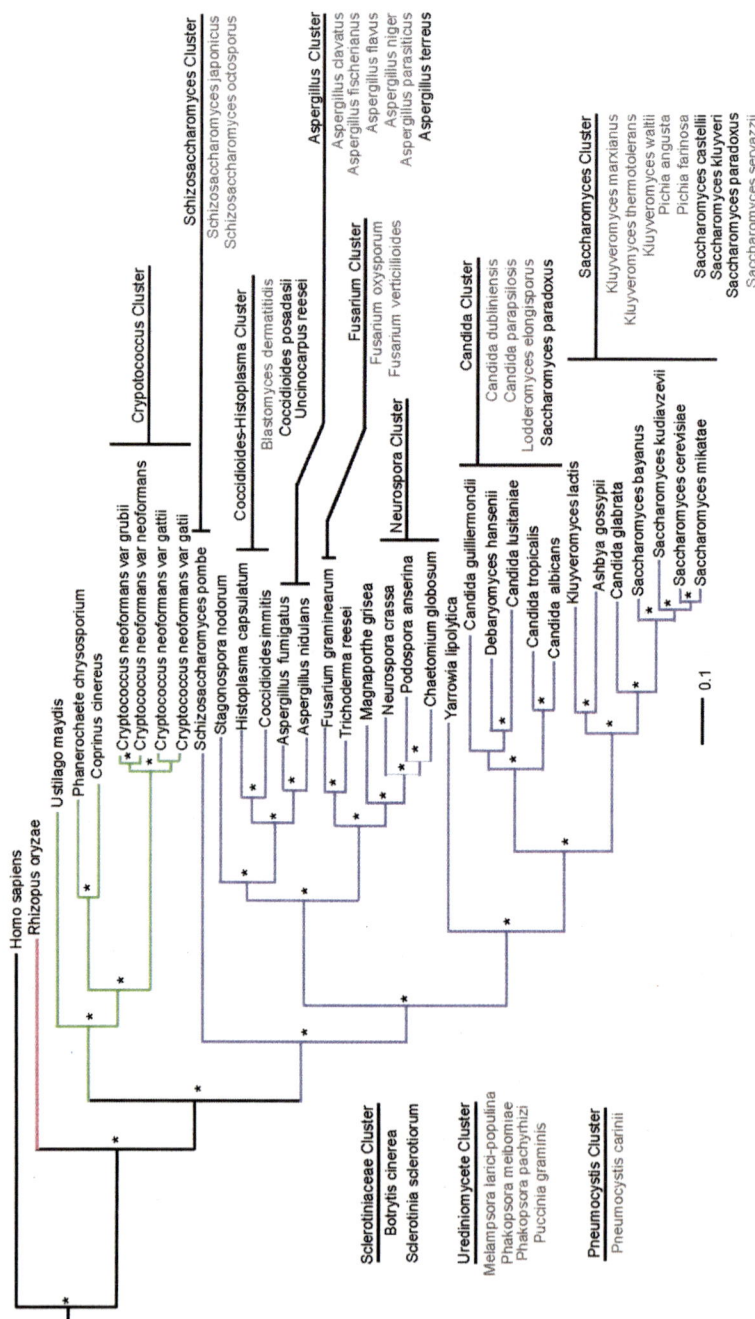

Figure 2. (*See facing page for legend.*)

FUNGAL GENOME SEQUENCING

The revolution in fungal genomics has been driven by the evolution of genome sequencing technology. Current whole genome shotgun (WGS) sequencing and assembly technologies produce fungal genome sequences with unparalleled accuracy and long-range contiguity at ever-reduced cost. These methods represent an advance over the clone-by-clone approaches used to sequence the first eukaryotic genomes. The clone-by-clone approach relied on labor-intensive clone-restriction mapping to pick sequencing templates, and required separate shotgun libraries for each clone to be prepared, tested, sequenced, and assembled. Ultimately, these maps were not sufficient to protect against both unnecessary overlap and errors originating both with the maps and sequencing. The adoption of more efficient high-throughput sequencing methods coupled with the simplicity of WGS strategies has greatly accelerated the pace of genome sequencing while dramatically reducing costs. Advances in assembly algorithms (Myers et al. 2000; Aparicio et al. 2002; Batzoglou et al. 2002; Jaffe et al. 2003; Mullikin and Ning 2003) and the inclusion of end sequences from large insert clones (e.g., Fosmids or BACs) routinely yield assemblies with high-sequence quality and continuity. For example, within the draft assembly of *Fusarium graminearum* an average base falls in a scaffold 5.4 Mb in length, while many scaffolds approach the length of intact chromosomes. Moreover, >99% of the individual bases in this assembly have consensus quality scores equivalent to that of a manually finished sequence.

Despite these advances, a number of challenges remain. Repetitive sequences present the single biggest difficulty in assembling WGS sequence data. The modest level of repetitive sequence ameliorates this problem in most fungi. However, the high identity repeats associated with telomeres, centromeres, and rDNA arrays remain difficult. Often these regions are not cloned in bacterial libraries, while in other cases these regions are cloned and sequenced but not correctly assembled. Although follow-up

---

Figure 2. Phylogeny of sequenced fungal genomes and genome clusters. Maximum likelihood tree of 33 fungi with available genome sequence. Additional genome sequences are shown to *right* with genomes in progress shown in gray. Clusters of related fungi are indicated. Phylogeny is rooted with *Dictyostelium discoideum* (data not shown), and major fungal groups are colored as in Fig. 1. Tree was generated based on protein sequence from 25 genes; asterisk indicates bootstrap values of >80 using PHYML with JTT model. Phylogeny was generated by Jason Stajich and modified with permission (http://fungal.genome.duke.edu/).

analyses (Farman and Leong 1995; Li et al. 2005) can accurately reconstruct telomeres, more robust automated methods are needed, as are independent mapping methods for assessing the size and position of these difficult to sequence regions.

A special case of repeated sequences are diploid genomes. In diploids, the extent of heterozygosity can vary dramatically across chromosomal regions. Regions of low polymorphism will be incorrectly merged during assembly, while highly polymorphic regions are separated. Consequently, allelic differences are difficult to distinguish from distinct paralogs. When possible, these complications have been avoided by sequencing a haploid form of the organism, or minimized by sequencing a closely related haploid as an aid. But in many cases, such as with *Candida albicans*, sequencing a diploid is unavoidable (Jones et al. 2004; Braun et al. 2005). New assembly algorithms are being developed to more accurately assemble whole-genome sequence data from diploid data sources (Vinson et al. 2005).

The challenges facing fungal genome sequencing are being met by new mapping and sequencing technologies. At least two different mapping approaches offer independent validation of genome assemblies without cloning, i.e., HAPPY mapping and optical mapping. HAPPY mapping is an established technique that determines the proximity of DNA markers through PCR assays using multiple pools of diluted, randomly broken genomic DNA (Dear and Cook 1989, 1993). This methodology is technically simple and does not produce large clone libraries. Optical mapping is a technology that has been newly applied to genome assembly (Zhou et al. 2004). The method produces genome-wide restriction maps based on images of single DNA molecules of megabase length. Comparing the order and distance between restriction sites to in silico digests of genome assemblies provides an independent assembly validation. Both HAPPY and optical mapping do not involve cloning and thus provide access to regions not present in WGS libraries. Both also allow sequences to be assigned to chromosomal locations.

Ongoing advances in sequencing technology also promise to further revolutionize fungal genomics. Although much work is still needed to optimize and fully validate these new approaches, their value is already apparent. For example, pyrosequencing methods implemented by 454 Life Sciences, have successfully generated sequence from *N. crassa* that could not be acquired through conventional sequencing methods. These sequences were found to be AT rich, which likely precluded efficient cloning in bacterial libraries. New instruments also provide the ability to inexpensively produce amounts of data, albeit consisting of short reads—tens to hundreds of base pairs per read compared with 500–1000 for conventional Sanger. The potential cost reduction enables 5–100× more strains or species to be sequenced for the current cost of producing a single genome. While early efforts have focused on producing high-quality reference sequences for individual strains or species, these new technologies will propel us to more fully describe the molecular diversity within related strains.

FUNGAL GENE ANNOTATION

Gene Prediction in Fungi

Gene annotation in the fungi is aided by the comparatively streamlined gene structures in these organisms. Fungal genomes display coding densities ranging from 37% to 61% and, as with other sequenced eukaryotes, gene density is inversely correlated with genome size. Coding sequence lengths average between 1.3 and 1.9 kb. Relative to metazoans, fungal genes are interrupted by few introns, although the fungi display a striking diversity of gene structures. Intron densities in fungi range from 5–6 introns per gene in basidiomycetes such as *Cryptococcus neoformans* (Loftus et al. 2005), to one to two introns per gene on average for many recently sequenced ascomycetes (e.g., *Neurospora, Magnaporthe*) (Galagan et al. 2003; Borkovich et al. 2004; Dean et al. 2005) to <300 introns in total in the hemiascomycete yeast *S. cerevisiae* (Goffeau et al. 1996). Introns are typically short in fungi, averaging between 80 and 150 bp in many ascomycetes. The basidiomycete *C. neoformans* is exceptional with regard to intron size, with an average intron length of 68 bp and possessing many introns as short as 35 bp. As described below, the structural diversity of introns in fungi provides a unique opportunity to study their evolution.

The relatively simple gene structures of most fungi facilitate accurate gene prediction. However, the majority of fungal species lack significant EST data. As a consequence, gene prediction in fungi relies heavily on de novo gene prediction. Given the significant differences in the characteristics of exons and introns between fungi, the training of gene prediction tools on organism-specific data is paramount. A growing number of de novo gene predictors provide this capability. These include GeneID (Guigo et al. 1992; Parra et al. 2000), FGenesh and FGenesh+ (Salamov and Solovyev 2000), SNAP (Korf 2004), Augustus (Stanke and Waack 2003; Stanke et al. 2004; Stanke and Morgenstern 2005), and GlimmerM (Salzberg et al. 1999). In addition to these tools, the programs GeneWise (http://www.ebi.ac.uk/Wise2/index.html) and Exonerate (Slater and Birney

2005) enable gene prediction based solely on the alignment of homologous protein or coding sequences.

Comparative gene prediction is a particularly attractive strategy for fungi, given clusters of related genomes. The utility of comparative gene prediction for fungi has been demonstrated through the comparative annotation of related *Saccharomyces* species (Kellis et al. 2003) and the adaptation of the program TWINSCAN for use with *C. neoformans* (Tenney et al. 2004). The latter program is a pairwise de novo gene prediction algorithm that utilizes homology from an informant genome to make predictions on a reference genome (Korf et al. 2001). TWINSCAN was trained on *C. neoformans* Serotype D (Loftus et al. 2005) and gene predictions were made using *C. neoformans* Serotype A (http://www.broad.mit.edu/) as the informant genome. Verification using known genes or RT–PCR indicated that 60%–72% of the predictions were exactly correct. Given the relative complexity of the *C. neoformans* genome as compared with other fungi, these results are encouraging. A challenge for the future will be to enable training of TWINSCAN and other comparative gene-prediction algorithms for the growing number of fungal genome clusters.

Alternative Splicing in Fungi

As with many other eukaryotes, a factor complicating gene annotation in fungi is the occurrence of alternative splicing. Examples of fungal genes with multiple alternative transcripts have been previously reported, but large-scale EST sequencing coupled with complete genome sequences is providing a more comprehensive view (Ebbole et al. 2004; Nelson 2004). Perhaps the most extensive genome-wide survey on alternative splicing in a fungus comes from the Basidiomycete *C. neoformans* Serotype D (Loftus et al. 2005). For this project, a high coverage of EST sequence was generated, resulting in alignments between at least one EST for the majority of predicted genes. These data revealed evidence of alternative splicing for 277 genes or 4.2% of the total. Although fungi appear to use alternative splicing less frequently than metazoans (estimates in human range from 40% of genes with alternative splicing to more than 80%) (Modrek and Lee 2002; Johnson et al. 2003; Kampa et al. 2004), these data represent the largest fraction of genes with alternative transcripts reported for any fungus so far, and likely represents a lower bound. Futhermore, the authors identified a variety of alternative splicing mechanisms including exon skipping and truncation, and extensions of both the 5′ and 3′ ends. The results for *C. neoformans* are noteworthy, as previous data from *S. cerevisiae* (Davis et al. 2000; Grate and Ares Jr. 2002;

Barrass and Beggs 2003) and *S. pombe* (Romfo et al. 2000) suggested that alternative splicing might only be prevalent in multicellular eukaryotes (Ast 2004). In addition, the few examples of alternative splicing in ascomycete yeasts (Romfo et al. 2000), as well as from other fungi (Ebbole et al. 2004), primarily involve intron retention. The results from *C. neoformans*, a basidiomycetous yeast, indicate that alternative splicing is likely more prevalent and richer than expected, even in single-celled organisms.

GENOME EVOLUTION

One noteworthy observation coming from the comparison of multiple genome sequences is how divergent fungi are at the genome level, despite apparent morphological and physiological similarities. For example, comparisons of the genomes of *Magnaporthe grisea* and *N. crassa*, related ascomycetes thought to have shared a common ancestor as recently as 200 million years ago (Mya) (Taylor et al. 1999; Berbee and Taylor 2000; Heckman et al. 2001), revealed an average amino acid identity of only 47% and virtually no conserved synteny (Dean et al. 2005). Only 113 regions were identified containing four or more genes in conserved colinearity. More generally, analyses of available complete fungal genomes reveal a rapid breakdown of conserved synteny over a relatively short evolutionary time span (data not shown). Even members of the same genus can display a remarkable divergence at the genomic level. A comparison of three species of *Aspergillus*—*A. nidulans*, *A. fumigatus*, and *A. oryzae*—revealed only 68% average amino acid identity between any pair of species (Galagan et al. 2005), an evolutionary distance comparable to that between human and fish (Dujon et al. 2004). At this distance, roughly 70% of *A. nidulans* could be mapped to a syntenic block with either *A. fumigatus* or *A. orzyae*, with roughly 50% of *A. nidulans* in conserved synteny across all three species (Galagan et al. 2005). Within these blocks of synteny, numerous microrearrangements were evident. These included many small inversions that have been shown to be a common pattern of rearrangement in eukaryotes (Seoighe et al. 2000; Aparicio et al. 2002; Kellis et al. 2003). But other patterns of breakage were equally prevalent, including translocations and segmental insertions, deletions, and duplications. Duplications and translocations in particular have been shown to be a common response of yeast undergoing experimental evolution (Dunham et al. 2002; Koszul et al. 2004) and thus are expected to contribute to the long-term evolution of fungal genomes. Moreover, a whole-genome duplication in yeast followed by massive gene loss, first predicted by Wolfe and colleagues (Wolfe and Shields 1997), has been confirmed by comparative analysis (Dujon et al. 2004; Kellis et al. 2004) and shown to have had a significant impact on yeast fermentation from carbon sources. Comparisons of fungi have also confirmed spatial patterns of rearrangement observed in other eukaryotes, namely, that rearrangements are far more common near telomeres and are frequently associated with repetitive sequence elements (Huynen et al. 2001; Carlton et al. 2002; Coghlan and Wolfe 2002; Kellis et al. 2003; Galagan et al. 2005; Lephart et al. 2005).

Together, these studies indicate that fungal genomes in particular, and eukaryotic genomes in general, are remarkably dynamic. In addition, studies in fungi are allowing us to investigate specific aspects of genome evolution in greater depth, and in some cases, connect genome evolutionary events with specific aspects of physiology. We review here two specific areas that have received considerable recent attention and for which the availability of genome sequence has led to new insights, i.e., intron evolution and genome defense. Owing to space, many other topics could not be considered, although several have been reviewed elsewhere (Archer and Dyer 2004; Borkovich et al. 2004; Fraser and Heitman 2004; Odds et al. 2004; Ryan and Smith 2004; Bell-Pedersen et al. 2005; Veneault-Fourrey and Talbot 2005; Yu and Keller 2005).

Intron Evolution

Although introns have been the object of intense study since their discovery over a quarter of a century ago (Sambrook 1977; Gilbert 1978), numerous questions remain concerning their origins, role, and evolutionary dynamics (for a review, see Lynch and Richardson 2002). Fungal genomes are particularly tractable for the study of introns for a number of reasons. First, the fundamental aspects of intron biology are shared between fungi and other eukaryotes, and thus, lessons learned from fungi are likely to have wide relevance. Second, as described above, fungal genomes are gene dense with relatively simple gene structures, facilitating the accurate prediction of intron boundaries. Finally, as described above, the fungi display a wider diversity of average intron density. This architectural diversity, coupled with the availability of fungal genomes spanning the kingdom, provides an opportunity to investigate intron dynamics. Several recent studies illustrate the utility of fungi for intron studies, and have provided new insight into the patterns and mechanisms of intron evolution.

In a study by Nielsen and colleagues (Nielsen et al. 2004), patterns of intron evolution were investigated in four Euascomycete fungal genomes (*A. nidulans*, *F. graminearum*, *M. grisea*, and *N. crassa*) spanning roughly

330 Myr. The conservation of predicted orthologous intron positions was determined in three of the fungi (using the fourth as an outgroup), and a probabilistic model was used to estimate the most likely rate of intron gain and loss giving rise to these observed conservation patterns. One immediate conclusion stemming from this work was the clear importance of intron gain within this group of Euascomycetes. All three non-outgroup lineages displayed significant numbers of predicted intron gains. In addition, even within this small set of organisms, differences in the pattern of intron dynamics were apparent, with the numbers of gained and lost introns approximately balanced in *M. grisea* and *F. graminearum*, but with roughly twice as many losses as gains in *N. crassa*. Rates of intron gain also varied substantially between gene families.

The subsequent sequencing of additional fungal genomes provided the opportunity to study intron dynamics over a wider evolutionary distance. With these additional data, Stajich and colleagues (J.E. Stajich, S.W. Roy, and F.S. Dietrich, in prep.) studied the patterns of intron gain and loss across 24 fungi spanning nearly the fungal kingdom. With *Homo sapiens* and *Arabidopsis thaliana* as outgroups, the authors developed a maximum likelihood approach to estimate intron loss and gain events and thereby calculate intron densities at various nodes in the fungal tree. Based on a set of more than 700 orthologous protein coding genes, the authors found numerous intron positions shared among plants, animal, and fungi, and they concluded both that these introns were common and present at the origin of the eukaryotic crown. Since the fungal last common ancestor, nearly all lineages were predicted to have suffered substantial intron loss, with particularly significant loss occurring at deeper branches and at the outset of the Hemiascomycete lineage. Interestingly, these authors also find intron gain to be as significant as intron loss in several recent lineages including the Euascomycetes (consistent with Nielsen et al. 2004) and the lineage leading to *C. neoformans*.

One intriguing characteristic of introns is the correlation between intron density and positional bias that has been observed in all eukaryotes sequenced to date (Mourier and Jeffares 2003). According to this 5' positional bias, introns are evenly distributed within the coding sequence of genes in intron-rich organisms, but are biased toward the 5' ends of genes in intron-poor organisms. It has been proposed that this bias may have arisen by intron loss through a mechanism of homologous recombination of spliced messages reverse transcribed from the 3' polyadenylated tail (Fink 1987; Mourier and Jeffares 2003; Roy and Gilbert 2005b). The plausibility of such a recombination-based mechanism has been demonstrated in experiments with intron-containing Ty elements in yeast

(Boeke et al. 1985). However, this hypothesis predicts that loss will be biased to introns in the 3' portions of coding sequences. The pattern of intron loss reported by Nielsen et al. (2004) did not reveal such a 3' bias. Instead, the rate of intron loss was lowest at the 3' end, while the highest rates of intron loss occurred in the middles of genes.

A similar pattern of positional intron loss was revealed by more recent work by Stajich and colleagues (J. Stajich and F.S. Dietrich, in prep.). In this analysis, recent intron loss was investigated in the genomes of four closely related *Cryptococcus* species (which diverged <37 Mya [Xu et al. 2000]). The authors identified several loci where multiple intron losses appear to have occurred from a single event. As these loci lacked close paralogs, these losses cannot be explained by gene conversion and suggest intron loss through recombination with RNA. Interestingly, these events all occur in the middle of genes, leaving introns intact at the 3' end. As with the results of Nielsen et al. (2004), these data suggest that intron loss alone, at least within the recent evolutionary history of the Cryptococci and the Euascomycetes, is not sufficient to explain the observed intron 5' positional bias in these species.

These and other data (Bon et al. 2003) have revealed fungal introns to be remarkably dynamic. Current gene architectures appear to reflect an interplay between intron gain and loss, with the balance between the two processes varying over evolutionary time. Based on existing data, intron loss appears to dominate in certain fungal clades. This is consistent with the results of an analysis of eight genomes—spanning plants, animals, protests, and including two fungi, *S. cerevisiae* and *S. pombe* (Roy and Gilbert 2005a)—that reported massive and net intron loss in six of the lineages examined. The role of loss appears diminished in more recent fungal evolution with certain lineages gaining nearly as many introns as lost. The mechanism by which introns are gained in any organism remains to be conclusively established, although several theories have been proposed (Logsdon Jr. et al. 1998; Coghlan and Wolfe 2002; Lynch and Richardson 2002; Fedorov et al. 2003). As additional fungal genomes are sequenced, these and other mysteries surrounding intron evolution may eventually be solved.

Genome Defense: Repeats, RIP, and RNAi

Repeat sequences are ubiquitous components of fungal genomes. In most genomes analyzed to date the majority of repeat sequences are associated with mobile genetic elements. Copies or remnants of both Class I (retroposons) or Class II (DNA transposons) have been identified, and the number of distinct families of mobile elements continues to grow (for review, see Kempken and Kuck 1998; see also Dean et al. 2005). Microsatellite repeats (Toth et al. 2000) and low-complexity sequence, as well as centromere and telomere associated repeats (Schechtman 1990) are also common. Fungal genomes contain varying amounts of repeat sequence, with "typical" repeat content ranging from between 3% (e.g., *A. nidulans, A. fumgatus,* and *A. oryzae*) to 10% (e.g., *Neurospora, Magnaporthe*). However, species outside of this range at either extreme have also been identified. Although repeat sequences may play a beneficial role in generating genetic diversity, their presence can also be detrimental, particularly in terms of genome stability. As a result, many organisms have developed "genome defense" systems that repress the activity of transposable elements. Two different genome defense mechanisms in fungi have received particular recent attention as a consequence of genomics—Repeat Induced Point Mutation and RNA silencing.

The first eukaryotic genome defense system described, discovered in the fungus *N. crassa*, is a process called Repeat Induced Point Mutation (RIP) (Selker 1990, 2002; Davis et al. 2000; Galagan and Selker 2004). RIP is a homology-based process that mutates repetitive DNA and frequently leads to epigenetic silencing through DNA methylation. Importantly, RIP has been shown to act on all duplicated sequence, including long segmental duplications, mobile element duplications, and gene duplications. The properties of RIP immediately suggest an impact on genome evolution, and the completion of the *N. crassa* genome sequence allowed the full extent of this impact to be determined (Galagan et al. 2003). Consistent with a role as a defense against mobile elements, the analysis of the *N. crassa* genome revealed a complete absence of intact transposons. However, this defense was shown to come at a price: Essentially, all paralogs in *N. crassa* appear to predate RIP, and since the emergence of RIP gene evolution through gene duplication has been arrested. Gene duplication is widely considered to be essential for the generation of new function. RIP thus illustrates the extent to which genomes can go to defend against mobile elements, and the impact this defense can have on genome structure and evolution. This impact has wide relevance in the fungi, RIP—albeit in a less severe form—has been observed in a growing number of other fungi (Galagan and Selker 2004)

The genome defenses of *N. crassa* are additionally fortified by two different RNA silencing mechanisms, quelling and meiotic silencing by unpaired DNA (MSUD). RNA silencing is a term that encompasses a range of phenomena found in many eukaryotic organisms. Fundamentally, these phenomena involve the repression of sequences with similarity to short RNA molecules. RNA silencing was originally described as "quelling" in the fungus *N. crassa* (Cogoni et al. 1996) and "cosuppression" in plants (Napoli et al. 1990). RNA silencing was subsequently described in *Caenorhabditis elegans* and other metazoans, where it is called RNA interference (RNAi) (Fire et al. 1998; Ketting and Plasterk 2000). The core machinery for RNA silencing—Argonautes, Dicers, and helicases—appears conserved across species (Hutvagner and Zamore 2002) and was first studied as part of the quelling pathway in Neurospora. More recently, it was shown that *N. crassa* possesses a second RNA silencing pathway called MSUD (Aramayo and Metzenberg 1996; Shiu et al. 2001; Shiu and Metzenberg 2002). The analysis of the genome sequence revealed that quelling and MSUD appear to be paralogous pathways derived through the duplication of a core set of genes (Galagan et al. 2003; Borkovich et al. 2004). These two pathways have evolved to operate during different parts of the Neurospora life cycle. Quelling acts during vegetative growth while MSUD acts during the meiosis phases of sexual reproduction. Together with RIP, which acts during sexual reproduction but premeiotically, these pathways effectively silence genes in aberrant copy number—and thus protect against genome instability—throughout the entire life cycle of *N. crassa* (Borkovich et al. 2004).

As in Metazoans, in fungi RNA silencing has emerged as a powerful experimental tool for manipulating gene expression. As described above, RNA silencing has been utilized in *C. albicans* to identify essential genes, and RNA-silencing experimental protocols have been developed for a host of other non-yeast fungi. Curiously, no endogenous microRNAs have yet been reported in fungi, although there is some evidence for antisense transcripts (Loftus et al. 2005). It has been suggested that these antisense transcripts may regulate gene expression through the RNAi pathway. But the extent, if at all, to which RNA silencing plays a gene regulatory role in fungi remains unknown.

PLANT PATHOGENESIS AND ENVIRONMENT

Fungi are central to the health of terrestrial ecosystems, and they have played a foundational role in the evolution of life on land. The colonization of land by eukaryotes is thought to have been established through the symbiosis of a fungus and a photosynthesizing organism (Gehrig et al. 1996; Heckman et al. 2001). The symbiosis between fungi and plants plays a crucial role in protecting plants from disease and facilitating nutrient uptake; 95% of all plant families have associated mycorrhizal fungi (Trappe 1987). Fungi also play a central role in degrading organic material. They are the dominant organisms in aerated soils (Frey et al. 1999), typically accounting for 10%–60% of the biomass in forest litter (Newell 1992; Metting 1993). In contrast to these beneficial roles, fungal plant pathogens have a devastating impact on agriculture. Fungi infect all major crop plants (Strange and Scott 2005) and lead to food contamination through the production of mycotoxins. In the United States alone, each year they are estimated to cause $33 billion dollars in damages (Madden and Wheelis 2003) and invoke expenditures of over $600 million on fungicides. Fungal pathogens have had a significant impact on human history. The dominance of tea over coffee in the British Empire can be traced to the failure of coffee in British Ceylon in the 1870s due to infection with leaf blight, caused by the fungus *Hemileia vastatrix*. These fields were planted instead with tea. Access to genome sequence promises to advance our knowledge of the underlying biology of fungal infection and the interaction of pathogen and host, as well as of the mechanisms by which fungi reproduce and persist in the environment. The reports on the genomes for *Phanerochaete chrysosporium* (Martinez et al. 2004) and *M. grisea* (Dean et al. 2005) illustrate this potential.

Lignin is a major component of plant cell walls and the second most abundant natural polymer (Martinez et al. 2004). Only a small group of fungi—termed white rot fungi—are able to degrade lignin, and as a consequence, these fungi play an important role in the global carbon cycle. The genome sequence of one white rot fungus, *P. chrysosporium*, has been generated and a preliminary analysis published (Martinez et al. 2004). The genome contains an extensive and highly redundant array of genes predicted to be involved in lignin degradation. Consistent with the ecosystem role of white rot, enzymes for carbohydrate catabolism outnumbered those for anabolism, the opposite of the pattern seen in other sequenced eukaryotes. The genome also revealed an extensive array of secondary metabolite gene clusters. The authors suggest that these genes may be attractive targets for bioprocess engineering. Functional studies that utilize the genome sequence may reveal the underlying cellular networks responsible for the important ecological role of white rot fungi.

M. grisea causes the most destructive disease of rice and has emerged as a central model organism for the study of fungal plant diseases. Rice blast, the disease caused by *M. grisea* is estimated to destroy enough rice annually to feed 60 million people (Zeigler et al. 1994). The generation and preliminary analysis of the *M. grisea* genome sequence has provided insight into the molecular basis of fungal plant pathogenicity (Dean et al. 2005). In particular, the genome revealed an expanded family of G-protein-coupled receptors (GPCRs), including a subfamily, one member of which had been previously shown to be required for pathogenesis. The other novel members of this subfamily were shown to be expressed during infection, with two genes specifically up-regulated during the development of a specialized infection structure called the appressorium. The genome also suggested significant diversity among different *M. grisea* strains. Of the seven known *M. grisea* avirulence genes, only four were found in the strain sequenced. With additional sequences, we will better understand this diversity and the role it may play in plant disease. Ultimately, these efforts may lead to improved methods for pathogen control and higher yields of staple foodstuffs worldwide.

MYCOSES AND MEDICINE

Fungal infections are the third most common hospital-acquired infection, and have emerged as a growing threat to human health (Beck-Sague and Jarvis 1993; Swartz 1994). They have lethal consequences for the growing population of patients immunocompromised with AIDS and leukemias or therapeutically immunosuppressed. The two most common fungal pathogens are *Candida* and *Aspergillus* species: Candidiasis is the most common HIV-related fungal infection with mortality reaching 49% (Gudlaugsson et al. 2003) while Aspergillosis has caused up to 10,000 hospitalizations per year with mortality as high as 20% (Dasbach et al. 2000). Emerging fungal infections represent an equally serious threat to healthy human populations. For example, in 2002, an outbreak of *C. neoformans* occurred on the east coast of Vancouver Island, British Columbia affecting at least 59, mostly immunocompetent, individuals and causing at least two deaths (Hoang et al. 2004). The incidence of Valley Fever caused by the dimorphic *Coccidioides* is increasing with more than 100,000 cases occurring each year in the United States alone (Chiller et al. 2003). Developing effective therapies against fungi has been more difficult than for bacterial pathogens, given the eukaryotic biology they share with humans; as a result, few effective antifungals are currently available. Most of the existing drugs have serious side effects, and resistance to these compounds is an increasing problem (Georgopapadakou 1998).

The analysis of the genomes of medically important fungi holds the potential to address these clinical issues. In particular, given the complete gene set for a pathogenic fungus, it becomes possible to predict genes necessary for fungal growth that lack human homologs. These may represent targets for antifungal drugs with fewer toxic side effects. This approach has been utilized to identify potential drug targets for *C. albicans* (Jones et al. 2004). Based on the human-curated gene set (see above), 228 genes were identified in the *C. albicans* genome sequence that were conserved in five other fully sequenced fungal genomes but that lacked significant sequence similarity in the human or mouse genomes (Braun et al. 2005). The authors suggested that, based on their predicted functions and localizations, these genes represent potential targets for small molecule inhibition.

The availability of complete genome sequence also facilitates genome-wide functional screens for drug targets. For example, De Backer and colleagues (De Backer et al. 2001) developed a method combining antisense RNA inhibition (see below) and promoter interference to identify genes critical for the growth of *C. albicans*, and subsequently used these genes as targets to identify new antifungals in a drug screen. The availability of an annotated genome sequence enabled the rapid identification of inhibited genes. Intriguingly, a significant fraction of *C. albicans* essential genes lacked homologs in *S. cerevisiae*, again highlighting the diversity of the fungal kingdom and the need for sequenced fungi beyond just a few models.

The growing complement of fungal genome sequences enables other strategies for investigating fungal infection. Comparing genomes from nonpathogenic species to related pathogenic organisms can identify genetic differences that contribute to infection and disease, while the comparison between strains with different host specificity may help clarify the genomic basis for differences in virulence and host interactions. Comparative analyses of these sorts are an exciting possibility arising from the sequencing of clusters of related genomes (as described above), often centering on a pathogenic fungus, but including related nonpathogenic fungi as in the case of *Coccidioides* spp. and *Uncinocarpus reesii*.

In addition to their role as pathogens, fungi also play a critical beneficial role in the development and production of pharmaceuticals through the production of secondary metabolites including Lovestatin and antibiotics such as penicillin, cephalosporins, and cyclosporine. The genomes of filamentous fungi have revealed an extensive—and occasionally unexpected—repertoire of secondary metabolites (Galagan et al. 2003; Kroken et al. 2003; Borkovich et al. 2004; Dean et al. 2005; Yu and Keller 2005). The burgeoning genomic resource available for fungi promises many further insights and discoveries into the friend and foe relationship between fungi and man.

THE FUTURE OF FUNGAL GENOMICS

The growing number of complete fungal genomes provides an unprecedented opportunity to study the biology and evolution of an entire eukaryotic kingdom. However, sequence is only the tip of the iceberg for fungal genomics. The availability of genome sequence has catalyzed the development of genome-wide functional studies for a growing number of fungal species. In particular, microarrays—both public and commercial—are available for numerous fungi, enabling not only expression studies, but also cross-genome hybridization, the identification of transcription-factor binding sites and chromatin modifications, and population genotyping. High-throughput proteomic methods are also increasingly being applied, providing insight into the protein modification and translational control. In addition, as highlighted above, comprehensive gene knock-out or knock-down projects are underway for several species. Ultimately, these data will enable a true systems biological approach to understanding fungal biology and evolution, and in particular the biology underlying the widespread medical, agricultural, and environmental impact of fungi.

ACKNOWLEDGMENTS

We are grateful to Jason Stajich for sharing his pre-publication results on intron evolution and the multigene phylogeny of sequenced fungi. This work was supported by grants from the NIH.

REFERENCES

Aparicio, S., Chapman, J., Stupka, E., Putnam, N., Chia, J.M., Dehal, P., Christoffels, A., Rash, S., Hoon, S., Smit, A., et al. 2002. Whole-genome shotgun assembly and analysis of the genome of *Fugu rubripes*. *Science* **297**: 1301–1310.

Aramayo, R. and Metzenberg, R.L. 1996. Meiotic transvection in fungi. *Cell* **86**: 103–113.

Archer, D.B. and Dyer, P.S. 2004. From genomics to post-genomics in *Aspergillus*. *Curr. Opin. Microbiol.* **7**: 499–504.

Ast, G. 2004. How did alternative splicing evolve? *Nat. Rev. Genet.* **5**: 773–782.

Barrass, J.D. and Beggs, J.D. 2003. Splicing goes global. *Trends Genet.* **19**: 295–298.

Batzoglou, S., Jaffe, D.B., Stanley, K., Butler, J., Gnerre, S., Mauceli, E., Berger, B., Mesirov, J.P., and Lander, E.S. 2002. ARACHNE: A whole-genome shotgun assembler. *Genome Res.* **12**: 177–189.

Beck-Sague, C. and Jarvis, W.R. 1993. Secular trends in the epidemiology of nosocomial fungal infections in the United States, 1980–1990. National Nosocomial Infections Surveillance System. *J. Infect. Dis.* **167**: 1247–1251.

Bell-Pedersen, D., Cassone, V.M., Earnest, D.J., Golden, S.S., Hardin, P.E., Thomas, T.L., and Zoran, M.J. 2005. Circadian rhythms from multiple oscillators: Lessons from diverse organisms. *Nat. Rev. Genet.* **6**: 544–556.

Berbee, M.L. and Taylor, J.W. 2000. *The Mycota, Vol. VIIB, systematics and evolution.* Springer-Verlag, New York.

———2001. Fungal molecular evolution: Gene trees and geologic time. In *The Mycota, Vol. VIIB, systematics and evolution* (eds. D.J. McLaughlin et al.), pp. 229–246. Springer, Berlin.

Boeke, J.D., Garfinkel, D.J., Styles, C.A., and Fink, G.R. 1985. Ty elements transpose through an RNA intermediate. *Cell* **40:** 491–500.

Bon, E., Casaregola, S., Blandin, G., Llorente, B., Neuveglise, C., Munsterkotter, M., Guldener, U., Mewes, H.W., Van Helden, J., Dujon, B., et al. 2003. Molecular evolution of eukaryotic genomes: Hemiascomycetous yeast spliceosomal introns. *Nucleic Acids Res.* **31:** 1121–1135.

Borkovich, K.A., Alex, L.A., Yarden, O., Freitag, M., Turner, G.E., Read, N.D., Seiler, S., Bell-Pedersen, D., Paietta, J., Plesofsky, N., et al. 2004. Lessons from the genome sequence of *Neurospora crassa*: Tracing the path from genomic blueprint to multicellular organism. *Microbiol. Mol. Biol. Rev.* **68:** 1–108.

Braun, B.R., van Het Hoog, M., d'Enfert, C., Martchenko, M., Dungan, J., Kuo, A., Inglis, D.O., Uhl, M.A., Hogues, H., Berriman, M., et al. 2005. A human-curated annotation of the *Candida albicans* genome. *PLoS Genet.* **1:** E1.

Carlton, J.M., Angiuoli, S.V., Suh, B.B., Kooij, T.W., Pertea, M., Silva, J.C., Ermolaeva, M.D., Allen, J.E., Selengut, J.D., Koo, H.L., et al. 2002. Genome sequence and comparative analysis of the model rodent malaria parasite *Plasmodium yoelii yoelii*. *Nature* **419:** 512–519.

Chiller, T.M., Galgiani, J.N., and Stevens, D.A. 2003. Coccidioidomycosis. *Infect. Dis. Clin. North Am.* **17:** 41–57, viii.

Coghlan, A. and Wolfe, K.H. 2002. Fourfold faster rate of genome rearrangement in nematodes than in *Drosophila*. *Genome Res.* **12:** 857–867.

Cogoni, C., Irelan, J.T., Schumacher, M., Schmidhauser, T.J., Selker, E.U., and Macino, G. 1996. Transgene silencing of the al-1 gene in vegetative cells of *Neurospora* is mediated by a cytoplasmic effector and does not depend on DNA-DNA interactions or DNA methylation. *EMBO J.* **15:** 3153–3163.

Dasbach, E.J., Davies, G.M., and Teutsch, S.M. 2000. Burden of aspergillosis-related hospitalizations in the United States. *Clin. Infect. Dis.* **31:** 1524–1528.

Davis, C.A., Grate, L., Spingola, M., and Ares Jr., M. 2000. Test of intron predictions reveals novel splice sites, alternatively spliced mRNAs and new introns in meiotically regulated genes of yeast. *Nucleic Acids Res.* **28:** 1700–1706.

De Backer, M.D., Nelissen, B., Logghe, M., Viaene, J., Loonen, I., Vandoninck, S., de Hoogt, R., Dewaele, S., Simons, F.A., Verhasselt, P., et al. 2001. An antisense-based functional genomics approach for identification of genes critical for growth of *Candida albicans*. *Nat. Biotechnol.* **19:** 235–241.

Dean, R.A., Talbot, N.J., Ebbole, D.J., Farman, M.L., Mitchell, T.K., Orbach, M.J., Thon, M., Kulkarni, R., Xu, J.R., Pan, H., et al. 2005. The genome sequence of the rice blast fungus *Magnaporthe grisea*. *Nature* **434:** 980–986.

Dear, P.H. and Cook, P.R. 1989. Happy mapping: A proposal for linkage mapping the human genome. *Nucleic Acids Res.* **17:** 6795–6807.

———. 1993. Happy mapping: Linkage mapping using a physical analogue of meiosis. *Nucleic Acids Res.* **21:** 13–20.

Dujon, B., Sherman, D., Fischer, G., Durrens, P., Casaregola, S., Lafontaine, I., De Montigny, J., Marck, C., Neuveglise, C., Talla, E., et al. 2004. Genome evolution in yeasts. *Nature* **430:** 35–44.

Dunham, M.J., Badrane, H., Ferea, T., Adams, J., Brown, P.O., Rosenzweig, F., and Botstein, D. 2002. Characteristic genome rearrangements in experimental evolution of *Saccharomyces cerevisiae*. *Proc. Natl. Acad. Sci.* **99:** 16144–16149.

Ebbole, D.J., Jin, Y., Thon, M., Pan, H., Bhattarai, E., Thomas, T., and Dean, R. 2004. Gene discovery and gene expression in the rice blast fungus, *Magnaporthe grisea*: Analysis of expressed sequence tags. *Mol. Plant Microbe Interact.* **17**: 1337–1347.

Farman, M.L. and Leong, S.A. 1995. Genetic and physical mapping of telomeres in the rice blast fungus, *Magnaporthe grisea*. *Genetics* **140**: 479–492.

Fedorov, A., Roy, S., Fedorova, L., and Gilbert, W. 2003. Mystery of intron gain. *Genome Res.* **13**: 2236–2241.

Fink, G.R. 1987. Pseudogenes in yeast? *Cell* **49**: 5–6.

Fire, A., Xu, S., Montgomery, M.K., Kostas, S.A., Driver, S.E., and Mello, C.C. 1998. Potent and specific genetic interference by double-stranded RNA in *Caenorhabditis elegans*. *Nature* **391**: 806–811.

Fraser, J.A. and Heitman, J. 2004. Evolution of fungal sex chromosomes. *Mol. Microbiol.* **51**: 299–306.

Frey, S.D., Elliott, E.T., and Paustian, K. 1999. Bacterial and fungal abundance and biomass in conventional and no-tillage agroecosystems along two climatic gradients. *Soil Biol. Biochem.* **31**: 573–585.

Galagan, J.E. and Selker, E.U. 2004. RIP: The evolutionary cost of genome defense. *Trends Genet.* **20**: 417–423.

Galagan, J.E., Calvo, S.E., Borkovich, K.A., Selker, E.U., Read, N.D., Jaffe, D., FitzHugh, W., Ma, L.J., Smirnov, S., Purcell, S., et al. 2003. The genome sequence of the filamentous fungus *Neurospora crassa*. *Nature* **422**: 859–868.

Galagan, J.E., Calvo, S.E., Cuomo, C., Ma, L.J., Wortman, J.R., Batzoglou, S., Lee, S.I., Basturkmen, M., Spevak, C.C., Clutterbuck, J., et al. 2005. Sequencing of *Aspergillus nidulans* and comparative analysis with *A. fumigatus* and *A. oryzae*. *Nature* **438**: 1105–1115.

Gehrig, H., Schussler, A., and Kluge, M. 1996. Geosiphon pyriforme, a fungus forming endocytobiosis with Nostoc (cyanobacteria), is an ancestral member of the Glomales: Evidence by SSU rRNA analysis. *J. Mol. Evol.* **43**: 71–81.

Georgopapadakou, N.H. 1998. Antifungals: Mechanism of action and resistance, established and novel drugs. *Curr. Opin. Microbiol.* **1**: 547–557.

Gilbert, W. 1978. Why genes in pieces? *Nature* **271**: 501.

Goffeau, A., Barrell, B.G., Bussey, H., Davis, R.W., Dujon, B., Feldmann, H., Galibert, F., Hoheisel, J.D., Jacq, C., Johnston, M., et al. 1996. Life with 6000 genes. *Science* **274**: 546, 563–567.

Grate, L. and Ares Jr., M. 2002. Searching yeast intron data at Ares lab Web site. *Methods Enzymol.* **350**: 380–392.

Gudlaugsson, O., Gillespie, S., Lee, K., Vande Berg, J., Hu, J., Messer, S., Herwaldt, L., Pfaller, M., and Diekema, D. 2003. Attributable mortality of nosocomial candidemia, revisited. *Clin. Infect. Dis.* **37**: 1172–1177.

Guigo, R., Knudsen, S., Drake, N., and Smith, T. 1992. Prediction of gene structure. *J. Mol. Biol.* **226**: 141–157.

Hawksworth, D.L. 1991. The fungal dimension of biodiversity—magnitude, significance, and conservation. *Mycol. Res.* **95**: 641.

Heckman, D.S., Geiser, D.M., Eidell, B.R., Stauffer, R.L., Kardos, N.L., and Hedges, S.B. 2001. Molecular evidence for the early colonization of land by fungi and plants. *Science* **293**: 1129–1133.

Hoang, L.M., Maguire, J.A., Doyle, P., Fyfe, M., and Roscoe, D.L. 2004. *Cryptococcus neoformans* infections at Vancouver Hospital and Health Sciences Centre (1997–2002): Epidemiology, microbiology and histopathology. *J. Med. Microbiol.* **53**: 935–940.

Hutvagner, G. and Zamore, P.D. 2002. RNAi: Nature abhors a double-strand. *Curr. Opin. Genet. Dev.* **12:** 225–232.

Huynen, M.A., Snel, B., and Bork, P. 2001. Inversions and the dynamics of eukaryotic gene order. *Trends Genet.* **17:** 304–306.

Jaffe, D.B., Butler, J., Gnerre, S., Mauceli, E., Lindblad-Toh, K., Mesirov, J.P., Zody, M.C., and Lander, E.S. 2003. Whole-genome sequence assembly for mammalian genomes: Arachne 2. *Genome Res.* **13:** 91–96.

Johnson, J.M., Castle, J., Garrett-Engele, P., Kan, Z., Loerch, P.M., Armour, C.D., Santos, R., Schadt, E.E., Stoughton, R., and Shoemaker, D.D. 2003. Genome-wide survey of human alternative pre-mRNA splicing with exon junction microarrays. *Science* **302:** 2141–2144.

Jones, T., Federspiel, N.A., Chibana, H., Dungan, J., Kalman, S., Magee, B.B., Newport, G., Thorstenson, Y.R., Agabian, N., Magee, P.T., et al. 2004. The diploid genome sequence of *Candida albicans*. *Proc. Natl. Acad. Sci.* **101:** 7329–7334.

Kampa, D., Cheng, J., Kapranov, P., Yamanaka, M., Brubaker, S., Cawley, S., Drenkow, J., Piccolboni, A., Bekiranov, S., Helt, G., et al. 2004. Novel RNAs identified from an in-depth analysis of the transcriptome of human chromosomes 21 and 22. *Genome Res.* **14:** 331–342.

Kellis, M., Patterson, N., Endrizzi, M., Birren, B., and Lander, E.S. 2003. Sequencing and comparison of yeast species to identify genes and regulatory elements. *Nature* **423:** 241–254.

Kellis, M., Birren, B.W., and Lander, E.S. 2004. Proof and evolutionary analysis of ancient genome duplication in the yeast *Saccharomyces cerevisiae*. *Nature* **428:** 617–624.

Kempken, F. and Kuck, U. 1998. Transposons in filamentous fungi—facts and perspectives. *Bioessays* **20:** 652–659.

Ketting, R.F. and Plasterk, R.H. 2000. A genetic link between co-suppression and RNA interference in *C. elegans*. *Nature* **404:** 296–298.

Korf, I. 2004. Gene finding in novel genomes. *BMC Bioinform.* **5:** 59.

Korf, I., Flicek, P., Duan, D., and Brent, M.R. 2001. Integrating genomic homology into gene structure prediction. *Bioinformatics* **17:** S140–S148.

Koszul, R., Caburet, S., Dujon, B., and Fischer, G. 2004. Eucaryotic genome evolution through the spontaneous duplication of large chromosomal segments. *EMBO J.* **23:** 234–243.

Kroken, S., Glass, N.L., Taylor, J.W., Yoder, O.C., and Turgeon, B.G. 2003. Phylogenomic analysis of type I polyketide synthase genes in pathogenic and saprobic ascomycetes. *Proc. Natl. Acad. Sci.* **100:** 15670–15675.

Lephart, P.R., Chibana, H., and Magee, P.T. 2005. Effect of the major repeat sequence on chromosome loss in *Candida albicans*. *Eukaryot. Cell* **4:** 733–741.

Li, W., Rehmeyer, C.J., Staben, C., and Farman, M.L. 2005. TERMINUS—Telomeric end-read mining in unassembled sequences. *Bioinformatics* **21:** 1695–1698.

Loftus, B.J., Fung, E., Roncaglia, P., Rowley, D., Amedeo, P., Bruno, D., Vamathevan, J., Miranda, M., Anderson, I.J., Fraser, J.A., et al. 2005. The genome of the basidiomycetous yeast and human pathogen *Cryptococcus neoformans*. *Science* **307:** 1321–1324.

Logsdon Jr., J.M., Stoltzfus, A., and Doolittle, W.F. 1998. Molecular evolution: Recent cases of spliceosomal intron gain? *Curr. Biol.* **8:** R560–R563.

Lowy, B. 1971. New records of mushroom stones from Guatemala. *Mycologia* **63:** 983–993.

Lynch, M. and Richardson, A.O. 2002. The evolution of spliceosomal introns. *Curr. Opin. Genet. Dev.* **12:** 701–710.

Madden, L.V. and Wheelis, M. 2003. The threat of plant pathogens as weapons against U.S. crops. *Annu. Rev. Phytopathol.* **41:** 155–176.

Martinez, D., Larrondo, L.F., Putnam, N., Gelpke, M.D., Huang, K., Chapman, J., Helfenbein, K.G., Ramaiya, P., Detter, J.C., Larimer, F., et al. 2004. Genome sequence of the lignocellulose degrading fungus *Phanerochaete chrysosporium* strain RP78. *Nat. Biotechnol.* **22:** 695–700.

Metting, F.B. 1993. Structure and physiological ecology of soil microbial communities. In *Soil microbial ecology: Applications in agricultural and environmental management*, pp. 3–25. Dekker, New York.

Modrek, B. and Lee, C. 2002. A genomic view of alternative splicing. *Nat. Genet.* **30:** 13–19.

Moss, M.O. 1987. Fungal biotechnology roundup. *Mycologist* **21:** 55–58.

Mourier, T. and Jeffares, D.C. 2003. Eukaryotic intron loss. *Science* **300:** 1393.

Mullikin, J.C. and Ning, Z. 2003. The phusion assembler. *Genome Res.* **13:** 81–90.

Myers, E.W., Sutton, G.G., Delcher, A.L., Dew, I.M., Fasulo, D.P., Flanigan, M.J., Kravitz, S.A., Mobarry, C.M., Reinert, K.H., Remington, K.A., et al. 2000. A whole-genome assembly of *Drosophila*. *Science* **287:** 2196–2204.

Napoli, C., Lemieux, C., and Jorgensen, R. 1990. Introduction of a chimeric chalcone synthase gene into petunia results in reversible co-suppression of homologous genes in trans. *Plant Cell* **2:** 279–289.

Nelson, M.A. 2004. EST evidence for alternative splicing in *Neurospora crassa*. In *Annual Neurospora Convention*.

Newell, S.Y. 1992. Estimating fungal biomass and productivity in decomposing litter. In *The fungal community: Its organization and role in the ecosystem* (eds. G.C. Carroll and D.T. Wicklow), pp. 521–561. Dekker, New York.

Nielsen, C.B., Friedman, B., Birren, B., Burge, C.B., and Galagan, J.E. 2004. Patterns of intron gain and loss in fungi. *PLoS Biol.* **2:** E422.

Odds, F.C., Brown, A.J., and Gow, N.A. 2004. *Candida albicans* genome sequence: A platform for genomics in the absence of genetics. *Genome Biol.* **5:** 230.

Parra, G., Blanco, E., and Guigo, R. 2000. GeneID in *Drosophila*. *Genome Res.* **10:** 511–515.

Rensberger, B. 1992. The Iceman: Now the research is on ice. *J. NIIT Res.* **4:** 25–27.

Romfo, C.M., Alvarez, C.J., van Heeckeren, W.J., Webb, C.J., and Wise, J.A. 2000. Evidence for splice site pairing via intron definition in *Schizosaccharomyces pombe*. *Mol. Cell. Biol.* **20:** 7955–7970.

Roy, S.W. and Gilbert, W. 2005a. Complex early genes. *Proc. Natl. Acad. Sci.* **102:** 1986–1991.

———. 2005b. The pattern of intron loss. *Proc. Natl. Acad. Sci.* **102:** 713–718.

Ryan, M.J. and Smith, D. 2004. Fungal genetic resource centres and the genomic challenge. *Mycol. Res.* **108:** 1351–1362.

Salamov, A.A. and Solovyev, V.V. 2000. Ab initio gene finding in *Drosophila* genomic DNA. *Genome Res.* **10:** 516–522.

Salzberg, S.L., Pertea, M., Delcher, A.L., Gardner, M.J., and Tettelin, H. 1999. Interpolated Markov models for eukaryotic gene finding. *Genomics* **59:** 24–31.

Sambrook, J. 1977. Adenovirus amazes at Cold Spring Harbor. *Nature* **268:** 101–104.

Schechtman, M.G. 1990. Characterization of telomere DNA from *Neurospora crassa*. *Gene* **88:** 159–165.

Selker, E.U. 1990. Premeiotic instability of repeated sequences in *Neurospora crassa*. *Annu. Rev. Genet.* **24:** 579–613.

———. 2002. Repeat-induced gene silencing in fungi. *Adv. Genet.* **46:** 439–450.

Seoighe, C., Federspiel, N., Jones, T., Hansen, N., Bivolarovic, V., Surzycki, R., Tamse, R., Komp, C., Huizar, L., Davis, R.W., et al. 2000. Prevalence of small inversions in yeast gene order evolution. *Proc. Natl. Acad. Sci.* **97:** 14433–14437.

Shiu, P.K. and Metzenberg, R.L. 2002. Meiotic silencing by unpaired DNA: Properties, regulation and suppression. *Genetics* **161:** 1483–1495.
Shiu, P.K., Raju, N.B., Zickler, D., and Metzenberg, R.L. 2001. Meiotic silencing by unpaired DNA. *Cell* **107:** 905–916.
Slater, G.S. and Birney, E. 2005. Automated generation of heuristics for biological sequence comparison. *BMC Bioinform.* **6:** 31.
Stanke, M. and Morgenstern, B. 2005. AUGUSTUS: A web server for gene prediction in eukaryotes that allows user-defined constraints. *Nucleic Acids Res.* **33:** W465–W467.
Stanke, M. and Waack, S. 2003. Gene prediction with a hidden Markov model and a new intron submodel. *Bioinformatics* **19:** II215–II225.
Stanke, M., Steinkamp, R., Waack, S., and Morgenstern, B. 2004. AUGUSTUS: A web server for gene finding in eukaryotes. *Nucleic Acids Res.* **32:** W309–W312.
Strange, R.N. and Scott, P.R. 2005. PLANT DISEASE: A threat to global food security. *Annu. Rev. Phytopathol.* **43:** 83–116.
Swartz, M.N. 1994. Hospital-acquired infections: Diseases with increasingly limited therapies. *Proc. Natl. Acad. Sci.* **91:** 2420–2427.
Taylor, T.N., Hass, H., and Kerp, H. 1999. The oldest fossil ascomycetes. *Nature* **399:** 648.
Tenney, A.E., Brown, R.H., Vaske, C., Lodge, J.K., Doering, T.L., and Brent, M.R. 2004. Gene prediction and verification in a compact genome with numerous small introns. *Genome Res.* **14:** 2330–2335.
Toth, G., Gaspari, Z., and Jurka, J. 2000. Microsatellites in different eukaryotic genomes: Survey and analysis. *Genome Res.* **10:** 967–981.
Trappe, J.M. 1987. Phylogenetic and ecologic aspects of mycotrophy in the angiosperms from an evolutionary standpoint. In *Ecophysiology of VA mycrorrhizal plants* (ed. G.R. Safir), pp. 5–25. CRC Press, Boca Raton, FL.
Veneault-Fourrey, C. and Talbot, N.J. 2005. Moving toward a systems biology approach to the study of fungal pathogenesis in the rice blast fungus *Magnaporthe grisea*. *Adv. Appl. Microbiol.* **57:** 177–215.
Vinson, J.P., Jaffe, D.B., O'Neill, K., Karlsson, E.K., Stange-Thomann, N., Anderson, S., Mesirov, J.P., Satoh, N., Satou, Y., Nusbaum, C., et al. 2005. Assembly of polymorphic genomes: Algorithms and application to *Ciona savignyi*. *Genome Res.* **15:** 1127–1135.
Wolfe, K.H. and Shields, D.C. 1997. Molecular evidence for an ancient duplication of the entire yeast genome. *Nature* **387:** 708–713.
Wood, V.R., Gwilliam, M.A., Rajandream, M., Lyne, R., Lyne, A., Stewart, J., Sgouros, N., Peat, J., Hayles, S., Baker, D., et al. 2002. The genome sequence of *Schizosaccharomyces pombe*. *Nature* **415:** 871–880.
Xu, J., Vilgalys, R., and Mitchell, T.G. 2000. Multiple gene genealogies reveal recent dispersion and hybridization in the human pathogenic fungus *Cryptococcus neoformans*. *Mol. Ecol.* **9:** 1471–1481.
Yu, J.H. and Keller, N. 2005. Regulation of secondary metabolism in filamentous fungi. *Annu. Rev. Phytopathol.* **43:** 437–458.
Zeigler, R.S., Leong, S.A., and Teeng, P.S. 1994. *Rice Blast Disease*. CAB International, Wallingford, UK.
Zhou, S., Kile, A., Kvikstad, E., Bechner, M., Severin, J., Forrest, D., Runnheim, R., Churas, C., Anantharaman, T.S., Myler, P., et al. 2004. Shotgun optical mapping of the entire Leishmania major Friedlin genome. *Mol. Biochem. Parasitol.* **138:** 97–106.

4

The *Arabidopsis* Genome: A Foundation for Plant Research

Michael Bevan
Cell and Developmental Biology Department
John Innes Centre
Norwich NR4 7UJ, United Kingdom

Sean Walsh
Computational Biology Department
John Innes Centre
Norwich NR4 7UJ, United Kingdom

*A*RABIDOPSIS THALIANA WAS THE FIRST PLANT, AND THE THIRD multicellular organism after *Caenorhabditis elegans* (The *C. elegans* Sequencing Consortium 1998) and *Drosophila melanogaster* (Adams et al. 2000), to be completely sequenced (The *Arabidopsis* Genome Initiative 2000). At the time, it was claimed that the *Arabidopsis* genome sequence "... creates the potential for direct and efficient access to a much deeper understanding of plant development and environmental responses, and permits the structure and dynamics of plant genomes to be assessed and understood." Five years on, how justified was this claim? Furthermore, a vision for the *Arabidopsis* research community was articulated based on the promise of the genome sequence. A noteworthy aspiration was to "determine the function of all *Arabidopsis* genes by 2010." What progress has been made toward this goal? This review takes a broad and necessarily shallow view of progress in *Arabidopsis* research and relates this to work in other reference organisms. Our analysis suggests that much of the extraordinary progress made in the past five years has drawn on the genome sequence, and shows that it has had a catalytic effect on the research community and on how plant science is conducted. However, the explosion of data has created unanticipated problems that the *Arabidopsis* and plant science community must address if it is to take successfully to the path of integrative biology.

PROGRESS IN SEQUENCING, REASSEMBLY, AND ANNOTATION OF THE GENOME

Since systematic sequencing was completed in late 2000, the genome sequence has undergone several rounds of reassembly, hole patching, and extension into unsequenced regions. The sequenced and analyzed regions of the genome cover ~119 Mb (million base pairs) of genome sequence. The most impressive accomplishment has been to extend BAC and YAC contigs further into pericentromeric regions (Hosouchi et al. 2002). The size of the remaining gap in each chromosome, estimated using gel electrophoresis, varied between 4Mb and 9 Mb, yielding an overall genome size of ~146 Mb. Several new genes were also identified in the pericentromeric heterochromatin. These regions are among the most comprehensively described (Hall et al. 2002) and are a key resource for understanding and using centromere functions, for example, to make minichromosomes, for understanding the biological functions of repeat sequences and how they originate and are maintained.

The initial set of gene models was generated by a combination of optimized ab initio gene-finding algorithms and supporting data such as EST and cDNA sequence. This was carried out in several locations (for maximum speed) and inevitably discrepancies occurred. Subsequently more systematic rounds of annotation incorporated a large number of full-length (FL) cDNA sequences (notably the RIKEN/SALK and Ceres resources) such that 16,000 of the 29,000 predicted genes are supported by experimental evidence (Wortman et al. 2003). Comparison of *Arabidopsis* sequences with genomic sequence from the closely related *Brassica oleracea* (Chinese cabbage) identified regions of high similarity that either identified putative new genes or extended existing gene models. About 30% of these new genes encoded a transcript. About 25% of the originally predicted genes had no supporting evidence such as an EST match or reasonable similarity of their putative peptide sequence to any other protein (the hypothetical genes), and consequently their biological relevance remains doubtful. Nevertheless, systematic analysis of this class of genes on Chromosome 2 revealed that the majority of these were expressed (Xiao et al. 2002). Other analyses of EST matches to the genome sequence found several hundred matches to putative noncoding regions and other noncanonical gene-like entities (Riano-Pachon et al. 2005).

The canonical genome sequence and annotation is currently represented by the TIGR (The Institute for Genome Research) v5 analysis (Haas et al. 2005) (http://www.tigr.org/tdb/e2k1/ath1/ath1.shtml). A comprehensive reassembly of the genome incorporated newly sequenced

BACs from centromeric regions and carefully examined clone overlaps to yield 119 Mb of analyzed sequence. Genes in this new assembly were then predicted using FL-cDNA and EST sequences, resulting in 19,117 genes matching assemblies of EST and FL-cDNA sequences. About 800 new genes were identified in intergenic space. A total of 26,207 protein-coding genes encode 27,885 distinctive proteins by alternative splicing, nearly all of which contain known protein domains. Approximately 36% of the predicted proteome is encoded by segmental and tandem genomic duplications. Approximately 1400 pseudogenes were identified, many of which had degenerated protein sequences and premature stop codons compared to the family members to which they were most closely related. Improved analysis of repeat and transposon sequences helped to identify 2355 transposon loci. Many of these had been defined as protein-coding genes in the original annotation. Table 1 summarizes some of the analyses of the TIGR v.5 annotation (Haas et al. 2005) and other data.

The new set of predicted genes was functionally annotated using conserved domain composition, not overall sequence homology as performed originally (The *Arabidopsis* Genome Initiative 2000). This permits protein families and relationships between proteins to be more thoroughly defined. Gene Ontology (Ashburner et al. 2000) terms were used to describe each gene in terms of the molecular function of the encoded protein, the biological process in which the protein functions, and the cellular component to which the protein may belong. All proteins were manually assigned to at least one of these categories by these authors (Haas et al. 2005). These manual and computational assignments will continue to be refined as further evidence of gene function is accumulated.

An Affymetrix whole-genome array (WGA) for the *Arabidopsis* genome was hybridized with cRNA to detect transcripts and their chromosomal

Table 1. Summary of *Arabidopsis* genome features from current analyses

Genome size	146 Mb (estimated)
Sequenced and annotated genome space	119 Mb
Predicted protein coding genes	26,207
Alternately spliced genes	2330
Protein coding genes with identified transcripts	19,117
Genes in protein families	18,641
Transposons and pseudogenes	3786
Distinct proteins	27,855
Nonredundant cloned ORFs	14,668
Genes with insertions in exon + intron space	24,589

location (Yamada et al. 2003). This landmark analysis supported ~5000 hypothetical gene models as actively transcribed genes, and identified transcripts arising from many intergenic regions and from 20% of the 1300 pseudogenes. Several transcriptional hotspots were identified in centromeric regions that corresponded to a variety of repeats and transposons as well as previously unrecognized genes. Finally, transcription units for most of the genes were accurately defined and used to identify full-length ORF clones for >30% of the genome. These ORFs have been cloned (by the RIKEN/SALK and Agrikola projects) and are critically important resources for functional genomics; their uses are described below.

THE DYNAMIC GENOME

One of the major features of the *Arabidopsis* genome revealed by the genome sequence was the extent of gene duplication and segmental duplications, which was surprising given the expectation of a functionally compact genome. Approximately 60% of the genome was thought to be derived from a single duplication event, possibly of the entire genome (The *Arabidopsis* Genome Initiative 2000). Subsequently, more detailed analysis (Ku et al. 2000; Blanc et al. 2003) proposed that the *Arabidopsis* lineage has undergone at least two duplications, the most recent being a polyploidization event during the early evolution of the crucifers (Blanc and Wolfe 2004b). These analyses support a model of *Arabidopsis* genome evolution involving cycles of gene duplication, gene loss, and gene divergence. Genes that remain duplicated tend to become specialized—for example, by different expression patterns (Blanc and Wolfe 2004a). The frequent observation of gene families as tandem arrays further supports the view of a dynamic genome driven by duplication events and specializing gene function from multiple copies of genes. The extensive work carried out based on the *Arabidopsis* genome sequence also supports interpretations of the evolution of the vertebrate lineage that propose a central role for genome duplications (Wolfe 2001). Future work aims to assess and understand genome dynamics, such as gene loss, and altered gene expression that occurs as a consequence of polyploidization. These studies draw on the *Arabidopsis* genome sequence to understand how genome duplications shape evolution and crop plant performance (Osborn et al. 2003).

A. thaliana is a wild species adapted to survive in a wide geographical range, and there is a long legacy of its use as a model for adaptation. Consequently there is an extensive range of natural variation in growth and environmental responsive traits that provide an exceptionally rich source of diversity. Several loci exhibiting variation in complex traits (Quantitative Trait Loci or QTL) have been cloned. Examples include using linkage disequilibrium (LD) to fine map the *FRI* and *FLC* loci controlling flowering time (Hagenblad et al. 2004). Natural variation in hypocotyls' responses to light was shown to be due to polymorphisms in phytochrome light receptors (Borevitz and Nordborg 2003). Frequent sequence polymorphisms were identified between the canonical sequenced strain Columbia (Co) and extensive sequence of another lab strain, Landsberg *erecta* (L*er*) (The *Arabidopsis* Genome Initiative 2000), that are a key resource for map-based cloning. High-throughput methods for identifying polymorphisms have been developed, such as capillary-based SNP detection, to exploit the useful polymorphisms. Affymetrix expression arrays have also been used for genotyping; total genomic DNA from Recombinant Inbred Lines (RILs) made from a cross of Col and L*er* was hybridized to the ATH1 Affymetrix array, and recombination events were identified (Borevitz and Nordborg 2003). Recently, Nordborg et al. (2005) sequenced small genomic regions of 96 accessions of *A. thaliana* to provide a systematic survey of a dense pattern of polymorphisms from a large sample of individuals. This allowed, for the first time, a view of patterns of genome-wide polymorphism and population structure. A clear relationship between geographical distance and genotype distance was shown by clustering of allele frequencies. This was quite surprising given that humans have transported *Arabidopsis* around the globe. There is widespread sharing of variation within these populations, and common accessions (ecotypes) share haplotypes though recombination, demonstrating that these lab strains must not be considered as "lineages." The authors also conclude that the statistical tests commonly used to analyze polymorphism data are not valid in *A. thaliana* because there is "too much local polymorphism." For example, there was a positive correlation between polymorphism levels and genome duplications, and a negative correlation with gene density. These factors require that studies to identify polymorphism frequencies that contribute to phenotypic variation will require surveys of the whole genome. Nevertheless, these authors propose that *A. thaliana* is a good model for evolutionary genomics. Recently, projects were announced by the DOE Joint Genome Institute (JGI) that aim to sequence the genomes of *Arabidopsis lyrata* and *Capsella rubella*, which are close relatives of *A. thaliana*, using a whole-genome shotgun strategy (http://www.jgi.doe.gov/sequencing/cspseqplans2006.html). These sequences will enable significant advances in many aspects of *A. thaliana* biology, for example, by defining ancestral polymorphisms and conserved regions or "footprints" that may have functional significance.

REPEATS

One of the reasons *Arabidopsis* was chosen for complete sequencing was its relative lack of repeat sequences compared to other experimentally tractable plants. Cytogenetic analysis (Fransz et al. 2000) revealed extensive tracts of pericentromeric heterochromatin and two rDNA loci at the northern ends of Chromosomes 2 and 4. Chromosome sequencing showed that their pericentromeric regions contained a complex mixture of retroelements, transposons, microsatellites, and middle-repetitive sequence. Unsequenced regions adjacent to and including centromeres contained homopolymeric tracts of characteristic 180-bp and 160-bp repeats. Interstitial heterochromatic regions (or knobs) have been completely sequenced within the context of surrounding low-copy sequences, and these regions have provided deep insights into how heterochromatin is initiated and maintained and how this chromatin state influences gene expression. One of these sequenced regions, hk4S on Chromosome 4, contained a tract of 22.5 tandem copies of the *AtEMSAT1* satellite repeats (Mayer et al. 1999; The Cold Spring Harbor Laboratory, Washington University Genome Sequencing Center, and PE Biosystems *Arabidopsis* Sequencing Consortium 2000). A tiling array was used to profile histone and DNA modifications across these repeats, and it was shown that these and other transposon repeats were methylated and marked by H3mK9 histone modification (Lippman et al. 2004). DNA methylation of these repeats was lost in mutants defective for the *DDM1* chromatin-modifying gene, and this was associated with loss of the H3mK9 mark and increased transcription (Gendrel et al. 2002). The tandem repeats encoded siRNAs, and the levels of these were strongly reduced in the *ddm1* mutant. Furthermore, genes located within the hk4S knob were transcriptionally silent and showed *DDM1*-dependent methylation. It was concluded, based on several well-characterized examples, that epigenetic gene silencing can be mediated by transposable elements inserted in or close to genes. This work was among the first to implicate RNAi in establishing and maintaining epigenetic marks on repeats and genes, and it was directly based on the careful assembly and analysis of complex repeat sequences on *Arabidopsis* Chromosome 4. It has established a new paradigm for understanding how epigenetic marks may be guided to appropriate genome sequences and stably inherited (Martienssen et al. 2005).

GENETIC RESOURCES

Before the genome sequence was completed, assembled, and annotated, scientists performed chromosome walks to identify mutant loci using YAC and cosmid clones. This approach was slow, uncertain, and had practical drawbacks. The precision and speed with which map-based cloning can now be conducted, given access to the genome sequence and polymorphism data, has greatly accelerated the rate of gene discovery and profitably extended the reach of genetic analysis into many research areas, such as cell biology and metabolism. A plethora of methods and resources have been established over the past five years that provide plant scientists with almost undreamt-of possibilities. Principal among these are the populations of *Arabidopsis* containing T-DNA and transposon insertions. The precise insertion sites of a vast number of these elements in the genome have been determined using PCR-based amplification of flanking sequences and sequencing (Table 2; Alonso et al. 2003; Rosso et al. 2003). These flanking sequences have been aligned with genome sequence assemblies in several databases that permit easy and usually unambiguous identification of genes harboring insertions. To date, there are ~320,000 sequenced insertions in the reference Columbia genome. Lines with sequenced insertion sites are freely available through the *Arabidopsis* Biological Resource Centre (ABRC) and the Nottingham *Arabidopsis* Stock Centre (NASC). These lines are an outstanding resource for functional genomics and should permit the functions of a large number of genes to be determined. This was one of the most important post-genomic objectives of the *Arabidopsis* research community. Future objectives that will further facilitate functional genomics include establishing a bulked set of reference lines with two verified insertion alleles in genes (J. Ecker, pers. comm.). This resource can then be used for systematic surveys of

Table 2. Functional genomics resources in *Arabidopsis thaliana*

Ecotype	Insertion type	Number of sequenced insertions
Wassilewskija	T-DNA	36,287
Landsberg *erecta*	Activation trap	270
Landsberg *erecta*	Enhancer trap	13
Landsberg *erecta*	Gene trap	8891
Landsberg *erecta*	Misexpression trap	899
Columbia	T-DNA	306,168
Columbia	SM	15,943

Data taken from http://www.atidb.org.

gene function, for establishing populations of crosses to examine genetic interactions, and so on. Other smaller populations containing specialized insertions for the mis-expression of genes, for detecting gene expression patterns as gene or enhancer traps, and for expressing reporter genes from enhancer traps have all been made (Table 2). A significant proportion of the insertions sites have been sequenced, and in most cases the lines have been deposited in the stock centers for distribution.

Many protein-coding genes (~1600) remain with no insertions within exons or introns. Simulations show that doubling the number of random insertions to 600,000 would raise the current proportion of protein-coding genes with insertions from 94% to 98% (Fig. 1). More directed approaches could be taken, for example, using TILLING (Till et al. 2003), RNAi (Lawrence and Pikaard 2003), or newly developed gene replacement strategies (Shaked et al. 2005). This latter method depends on expression of a yeast *RAD54* gene, encoding a member of the *SWI/SNF2* chromatin-remodeling gene family. The gene promotes strand invasion in yeast and is required in that organism for efficient recombination. Expression in *Arabidopsis* increased homologous recombination frequencies by 27-fold, yielding a useful frequency of between 0.01 and 0.1, making it feasible to screen transformants directly by PCR. This method will prove to be very useful and will undoubtedly be further enhanced. Finally, fast neutrons are emerging as the mutagen of choice, as calibrated doses make deletions of corresponding size and frequency (Li et al. 2001).

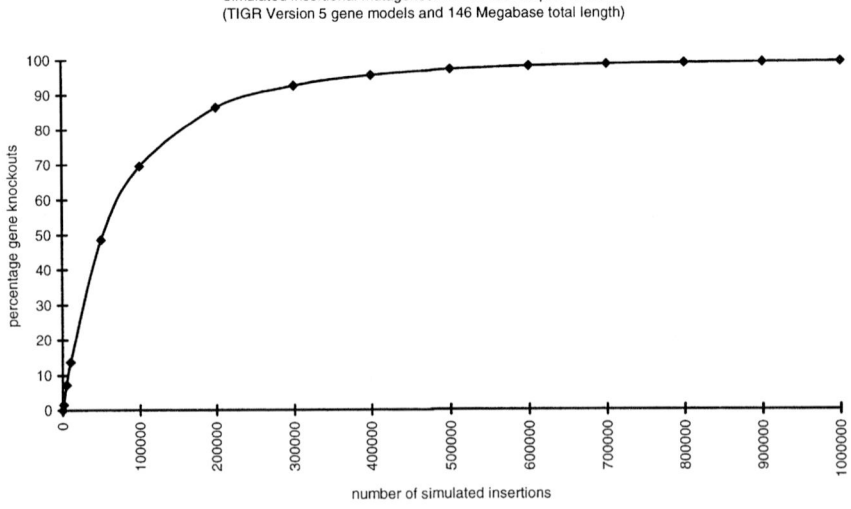

Figure 1. Simulation of random insertions in the *Arabidopsis* genome.

Small deletions targeting loci can be detected in multiplexed samples using PCR, and populations with finely mapped deletions providing coverage of the genome can be made. It is also efficient and feasible to screen deletion lines by hybridization of genomic DNA to Affymetrix chips to map deletions. This strategy has been used in *Medicago truncatula* to identify genes with reduced transcription caused by nonsense codons generated by EMS mutagenesis (Mitra et al. 2004). Taken together, the resources for functional genomics that have been developed and made available provide the entire plant research community with many of the resources needed for progress toward the goal of describing the functions of every gene by 2010.

GENE EXPRESSION

Array-based transcript detection methods were devised soon after the completion of the genome sequence. Perhaps the most widely used system is the Affymetrix ATH1 array, which has probe sets representing ~24,000 genes as 11 25-mer probe pairs per gene on a single chip. Full-length cDNAs and long gene-specific oligonucleotides have also been printed onto microarrays and used to analyze gene expression. A comprehensive set of genome sequence tags (GSTs) has also been created (Hilson et al. 2004) in the CATMA (Complete *Arabidopsis* Transcriptome Analysis) project (http://www.catma.org/). These are PCR-amplified regions of genomic DNA between 150 and 300 bp long that provide gene-specific probes to ~21,000 *Arabidopsis* genes. These have been used in transcript profiling experiments with printed microarrays (Hilson et al. 2004), and PCR amplicons and cloned GSTs are available from NASC. GSTs were also configured in GATEWAY vectors and used in RNAi experiments to down-regulate gene expression. These reagents provide a key resource for functional genomic experiments.

New insights into the role of RNA species in regulating gene expression and genome dynamics are being gained rapidly based on reference to the genome sequence and annotation. Massively parallel signature sequencing (MPSS) identified a large number of noncoding RNAs (ncRNA) (Meyers et al. 2004), as did whole-genome arrays (Yamada et al. 2003). These data also revealed extensive antisense transcription of genes (Wang et al. 2005) and microRNAs, most of which appear to be encoded in non-gene space and may act as regulators of gene expression (Reinhart et al. 2002). They may serve as sequence templates to direct chromatin protein complexes to specific locations, as has been suggested for repeats (Martienssen and Colot 2001; Gendrel et al. 2002; Lippman et al. 2004). One of the exciting challenges in this area, to which an accurately annotated genome sequence can contribute, is establishing the role of small RNA species in growth and development by determining their roles in controlling gene expression through chromatin remodeling and translational control. The machinery producing ncRNAs is starting to be understood—for example, one of the surprises arising from analysis of the genome sequence was an outlier RNA polymerase dubbed RNA Pol IV (The *Arabidopsis* Genome Initiative 2000). This has recently been shown to be a silencing-specific RNA polymerase required for production and maintenance of small interfering RNA (siRNA) (Herr et al. 2005).

Many microarray experiments have been conducted that provide a large amount of quantitative data on gene expression in different *Arabidopsis* tissues and in response to different treatments and experimental conditions. Data from purified cell types (Birnbaum et al. 2003) and microdissected tissues (Casson et al. 2005) continue to extend the resolution of array analysis to the cellular level. Recently, Zanetti et al. (2005) immunopurified polysomes using an epitope-tagged ribosomal protein and showed that transcript profiles obtained by RNA isolated from these polysomes yielded highly reproducible data. Thus it may be possible to express epitope-tagged ribosome proteins from cell-specific promoters (e.g., from enhancer trap lines) to fractionate cell-specific mRNA populations. Finally, it is conceivable that microarray data can be integrated with reporter gene expression patterns, for example, derived from gene trap lines (Sundaresan et al. 1995), to establish a comprehensive and cell-specific analysis of gene expression.

Microarray data from *Arabidopsis* experiments are accumulating rapidly in different databases such as ArrayExpress and NASC. For example, the ATGENEXPRESS network (http://web.uni-frankfurt.de/fb15/botanik/mcb/AFGN/atgenex.htm) recently integrated the results from ~1300 microarray experiments in databases at TAIR and NASC. Data from different experiments can be analyzed because data submitted to these databases are MIAME compliant—that is, there are standard descriptions of experimental conditions, hybridization conditions, and so on that permit statistically sound comparisons. Tools such as GENEVESTIGATOR have been developed to analyze sets of *Arabidopsis* microarray data (Zimmermann et al. 2004). This system provides sophisticated data-mining and other analytical tools to identify expression profiles of sets of genes in different conditions and tissues. These data can also be downloaded for further analyses. So far, array experiments have mostly been interpreted qualitatively—for example, to describe the classes of genes regulated by light, stress, and so on.

The comprehensive and quantitative data obtained from multiple chip experiments can also be used to establish functional modules based on coregulation and tissue specificity and to model gene expression networks. For example, gene expression data from ATH1 chips of genes involved in metabolism have been mapped to corresponding pathways to interpret metabolic consequences of gene expression (Thimm et al. 2004). Recently a larger study (Schmid et al. 2005) integrated data from 79 experiments that assayed RNA levels in different tissues and organs of *Arabidopsis* plants. An expression map of the classes of genes expressed in different organs and tissues was established by reference to GO classifications of gene function. This type of meta-analysis is an initial step in establishing a quantitative framework for interpreting the influence of mutations on gene expression and for defining functional modules. Similar expression maps have been made for yeast (Hughes et al. 2000), *C. elegans* (Kim et al. 2001), and mouse (Zhang et al. 2004). These studies associated coregulated genes and GO descriptions of gene expression to establish functional modules using clustering analysis. Coclustering of genes with known and unknown functions ("guilt by association") can provide important clues to the functions of many genes, which can subsequently be experimentally defined. Such analyses, when coupled to comprehensive GO annotations and interactome data from *Arabidopsis* (see below), will provide important quantitative information about cellular processes. These analyses currently provide an "averaged" view of tissues that are composed of several cell types, and resolution at the cellular level is required to provide key data on the subfunctionalization of cells and how cell functions are integrated to generate tissue and ultimately whole plant responses. This issue is discussed further below in the context of cell and systems biology in *Arabidopsis*.

Array data have also started to be used for determining transcriptional networks, that is, the interplay of transcription factors and promoter sequences that directs the place, time, and level of gene transcription. Establishing models of transcription networks is critically important for defining the logic of regulatory sequences. Experimental approaches in yeast (Lee et al. 2002) used chromatin immunoprecipitation to identify the genome localization of 106 epitope-tagged transcriptional regulators. The promoter regions of ~2300 genes were tagged by at least one regulator, many were tagged by several regulators, and each regulator bound on average 38 promoter regions. Models of regulatory networks such as feedforward loops and regulatory chains were established based on these data. Currently, it is challenging to establish regulatory networks using this strategy in *Arabidopsis*, as it has ~1700 putative DNA-binding transcription factors and a much larger genome, and multicellularity will complicate interpretation. Nevertheless, the promise of this type of analysis suggests it will be increasingly used in *Arabidopsis* to establish transcriptional regulatory networks.

Computational methods can also generate reasonable models of promoter functions that promise to reduce the complexity of experiments that involve vast numbers of transcription factors such as found in *Arabidopsis*. Algorithms have been developed that correlate promoter motifs with expression patterns. One method involves clustering genes according to their expression levels in multiple conditions and identifying promoters of genes in coregulated clusters. Global alignment, frequency analysis, and other methods have then been used to identify sequence motifs common to each cluster (Tavazoie et al. 1999; Bussemaker et al. 2001). The motifs can then be related to the functions of known promoter motifs (e.g., those in the PLACE database) (Higo et al. 1999) and experimentally tested to determine their function. The application of these strategies in *Arabidopsis* holds great promise in interpreting gene expression data and generating models of gene expression networks. Of particular relevance to this type of analysis is the identification of promoter sequences conserved between closely related species. These phylogenetic footprints may identify functionally relevant promoter domains, for example, comparison of the *CHS* and *AP3* promoter sequences across relatives of *Arabidopsis* identified conserved domains that were functionally significant (Koch et al. 2001). Whole-genome sequencing and assembly of *A. lyrata* and *C. rubella*, together with the planned sequence of *Brassica* species, will provide new opportunities for the systematic determination of gene regulatory architecture in *A. thaliana* as has been proposed for yeast (Kellis et al. 2003).

DETERMINING GENE FUNCTION

The foundations for systematic determination of gene function have been laid by the reannotation of the genome (Haas et al. 2005), the generation of >14,000 nonredundant full-length cDNA clones (Seki et al. 2002) (http://signal.salk.edu/cgi-bin/tdnaexpress; http://rarge.gsc.riken.go.jp/cdna/cdna.pl), the definition of transcription units (Yamada et al. 2003; Meyers et al. 2004), and the development of mathematically structured and systematic descriptions of gene functions (http://www.arabidopsis.org/tools/bulk/go/index). As described above, the initial annotation of the genome was based on limited full-length cDNA sequences. Currently there are 14,668 nonredundant cDNAs available from the RIKEN Bioresource Center (BRC) database made in collaboration between the SSP (Salk, Stanford and Plant Gene Expression Laboratory) and RIKEN groups. Many of these ORFs are cloned into a versatile vector that permits recombination-based manipulation of ORFs into epitope tags, yeast two-hybrid baits, and so on. These collections provide an indispensable resource for large-scale screens, for example, for determining the interactome, systematic determination of biochemical functions, and protein localization. Two large-scale studies have pioneered the systematic determination of biochemical activities of many members of the glycosyl transferase (Bowles et al. 2005) and cytochrome P450 protein families (Schuler and Werck-Reichhart 2003) that catalyze many diverse biochemical reactions in plants. Smaller-scale studies have shown the utility of using ORF-GFP fusions to determine the subcellular localization of proteins (Cutler et al. 2000; Koroleva et al. 2005), and these can be scaled up for cell-based assays using transient expression systems.

Several studies have described in some detail the relationships and functions of genes in large families and have incorporated experimental data from the literature. These include important studies of the MYB (Stracke et al. 2001), bZIP (Jakoby et al. 2002), and b-HLH (Toledo-Ortiz et al. 2003) transcription factor families, the CDPK-SnRK superfamily of protein kinases (Hrabak et al. 2003), pentatricopeptide repeat proteins (Lurin et al. 2004), and receptor-like kinases (Shiu et al. 2004). Many enzymes of primary metabolism have been carefully annotated and incorporated into specialist databases such as AraCYC (Mueller et al. 2003). Enzymes involved in carbohydrate metabolism have been thoroughly analyzed and annotated (http://afmb.cnrs-mrs.fr/CAZY/). These studies and reviews are critically important for functional genomics studies, as they engage experts in systematic analyses of large numbers of genes. These studies would have even more value if the data were expressed in standardized ontological terms and incorporated into a common framework such as offered by genome databases.

Proteomics strategies are increasingly used in *Arabidopsis* research, and the power of these methods has been greatly improved by the careful reannotation of the genome. Proteins found in the chloroplast, mitochondria, peroxisome, cell wall, nucleus, vacuoles, nucleolus, and cytoskeleton have all been carefully cataloged (Baginsky and Gruissem 2004; Heazlewood and Millar 2005; Pendle et al. 2005) and provide many important clues for future functional studies. Purification and mass-spectrometry of epitope-tagged proteins are increasingly used to identify complexes and post-translational modifications. In other model eukaryotes, notably yeast and *C. elegans*, "interactomes," or networks of protein–protein interactions, have been established by massive yeast two-hybrid screens using ORFs cloned in recombinational systems. Analysis of *C. elegans* interactomes (Li et al. 2004) identified functionally significant interactions that suggested functions for many proteins. Clearly, the availability of a large and well-characterized ORF collection in *Arabidopsis* makes the generation and analysis of *Arabidopsis* interactomes an important and feasible priority. The integration of protein interaction, genetic, and transcriptional networks has revealed global patterns of connections that have generated important biological insights. In yeast (Zhang et al. 2005) networks of interactions, such as protein–protein interactions and gene co-expression data, corresponded with known complexes (replication, actin associated, chromatin remodeling, etc.) and interactions between complexes (actin-associated proteins, motor proteins, protein translocation, etc.). Regulonic complexes linking transcription factors and their regulation of components of complexes were also established. These powerful types of analyses will be possible in *Arabidopsis* once interactome data are generated.

COMPARATIVE GENOMICS AND CROP PLANT RESEARCH

Table 3 shows the current state of sequencing in plants, mosses, and green algae. This includes the first planned sequence of a member of the asterid family, tomato (*Lycopersicon esculentum*), which will complement the five rosid species being sequenced. A broader sequence survey of plant taxa, including key species representing important nodes in plant evolution, is also under way (Pryer et al. 2002). This sampling of taxa strongly aids taxonomy and evolutionary studies and is also relevant to identifying plant biodiversity (Wheeler et al. 2004).

Recently the map-based sequence and analysis of the rice genome has been published (The International Rice Genome Sequence Project 2005) and is described in Chapter 5. Comparison of the rice and *Arabidopsis* proteomes showed that 71% of predicted rice proteins were reasonably similar to *Arabidopsis* proteins. While much more analysis needs to be done to determine relationships, this promising and somewhat unexpected high similarity suggests that the cellular and biochemical functions of many rice genes can be interpreted according to experiments conducted in *Arabidopsis*. Thus, determining orthologous relationships between *Arabidopsis* and crop plant genes is a sound way of translating information on *Arabidopsis* gene function and is therefore a high priority. The same strategy applied to the current annotation of the *Arabidopsis* genome (Haas et al. 2005), based on defining common domain architectures and

The *Arabidopsis* Genome 85

Table 3. Public plant genome sequencing projects

Group	Organism	EST	Gen.	Comment
Plants and mosses				
Asterids	Tomato	+	+	Map-based sequencing initiated
	Tobacco	+		
	Potato	+	+	Map-based sequencing planned
Rosids	*Arabidopsis thaliana*	+	+	Map-based sequencing completed 2000
	Arabidopsis lyrata		+	Whole-genome shotgun starts 2006
	Capsella rubella (Shepherd's Purse)		+	Whole-genome shotgun starts 2006
	Brassica	+	+	Map-based sequencing planned
	Cotton	+		
	Soybean	+		
	Lotus	+	+	Map-based sequencing underway
	Mimulus (Monkey Flower)		+	Whole-genome shotgun starts 2006
	Medicago (barrel medic)	+	+	Map-based sequencing underway
	Cottonwood/aspen	+	+	Whole-genome shotgun completed 2004
Monocots	Barley	+		
	Rice	+	+	Map-based sequencing completed 2005
	Sorghum	+	+	Whole-genome shotgun starts 2006
	Wheat	+		
	Maize	+	+	Map-based sequencing starts 2006
Conifers	Pinus	+		
	Gnetum	+		
Club moss	Selaginella	+	+	Whole-genome shotgun underway
Moss	Physcomitrella	+	+	Whole-genome shotgun underway
Green algae	Chlamydomonas		+	Whole-genome shotgun completed 2005

their conservation, could be applied to sequenced plant genomes as part of a concerted comparative genomics study. The extensive synteny observed between *Arabidopsis*, *M. truncatula*, and soybean (Mudge et al. 2005) suggests that novel bioinformatics approaches can be developed to take into account the chromosomal context of genes in determining ancestral relationships and relationships between paralogous families.

Several aspects of *Arabidopsis* biology have provided unexpected yet important knowledge about crop plant biology. Its small stature and annual growth habit suggested *Arabidopsis* had little to offer to understanding wood formation in trees. Nevertheless, inflorescence stems undergo secondary thickening and a bona fide cambium forms, and many genes expressed during secondary thickening and associated with cambial activity are highly conserved between *Arabidopsis* and *Populus* (Hertzberg et al. 2001). Furthermore, the systematic characterization of genes involved in cell wall formation, lignification, and vascular cell identity in *Arabidopsis* provides key leads for an important biotechnology sector. Although *Arabidopsis* and other members of the Brassicaceae do not engage in symbiotic interactions, studies in the legume species *Lotus japonicus*, *M. truncatula*, and pea have identified classes of proteins with conserved functions such as those involved in Ca^{2+}-mediated signaling, polar cell growth, and trimeric G-protein-mediated signaling (Limpens and Bisseling 2003). Extensive analysis in *Arabidopsis* of the function of this class of protein in responses to growth regulators and pathogens provides a strong framework for dissecting their role in early nodulation events.

BIOINFORMATICS AND MODELING

Arabidopsis researchers are supported by the hub provided by TAIR (The *Arabidopsis* Information Resource) and an extensive set of specialist genome databases that distribute *Arabidopsis* data (Rhee et al. 2003). Currently, a dedicated group is annotating *Arabidopsis* and other plant proteins centrally (ftp://ftp.arabidopsis.org/home/tair/Genes/Gene_Ontology/), and these GO terms describing gene functions are starting (albeit quite slowly) to replace the ad hoc system of gene descriptions currently widely used. GO terms describe proteins according to molecular function, biological process, and cellular component (Ashburner et al. 2000). These generic GO classifications provide a strong unifying concept in genome analysis within *Arabidopsis* and other plant species and between all genomes (http://www.godatabase.org/cgi-bin/amigo/go.cgi). Plant Ontology and Trait Ontology have also established controlled vocabularies describing plant anatomy and development, and plant traits and phenotypes (see http://www.gramene.org/plant_ontology/), while Structure Ontology aims to unify the description of genome features allowing for rapid feature transfer to newly sequenced genomes (http://song.sourceforge.net/).

An increasing number of large-scale studies relate biological data to genome features. To manage this work, powerful and useful generic software and database schemas have been developed by the Generic Model Organism Project (GMOD; http://www.gmod.org/). Chief among these is the Genome Browser. It provides a powerful application programming interface that allows sophisticated queries and reports to be constructed with relative ease. The system also provides a facility to allow nonprogrammers to overlay and visualize private data (Stein et al. 2002). Genome Browser has been used for large-scale functional genomics studies (e.g., http://www.atidb.org) and research on epigenetic control of chromosome-scale features (Martienssen et al. 2005). The benefits of data capture and dissemination from distributed sources such as specialist genome databases include direct access to up-to-date data generated by experts, but this creates problems for users who need to know where information can be found, how to relate different data sets, and how to overcome incompatibilities between different systems. The BioMOBY system enables interoperability between database providers through a Web service–like architecture. Services are registered at a central repository that allows bespoke reports to be generated with the visual programming tool Taverna (http://taverna.sourceforge.net/). This method has been successfully implemented by the PlaNet Database Consortium (Wilkinson et al. 2005), who provide a comprehensive array of *Arabidopsis* genome-related data resources through BioMOBY services (http://mips.gsf.de/projects/plants/PlaNetPortal/index_html).

One of the major challenges and responsibilities of researchers working on reference organisms such as *Arabidopsis* is the need systematically to capture the large amount of functional genomics data being generated. This permits all plant researchers to make links and integrate diverse types of data to obtain a global perspective of gene functions, as it is now well understood that networks of interactions rather than individual proteins themselves provide a better description of how biological processes are regulated. Currently 24 papers per working day are produced that mention *Arabidopsis*. This vast amount of information is not able to be captured by manual curation strategies alone, thus these data are fragmented, dispersed, and buried in the literature. Furthermore, the data are represented by a myriad of unrelated terms so that they cannot be accessed or manipulated computationally. In response to the problems created by traditional publication methods, text-mining tools are being developed to resurrect data for computational manipulation. However, these approaches are currently limited by restricted access to the full text of journal articles (Krallinger and Valencia 2005). The adoption of semantic tags by learned journals, which enable computers to mine text (Mons 2005), would also be a huge advance in this respect.

This fragmentation occurs at a time when many aspects of *Arabidopsis* research increasingly require individual laboratories to manipulate a common set of data and query all the data at once. The current display of data from several sources such as SIGnAL, TAIR, MIPS, ATIDB, and AtEnsembl further exacerbates the fragmentation problem. A strategy that promotes and rewards a division of labor between databases could help to overcome one aspect of this problem, that is, the collection, processing, analysis, and redistribution of major data sets. This requires free and open access to published community data, including software and backend database systems.

Once data can be successfully captured, they need to be integrated with existing knowledge to generate new hypotheses. Pathways are well understood biologically and computationally in graph theory and provide a strong framework upon which to build this integrated knowledge (Cary et al. 2005). The AraCYC (Mueller et al. 2003) database is a carefully curated database of metabolic pathways that could be expanded to include a more extensive network of metabolic reactions. However, it is difficult for such a system to form the basis for a community effort because its licensing terms are currently inconsistent with the need to share and develop data and databases in an open source environment. The Reactome system (http://www.reactome.org) does provide a framework for describing the molecular biology of an organism (the reactome). The initial implementation is for the human genome (Joshi-Tope et al. 2005) and incorporates a data model based on the concepts of reaction and pathway to describe molecular processes, such as signal transduction, gene regulation, and metabolism up to higher-level processes. Data are captured by expert curators, and pathways are projected onto other organisms to provide a framework for inferring gene functions. Reactome captures all actors in molecular processes as instances of database entities. This goes closer to modeling biological reality than is possible with systems like Gene Ontology or MapMan (Thimm et al. 2004) in their current invocations, allowing data sets from hybridization arrays and mass-spectroscopy experiments in metabolomics and proteomics to be overlaid on these pathways. In this way high-throughput and highly parallel data sets can be interpreted as correlations with existing knowledge. *Arabidopsis* data are currently being annotated into Reactome, and individuals can also use the curatorial tools to establish pathways in which they are experts, but the degree of evidence for particular data sets requires careful consideration or the adoption of specific evidence ontologies.

One of the most important and challenging aspects of *Arabidopsis* research is to translate knowledge for use in crop plants and to establish evolutionary relationships. Establishing orthologous relationships among the proteomes of the sequenced genomes of crop plants is therefore becoming a high priority. Genome and segmental duplication events during the evolution of flowering plants, and the abundance of large gene families, indicates that defining gene relationships will be as demanding as it will be rewarding. OrthoMCL (Li et al. 2003) establishes similarity matrices using BLASTP within and between species, and then uses Markov clustering to resolve multiple relationships in similarity space. Complete annotations are available for several eukaryotes, including *Arabidopsis* (http://www.cbil.upenn.edu/gene-family). The recent completion of a high-quality rice genome sequence provides the first major opportunity to establish orthologous relationships between rice and *Arabidopsis* proteomes.

The availability of gene catalogs and systematic descriptions of *Arabidopsis* gene function provide plant scientists with the opportunity to develop high-throughput assays to determine cellular readouts from the activities of multiple genes (Gibon et al. 2004). These studies can then be used to develop quantitative models of pathways and other cellular phenotypes to test predictions and develop new experiments. L-Systems have been developed to provide a quantitative framework for describing plant architecture and how this changes during growth and development (Prusinkiewicz 2004), and for describing physiological responses (Allen et al. 2005). This approach describes plants as a set of modules, each of which has a single mathematical description. Variables such as growth rate, genetic regulatory networks, and so on can be incorporated into these modules and the model run to "grow" the virtual plant (Fig. 2). The effects of mutations and gene interactions on growth can then be described according to the growth model. This provides a robust universal framework for describing growth and development phenotypes. At a cellular level, models of gene action in the shoot apical meristem have been established (Jonsson et al. 2005). These incorporate cellular lineage data (Reddy et al. 2004) and information about the spatial and temporal interactions of *CLAVATA1*, *CLAVATA3*, and *WUSCHEL* that maintain expression zones in the shoot apical meristem. This model was able to simulate changes in the *WUSCHEL* domain seen in experiments. These

Figure 2. L-Systems model of *Arabidopsis* inflorescence growth. This graphic shows a rendering of an L-systems model of the flowering shoot apex of *Arabidopsis*. This figure was generated by Przemek Prusinkiewicz, Enrico Coen, and their colleagues.

pioneering efforts in quantitative modeling of whole-plant and cell-based growth and development provide foundations for more extensive work aimed at modeling organ development at a cellular level. The planar shape of leaves (Rolland-Lagan et al. 2003) is a promising area in which significant progress is already being made.

PERSPECTIVES

Looking forward, the prospects for plant science appear to have never been better. The *Arabidopsis* and rice genomes, and the genome sequences currently being generated and analyzed (Table 3), provide a strong platform for supporting integrative plant science across model and crop species. The complete and accurate sequence of reference genomes from the major groups provides a framework for using sequence from promising high-throughput methods (Margulies et al. 2005) for gene discovery in many new groups of plants. Natural variation can be explored to understand adaptation (Weigel and Nordborg 2005) and broaden the scope of plant breeding (Gur and Zamir 2004). Access to extensive *Arabidopsis* functional genomics resources (Alonso et al. 2003) promotes all plant researchers to consider developing research programs with genetics at their core. New types of screens can be developed in plants using cell-based systems to dissect regulatory pathways and high-throughput RNAi-based methods (DasGupta et al. 2005). Stomatal and trichome cells are good candidates for systematically determining sets of genes involved in signal transduction and cell shape.

These unprecedented opportunities are coupled to socioeconomic trends suggesting a greater need for plant research. For example, more people deserve access to higher-quality food, and plant research can help promote improved plant productivity. Agriculture consumes most of the available high-quality fresh water, and plant research may be able to promote more efficient use of this precious resource. Currently only a small proportion of plant biomass is directly used for fuel and fiber. Increased atmospheric CO_2 levels, dwindling fossil fuel reserves, and their increasing costs suggest that we now need to accelerate research plans to make greater use of plant-based biomass as a renewable chemical feedstock and for energy production.

ACKNOWLEDGMENTS

We thank Przemek Prusinkiewicz, Enrico Coen, and colleagues for Figure 2. This work was funded by the Core Strategic Grant to the John Innes Centre and EC grant QRL1-CT-2001-00006 (PlaNet) to M.B. and S.W.

REFERENCES

Adams, M.D., Celniker, S.E., Holt, R.A., Evans, C.A., Gocayne, J.D., Amanatides, P.G., Scherer, S.E., Li, P.W., Hoskins, R.A., Galle, R.F., et al. 2000. The genome sequence of *Drosophila melanogaster*. *Science* **287:** 2185–2195.

Allen, M.T., Prusinkiewicz, P., and DeJong, T.M. 2005. Using L-systems for modeling source-sink interactions, architecture and physiology of growing trees: The L-PEACH model. *New Phytol.* **166:** 869–880.

Alonso, J.M., Stepanova, A.N., Leisse, T.J., Kim, C.J., Chen, H., Shinn, P., Stevenson, D.K., Zimmerman, J., Barajas, P., Cheuk, R., et al. 2003. Genome-wide insertional mutagenesis of *Arabidopsis thaliana*. *Science* **301:** 653–657.

The *Arabidopsis* Genome Initiative. 2000. Analysis of the genome sequence of the flowering plant *Arabidopsis thaliana*. *Nature* **408:** 796–815.

Ashburner, M., Ball, C.A., Blake, J.A., Botstein, D., Butler, H., Cherry, J.M., Davis, A.P., Dolinski, K., Dwight, S.S., Eppig, J.T., et al. 2000. Gene ontology: Tool for the unification of biology. *Nat. Genet.* **25:** 25–29.

Baginsky, S. and Gruissem, W. 2004. Chloroplast proteomics: Potentials and challenges. *J. Exp. Bot.* **55:** 1213–1220.

Birnbaum, K., Shasha, D.E., Wang, J.Y., Jung, J.W., Lambert, G.M., Galbraith, D.W., and Benfey, P.N. 2003. A gene expression map of the *Arabidopsis* root. *Science* **302:** 1956–1960.

Blanc, G. and Wolfe, K.H. 2004a. Functional divergence of duplicated genes formed by polyploidy during *Arabidopsis* evolution. *Plant Cell* **16:** 1679–1691.

———. 2004b. Widespread paleopolyploidy in model plant species inferred from age distributions of duplicate genes. *Plant Cell* **16**: 1667–1678.

Blanc, G., Hokamp, K., and Wolfe, K.H. 2003. A recent polyploidy superimposed on older large-scale duplications in the *Arabidopsis* genome. *Genome Res.* **13**: 137–144.

Borevitz, J.O. and Nordborg, M. 2003. The impact of genomics on the study of natural variation in *Arabidopsis*. *Plant Physiol.* **132**: 718–725.

Bowles, D., Isayenkova, J., Lim, E.K., and Poppenberger, B. 2005. Glycosyltransferases: Managers of small molecules. *Curr. Opin. Plant Biol.* **8**: 254–263.

Bussemaker, H.J., Li, H., and Siggia, E.D. 2001. Regulatory element detection using correlation with expression. *Nat. Genet.* **27**: 167–171.

Cary, M.P., Bader, G.D., and Sander, C. 2005. Pathway information for systems biology. *FEBS Lett.* **579**: 1815–1820.

Casson, S., Spencer, M., Walker, K., and Lindsey, K. 2005. Laser capture microdissection for the analysis of gene expression during embryogenesis of *Arabidopsis*. *Plant J.* **42**: 111–123.

The *C. elegans* Sequencing Consortium. 1998. Genome sequence of the nematode *C. elegans*: A platform for investigating biology. *Science* **282**: 2012–2046.

The Cold Spring Harbor Laboratory, Washington University Genome Sequencing Center, and PE Biosystems *Arabidopsis* Sequencing Consortium. 2000. The complete sequence of a heterochromatic island from a higher eukaryote. *Cell* **100**: 377–386.

Cutler, S.R., Ehrhardt, D.W., Griffitts, J.S., and Somerville, C.R. 2000. Random GFP:cDNA fusions enable visualization of subcellular structures in cells of *Arabidopsis* at a high frequency. *Proc. Natl. Acad. Sci.* **97**: 3718–3723.

DasGupta, R., Kaykas, A., Moon, R.T., and Perrimon, N. 2005. Functional genomic analysis of the Wnt-wingless signaling pathway. *Science* **308**: 826–833.

Fransz, P.F., Armstrong, S., de Jong, J.H., Parnell, L.D., van Drunen, C., Dean, C., Zabel, P., Bisseling, T., and Jones, G.H. 2000. Integrated cytogenetic map of chromosome arm 4S of *A. thaliana*: Structural organization of heterochromatic knob and centromere region. *Cell* **100**: 367–376.

Gendrel, A.V., Lippman, Z., Yordan, C., Colot, V., and Martienssen, R.A. 2002. Dependence of heterochromatic histone H3 methylation patterns on the *Arabidopsis* gene DDM1. *Science* **297**: 1871–1873.

Gibon, Y., Blaesing, O.E., Hannemann, J., Carillo, P., Hohne, M., Hendriks, J.H., Palacios, N., Cross, J., Selbig, J., and Stitt, M. 2004. A robot-based platform to measure multiple enzyme activities in *Arabidopsis* using a set of cycling assays: Comparison of changes of enzyme activities and transcript levels during diurnal cycles and in prolonged darkness. *Plant Cell* **16**: 3304–3325.

Gur, A. and Zamir, D. 2004. Unused natural variation can lift yield barriers in plant breeding. *PLoS Biol.* **2**: e245.

Haas, B.J., Wortman, J.R., Ronning, C.M., Hannick, L.I., Smith Jr., R.K., Maiti, R., Chan, A.P., Yu, C., Farzad, M., Wu, D., et al. 2005. Complete reannotation of the *Arabidopsis* genome: Methods, tools, protocols and the final release. *BMC Biol.* **3**: 7.

Hagenblad, J., Tang, C., Molitor, J., Werner, J., Zhao, K., Zheng, H., Marjoram, P., Weigel, D., and Nordborg, M. 2004. Haplotype structure and phenotypic associations in the chromosomal regions surrounding two *Arabidopsis thaliana* flowering time loci. *Genetics* **168**: 1627–1638.

Hall, A.E., Fiebig, A., and Preuss, D. 2002. Beyond the *Arabidopsis* genome: Opportunities for comparative genomics. *Plant Physiol.* **129**: 1439–1447.

Heazlewood, J.L. and Millar, A.H. 2005. AMPDB: The *Arabidopsis* mitochondrial protein database. *Nucleic Acids Res.* **33:** D605–D610.

Herr, A.J., Jensen, M.B., Dalmay, T., and Baulcombe, D.C. 2005. RNA polymerase IV directs silencing of endogenous DNA. *Science* **308:** 118–120.

Hertzberg, M., Aspeborg, H., Schrader, J., Andersson, A., Erlandsson, R., Blomqvist, K., Bhalerao, R., Uhlen, M., Teeri, T.T., Lundeberg, J., et al. 2001. A transcriptional roadmap to wood formation. *Proc. Natl. Acad. Sci.* **98:** 14732–14737.

Higo, K., Ugawa, Y., Iwamoto, M., and Korenaga, T. 1999. Plant *cis*-acting regulatory DNA elements (PLACE) database: 1999. *Nucleic Acids Res.* **27:** 297–300.

Hilson, P., Allemeersch, J., Altmann, T., Aubourg, S., Avon, A., Beynon, J., Bhalerao, R.P., Bitton, F., Caboche, M., Cannoot, B., et al. 2004. Versatile gene-specific sequence tags for *Arabidopsis* functional genomics: Transcript profiling and reverse genetics applications. *Genome Res.* **14:** 2176–2189.

Hosouchi, T., Kumekawa, N., Tsuruoka, H., and Kotani, H. 2002. Physical map-based sizes of the centromeric regions of *Arabidopsis thaliana* chromosomes 1,2, and 3. *DNA Res.* **9:** 117–121.

Hrabak, E.M., Chan, C.W., Gribskov, M., Harper, J.F., Choi, J.H., Halford, N., Kudla, J., Luan, S., Nimmo, H.G., Sussman, M.R., et al. 2003. The *Arabidopsis* CDPK-SnRK superfamily of protein kinases. *Plant Physiol.* **132:** 666–680.

Hughes, T.R., Marton, M.J., Jones, A.R., Roberts, C.J., Stoughton, R., Armour, C.D., Bennett, H.A., Coffey, E., Dai, H., He, Y.D., et al. 2000. Functional discovery via a compendium of expression profiles. *Cell* **102:** 109–126.

The International Rice Genome Sequence Project. 2005. The map-based sequence of the rice genome. *Nature* **436:** 793–800.

Jakoby, M., Weisshaar, B., Droge-Laser, W., Vicente-Carbajosa, J., Tiedemann, J., Kroj, T., and Parcy, F. 2002. bZIP transcription factors in *Arabidopsis*. *Trends Plant Sci.* **7:** 106–111.

Jonsson, H., Heisler, M., Reddy, G.V., Agrawal, V., Gor, V., Shapiro, B.E., Mjolsness, E., and Meyerowitz, E.M. 2005. Modeling the organization of the WUSCHEL expression domain in the shoot apical meristem. *Bioinformatics* **21 Suppl 1:** i232–i240.

Joshi-Tope, G., Gillespie, M., Vastrik, I., D'Eustachio, P., Schmidt, E., de Bono, B., Jassal, B., Gopinath, G.R., Wu, G.R., Matthews, L., et al. 2005. Reactome: A knowledgebase of biological pathways. *Nucleic Acids Res.* **33:** D428–D432.

Kellis, M., Patterson, N., Endrizzi, M., Birren, B., and Lander, E.S. 2003. Sequencing and comparison of yeast species to identify genes and regulatory elements. *Nature* **423:** 241–254.

Kim, S.K., Lund, J., Kiraly, M., Duke, K., Jiang, M., Stuart, J.M., Eizinger, A., Wylie, B.N., and Davidson, G.S. 2001. A gene expression map for *Caenorhabditis elegans*. *Science* **293:** 2087–2092.

Koch, M.A., Weisshaar, B., Kroymann, J., Haubold, B., and Mitchell-Olds, T. 2001. Comparative genomics and regulatory evolution: Conservation and function of the Chs and Apetala3 promoters. *Mol. Biol. Evol.* **18:** 1882–1891.

Koroleva, O.A., Tomlinson, M.L., Leader, D., Shaw, P., and Doonan, J.H. 2005. High-throughput protein localization in *Arabidopsis* using *Agrobacterium*-mediated transient expression of GFP-ORF fusions. *Plant J.* **41:** 162–174.

Krallinger, M. and Valencia, A. 2005. Text-mining and information-retrieval services for molecular biology. *Genome Biol.* **6:** 224.

Ku, H.M., Vision, T., Liu, J., and Tanksley, S.D. 2000. Comparing sequenced segments of the tomato and *Arabidopsis* genomes: Large-scale duplication followed by selective gene loss creates a network of synteny. *Proc. Natl. Acad. Sci.* **97:** 9121–9126.

Lawrence, R.J. and Pikaard, C.S. 2003. Transgene-induced RNA interference: A strategy for overcoming gene redundancy in polyploids to generate loss-of-function mutations. *Plant J.* **36:** 114–121.

Lee, T.I., Rinaldi, N.J., Robert, F., Odom, D.T., Bar-Joseph, Z., Gerber, G.K., Hannett, N.M., Harbison, C.T., Thompson, C.M., Simon, I., et al. 2002. Transcriptional regulatory networks in *Saccharomyces cerevisiae. Science* **298:** 799–804.

Li, X., Song, Y., Century, K., Straight, S., Ronald, P., Dong, X., Lassner, M., and Zhang, Y. 2001. A fast neutron deletion mutagenesis-based reverse genetics system for plants. *Plant J.* **27:** 235–242.

Li, L., Stoeckert Jr., C.J., and Roos, D.S. 2003. OrthoMCL: Identification of ortholog groups for eukaryotic genomes. *Genome Res.* **13:** 2178–2189.

Li, S., Armstrong, C.M., Bertin, N., Ge, H., Milstein, S., Boxem, M., Vidalain, P.O., Han, J.D., Chesneau, A., Hao, T., et al. 2004. A map of the interactome network of the metazoan *C. elegans. Science* **303:** 540–543.

Limpens, E. and Bisseling, T. 2003. Signaling in symbiosis. *Curr. Opin. Plant Biol.* **6:** 343–350.

Lippman, Z., Gendrel, A.V., Black, M., Vaughn, M.W., Dedhia, N., McCombie, W.R., Lavine, K., Mittal, V., May, B., Kasschau, K.D., et al. 2004. Role of transposable elements in heterochromatin and epigenetic control. *Nature* **430:** 471–476.

Lurin, C., Andres, C., Aubourg, S., Bellaoui, M., Bitton, F., Bruyere, C., Caboche, M., Debast, C., Gualberto, J., Hoffmann, B., et al. 2004. Genome-wide analysis of *Arabidopsis* pentatricopeptide repeat proteins reveals their essential role in organelle biogenesis. *Plant Cell* **16:** 2089–2103.

Margulies, M., Egholm, M., Altman, W.E., Attiya, S., Bader, J.S., Bemben, L.A., Berka, J., Braverman, M.S., Chen, Y.J., Chen, Z., et al. 2005. Genome sequencing in microfabricated high-density picolitre reactors. *Nature* **437:** 376–380.

Martienssen, R.A. and Colot, V. 2001. DNA methylation and epigenetic inheritance in plants and filamentous fungi. *Science* **293:** 1070–1074.

Martienssen, R.A., Doerge, R.W., and Colot, V. 2005. Epigenomic mapping in *Arabidopsis* using tiling microarrays. *Chromosome Res.* **13:** 299–308.

Mayer, K., Schuller, C., Wambutt, R., Murphy, G., Volckaert, G., Pohl, T., Dusterhoft, A., Stiekema, W., Entian, K.D., Terryn, N., et al. 1999. Sequence and analysis of Chromosome 4 of the plant *Arabidopsis thaliana. Nature* **402:** 769–777.

Meyers, B.C., Vu, T.H., Tej, S.S., Ghazal, H., Matvienko, M., Agrawal, V., Ning, J., and Haudenschild, C.D. 2004. Analysis of the transcriptional complexity of *Arabidopsis thaliana* by massively parallel signature sequencing. *Nat. Biotechnol.* **22:** 1006–1011.

Mitra, R.M., Gleason, C.A., Edwards, A., Hadfield, J., Downie, J.A., Oldroyd, G.E., and Long, S.R. 2004. A Ca^{2+}/calmodulin-dependent protein kinase required for symbiotic nodule development: Gene identification by transcript-based cloning. *Proc. Natl. Acad. Sci.* **101:** 4701–4705.

Mons, B. 2005. Which gene do you mean? *BMC Bioinformatics* **6:** 142.

Mudge, J., Cannon, S.B., Kalo, P., Oldroyd, G.E., Roe, B.A., Town, C.D., and Young, N.D. 2005. Highly syntenic regions in the genomes of soybean, *Medicago truncatula*, and *Arabidopsis thaliana. BMC Plant Biol.* **5:** 15.

Mueller, L.A., Zhang, P., and Rhee, S.Y. 2003. AraCyc: A biochemical pathway database for *Arabidopsis. Plant Physiol.* **132:** 453–460.

Nordborg, M., Hu, T.T., Ishino, Y., Jhaveri, J., Toomajian, C., Zheng, H., Bakker, E., Calabrese, P., Gladstone, J., Goyal, R., et al. 2005. The pattern of polymorphism in *Arabidopsis thaliana. PLoS Biol.* **3:** e196.

Osborn, T.C., Pires, J.C., Birchler, J.A., Auger, D.L., Chen, Z.J., Lee, H.S., Comai, L., Madlung, A., Doerge, R.W., Colot, V., et al. 2003. Understanding mechanisms of novel gene expression in polyploids. *Trends Genet.* **19:** 141–147.

Pendle, A.F., Clark, G.P., Boon, R., Lewandowska, D., Lam, Y.W., Andersen, J., Mann, M., Lamond, A.I., Brown, J.W., and Shaw, P.J. 2005. Proteomic analysis of the *Arabidopsis* nucleolus suggests novel nucleolar functions. *Mol. Biol. Cell* **16:** 260–269.

Prusinkiewicz, P. 2004. Modeling plant growth and development. *Curr. Opin. Plant Biol.* **7:** 79–83.

Pryer, K.M., Schneider, H., Zimmer, E.A., and Banks, J. 2002. Deciding among green plants for whole genome studies. *Trends Plant Sci.* **7:** 550–554.

Reddy, G.V., Heisler, M.G., Ehrhardt, D.W., and Meyerowitz, E.M. 2004. Real-time lineage analysis reveals oriented cell divisions associated with morphogenesis at the shoot apex of *Arabidopsis thaliana*. *Development* **131:** 4225–4237.

Reinhart, B.J., Weinstein, E.G., Rhoades, M.W., Bartel, B., and Bartel, D.P. 2002. MicroRNAs in plants. *Genes & Dev.* **16:** 1616–1626.

Rhee, S.Y., Beavis, W., Berardini, T.Z., Chen, G., Dixon, D., Doyle, A., Garcia-Hernandez, M., Huala, E., Lander, G., Montoya, M., et al. 2003. The *Arabidopsis* Information Resource (TAIR): A model organism database providing a centralized, curated gateway to *Arabidopsis* biology, research materials and community. *Nucleic Acids Res.* **31:** 224–228.

Riano-Pachon, D.M., Dreyer, I., and Mueller-Roeber, B. 2005. Orphan transcripts in *Arabidopsis thaliana*: Identification of several hundred previously unrecognized genes. *Plant J.* **43:** 205–212.

Rolland-Lagan, A.G., Bangham, J.A., and Coen, E. 2003. Growth dynamics underlying petal shape and asymmetry. *Nature* **422:** 161–163.

Rosso, M.G., Li, Y., Strizhov, N., Reiss, B., Dekker, K., and Weisshaar, B. 2003. An *Arabidopsis thaliana* T-DNA mutagenized population (GABI-Kat) for flanking sequence tag-based reverse genetics. *Plant Mol. Biol.* **53:** 247–259.

Schmid, M., Davison, T.S., Henz, S.R., Pape, U.J., Demar, M., Vingron, M., Scholkopf, B., Weigel, D., and Lohmann, J.U. 2005. A gene expression map of *Arabidopsis thaliana* development. *Nat. Genet.* **37:** 501–506.

Schuler, M.A. and Werck-Reichhart, D. 2003. Functional genomics of P450s. *Annu. Rev. Plant Biol.* **54:** 629–667.

Seki, M., Narusaka, M., Kamiya, A., Ishida, J., Satou, M., Sakurai, T., Nakajima, M., Enju, A., Akiyama, K., Oono, Y., et al. 2002. Functional annotation of a full-length *Arabidopsis* cDNA collection. *Science* **296:** 141–145.

Shaked, H., Melamed-Bessudo, C., and Levy, A.A. 2005. High-frequency gene targeting in *Arabidopsis* plants expressing the yeast RAD54 gene. *Proc. Natl. Acad. Sci.* **102:** 12265–12269.

Shiu, S.H., Karlowski, W.M., Pan, R., Tzeng, Y.H., Mayer, K.F., and Li, W.H. 2004. Comparative analysis of the receptor-like kinase family in *Arabidopsis* and rice. *Plant Cell* **16:** 1220–1234.

Stein, L.D., Mungall, C., Shu, S., Caudy, M., Mangone, M., Day, A., Nickerson, E., Stajich, J.E., Harris, T.W., Arva, A., et al. 2002. The generic genome browser: A building block for a model organism system database. *Genome Res.* **12:** 1599–1610.

Stracke, R., Werber, M., and Weisshaar, B. 2001. The R2R3-MYB gene family in *Arabidopsis thaliana*. *Curr. Opin. Plant Biol.* **4:** 447–456.

Sundaresan, V., Springer, P., Volpe, T., Haward, S., Jones, J.D., Dean, C., Ma, H., and Martienssen, R. 1995. Patterns of gene action in plant development revealed by enhancer trap and gene trap transposable elements. *Genes & Dev.* **9:** 1797–1810.

Tavazoie, S., Hughes, J.D., Campbell, M.J., Cho, R.J., and Church, G.M. 1999. Systematic determination of genetic network architecture. *Nat. Genet.* **22:** 281–285.

Thimm, O., Blasing, O., Gibon, Y., Nagel, A., Meyer, S., Kruger, P., Selbig, J., Muller, L.A., Rhee, S.Y., and Stitt, M. 2004. MAPMAN: A user-driven tool to display genomics data sets onto diagrams of metabolic pathways and other biological processes. *Plant J.* **37:** 914–939.

Till, B.J., Reynolds, S.H., Greene, E.A., Codomo, C.A., Enns, L.C., Johnson, J.E., Burtner, C., Odden, A.R., Young, K., Taylor, N.E., et al. 2003. Large-scale discovery of induced point mutations with high-throughput TILLING. *Genome Res.* **13:** 524–530.

Toledo-Ortiz, G., Huq, E., and Quail, P.H. 2003. The *Arabidopsis* basic/helix–loop–helix transcription factor family. *Plant Cell* **15:** 1749–1770.

Wang, X.J., Gaasterland, T., and Chua, N.H. 2005. Genome-wide prediction and identification of *cis*-natural antisense transcripts in *Arabidopsis thaliana*. *Genome Biol.* **6:** R30.

Weigel, D. and Nordborg, M. 2005. Natural variation in *Arabidopsis*. How do we find the causal genes? *Plant Physiol.* **138:** 567–568.

Wheeler, Q.D., Raven, P.H., and Wilson, E.O. 2004. Taxonomy: Impediment or expedient? *Science* **303:** 285.

Wilkinson, M., Schoof, H., Ernst, R., and Haase, D. 2005. BioMOBY successfully integrates distributed heterogeneous bioinformatics Web Services. The PlaNet exemplar case. *Plant Physiol.* **138:** 5–17.

Wolfe, K.H. 2001. Yesterday's polyploids and the mystery of diploidization. *Nat. Rev. Genet.* **2:** 333–341.

Wortman, J.R., Haas, B.J., Hannick, L.I., Smith Jr., R.K., Maiti, R., Ronning, C.M., Chan, A.P., Yu, C., Ayele, M., Whitelaw, C.A., et al. 2003. Annotation of the *Arabidopsis* genome. *Plant Physiol.* **132:** 461–468.

Xiao, Y.L., Malik, M., Whitelaw, C.A., and Town, C.D. 2002. Cloning and sequencing of cDNAs for hypothetical genes from Chromosome 2 of *Arabidopsis*. *Plant Physiol.* **130:** 2118–2128.

Yamada, K., Lim, J., Dale, J.M., Chen, H., Shinn, P., Palm, C.J., Southwick, A.M., Wu, H.C., Kim, C., Nguyen, M., et al. 2003. Empirical analysis of transcriptional activity in the *Arabidopsis* genome. *Science* **302:** 842–846.

Zanetti, M.E., Chang, I.F., Gong, F., Galbraith, D.W., and Bailey-Serres, J. 2005. Immunopurification of polyribosomal complexes of *Arabidopsis* for global analysis of gene expression. *Plant Physiol.* **138:** 624–635.

Zhang, W., Morris, Q.D., Chang, R., Shai, O., Bakowski, M.A., Mitsakakis, N., Mohammad, N., Robinson, M.D., Zirngibl, R., Somogyi, E., et al. 2004. The functional landscape of mouse gene expression. *J. Biol.* **3:** 21.

Zhang, L.V., King, O.D., Wong, S.L., Goldberg, D.S., Tong, A.H., Lesage, G., Andrews, B., Bussey, H., Boone, C., and Roth, F.P. 2005. Motifs, themes and thematic maps of an integrated *Saccharomyces cerevisiae* interaction network. *J. Biol.* **4:** 6.

Zimmermann, P., Hirsch-Hoffmann, M., Hennig, L., and Gruissem, W. 2004. GENEVESTIGATOR. *Arabidopsis* microarray database and analysis toolbox. *Plant Physiol.* **136:** 2621–2632.

5

Grains of Knowledge: Genomics of Model Cereals

Andrew H. Paterson
Plant Genome Mapping Laboratory
University of Georgia
Athens, Georgia 30602

Michael Freeling
Department of Plant and Microbial Biology
University of California
Berkeley, California 94704

Takuji Sasaki
National Institute of Agrobiological Sciences
Tsukuba, Ibaraki 305-8602, Japan

THE CULTIVATED CEREALS, MEMBERS OF THE POACEAE FAMILY of the angiosperms, provide about half of the calories consumed by humans and a growing share of biofuel. Together with their economic importance, the Poaceae are an attractive group for comparative genomics because they include many important crops with diverse native distributions and at least 35-fold variation in genome size (e.g., rice = 420 Mb; wheat = ~15,000 Mb). The independent domestication of rice in both Africa and Asia, sorghum in Africa, maize in America, and wheat in the Near East has provided an excellent study system in which to explore the genetic complexity of adapting plants to human use (for example, see Paterson et al. 1995; Paterson 2002).

Recent efforts to characterize Poaceae genomes better are reflected in their expansion from 1% to about 6% of the DNA sequence resources in GenBank (Paterson et al. 2003). This is exemplified by the nearly finished sequencing of each of two *Oryza* subspecies (see below), supplemented

Genomes, ©2006 Cold Spring Harbor Laboratory Press 0-87969-806-3   97

by exploratory genome-wide efforts in maize (Whitelaw et al. 2003) and sorghum (Bedell et al. 2005), and large EST and STS-based DNA marker collections for many others.

Collectively, genomic resources for diverse Poaceae promise new insights into molecular evolution, botanical diversity, and agricultural productivity. The power of any family or clade as a system to answer fundamental questions depends largely on the number of whole-genome sequences available, the exact branch lengths and positions of these data sets in the phylogenetic tree, and the positions of whole-genome duplications in the tree. For efficient comparison, orthologous genes or regions cannot be too closely or too distantly related; grass divergence events happened at useful times. Even if the grass family were represented solely by the whole genomes of rice, sorghum, and maize, their relationship just happens to confer tremendous analytic power to unravel much of the evolutionary history of both entire genomes (Fig. 1A,B) and individual genes (Fig. 2) in this important family. In grasses is the happy union of economic and scientific interests.

RICE

Rice is considered a model cereal crop because it has a relatively small genome size as compared with other cereals, a vast germplasm collection,

---

Figure 1. Unraveling the history of cereal gene and genome evolution. (*A*) A whole-genome duplication **A** and associated divergence and/or loss **B** of some members of duplicated gene pairs, determined the gene set that was inherited by all cereals **C**. After divergence of the cereals from this common ancestor **D**, continuing gene loss (**E**: note that locus e has been preserved on one homoeolog in the indicated lineage and the other homoeolog in the alternate lineage, thus changing its linkage relationship to flanking genes) together with the effects of other rearrangement mechanisms such as transposition **F**, led to incongruities in gene arrangement of modern cereal crops. Additional whole-genome duplications in some lineages such as maize, sugarcane, and wheat **G**, accompanied by continuing activity of transposition mechanisms **H**, resulted in further differentiation of the modern gene repertoire and order from that of the ancestral cereal order. Finished sequences of genomes representing the major taxonomic groups within the cereals will permit unraveling of the nature and timing of many of the events that account for such differences, as well as inference of the general organization of the genome of their common ancestor. (*B*) Phylogenetic relationships among selected Poaceae lineages. Current thinking on the approximate relationships among the major lineages discussed are illustrated. (*) The *Sorghum* genus includes a recently formed polyploid, Johnson grass (*S. halepense*). (**) The *Triticum* lineage includes several recently formed polyploids, most notably tetraploid *T. durum* (durum wheat) and hexaploid *T. aestivum* (bread wheat).

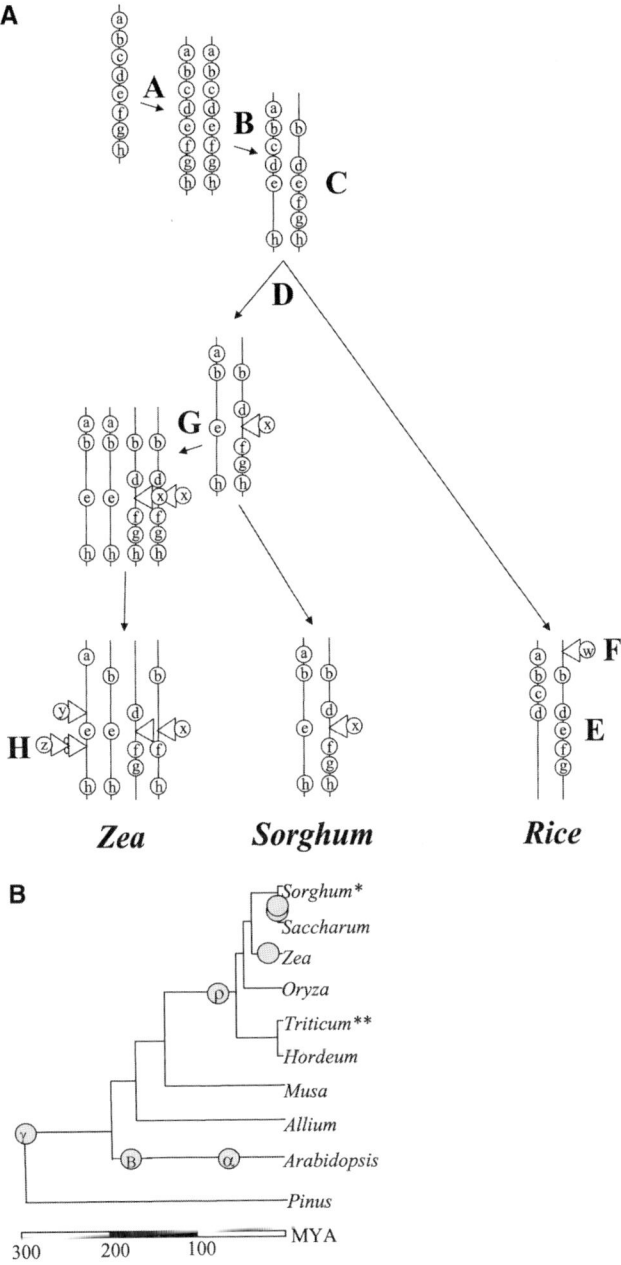

Figure 1. (*See facing page for legend.*)

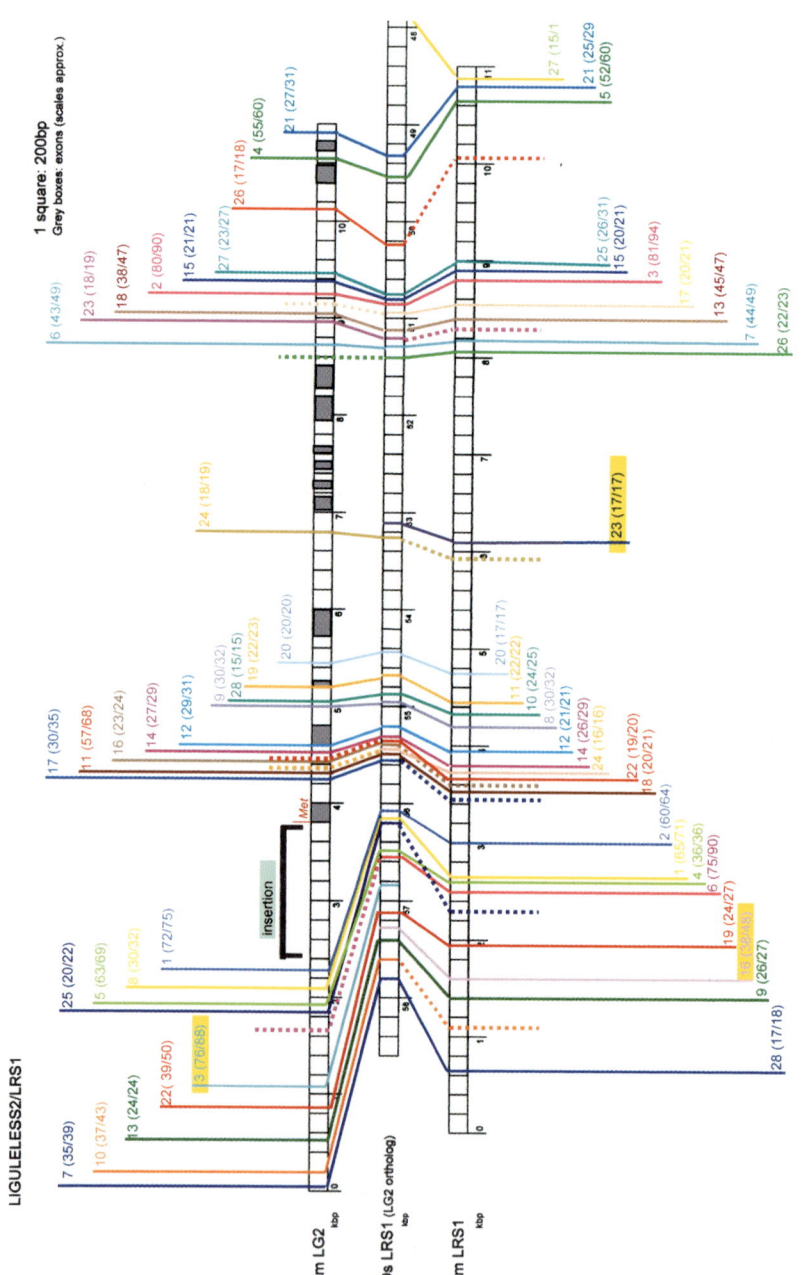

Figure 2. (*See facing page for legend.*)

an enormous repertoire of molecular genetic resources, and an efficient transformation system. The scientific value of rice is further enhanced with the elucidation of the genome sequence of the two major subspecies of cultivated rice, *Oryza sativa* ssp. *japonica* and ssp. *indica*. The sequence of the *japonica* cultivar Nipponbare was recently completed by a consortium of 10 countries, which comprised the International Rice Genome Sequencing Project (IRGSP), and represents a map-based finished sequence of the entire genome obtained using the hierarchical clone-by-clone sequencing strategy (Sasaki and Burr 2000). The sequence of the *indica* cultivar was derived by a whole-genome shotgun sequencing approach (Yu et al. 2002, 2005). These genome sequences are invaluable resources not only in understanding the structure and function of the rice plant itself but also in deciphering the organization of other cereal genomes, which share an appreciable degree of synteny with rice (e.g., Paterson et al. 2004; Devos 2005).

Genome Organization and Sequence

A total of 370 Mb of finished sequence of PAC and BAC clones from *japonica* rice, including virtually all of the euchromatic regions, revealed several characteristic features of the rice genome. A total of 57,000 protein-encoding sequences were inferred by computational gene predictions from all finished sequences (Yu et al. 2005). This is definitely an overestimate because we know that a large fraction of these (13% in chr. 1 and 18% in chr. 10) are various kinds of transposable elements (TEs) (Sasaki et al. 2002; The Rice Chromosome 10 Sequencing Consortium 2003). Exclusion of

---

Figure 2. Two independent BLAST comparisons plotted coordinately: maize gene *lg2* (AY180106 from a BAC) with its rice ortholog and maize gene *lrs1* (AY180107 from a BAC) with the same rice ortholog (RGP PAC AP003287). The purple alignment line of lg2CNS17 represents how sequences near 5′ exons align. Exons were identified in genomic DNA for all three genes using the complete cDNA of *lg2*-mRNA (AF036949) and then masked. Bl2seq conditions were modified from those of Kaplinsky et al. (2002b). The BLAST result is represented by the solid, multicolored lines connecting the maize *lg2* (Zm LG2) and rice *lrs1* (Os LRS1) gene diagrams. The identity match is indicated parenthetically. Each color connects CNSs that are essentially the same. A broken line reflects a maize lg2-lrs1 CNS retention, but just below the 15/15 cutoff. Yellow highlighting denotes those rare CNSs that are fractionated. Fractionation at the DNA level could, hypothetically, explain subfunctionalization at the phenotypic level. An insertion into *lg2* promoter is noted. Figure was reproduced with permission from the Genetics Society of America © 2004, from Langham et al. (2004).

all TE-related predicted genes would reduce the estimated gene number to about 43,000, implying a much lower gene density than the 4.5/kb value for *Arabidopsis* (*Arabidopsis* Genome Initiative 2000).

The map-based sequence also provides accurate positional information of genome components characterized by the presence of repeat or exogenous sequences. Widespread gene transfer from the organelles is suggested by the presence in the nuclear genome of many chloroplast and mitochondrial DNA fragments, including some nearly intact copies. The sequence of the centromeres revealed clusters of highly repetitive CentO satellite DNA located within the functional domain and flanked by centromere-specific retrotransposons.

A draft sequence of the *indica* subspecies representing 4.2× genome coverage and comprising 103,044 scaffolds corresponding to 466 Mb (Yu et al. 2002) was recently increased to 6.28× coverage and further revised using the Syngenta Nipponbare whole-genome shotgun sequence data (Goff et al. 2002) resulting in a total genome assembly of 466.3 Mb (Yu et al. 2005). A total of 19,079 nonredundant Nipponbare full-length cDNAs (97.7%) were localized and 49,088 genes were predicted. The discrepancies in gene numbers between *japonica* and *indica* genomes illustrate the need for more uniform and accurate methods of gene identification. In the *japonica* genome, this concern is being addressed through manual curation of annotation using full-length cDNA information.

Gene Functional Assignments and Resources

In addition to the assembled genomic sequences, public databases currently contain 386,487 rice ESTs (dbEST release 031105). A collection of more than 32,000 nonredundant rice full-length cDNA sequences is also publicly available (Kikuchi et al. 2003). These resources together with the accurate map-based genome sequence will be indispensable for further functional characterization of the rice genome. To address the huge task of determining the function of all predicted genes in rice, the International Rice Functional Genomics Consortium has been established using much the same organizational model as the IRGSP (Hirochika et al. 2004). The functional genomics resources currently available and which can be used for both forward and reversed genetics methods include insertion mutant lines generated using *Tos17*, T-DNA, and *Ac/Ds* elements, as well as populations of deletion mutants.

The map-based sequence has proven especially useful for the identification of genes underlying diverse agronomic traits such as flowering time, plant architecture and development, fertility restoration, and disease resistance. Many of these traits are governed by multiple genetic loci or QTLs. The genome sequence facilitates the development of DNA markers for this analysis. At least three genomewide SNP discovery studies have been conducted, with marked differences in numbers and rates of inferred SNPs due to the use of different target sequences (low-copy vs. total genomic vs. BAC-end) and different approaches to filter paralogs and artifactual SNPs from the respective data sets. Comparison of low-copy DNA across the entire *japonica* and *indica* sequences using short sequence alignments and stringent filtering criteria resulted in inference of 408,898 (Feltus et al. 2004) SNPs or about 1 per kb across the genome (but 1.7 per kb in the low-copy DNA that was screened), of which about 80% could be empirically verified in a sample. Analysis of total genomic DNA (Zhao et al. 2004) suggested ~5 million SNPs using longer alignments but looser stringencies to filter paralogs and false positives. Comparison of the *japonica* sequence with the end sequences of BAC clones from another *indica* cultivar, Kasalath, revealed a SNP frequency of 0.71% (Katagiri et al. 2004), intermediate between the other two studies.

Mapping populations such as recombinant inbred lines (RILs), backcross inbred lines (BILs), doubled haploid lines (DHLs), and chromosome segment substitution lines have been developed for rice to facilitate identification of target genes (Yano and Sasaki 1997).

SORGHUM

The #5 grain crop worldwide based on tonnage (after wheat, rice, maize, and barley; see http://www.fao.org), sorghum is unusually tolerant of low input levels, an essential trait for agriculture in areas that receive little rainfall, such as Northeast Africa and the U.S. Southern Plains. The *Sorghum* genus also includes one of the world's most noxious weeds. Many features that make "Johnson grass" (*S. halepense*) such a troublesome weed are actually desirable in forage, turf, and biomass crops, adding further value to enhanced knowledge of the *Sorghum* genus.

As a model for the tropical grasses, sorghum is a logical complement to *Oryza* (rice: Fig. 1B). Sorghum is representative of tropical grasses in that it has C4 photosynthesis, biochemical and morphological specializations that improve net carbon assimilation at high temperatures. By contrast, rice uses C3 photosynthesis, more typical of temperate grasses. Sorghum and maize shared a common ancestor about 12 million years ago (Mya)—however, the sorghum genome is much smaller (~736 Mbp), due to both recent polyploidization and repetitive DNA propagation in maize (see below). Sorghum is an even closer relative of sugarcane, arguably the most important biomass/biofuel crop worldwide (http://www.fas.usda.gov). However, sugarcane has outdone maize, undergoing at least two whole-genome duplications in the ~5 Myr since divergence from sorghum (Ming et al. 1998), resulting in a genome larger than that of human and with 4-fold or greater redundancy of most genes. While this article was in review, the U.S. Department of Energy Joint Genome Institute Community Sequencing Program announced its intention to perform 8× whole-genome shotgun coverage of the *Sorghum bicolor* genome, for public release.

Genome Organization and Sequence

The most detailed sorghum sequence-tagged site (STS)-based map is from a cross between *Sorghum bicolor* (*SB*) and *S. propinquum* (*SP*), comprising 2512 restriction fragment length polymorphism loci that span 1059.2 cM (Bowers et al. 2003). A total of 865 heterologous probes link the sorghum map to those of *Saccharum* (sugarcane: Ming et al. 1998), *Zea* (maize: Bowers et al. 2003), *Oryza* (rice: Paterson et al. 1995, 2004), *Pennisetum* (millet, buffelgrass: Jessup et al. 2003), the Triticeae (wheat, barley, oat, rye), *Panicum* (switchgrass: Missaoui et al. 2005), and *Cynodon* (bermudagrass: C. Bethel, E. Sciara, J. Estill, W. Hanna, and A.H. Paterson, in prep.).

Sorghum was the first plant for which a BAC library was reported (Woo et al. 1994). Physical maps of both *SB* and *SP* have been constructed and genetically anchored by hybridization to the mapped STS loci (Bowers et al. 2005). In a recent assembly (available at http://cggc.agtec.uga.edu), 40,957 *SP* BAC fingerprints (7× genome coverage) yielded 1541 contigs (averaging 24.4 BACs and 12.5 hybridization loci), while 69,545 *SB* BACs (11× coverage) yielded 1869 contigs (35.6 BACs, 10.1 hybridization loci). These assemblies exclude 100% of "Q" (questionable) clones. About 200 contigs exceed 1 million bp, and another ~500 are >500 kb. The contigs appear to cover >90% of the sorghum genome.

Sorghum was the first organism for which Cot-based reduced-representation sequencing approaches were reported (Peterson et al. 2002a), yielding extensive coverage of its repetitive DNA sequence and a promising means to eventually obtain most of its low-copy DNA (Peterson et al. 2002b). A complementary resource providing about 1× coverage of the hypomethylated DNA of sorghum has been reported recently (Bedell et al. 2005).

Gene Functional Assignments and Resources

As of this writing, about 170,000 sorghum ESTs were available in GenBank, many from a carefully structured experiment that produced ~5000 ESTs from each of 10 libraries representing diverse tissues and developmental stages (Cordonnier-Pratt et al. 2005). Nearly 300,000 ESTs for sugarcane are also available (Vettore et al. 2003).

While its minimal level of gene duplication makes sorghum, like rice, an attractive system for many approaches to determining gene function, this opportunity has been underexploited. Past efforts by the U.S. Department of Agriculture Agricultural Research Service have resulted in a collection of several hundred mutants (C. Franks, pers. comm.), but only a small subset have been studied formally. Identification of an active transposable element in sorghum (Chopra et al. 1999) has laid the foundation for insertional mutagenesis in sorghum.

MAIZE

Maize (n = 10) is a recent domesticate of the tropical grass teosinte (Doebley 2004). The most recent maize whole-genome duplication happened approximately 12 Mya (Gaut and Doebley 1997). Sorghum (n = 10), with its clearly diploid genome (Chittenden et al. 1994; Moore et al. 1995), differentiated from a maize ancestor just before this tetraploidy (Swigonova et al. 2004). Over the 12 Myr since tetraploidy, the maize genome "diploidized" by deleting most of the duplicated centromere regions and also deleting or tolerating degeneration of one member of most of its paired gene sets, sometimes fragmenting ancestral gene orders across multiple chromosomes and obscuring similarities in gene order that existed among its ancestors (Fig. 1A). The results of fractionating a tetraploid back toward a diploid can be complicated (e.g., Song and Messing 2003; Brunner et al. 2005). Comparisons of sorghum and maize chromosomal details should facilitate condensing maize back onto the tribal (*Andropogonae*) and eventually family (grass) ancestral genome. While all flowering plants are paleopolyploids, maize has largely been restored to diploidy.

Few if any plants rival the contributions of maize as a Mendelian genetic model during the 20th century (MaizeGDB [http://maizegdb.org] serves as a clearinghouse for maize genetic information). Today, maize continues to be a leading botanical model system for analysis in several areas, including transposons (Lisch 2002; Kolkman et al. 2005), meiosis where visualization of chromosomes is important (e.g., Harper et al. 2004), and the limits of fine mapping of phenotypes (Wright et al. 2005). Not all maize inbred lines carry identical gene contents (Fu and Dooner 2002); heterosis could well be a result. Transposons have been implicated as a mechanism generating such rapid intraspecies movement and diversity (Lai et al. 2004; H. Dooner, pers. comm.).

Genome Organization and Sequence

The 21st century finds maize in the process of being sequenced. With an estimated 2300–2600 Mb of chromosomal DNA (6× rice and 20× *Arabidopsis*), of which at least 60% is retrotransposon, the maize genome has initially been "filtered" to enhance for unmethylated (Rabinowicz et al. 1999) or low-repeat (Peterson et al. 2002a; Yuan et al. 2003) sequence—before shotgun sequencing (Whitelaw et al. 2003). The up-to-date result, best seen at the appropriate TIGR Web page (http://maize.tigr.org/), is about 1 million reads of about 700 bp each from inbred line B73. Additional reads, from other genotypes, are also available in Gen-Bank. Three independent groups are assembling these sequences into contigs, defining what has come to be called "gene space." Without doubt, this filtration method could miss particular genes, genes that will only be found in unbiased genomic sequence. Even so, the vast majority of nonrepetitive DNA in maize is represented by the current genome survey sequences. At press time, the U.S. National Science Foundation is considering proposals to "finish" the maize genome.

Gene Functional Assignments and Resources

Maize genetics enjoys a host of both public and private resources. MaizeGDB holds the current maps, nomenclatures, gene content data, descriptions and photographs of mutant phenotypes, and references, and lists the positions of thousands of BACs and mapping markers, markers that are often also useful in grasses other than maize. The 407,423 maize ESTs have been used by both TIGR and MaizeGDB to form gene lists; TIGR finds about 58,500 gene-like units in maize. Based on these or similar gene sequences, two projects—Pioneer Hi-Bred's TUSC (Trait Utility System for Corn) and Maize Targeted Mutagenesis (MTM)—permit users to order seeds homozygous for *Mu* transposon insertions potentially knocking out any gene that has been sequenced. Two public projects generated ears segregating for mutants that are likely caused by *Mu* transposon insertion. These phenotypes are often photographed or described at the MTM or MaizeGDB Web sites; phenotypes may also be screened and seed may be ordered from the Maize Genetics Cooperation Stock Center's Web site (http://w3.aces.uiuc.edu/maize-coop/). Transposon

Mu has been a particularly useful mutagen in maize (Lisch 2002) and allows for subsequent tagging and cloning by a variety of methods (Brutnell 2002). Forward genetics on essential genes can be challenging: A maize "TILLING" service (Till et al. 2004) delivers point mutants given gene sequence (http://genome.purdue.edu/maizetilling). Adding genes to maize is possible but time-consuming; alternative public projects can help add a user's gene to maize within a T-DNA (http://www.biotech. aistate. edu/service_facilities/plant_transformation.html) or by biolistics (http://www.psu.missouri.edu/muptcf/service/customized_service.htm). The maize ESTs, or derived synthetic DNAs, have been used to make hybridization microarrays. Transcript profiling experiments may be found at MaizeGDB and links therein.

CROSS-TAXON MESSAGES

For several reasons, the cereals are a particularly promising family in which to answer fundamental questions about many aspects of plant genome evolution. Their ~35-fold variation in genome size (c-value), while sharing a largely common set of genes, invites questions into evolution of genome size and structure. The fact that they all share a whole-genome duplication event that occurred "shortly" (ca. 20 Myr) before their divergence invites inquiry into the range of fates of ancient homeologs in reproductively isolated lineages. Several recent duplications (sugarcane, maize) or polyploidizations (wheat, Johnson grass) provide potential systems in which to explore repeating patterns in the fates of genes and gene families, as exemplified in Figure 1. Correspondence in the chromosomal locations of genes controlling many traits, for example, key features of domestication (Paterson et al. 1995) and weediness (Hu et al. 2003), highlights the many leveraging opportunities for information about model cereals. Finally, their economic importance makes a strong case for sequencing of the genomes of additional cereals and provides a large community to help translate such investments into agricultural benefits. We elaborate on a few promising early opportunities below.

Comparative Gene Fates

A natural first question posed of the *second* genome sequenced in a taxonomic group is whether taxon-specific genes can be discerned. That is, will the analysis of well-annotated versions of their completed genomes reveal genes that are present in rice and maize (for example) but absent (or at least unrecognizable) in *Arabidopsis, Populus, Medicago*, or other flowering plants? At present, approximately 71% of predicted rice proteins have a homolog in *Arabidopsis* whereas a total of 2859 genes appear to be unique to rice and other cereals. These numbers must be considered preliminary estimates, as inferences of the existence of taxon-specific genes are complicated by differences in sequence annotation methods, evolutionary rates of entire genes and specific motifs, chronic problems with sequencing of heterochromatic regions, and other obstacles. Nonetheless, the finding that specific transposable element families in rice may occasionally "stitch together" new genes from parts of existing ones (Jiang et al. 2004) makes the cereals an especially interesting group to explore for such events.

While positional correspondence of many genes across the cereals is well known, functional correspondence among the vast majority of these genes remains a hypothesis to be tested. All cereals are ancient polyploids, most recently as a result of a duplication event thought to have occurred perhaps 20 Myr prior to their divergence from common ancestors. This 20-million-year lag appears to contraindicate the possibility that polyploidization contributed directly to species divergence (e.g., Lynch and Force 2000) in the cereals but opens the door to the possibility of extensive "subfunctionalization" (Lynch and Conery 2000), with pairs of ancient homoeologs each undergoing compensatory mutations that result in subdivision of the ancestral gene's function. Further, post-polyploidization gene loss appears to have still been happening 20 million years later, in that loss of different members of homoeologous sets accounts for a sizable portion of apparent deviations from gene order conservation in rice and sorghum (Paterson et al. 2004). If subfunctionalization or even neofunctionalization (the acquisition of entirely new function) had preceded loss of different members of homoeologous sets in the two lineages, then the phenotypic consequences could be substantial. A host of genomic data will contribute to investigation of the degree to which functional correspondence parallels positional correspondence of individual genes, particularly including genomewide expression profiling, and saturation mutagenesis, providing empirical data with which to develop and test hypotheses about readily observable differences among cereals in sequence and structure of specific genes and their flanking regions.

Conserved Noncoding Sequences: Approaching Gene Regulation Computationally

All but a few of the most basal grasses radiated into subfamilies about 50 Mya (Kellogg 2001). An estimated 20 Myr prior to this radiation, the grass lineage had a whole-genome duplication, with about 20% of genes being retained as pairs within modern rice and many other modern grasses considered to be diploids (Paterson et al. 2003, 2004, 2005; Vandepoele et al. 2003; Wang et al. 2005; Yu et al. 2005). By multiplying the published rate of neutral base-pair substitution in grasses by the divergence time, it was calculated that a 15/15-bp exact positional match between orthologous genes in maize versus rice (or, coincidentally, man vs. mouse) would occur without selection only four times in a million; pairwise noncoding alignments with e-values equal to or more significant than 15/15 within a gene space define a grass conserved noncoding sequence (CNS) (Kaplinsky et al. 2002). So, alignments between retained duplicates within all grasses and comparisons between orthologs representing different grass subfamilies have diverged for enough time to insure CNS function but have not diverged so much that alignments are lost. As with mammals, the number of sequenced or in-progress grass genomes and their exact placement in the phylogenetic tree provide a powerful experimental system for studying the relationships between CNSs, gene regulation, and phenotype (Lockton and Gaut 2005). Grasses have far fewer and much smaller CNSs than do mammals, a possible indication of relative developmental simplicity (Inada et al. 2003). This relative simplicity has experimental advantages. For one case, CNS function involves a CNS-rich 1.5-Kb region of intron within the *knotted* Class I homeobox gene in grasses (Inada et al. 2003); mutant phenotypes in maize indicate that transposon insertions in this region can prevent a negative function keeping this gene "off" in wild-type leaves (Greene et al. 1994), which is the sort of regulatory switch often accomplished at the chromatin level. See Lockton and Gaut (2005) for a review of plant CNS research and on the power of having usefully placed outgroup genes for CNS research. A graphic example of maize–rice CNS alignments and possible subfunctionalization is illustrated in Figure 2. CNSs found in maize–rice comparisons have been shown to work well as specific PCR primer sites. These sites are conserved in grasses representing the breadth of grass diversity (Kaplinsky et al. 2002). Thus, CNS research between any two usefully diverged grasses delivers pangrass PCR mapping and sequence-extraction tools.

Heterochromatin and Its Importance

In large repeat-rich plant genomes, sequencing plans naturally prioritize relatively gene-rich euchromatin, both because of its information yield and relative ease of assembly. It is important, however, that heterochromatic (particularly pericentromeric) regions be sequenced to the degree possible at least in a few models to permit exploration of their unique features. It is a widespread observation that recombination (in the sense of reciprocal chromosomal exchange) is rare in pericentromeric regions. Suppression of recombination, together with a high tolerance for repetitive DNA, may create the sort of genomic environment that would favor the evolution of "co-adapted gene complexes," which are predicted to be favored by domestication (D'Ennequin et al. 1999) and have been observed in several cases (Paterson 2002). Identification and characterization of such complexes would be very important in a wide range of contexts. Early progress toward this goal has been accomplished for individual centromeres (Nagaki et al. 2004).

NEEDS AND OPPORTUNITIES

The availability of so many economically important taxa within the Poaceae family has made it a logical focus of monocot genomics. Finishing the maize genome, together with the sequencing of sorghum and at least one member of the Triticeae group including wheat and barley (a very important group that was not otherwise addressed in this manuscript), is a natural priority. A good case can also be made for several other family members—for example, foxtail millet (*Setaria* spp.) has a small genome and represents the sister tribe to the one containing maize–sorghum–sugarcane (thus, it is an appropriate outgroup for studies of these tropical grasses). It also contains some of the toughest, weediest species and landraces in the plant kingdom, offering new genes to help confer on our major crops the ability to remain productive under desertification.

It would be of direct value to Poaceae genomics to have extensive resources (preferably a complete genome) for a closely related taxon that could provide an "outgroup" for phylogenetic triangulation of events at a wide range of levels ranging from gain/loss of individual genes and parts of genes to chromosomal-level rearrangements. Within the two orders closest to the monocot suborder Poales, the Zingiberales includes banana (*Musa* spp.), a genome with $2n = 18$ and $1C = 600$ Mbp, which is of major global importance in terms of food and income security to millions of smallholder farmers throughout the developing countries of the tropics and subtropics. Further, the Commelinales includes pineapple, *Ananas comosus* L., a diploid, self-incompatible, with $2n = 50$, and $1C = 526$ Mbp (Arumuganathan and Earle 1991). Besides being an important fruit crop, pineapple represents still another strategy for carbon fixation (CAM). Its abnormally high somatic mutation rate (Collins 1936, 1960) suggests that it, like sorghum and rice, may prove to be a facile system for modern approaches to relate gene sequence to function.

While most Poaceae genes are now identifiable from current or pending genomic sequences, there remain many gaps in knowledge of the patterns of intragenic variation across the major branches of the family. Particularly glaring is the lack of knowledge of the Bambusoids (bamboos), for which only 295 sequences are found in GenBank as of this writing; the Chloridoids including major turfgrasses such as Bermuda (*Cynodon*) and zoysia (*Zoysia*) and orphan crops such as tef (*Eragrostis tef*), with 5367 sequences as of this writing; and Arundinoids, including the reeds, with only 178 sequences. One can also make good cases for more information about several lower-level taxonomic groups within the better-studied branches of the Poaceae.

Much of the value of whole-genome sequences in the cereals is likely to be realized by analysis of levels and patterns of allelic diversity in cultivars and their wild relatives. Germplasm collections are the underpinning of crop improvement—however, the 300,000+ accessions in global cereal genetic resources collections are woefully underexplored. Association genetics approaches (Thornsberry et al. 2001) are a promising tactic by which to begin to realize the value that lies latent in these resources; however strategies will have to accommodate species-specific features of the structuring of diversity. For example, linkage disequilibrium (LD) decays rapidly over a few hundred base pairs in maize (Remington et al. 2001), but in sorghum average LD (r^2) in six unlinked genomic regions only falls below 0.2 for pairs of sites >15 kb apart (Hamblin et al. 2005). Thus, where gene-by-gene approaches may be necessary in maize, LD-based association genetics appears more promising in sorghum.

REALIZING THE OPPORTUNITIES: DATABASE RESOURCES

Realization of the cross-taxon messages that lie latent in the Poaceae will require still more attention to integrative computational resources. Increasing complexity of data and sophistication of user queries make the appropriate content and organization of such databases a moving target, and in addition complementary resources are needed for additional taxa. Gramene (Ware et al. 2002; http://www.gramene.org) represents a valuable early effort toward a pan-Poaceae genomic database from a rice-centric perspective, providing a platform for comparison of rice mapping and sequence information with corresponding data in other cereal crops.

The rice genome sequencing effort has been accompanied by generation of several databases. The Rice Genome Automated Annotation System

(RiceGAAS: http://ricegaas.dna.affrc.go.jp) is both an annotation tool and a database of the annotation of all rice genome sequences ranging from 10 kb to 1 Mb submitted to GenBank. Manually curated annotation of the Nipponbare genome sequence can be accessed through the Rice Annotation Database (RAD: http://rad.dna.affrc.go.jp) and the Integrated Rice Genome Explorer (INE: http://rgp.dna.affrc.go.jp/giot/INE.html). Furthermore, INE integrates the genome sequence information with the genetic map, physical map, and transcript map of rice. The TIGR Rice Genome Annotation Database (http://www.tigr.org/tdb/e2k1/osa1/) also features the sequence and annotation data for the rice genome. Finally, Oryzabase (http://www.shigen.nig.ac.jp/rice/oryzabase/top/top.jsp) provides information on classical rice genetics as well as current advances in rice genomics.

While there are many online resources supporting research on maize, most of them may be accessed through the "links" button at the top of the MaizeGDB homepage. MaizeGDB is "information central" for maize research, both basic and commodity driven. The Maize Page also carefully links maize resources (http://maize.agron.iastate.edu). For example, while the INRA Maize Genomic Database (http://moulon.moulon.inra.fr/imgd)—a must visit for all QTL studies—was not mentioned in the paragraph on maize above, its link may be found readily.

For sorghum, an online resource (http://cggc.agtec.uga.edu) provides access to sorghum EST, genetic/physical map, and polymorphism data via Web interfaces and bulk downloads. Underlying this resource is an Oracle relational database containing tables for data storage, display parameters, and links to external Web sites as well as database views to implement on-line statistical calculations, with flexibility for a wide range of query options. Examples of Web-accessible displays are the IntegratedMap display that provides a graphic view of the most detailed sorghum STS-based genetic map along with mapped polymorphisms and anchored BAC contig data, the OxfordGrid display for comparative genetic map views, and the ESTminer display that shows contigs and clusters as color-coded tree nodes with expandable cDNA library statistics. The Web displays are compatible with a wide range of browsers and operating systems. Additional complementary Web-based resources focus on functional genomics of the transcriptome (http://fungen.botany.uga.edu) and genomics of the unique abiotic stress responses of sorghum (http://sorgblast2.tamu.edu). Finally, online resources from SUCEST (http://sucest.lbi.dcc.unicamp.br/en/), an extensive EST project in closely related sugarcane (Vettore et al. 2003), are also often of value for sorghum genomics.

REFERENCES

Arabidopsis Genome Initiative. 2000. Analysis of the genome sequence of the flowering plant *Arabidopsis thaliana*. *Nature* **408**: 796–815.
Arumuganathan, K. and Earle, E. 1991. Estimation of nuclear DNA content of plants by flow cytometry. *Plant Mol. Biol. Reptr.* **9**: 208–218.
Bedell, J.A., Budiman, M.A., Nunberg, A., Citek, R.W., Robbins, D., Jones, J., Flick, E., Rohlfing, T., Fries, J., Bradford, K., et al. 2005. Sorghum genome sequencing by methylation filtration. *PLoS Biol.* **3**: 103–115.
Bowers, J.E., Abbey, C., Anderson, S., Chang, C., Draye, X., Hoppe, A.H., Jessup, R., Lemke, C., Lennington, J., Li, Z., et al. 2003. A high-density genetic recombination map of sequence-tagged sites for sorghum, as a framework for comparative structural and evolutionary genomics of tropical grains and grasses. *Genetics* **165**: 367–386.
Bowers, J.E., Arias, M.A., Asher, R., Avise, J.A., Ball, R.T., Brewer, G.A., Buss, R.W., Chen, A.H., Edwards, T.M., Estill, J.C., et al. 2005. Comparative physical mapping links conservation of microsynteny to chromosome structure and recombination in Grasses. *Proc. Natl. Acad. Sci.* **102**: 13206–13211.
Brunner, S., Fengler, K., Morgante, M., Tingey, S., and Rafalski, A. 2005. Evolution of DNA sequence nonhomologies among maize inbreds. *Plant Cell* **17**: 343–360.
Brutnell, T.P. 2002. Transposon tagging in maize. *Funct. Integr. Genomics* **2**: 4–12.
Chittenden, L.M., Schertz, K.F., Lin, Y.R., Wing, R.A., and Paterson, A.H. 1994. A detailed Rflp map of sorghum-bicolor X S-Propinquum, suitable for high-density mapping, suggests ancestral duplication of sorghum chromosomes or chromosomal segments. *Theor. Appl. Genet.* **87**: 925–933.
Chopra, S., Brendel, V., Zhang, J.B., Axtell, J.D., and Peterson, T. 1999. Molecular characterization of a mutable pigmentation phenotype and isolation of the first active transposable element from sorghum bicolor. *Proc. Natl. Acad. Sci.* **96**: 15330–15335.
Collins, J.I. 1936. A frequently mutating gene in the pineapple *Ananas comosus* (L.) Merr. *Am. Nat.* **70**: 467–476.
———. 1960. *The Pineapple*. Leonard Hill, Ltd., London.
D'Ennequin, M.L.T., Toupance, B., Robert, T., Godelle, B., and Gouyon, P. 1999. Plant domestication: A model for studying the selection of linkage. *J. Evol. Biol.* **12**: 1138–1147.
Devos, K. 2005. Updating the "Crop Circle." *Curr. Opin. Plant Biol.* **8**: 155–162.
Doebley, J. 2004. The genetics of maize evolution. *Annu. Rev. Genet.* **38**: 37–59.
Feltus, F.A., Wan, J., Schulze, S.R., Estill, J.C., Jiang, N., and Paterson, A.H. 2004. An SNP resource for rice genetics and breeding based on subspecies *Indica* and *Japonica* genome alignments. *Genome Res.* **14**: 1812–1819.
Fu, H.H. and Dooner, H.K. 2002. Intraspecific violation of genetic colinearity and its implications in maize. *Proc. Natl. Acad. Sci.* **99**: 9573–9578.
Gaut, B.S. and Doebley, J.F. 1997. DNA sequence evidence for the segmental allotetraploid origin of maize. *Proc. Natl. Acad. Sci.* **94**: 6809–6814.
Goff, S.A., Ricke, D., Lan, T.H., Presting, G., Wang, R.L., Dunn, M., Glazebrook, J., Sessions, A., Oeller, P., Varma, H., et al. 2002. A draft sequence of the rice genome (*Oryza sativa* L. ssp *japonica*). *Science* **296**: 92–100.
Greene, B., Walko, R., and Hake, S. 1994. Mutator insertions in an intron of the maize *Knotted1* gene result in dominant suppressible mutations. *Genetics* **138**: 1275–1285.

Hamblin, M.T., Salas Fernandez, M.G., Casa, A.M., Mitchell, S.E., Paterson, A.H., and Kresovich, S. 2005. Patterns of short- and medium-range linkage disequilibrium in the domesticated grass *Sorghum bicolor*. *Genetics* **171**: 1247–1256.

Harper, L., Golubovskaya, I., and Cande, W.Z. 2004. A bouquet of chromosomes. *J. Cell Sci.* **117**: 4025–4032.

Hirochika, H., Guiderdoni, E., An, G., Hsing, Y.I., Eun, M.Y., Han, C.D., Upadhyaya, N., Ramachandran, S., Zhang, Q.F., Pereira, A., et al. 2004. Rice mutant resources for gene discovery. *Plant Mol. Biol.* **54**: 325–334.

Hu, F.Y., Tao, D.Y., Sacks, E., Fu, B.Y., Xu, P., Li, J., Yang, Y., McNally, K., Khush, G.S., Paterson, A.H., et al. 2003. Convergent evolution of perenniality in rice and sorghum. *Proc. Natl. Acad. Sci.* **100**: 4050–4054.

Inada, D.C., Bashir, A., Lee, C., Thomas, B.C., Ko, C., Goff, S.A., and Freeling, M. 2003. Conserved noncoding sequences in the grasses. *Genome Res.* **13**: 2030–2041.

Jessup, R.W., Burson, B.L., Burow, G., Wang, Y.W., Chang, C., Li, Z., Paterson, A.H., and Hussey, M.A. 2003. Segmental allotetraploidy and allelic interactions in buffelgrass (*Pennisetum ciliare* (L.) Link syn. *Cenchrus ciliaris* L.) as revealed by genome mapping. *Genome* **46**: 304–313.

Jiang, N., Bao, Z.R., Zhang, X.Y., Eddy, S.R., and Wessler, S.R. 2004. Pack-MULE transposable elements mediate gene evolution in plants. *Nature* **431**: 569–573.

Kaplinsky, N.J., Braun, D.M., Penterman, J., Goff, S.A., and Freeling, M. 2002. Utility and distribution of conserved noncoding sequences in the grasses. *Proc. Natl. Acad. Sci.* **99**: 6147–6151.

Katagiri, S., Wu, J., Ito, Y., Karasawa, W., Shibata, M., Kanamori, H., Katayose,Y., Namiki, N., Matsumoto, T., and Sasaki, T. 2004. End sequencing and chromosomal in silico mapping of BAC clone derived from an indica rice cultivar, Kasalath. *Breeding Sci.* **54**: 273–279.

Kellogg, E.A. 2001. Evolutionary history of the grasses. *Plant Physiol.* **125**: 1198–1205.

Kikuchi, S., Satoh, K., Nagata, T., Kawagashira, N., Doi, K., Kishimoto, N., Yazaki, J., Ishikawa, M., Yamada, H., Ooka, H., et al. 2003. Collection, mapping, and annotation of over 28,000 cDNA clones from japonica rice. *Science* **301**: 376–379.

Kolkman, J.M., Conrad, L.J., Farmer, P.R., Hardeman, K., Ahern, K.R., Lewis, P.E., Sawers, R.J.H., Lebejko, S., Chomet, P., and Brutnell, T.P. 2005. Distribution of activator (Ac) throughout the maize genome for use in regional mutagenesis. *Genetics* **169**: 981–995.

Lai, J., Ma, J., Swigonova, Z., Ramakrishna, W., Linton, E., Llaca, V., Tanyolac, B., Park, Y.J., Jeong, O.Y., Bennetzen, J.L., et al. 2004. Gene loss and movement in the maize genome. *Genome Res.* **14**: 1924–1931.

Langham, R.J., Walsh, J., Dunn, M., Ko, C., Goff, S., and Freeling, M. 2004. Genomic duplication, fractionation and the origin of regulatory novelty. *Genetics* **166**: 935–945.

Lisch, D. 2002. Mutator transposons. *Trends Plant Sci.* **7**: 498–504.

Lockton, S. and Gaut, B.S. 2005. Plant conserved non-coding sequences and paralogue evolution. *Trends Genet.* **21**: 60–65.

Lynch, M. and Conery, J.S. 2000. The evolutionary fate and consequences of duplicate genes. *Science* **290**: 1151–1155.

Lynch, M. and Force, A.G. 2000. The origin of interspecific genomic incompatibility via gene duplication. *Am. Nat.* **156**: 590–605.

Ming, R., Liu, S.C., Lin, Y.R., da Silva, J., Wilson, W., Braga, D., van Deynze, A., Wenslaff, T.F., Wu, K.K., Moore, P.H., et al. 1998. Detailed alignment of saccharum and sorghum chromosomes: Comparative organization of closely related diploid and polyploid genomes. *Genetics* **150**: 1663–1682.

Missaoui, A., Paterson, A.H., and Bouton, J.H. 2005. Investigation of genome organization in switchgrass (Panicum virgatum L.) using DNA markers. *Theor. Appl. Genet.* **110**: 1372–1383.

Moore, G., Devos, K.M., Wang, Z., and Gale, M.D. 1995. Cereal genome evolution—Grasses, line up and form a circle. *Curr. Biol.* **5**: 737–739.

Nagaki, K., Cheng, Z.K., Ouyang, S., Talbert, P.B., Kim, M., Jones, K.M., Henikoff, S., Buell, C.R., and Jiang, J.M. 2004. Sequencing of a rice centromere uncovers active genes. *Nature Gen.* **36**: 138–145.

Paterson, A.H. 2002. What has QTL mapping taught us about plant domestication? *New Phytologist* **154**: 591–608.

Paterson, A., Lin, Y., Li, Z., Schertz, K., Doebley, J., Pinson, S., Liu, S., Stansel, J., and Irvine, J. 1995. Convergent domestication of cereal crops by independent mutations at corresponding genetic loci. *Science* **269**: 1714–1718.

Paterson, A.H., Bowers, J.E., Peterson, D.G., Estill, J.C., and Chapman, B.A. 2003. Structure and evolution of cereal genomes. *Curr. Opin. Genet. Devel.* **13**: 644–650.

Paterson, A.H., Bowers, J.E., and Chapman, B.A. 2004. Ancient polyploidization predating divergence of the cereals, and its consequences for comparative genomics. *Proc. Natl. Acad. Sci.* **101**: 9903–9908.

Paterson, A.H., Bowers, J.E., Vandepoele, K., and Van de Peer, Y. 2005. Ancient duplication of cereal genomes. *New Phytologist* **165**: 658–661.

Peterson, D.G., Schulze, S.R., Sciara, E.B., Lee, S.A., Bowers, J.E., Nagel, A., Jiang, N., Tibbitts, D.C., Wessler, S.R., and Paterson, A.H. 2002a. Integration of Cot analysis, DNA cloning, and high-throughput sequencing facilitates genome characterization and gene discovery. *Genome Res.* **12**: 795–807.

Peterson, D.G., Wessler, S.R., and Paterson, A.H. 2002b. Efficient capture of unique sequences from eukaryotic genomes. *Trends Genet.* **18**: 547–550.

Rabinowicz, P.D., Schutz, K., Dedhia, N., Yordan, C., Parnell, L.D., Stein, L., McCombie, W.R., and Martienssen, R.A. 1999. Differential methylation of genes and retrotransposons facilitates shotgun sequencing of the maize genome. *Nat. Genet.* **23**: 305–308.

Remington, D.L., Thornsberry, J.M., Matsuoka, Y., Wilson, L.M., Whitt, S.R., Doebley, J., Kresovich, S., Goodman, M.M., and Buckler, E.S. 2001. Structure of linkage disequilibrium and phenotypic associations in the maize genome. *Proc. Natl. Acad. Sci.* **98**: 11479–11484.

The Rice Chromosome 10 Sequencing Consortium. 2003. In-depth view of structure, activity, and evolution of rice chromosome 10. *Science* **300**: 1566–1569.

Sasaki, T. and Burr, B. 2000. International Rice Genome Sequencing Project: The effort to completely sequence the rice genome. *Curr. Opin. Plant Biol.* **3**: 138–141.

Sasaki, T., Matsumoto, T., Yamamoto, K., Sakata, K., Baba, T., Katayose, Y., Wu, J.Z., Niimura, Y., Cheng, Z.K., Nagamura, Y., et al. 2002. The genome sequence and structure of rice chromosome 1. *Nature* **420**: 312–316.

Song, R. and Messing, J. 2003. Gene expression of a gene family in maize based on noncollinear haplotypes. *Proc. Natl. Acad. Sci.* **100**: 9055–9060.

Swigonova, Z., Lai, J.S., Ma, J.X., Ramakrishna, W., Llaca, V., Bennetzen, J.L., and Messing, J. 2004. Close split of sorghum and maize genome progenitors. *Genome Res.* **14**: 1916–1923.

Thornsberry, J.M., Goodman, M.M., Doebley, J., Kresovich, S., Nielsen, D., and Buckler, E.S. 2001. Dwarf8 polymorphisms associate with variation in flowering time. *Nat. Genet.* **28**: 286–289.

Till, B., Reynolds, S., Weil, C., Springer, N.M., Burtner, C., Young, K., Bowers, E., Codomo, C., Enns, L., Odden, A., et al. 2004. Discovery of induced point mutations in maize genes by TILLING. *BMC Plant Biol.* **4:** 12.

Vandepoele, K., Simillion, C., and Van de Peer, Y. 2003. Evidence that rice and other cereals are ancient aneuploids. *Plant Cell* **15:** 2192–2202.

Vettore, A.L., da Silva, F.R., Kemper, E.L., Souza, G.M., da Silva, A.M., Ferro, M.I.T., Henrique-Silva, F., Giglioti, E.A., Lemos, M.V.F., Coutinho, L.L., et al. 2003. Analysis and functional annotation of an expressed sequence tag collection for tropical crop sugarcane. *Genome Res.* **13:** 2725–2735.

Wang, X., Shi, X., Hao, B.L., Ge, S., and Luo, J. 2005. Duplication and DNA segmental loss in rice genome and their implications for diploidization. *New Phytologist* **165:** 937–946.

Ware, D.H., Jaiswal, P.J., Ni, J.J., Yap, I., Pan, X.K., Clark, K.Y., Teytelman, L., Schmidt, S.C., Zhao, W., Chang, K., et al. 2002. Gramene, a tool for grass Genomics. *Plant Physiol.* **130:** 1606–1613.

Whitelaw, C.A., Barbazuk, W.B., Pertea, G., Chan, A.P., Cheung, F., Lee, Y., Zheng, L., van Heeringen, S., Karamycheva, S., Bennetzen, J.L., et al. 2003. Enrichment of gene-coding sequences in maize by genome filtration. *Science* **302:** 2118–2120.

Woo, S.-S., Jiang, J., Gill, B., Paterson, A., and Wing, R. 1994. Construction and characterization of a bacterial artificial chromosome library of *Sorghum bicolor*. *Nucleic Acids Res.* **22:** 4922–4931.

Wright, S.I., Bi, I.V., Schroeder, S.G., Yamasaki, M., Doebley, J.F., McMullen, M.D., and Gaut, B.S. 2005. The effects of artificial selection of the maize genome. *Science* **308:** 1310–1314.

Yano, M. and Sasaki, T. 1997. Genetic and molecular dissection of quantitative traits in rice. *Plant Mol. Biol.* **35:** 145–153.

Yu, J., Hu, S.N., Wang, J., Wong, G.K.S., Li, S.G., Liu, B., Deng, Y.J., Dai, L., Zhou, Y., Zhang, X.Q., et al. 2002. A draft sequence of the rice genome (*Oryza sativa* L. ssp *indica*). *Science* **296:** 79–92.

Yu, J., Wang, J., Lin, W., Li, S.G., Li, H., Zhou, J., Ni, P.X., Dong, W., Hu, S.N., Zeng, C.Q., et al. 2005. The genomes of *Oryza sativa*: A history of duplications. *PLoS Biol.* **3:** 266–281.

Yuan, Y.N., SanMiguel, P.J., and Bennetzen, J.L. 2003. High-Cot sequence analysis of the maize genome. *Plant J.* **34:** 249–255.

Zhao, W.M., Wang, J., He, X.M., Huang, X.B., Jiao, Y.Z., Dai, M.T., Wei, S.L., Fu, J., Chen, Y., Ren, X.Y., et al. 2004. BGI-RIS: An integrated information resource and comparative analysis workbench for rice genomics. *Nucleic Acids Res.* **32:** D377–D382.

6

Genomics in *Caenorhabditis elegans*: So Many Genes, Such a Little Worm

LaDeana W. Hillier
Genome Sequencing Center, Washington University School of Medicine
St. Louis, Missouri 63108

Alan Coulson
MRC Laboratory of Molecular Biology, Cambridge CB2 2QH, and
The Wellcome Trust Sanger Institute, Wellcome Trust Genome Campus
Hinxton, Cambridge CB10 1SA, United Kingdom

John I. Murray and Zhirong Bao
Department of Genome Sciences, University of Washington
Seattle, Washington 98195

John E. Sulston
The Wellcome Trust Sanger Institute, Wellcome Trust Genome Campus
Hinxton, Cambridge CB10 1SA, United Kingdom

Robert H. Waterston
Department of Genome Sciences, University of Washington
Seattle, Washington 98195

IN 1965 SYDNEY BRENNER SELECTED *CAENORHABDITIS ELEGANS* for his studies of development and the nervous system because of its simple anatomy, its stereotyped behavior, and the ease of genetic manipulation. Even at inception, the goal of studying the worm was an understanding of how genes dictated form and behavior. This holistic view of the organism (now dubbed "systems biology") stimulated the collection of comprehensive data sets. The anatomy was described through serial electron microscopic reconstruction with the nervous system defined at the level of the synapse (White et al. 1986). The complete cell lineage of the 959 adult somatic cells was determined (Sulston and Horvitz 1977; Kimble and Hirsh 1979; Sulston et al. 1983) and found to be remarkably consistent animal to animal. Investigators commonly sought to collect all genes affecting a certain trait through mutations (however illusory that completeness might be in retrospect).

The construction of a clone-based physical map (Coulson et al. 1986, 1995; Sulston et al. 1988), one of the earliest genome projects, was undertaken in the early 1980s in the same spirit. The map of overlapping cosmids and later Yeast Artificial Chromosomes (YACs) (Coulson et al. 1988, 1991), along with efficient means of transformation, provided the community with the wherewithal to recover the DNA for any well-mapped mutant readily and rapidly. But perhaps more importantly, the existence of a nearly complete physical map in 1989 helped convince James D. Watson, head of the National Center for Human Genome Research at the time, that the worm should be included in the select set of model organisms to be targeted by the Human Genome Project (HGP), the so-called Security Council of the HGP (Sulston and Ferry 2002). We, in turn, were drawn to the project by the vision of a complete genome sequence, whose catalytic effect would drive research on the worm forward.

This review begins with an update on the genome sequence since our last report in 1998 (The *C. elegans* Genome Sequencing Consortium 1998). We describe the current state of the genome annotation of the sequence and then consider the collection of systematic data sets and analyses that the genome sequence has enabled and stimulated. All of these data and more are collected in WormBase (Chen et al. 2005a), which is briefly summarized (see Table 1 for Web sites). In conclusion, we discuss the challenges ahead as we strive for a molecular explanation of how the genome sequence produces a worm.

Table 1. *Caenorhabditis elegans* online repositories

Web address	Description
www.wormbase.org	Biology and genome database
elegans.swmed.edu	*C. elegans* WWW server
www.wormatlas.org	Behavioral and structural anatomy
www.wormbook.org	Online review of *C. elegans* biology
www.wormclassroom.org	Education and online learning community
www.rnai.org	Phenotypic data from RNAi studies

Additional Web sites are available in the Supplemental material.

GENOME SEQUENCES

The *C. elegans* Genome Sequence Is Complete

When the sequence of the 100-Mb genome of *C. elegans* was published in 1998 (The *C. elegans* Genome Sequencing Consortium 1998), very little important information was believed to be missing. Nonetheless, several recalcitrant gaps remained, and we had aimed from the start for a complete description of the content and structure of this benchmark genome. With persistence, we have now accumulated, by a variety of methods, the mapping and sequence information that completes the genome. The work behind this achievement is summarized in Text Box 1 and described in more detail in the Supplemental material.

BOX 1 ■ Completing the *Caenorhabditis elegans* Genome Sequence

At publication in 1998, there were tens of unfinished YACs and three unfinished cosmids and fosmids. These clones were all completed over the next year or two using the array of methods available for clone finishing (International Human Genome Sequencing Consortium 2004). In addition, we corrected ~20 misassembled, ambiguous, or deleted regions along with ~200 single base corrections (mostly in early projects) stemming from detailed analysis of Expressed Sequence Tags (ESTs) (McCombie et al. 1992; Waterston et al. 1992; Kohara 1996; The *C. elegans* Genome Sequencing Consortium 1998) and other data including community feedback.

More significantly, there remained two internal map gaps on Chromosomes III and IV, respectively, where no spanning clones were available, and three telomeric (Chromosome II right, where left and right are with reference to the genetic map) or subtelomeric (Chromosome I left and Chromosome X left) gaps. The telomere clone cTel33B (one from a set of eleven isolated by Wicky et al. 1996) eventually overlapped Y74C9 as its sequence was completed, capping the left end of Chromosome I. Plasmid cTel7X was linked to Y35H6 on the left end of Chromosome X through three PCR fragments, capping that chromosome end.

The internal gaps persisted despite the high redundancy of the initially mapped clones (some 30-fold from YAC, cosmid, and fosmid clones) and after screening a new BAC library (Exelixis, http://www.exelixis.com, pers. comm.). Given the rarity of these regions in large insert clone libraries, we turned to a strategy of directly subcloning and shotgun-sequencing a restriction fragment from whole genomic DNA for these internal gaps and the uncloned telomere from Chromosome II right.

The regions containing the internal gaps and the remaining telomere were mapped by macrorestriction Southern-blot analysis, using probes derived from the known flanking sequence. To obtain useful purity of the fragments, we

(Continued)

> **BOX 1 ■ (Continued)**
>
> adopted a successive digest scheme, using pulsed field gel electrophoresis (PFGE) to isolate the product of the first digest, digesting this in situ with a second enzyme, and subcloning the isolated DNA from a second PFGE purification. Inevitably these libraries were contaminated with copurifying DNA (50%–95% contaminated), but the dominant contig was easily identified in each case and the rest accounted for with known sequence.
>
> The spanning sequence for the internal gaps was in each case a small fraction of the size predicted by Southern blots (6 kb vs. the predicted 250 kb and 20 kb vs. 70 kb for Chromosomes III and IV, respectively). Perhaps the fragment mobility in PFGE can be anomalous at high concentrations (Doggett et al. 1992) (we used 50–100 μ/mL) or result from unusual sequence features, which might also account for the poor representation of the regions in libraries. The telomere segment was in better agreement (82 kb vs. 90 kb predicted), with the difference accounted for at least in part by exclusion of the telomere repeat from the assembled sequence.

As a result, the *C. elegans* sequence is fully contiguous telomere to telomere and with the mitochondrial genome totals 100,291,840 bp. A few problems may remain, such as undetected deletions within the clones or minor misassemblies. Some long multicopy tandem repeats, where not completely sequenced, have been characterized with respect to sequence content and tagged as such in sequence entries. But because of the hierarchical (clone) based shotgun methods used, all larger genomic duplications should be resolved (including one tandem repeat of 108 kb with only 10 sequence differences between the two copies). The per base error rate has been estimated at $<10^{-5}$. Reports from the community of problems with the sequence are now exceedingly infrequent, suggesting that remaining problems are rare, indeed. The genome seems in good shape!

Other *Caenorhabditis* Genomes

The comparison of related genomes provides a powerful tool for genome interpretation. In support of this objective, a draft sequence of the *Caenorhabditis briggsae* genome was produced (Stein et al. 2003). This whole-genome shotgun project produced a sequence with just 899 supercontigs (ordered and oriented contiguous sequence segments) spanning 106 Mb of DNA sequence with ~3 Mb of undetected overlaps and another ~2Mb of inferred gaps. When combined with the physical map, 102 Mb was placed in 142 ultracontigs ("supercontigs" ordered and oriented by their position within the physical map). More recently, the construction of a genetic map using single nucleotide polymorphic (SNP) markers has positioned 100 Mb along the six chromosomes and refined the sequence map (R.H. Waterston, S. Baird, L. Hillier, and R. Miller, unpubl.).

The *C. briggsae* sequence has proven useful in gene prediction (Wei et al. 2005), definition of regulatory elements (Luersen et al. 2004; Teng et al. 2004), and recognition of microRNAs (miRNAs; see below). But with only two species to compare, the signals of selection are often difficult to tease out from the noise of neutral change. To add power to the analysis, additional nematode genomes are currently under way (http://www.genome.gov/11007952), including the three closest of the known *Caenorhabditis* genomes, *Caenorhabditis remanei*, *Caenorhabditis japonica*, and *Caenorhabditis n. sp. PB2801*, and the more distantly related species *Pristionchus pacificus* and *Brugia malayi* (http://www.genome.gov/10002154). All are based on whole-genome shotgun assemblies. The three additional *Caenorhabditis* sequences should refine the definition of conserved features and may reveal sequences that have changed more rapidly in one lineage but not in others. The sequence of multiple species may be particularly critical in defining regulatory elements and noncoding RNA genes. The multiple *Caenorhabditis* species combined with the more distantly related nematodes should also provide insights into structure–function relationships at the protein level. As sequencing costs continue to drop, complete sequencing of other *C. elegans* isolates will undoubtedly be undertaken and add to our knowledge of the functional elements and their evolution.

GENE ANNOTATION

Protein-Coding Genes

The identification of the full set of *C. elegans* protein-coding genes is approaching completion. WormBase (release WS140) (Chen et al. 2005a) currently lists 19,735 genes with 2685 alternative splice forms, bringing the predicted protein count to 22,420 (producing 22,269 unique peptide sequences). More than 90% of the alternatively spliced genes have only one or two alternative spliced forms (Spieth and Lawson 2005). Trans-splicing is common in the worm, with more than half of *C. elegans* pre-mRNAs receiving an SL1 leader sequence and 20% an SL2 (Blumenthal 2005). More than 90% of the genes are directly supported by experimental evidence.

Nematodes are unusual among animals in having operons, polycistronic gene clusters containing two or more genes (Blumenthal and Gleason 2003; Blumenthal 2005). Currently, there are >1000 operons identified, each containing between two and eight genes, and accounting for ~15% of all *C. elegans* genes. Those genes that encode the basic machinery of gene expression are more frequently included in operons, while tissue-specific genes tend not to be part of operons (Blumenthal and Gleason 2003).

The protein-coding gene set was based initially on predictions by GeneFinder (P. Green, unpubl.), a gene prediction program developed in conjunction with the *C. elegans* genome project (The *C. elegans* Genome Sequencing Consortium 1998). The accuracy of individual exon prediction was high, but the prediction of complete genes was less reliable because of the combinatorics of multiexon genes and the challenges in detecting the start and stop of genes, especially in an organism with operons. Nonetheless, the GeneFinder predictions have been an excellent point of departure and have served the worm community well.

The computer predictions have been validated and modified by experimental data. Expressed sequence tags (ESTs) aligned with the genome now number more than a quarter of a million (McCombie et al. 1992; Waterston et al. 1992; Kohara 1996). Most ESTs come from the Kohara lab, which used methods to reduce the prevalence of abundant messages. In most cases, ESTs were derived from both 5'- and 3'-ends of cDNA clones, with the 3'-end establishing the 3'-UTR and the polyadenylation site and the 5'-end sampling the coding region or establishing the 5'-UTR for full-length clones. In turn, these clones provided representatives for full-length cDNA sequencing, with >2800 full-length sequences currently in the database. SAGE (Serial Analysis of Gene Expression) (Velculescu et al. 1995) of more than 30 libraries (http://elegans.bcgsc.ca/home/ge_consortium.html) from worms of a variety of stages, growth conditions, tissues, and cell types has yielded >2.5 million high-quality tags (McKay et al. 2003). These tags provide additional support for 16,212 genes, of which 2682 only have SAGE support. In addition, SAGE tags reveal ~500 open reading frames (ORFs) with *C. briggsae* homology that are not in the present gene predictions (G. Vatcher and D. Moerman, pers. comm.). More recently, a method was developed to obtain 5'-end SAGE-like tags for messages with SL1 or SL2 transpliced leaders (Hwang et al. 2004). An initial set of 13,525 tags identified the 5'-end of 2012 genes, confirming the 5'-end of 1512 known or predicted genes and modifying the end of another 401 genes. The 5'-ends of 99 previously unknown genes were also found. A larger sampling of 5'-end tags, now under way, identifies some 6500 5'-ends with 330 not associated with known or predicted genes (B.J. Hwang, H. Muller, S. McKay, P. Huang, S. Gharib, S. Jones, M. Marra, D. Moerman, D. Baillie and P.W. Sternberg, pers. comm.).

As these random-sampling-based methods become less efficient at gene confirmation/discovery, directed methods that begin with the predicted gene models became more useful. As part of an effort to obtain full-length cDNA clones for all *C. elegans* genes (the ORFeome Project) (Lamesch et al. 2004), >12,500 ORFs have been cloned in Gateway vectors, using RT-PCR starting from the gene models. Beyond confirming the transcription of these models, the data also modify the predicted gene models. Together with the EST libraries, OSTs (ORFeome sequence tags) (Lamesch et al. 2004) define 46,830 exon/intron boundaries. Green and colleagues have also been using RT-PCR to test systematically all unconfirmed intron–exon boundaries (see below) (P. Green, pers. comm).

Many of the remaining unsupported gene models and any as-yet-undetected genes in the genome are likely to be poorly expressed, may have weaker statistical signals, and may be less well conserved across species, making their identification by either computational or experimental means more difficult. Improvements in gene prediction programs may help tease out these signals. Twinscan (Korf et al. 2001), an HMM-based program derived from GenScan (Burge and Karlin 1997) that can use comparative sequence in predictions, has used a more realistic model of intron length, added a minor splice variant to splice tables and the *C. briggsae* sequence to produce an improved gene set over current WormBase predictions (Wei et al. 2005). While most Twinscan predictions overlap at least in part with existing predictions, >2000 are unique to Twinscan. RT-PCR experiments suggest that more than half of these may be transcribed (Wei et al. 2005). In a broad assault on the remaining unconfirmed exons and genes, P. Green (unpubl.) has used a substantially improved GeneFinder with relaxed constraints in order to capture most real genes at the cost of false positives. All the unconfirmed exon–intron boundaries are being tested by RT-PCR across the genome. In addition, SL1 and SL2 primers are being used in combination with internal primers to identify the 5′-ends of transpliced messages. Preliminary analysis of the data indicates that the gene set may rise to >21,000 confirmed protein-coding genes. The drive to complete the gene set will undoubtedly begin to challenge our notions of a gene.

Noncoding RNA Genes

Many transcripts function at the RNA level, including rRNAs, tRNAs, snRNAs, and snoRNAs. *C. elegans* contains all the major types of eukaryotic RNA genes: >1300 (Stricklin et al. 2005) of these genes have been identified, including 630 tRNAs, 78 snRNAs, and 17 snoRNAs. Of the rRNA genes, the 18S, 28S, and 5.8S are transcribed separately by RNA polymerase I in the ~55 copies of the 7.2-kb rDNA repeat on I (Sulston and Brenner 1974; The *C. elegans* Sequencing Consortium 1998). The 5S gene along with the SL1 spliced leader gene lies in a 1-kb tandem repeat with ~110 copies on V (Sulston and Brenner 1974; Nelson and Honda 1985). (With uncertainty about the exact copy number of these large tandem repeats, only representative members of each are included in the sequence.) There are also 20 copies of the SL2 repeat dispersed in the genome. In addition to these well-known genes, the *lin-4* and *let-7* genes provided the first examples of functional miRNAs (Lee et al. 1993; Wightman et al. 1993; Reinhart et al. 2000), which are now recognized to be common features of eukaryotic genomes, including human. Indeed, many worm miRNA genes have clear homologs in mammalian genomes. Methods are now being developed for large-scale in vivo validation of predicted miRNA targets in *C. elegans*; for example, a dozen novel predicted targets of *let-7* have been tested using comparative expression analyses in transgenic worms (N. Rajewsky, S. Lall, and F. Piano, unpubl.). Computational and experimental methods have identified at least 114 miRNA genes (Ambros et al. 2003; Griffiths-Jones 2004; http://microrna.sanger.ac.uk/sequences/), and intriguing new work is providing evidence about the roles of these RNAs in cell and developmental processes.

Other novel RNA genes and gene families may well exist in the worm genome. Current computational methods to identify such genes and families use conservation of secondary structure across species but are subject to high false-positive rates (Rivas and Eddy 2001; Lim et al. 2003), obscuring real genes. With the sequencing of additional related species (Rivas and Eddy 2001; Coventry et al. 2004; Washietl et al. 2005) the false-positive rate may drop sufficiently to allow the emergence of additional RNA genes. SAGE can provide evidence for some RNA genes (Jones et al. 2001), and the development of tiling microarrays covering essentially all of the genome may well point to additional possible genes for more detailed study.

GLOBAL STUDIES ENABLED BY THE GENOME SEQUENCES

The genome sequence, by providing a comprehensive view of the information needed to specify the animal and its behavior, has stimulated a variety of systematic studies to define the functional elements of the genome and to capture functional information about those elements more effectively. Occasionally these data sets provide direct insight into biological mechanism; more often they provide resources that enable investigators focused on specific mechanisms to speed their work. Increasingly these more systematic approaches are being integrated into the more specific studies. We provide examples of these data sets and their use below.

Gene Expression

In a multicellular organism a major insight into gene function comes from when, where, and under what conditions a gene is expressed. Approaches that yield expression data on many genes in parallel and other systematic efforts have been enabled by the genome sequence. Many of these approaches are shared with other organisms; others exploit the comprehensive knowledge of the worm's simple anatomy and the cell lineage to provide high temporal and anatomic resolution.

Large data sets measuring RNA levels in specific worm populations are available for both microarray analysis and SAGE. Microarrays provide data on many genes at once but depend on the current state of gene models, while SAGE and related approaches give a potentially unbiased sampling but are more expensive. Microarray data have been acquired from hundreds of experiments using populations of worms, including various stages, different sexes and mutants, and various growth conditions. Early on, much of the data were generated using spotted DNA arrays, and these continue to be widely used (http://www.genome.wustl.edu/genome/celegans/microarray/ma_gen_info.cgi). These resources have been augmented by arrays from commercial suppliers. For example, Affymetrix offers a chip representing an estimated 22,500 transcripts from almost 19,000 gene models (http://www.affymetrix.com/products/arrays/specific/celegans.affx), and NimbleGen offers a chip with 390,000 probes covering 21,121 genes with a minimum of 17 probes per gene (http://www.nimblegen.com/products). Clustering the resultant expression data reveals sets of genes that respond similarly within the populations examined, and based on the presence of previously characterized genes within those clusters, inferences can be drawn about the role of the genes in the group. For example, in a pioneering study, Kim and colleagues (Kim et al. 2001) found 44 different clusters and were able to associate 30 of these with possible functions. Early SAGE analysis targeted differences in gene expression patterns between dauer and non-dauer worms, highlighting the substantial transcriptional differences in the specialized dauer stage and identifying noncoding transcripts with sequence related to the telomere repeat (Jones et al. 2001). In another application, SAGE was used to compare long-lived mutants with control populations to reveal genes and pathways potentially involved in life-span extension (Holt and Riddle 2003). These experiments also demonstrated the potential of SAGE to reveal previously unknown genes and alternative splice and polyadenylation variants.

Using amplification, with the caveats this introduces, gene expression has been measured in small populations of purified cell types and in carefully staged embryos. Specific cell types can be labeled using tissue-specific promoters driving GFP (Green Fluorescent Protein), and until about the 400-min stage, embryonic cells can be dissociated with the labeled cells recovered by FACS (Fluorescence Activated Cell Sorting). Cells can be harvested immediately or placed in culture to allow further differentiation (Christensen et al. 2002; Zhang et al. 2004; Blacque et al. 2005; Fox et al. 2005) and analyzed for mRNA content by either microarray analysis or SAGE. In a variant of this, a tagged poly(A) binding protein (PABP) has been expressed in specific cell types, and mRNAs from these cells have been recovered by immunoprecipitation (Roy et al. 2002; Kunitomo et al. 2005). To obtain information about the temporal progression of gene expression in early embryogenesis, Baugh and colleagues (Baugh et al. 2003, 2005) staged small cohorts of embryos by visual selection of embryos at the four-cell stage, which were then allowed to develop. Samples were taken at intervals approximating the successive rounds of cell division of the embryo. Quantitative analysis of the resultant data showed successive sets of gene expression, suggesting a causal relationship. This relationship was confirmed for a few examples, revealing several potential regulatory networks.

In contrast to methods that extract RNAs, gene products (mRNA or protein) can be assayed directly in the animal to determine the site and time of gene expression. Both RNA hybridization and antibody have been used traditionally for this purpose. RNA in situ methods are more readily carried out systematically, and Kohara (http://www.nig.ac.jp/section/kohara/kohara-e.html; nematode.lab.nig.ac.jp/db2/index.php) currently displays whole-mount in situ images of 11,237 cDNA clones with various stages available for inspection. Certain tissue patterns are readily recognized, but individual cell identity is difficult to determine. Antibody methods have been more difficult to scale up, but new methods for generating high-affinity reagents may change this.

The advent of in vivo GFP labeling methods allows gene expression patterns to be visualized in living worms. Promoter::GFP fusions are being generated on a genome scale in conjunction with the Promoterome project (Dupuy et al. 2004)—the effort has already released promoter fusions (up to 2 kb) from ~6500 *C. elegans* genes, and plans are under way for a more comprehensive set. Two groups are systematically transforming these constructs or related ones using PCR and imaging the resultant worms with fluorescent microscopy. The Hope Lab Web site (http://129.11.204.86:591/default.htm) provides descriptions and images for >300 genes, and the BC Genome Center site (http://www.bcgsc.ca/gc/celegans/) provides information on some 1750 genes, with images available on a subset of these. The former group has focused on transcription factors, while the latter has targeted *C. elegans* genes with human homologs. The fidelity of the transgene patterns to native genes is, of course, a central issue with such approaches. Transgenes introduced by injection typically are incorporated into large extrachromosomal arrays and are subject to somatic loss and germ-line silencing; nevertheless, the observed expression patterns have been generally reliable. In addition, promoters and other regulatory sequences are not defined for most genes in *C. elegans*, so that as an expedient both projects use the upstream region of arbitrary length to drive expression. Since intergenic regions in *C. elegans* are usually small, often these constructs extend to the adjacent gene. Despite these obvious limitations, the available gene expression patterns are highly valuable.

A challenge in using the in vivo expression data is the need for an expert to interpret the patterns. To circumvent this, our laboratory (Z. Bao, J. Murray, T. Boyle, and R. Waterston, unpubl.) has embarked on a project that will automate the assignment of gene expression to individual cells throughout early development. The method uses four-dimensional images of worms with nuclei labeled with GFP-histone fusions to follow cell divisions throughout embryogenesis, thereby automating the determination of the cell lineage. Because the lineage in wild type is highly reproducible and the fate of every daughter cell is known, knowledge of the lineage history of an animal is tantamount to knowledge of its anatomy. Introduction of a second reporter gene driven by a promoter sequence of interest into this background thus holds the promise of providing expression data with single-cell resolution and high temporal fidelity automatically. Introduction of the constructs via bombardment also may yield single-copy integrants and circumvent germ-line silencing in many cases. The current implementation traces the lineage through 250 cells with only minor editing and thus is already useful for early embryonic events.

Gene Disruption

A second powerful insight into gene function comes from analysis of the phenotype of animals carrying mutant forms of a gene. Traditional methods, including chemical mutagenesis, irradiation, and transposon insertion, have produced mutant alleles in fewer than 1000 genes. Furthermore, homologous recombination, so powerful in yeast and mammals, is relatively ineffective in *C. elegans*.

Fortunately other methods have emerged that allow systematic disruption of gene function. Since its discovery in the worm (Fire et al. 1998; Piano et al. 2000; Sonnichsen et al. 2005), RNA interference (RNAi), where double-stranded RNA induces sequence-specific degradation of homologous mRNAs, has become the most widely used means of inhibiting gene function. The double-stranded RNA can be introduced by injection, soaking, and even by feeding worms bacteria expressing the dsRNA (Timmons and Fire 1998). Inhibition is rarely complete, and neuronally expressed genes are particularly resistant to RNAi effects. Nonetheless, the ease of use of feeding libraries and other modes of delivery has facilitated systematic genome-wide RNAi screens by several groups (Fraser et al. 2000; Maeda et al. 2001; Kamath and Ahringer 2003; Sonnichsen et al. 2005), and currently >18,000 *Escherichia coli* strains have been constructed and have been widely distributed. Initial screens were for easily scored phenotypes such as viability, slow growth, or altered movement and body shape. These screens and others have produced phenotypes for >3300 genes (the *E. coli* RNAi library covers 86% of all *C. elegans* genes) (http://www.gurdon.cam.ac.uk/ ~ahringerlab/ pages/rnai.html), including 721 genes required for embryogenesis (Vidalain et al. 2004). To examine genetic robustness at a functional level, a double RNAi feeding screen is being carried out to test 2000 putative duplicate gene pairs for redundant function (S. Woods and J. Ahringer, unpubl.). The RNAi library is being increasingly used to screen for more specific phenotypes or in certain mutant backgrounds, including backgrounds that appear to enhance RNAi effects (Wang et al. 2005). The success of these has, in turn, stimulated efforts to automate various aspects of phenotype analysis.

To complement RNAi and to provide permanent lines with transmissible defects, projects are under way to knock out genes, using either chemically induced deletions or transposons. Both methods use PCR to detect length differences in populations of treated animals. At present, the Gene Knockout Consortium (http://www.celeganskoconsortium.omrf.org/) and the National Bioresource Center (http://shigen.lab.nig.ac.jp/c.elegans/index.jsp) have each generated gene deletions. The former has generated deletions in some 1800 genes, with >1300 of these stabilized and archived, while the latter lists ~1600 gene deletions. The NemaGENETAG Consortium (http://elegans.imbb.forth.gr/nemagenetag/home.html) has produced >150

Mos1-tagged strains and plans to do more (P. Kuwabara, unpubl.). The TILLING (Targeting Induced Local Lesions IN Genomes) approach (McCallum et al. 2000), because it is adaptable to any organism that can be chemically mutagenized, has been used in *C. elegans* and proven to be successful at generating point mutations including stop codons (R. Plasterk, unpubl.). TILLING has the potential advantage of producing an allelic series (mutations of varying severity). As sequencing costs fall, direct sequencing of mutagenized lines may become the method of choice (R. Plasterk, pers. comm.).

Gene Regulation

The signals that control gene activity in time and space are also embedded in the genome. They act at the DNA level as promoters and other *cis*-regulatory elements; at the RNA level as elements that govern translation and stability; and at the protein level through post-translational modification and turnover. In contrast to protein-coding regions, no algorithms currently exist that can effectively recognize these signals ab initio in genome sequence. Early work in the area focused on individual genes and through traditional methods established the precise sequences driving gene expression (Okkema and Fire 1994; Fukushige et al. 1996; Okkema et al. 1997). But with the genome sequence, a combination of gene expression data, comparative sequence analysis, and improving computer programs, there is progress in the recognition of the DNA elements and to some extent the RNA elements.

At the DNA level the gene expression sets described above have been critical, allowing genes to be grouped or stratified by time and tissue. Candidate elements have been identified associated with genes expressed in heat shock (Nikolaidis and Nei 2004), muscle (GuhaThakurta et al. 2004), and the gut (Gaudet et al. 2004), particularly the pharynx. For example, Mango and her colleagues (Gaudet et al. 2004) identified genes expressed in the pharynx by comparing mutant embryos enriched and depleted of pharyngeal cells using microarrays. They grouped the genes by early or late expression and then looking between species and across genes, they identified nine candidate regulatory motifs, two of which were previously known. They confirmed several of these for activity in vivo and, in turn, used the motifs to search for additional genes with the motifs. The resultant sets were significantly enriched for genes expressed in the pharynx. This strategy should become more powerful as additional

Caenorhabditis genome sequences become available and as gene expression data are refined.

Parallel to expression data and comparative genome analysis, investigators have attempted to identify the target sequence for known transcription factors. Using the yeast one-hybrid system, the motifs recognized by the DNA-binding domains of the worm's ~600 transcription factors are being systematically dissected (Deplancke et al. 2004). Others are exploring ways to apply chromatin precipitation to discover the in vivo sites of protein–DNA interaction and to use DNase I hypersensitivity to find regions of open chromatin. SELEX (systematic evolution of ligands by exponential amplification) offers another approach to identify binding motifs that might be applied at scale (Roulet et al. 2002). Combining knowledge of transcription-factor-binding sites and the identification of functional sites associated with genes could provide powerful insights into the networks of gene regulation that underlie development.

Many motifs encoded in DNA within genes act at the RNA level to regulate splicing, localization, translation, RNA editing, or other processes. These RNA regulatory elements can be studied in largely the same fashion as the transcriptional regulatory elements: sequence conservation can be used to identify candidate elements; pull-down experiments can link RNA-binding proteins to their target genes and candidate motifs; function can be assayed by fusions with reporter proteins. *C. elegans* has ~500 RNA-binding proteins, and genetic, biochemical, and computational analyses have revealed critical roles of protein–RNA complexes, 3′-untranslated regions, RNA-binding proteins (and their targets), and RNA–RNA interactions in development.

A complication for defining the RNA regulatory elements is that the regulatory information often resides inside the three-dimensional RNA secondary structure rather than be encoded directly in the primary sequence. This makes it more difficult to predict regulatory elements computationally. The computational prediction of RNA regulatory elements must proceed hand in hand with the structural analysis of the RNA genes that regulate them.

Proteomics

With the well-annotated *C. elegans* genome in hand, both the study of individual proteins and the study of interactions among those proteins can proceed. In a high-throughput proteomic effort to confirm protein-coding genes, G. Merrihew, J.H. Thomas, and M.J. MacCoss (unpubl.) are using mass spectrometry to validate experimentally even small predicted

ORFs. They currently have identified 3363 proteins, 121 of which previously had no experimental support (39 of these were identified based on a translated intergenic ORF set, and the remainder from GeneFinder predictions) (P. Green, unpubl.). Others are finding success using mass spectrometry to quantify relative protein levels in *C. elegans* embryos and adults (Venable et al. 2004). Mass spectrometry approaches should also reveal post-translational modifications that may alter activity.

The study of individual protein structures is also well under way. A *C. elegans* structural genomics group has formed a high-throughput protein-to-structure pipeline (Liu et al. 2005b). They have determined the crystal structure of 78 proteins or protein fragments (http://sgce.cbse.uab.edu/index.php) and solved 19 structures (e.g., Symersky et al. 2003; Lu et al. 2004). Another structural genomics effort (http://www.nesg.org; Wunderlich et al. 2004) identified seven structures.

Identifying protein–protein interactions and the effects of any modifications on those interactions will be key to any molecular understanding of the worm. Computational-aided methods, some using comparative data (Liu et al. 2005a; Sharan et al. 2005), have the potential for revealing these, but large-scale studies depend on experimental data. Armed with the set of 11,000 cloned ORFs (*C. elegans* ORFeome project) (Lamesch et al. 2004), researchers have generated a *C. elegans* interactome network map that contains >5500 potential interactions (Li et al. 2004) and are moving toward defining the entire set. Along with another map for *Drosophila melanogaster* (Sanchez et al. 1999), these data sets, although containing high proportions of false positives and negatives, nevertheless represent the first of their kind for metazoan organisms. Critically, the interactome map serves as a foundation for integration of studies of development and disease, both for individual proteins and at the level of networks of interactions.

Population Biology and Evolution

Beyond aiding in a molecular understanding of the form and behavior of the worm, the genome sequence has also facilitated studies of the evolutionary processes acting on the worm genome. While we cannot access *C. elegans* ancestors, comparative analysis allows inferences about that ancestral state and the events that have occurred since the divergence of two species. With the sequence of the laboratory strain N2 in hand, the study of variation in different isolates of *C. elegans* from around the world became straightforward. Variation could be readily determined either through using PCR to recover specific areas or from random whole-genome sequence reads from these different isolates aligned with the N2 sequence. A patchwork pattern of variation within most isolates suggested that most isolates had resulted from an interbreeding event followed by isolation, perhaps facilitated by hermaphroditic reproduction (Koch et al. 2000). Surprisingly, there is high population diversity at the local level—on the scale of centimeters—but the diversity levels off very quickly so that there is about the same amount of diversity among isolates from different countries as among isolates from the same compost heap (Fitch 2005).

Among the different isolates, the Hawaiian strain, CB4856, however, proved to have widely and more uniformly dispersed sequence differences (Wicks et al. 2001). A difference was observed once every 850 bases, with transitions outnumbering transversions (57% vs. 43%) and indels (one or more bases added or removed) accounting for more than one-quarter of the differences. Somewhat surprisingly, this rate of difference suggests an effective population size not much different from that of humans. Recent comparison of the genomes using comparative genome hybridizations with microarrays reveals a surprising number of larger deletions in the Hawaiian strain (D. Moerman, pers. comm.). The single nucleotide polymorphisms (SNPs) have also provided the basis for an effective genetic mapping strategy (Wicks et al. 2001; Swan et al. 2002).

C. elegans autosomes have an unusual organization, with recombination significantly elevated on the terminal thirds compared to the centers. Essential genes are more frequently located in the centers in contrast to gene families, which are overrepresented on the arms. This has led to speculation that the arms are sites of high gene death and birth. Consistent with this notion, SNP density appears to be elevated on the arms (Koch et al. 2000). Comparison of the *C. elegans* and *C. briggsae* genomes has shown dramatic differences in expansion of chemosensory genes on the arms in the two species (Chen et al. 2005b) and for positive selection of members of the srz family of G-protein-coupled receptors (Thomas et al. 2005) also clustered on the arms. Furthermore, protein and regulatory evolution is weakly coupled in orthologs but not paralogs, and duplicates of both species show acceleration of both regulatory and protein evolution compared to orthologs (Castillo-Davis et al. 2004). Strikingly, the *C. briggsae* genome shows the same pattern of high recombination on the autosome arms, showing that this is a well-established feature of genome architecture (R.H. Waterston, L.W. Hillier, S. Baird, and R. Miller, unpubl.).

Comparative studies of the five *Caenorhabditis* genomes may also shed light on the evolution of the hermaphrodite–male mode of reproduction, which is believed to have evolved independently in *C. elegans* and *C. briggsae*. The other three *Caenorhabditis* species have female–male sexual systems. Just comparison of the genomes of the two self-fertilizing species have yielded insights into the dynamics of sex and gamete-specific gene evolution (Kiontke et al. 2004; Cutter and Ward 2005; Nayak et al. 2005) and the genomic organization of reproductive genes (Miller et al. 2004). Intriguingly, the genomes of both *Caenorhabditis remanei* and *Caenorhabditis n. sp. PB2801* are significantly larger than the genomes of the self-fertilizing species (J.S. Johnston, pers. comm.).

WormBase

Central to making all this information available to the community has been the ongoing development of WormBase (Chen et al. 2005a; http://www.wormbase.org), an outgrowth of ACeDB (A *C. elegans* database; http://www.acedb.org). ACeDB was developed in conjunction with the genome project to coordinate the effort to integrate the sequence with the genetic and physical maps and to provide public access to the project and its data.

WormBase contains a wide range of information about the biology and genomics of the worm. It acts as the repository of all the genome annotation for *C. elegans* as well as *C. briggsae* and related nematodes. It curates gene models, reconciling the predictions and the various experimental data sets. It acquires associated functional information from high-throughput experiments and more traditional experiments reported in the literature. WormBase also contains an extensive bibliography of papers published on *C. elegans* along with unpublished abstracts from regional meetings and the biennial International Worm Meetings and the brief reports in the *Worm Breeder's Gazette*.

WormBase supports five different methods of access through its interactive Web interface, with each adapted to specific purposes. These are

1. Web browsing for the casual user, with simple queries and navigation through a variety of displays;
2. batch retrieval for gene and sequence fields;
3. query language searching allowing ad hoc queries for more sophisticated users;
4. bulk downloads of gene sets, other data sets, or even the entire database to provide local access; and
5. scripting to allow formatting and processing of query results for those with some programming skills.

WormBase also supports the Distributed Annotation System (DAS, also developed in conjunction with the worm genome project) (Dowell et al. 2001) allowing users to add their own data tracks to browser displays.

WormBase continues to evolve, improving user interfaces and adding new data sets, such as movies, protein structures, and new genome sequences.

CONCLUSIONS

The *C. elegans* genome sequence, now complete, has spurred research on the worm to an extent only dimly foreseen by the early advocates of the genome project. The impact extends beyond the large data sets, the sequencing of additional nematode genomes, and the development of WormBase. The sequence and the associated resources have empowered individual worm labs to investigate central biological issues, rather than the process of cloning and sequencing. It also places their work in a larger context. The abundance of resources has also drawn very talented new investigators into the field. The worm leads the field in studies of apoptosis, aging, development, neurobiology, and other areas.

But the impact extends well beyond the worm field itself. Stimulated by successes in *C. elegans*, ESTs have been generated for almost every major class of nematode parasites of humans (Mitreva et al. 2005), and with the *C. elegans* genome as a point of reference, these data sets are opening new avenues to conquer these insidious diseases. Nematode-specific genes provide potential drug targets, with *C. elegans* able to serve as an initial testbed for evaluating candidate compounds.

More broadly, the worm sequence, through GenBank and the browsers (UCSC ENSEMBL, NCBI), provides a portal to the worm for investigators of other organisms. Either through direct homology searches or through established orthology tables, scientists can rapidly learn that *C. elegans* has a gene related to their gene of interest and then from WormBase and the literature learn what is known about that gene. They may well be drawn into the field to study the gene in worms, because of the ease of experimentation and wealth of resources. Many a worm researcher has had colleagues appear in their office asking about how to do experiments with the worm. New collaborations result, with the worm field enormously enriched by these "outsiders" perspectives, opening up possibilities for impact on human health and well being that otherwise might have been missed.

The impact of the *C. elegans* genome project extends in other directions. The early success of the *C. elegans* EST project was the direct forerunner of the large-scale public domain human and mouse EST projects, without which mammalian microarray and proteomic investigations in the 1990s would have been extremely limited. In genome sequencing, the worm project demonstrated the feasibility of using Sanger-based sequencing methods and a hierarchical (clone-based) shotgun strategy for the Human Genome Project. Significantly, the worm project also provided the model for the data release policies of the Human Genome Project. The worm genome project had adopted from the start a policy of rapid and open data release, extending the practice of early data sharing of the worm community. This policy drew the worm labs into the project, led to a clear delineation of tasks (the genome centers provided the sequence and the individual labs gave it biological meaning), and accelerated the impact of the sequence.

But the task of understanding the worm at a molecular level has just begun. Having captured the large but finite information of the genome, we can now begin to see the enormity of the task before us. We need a full parts list, not just the protein-coding genes, but the RNA genes, the regulatory elements, and any other functional elements of the genome. We need to know the motifs that transcription factors bind in vivo, and that has to be coupled with a precise knowledge of when, where, and at what level each gene is expressed. *C. elegans* is probably the only experimental animal in which the resolution can be at the single-cell level throughout development; we should exploit this. With this information, the regulatory networks that control development should emerge, yielding circuit diagram models of development. Success with this lowly nematode will again have profound impact on the efforts to extend this knowledge to human biology, with all its implications for human health and well-being.

But we need to move beyond this network view to achieve a true molecular understanding of worm biology. Protein function will have to be defined in detail. RNAi knock-downs, gene knockouts, and protein–protein interaction networks will be a start, but our knowledge of function will have to go much deeper. Undoubtedly we will need to understand the small molecule component of cells and their flux as well.

These will be challenging studies as we delve deeper and deeper into the molecular description. It will take common resources, new methods, and perseverance. But the synergy of hypothesis-driven and data-driven science of the past decade combined with the spirit of the community are major assets. These, with the inherent advantages of the worm so presciently recognized by Brenner more than 40 years ago, make *C. elegans* the prime candidate for achieving such a grandiose goal. We can't let the opportunity pass.

ACKNOWLEDGMENTS

The authors gratefully acknowledge the members of the *C. elegans* Sequencing Consortium as well as the team members of WormBase. The authors also thank Heidi Browning, Cindi Madej, and Susan Strome for their advice and gifts of macrorestriction Southern blots. We also thank Tim Schedl, Don Moerman, Lincoln Stein, and Paul Sternberg for their comments on the manuscript. This work has been funded by the UK Medical Research Council, the Wellcome Trust, and the National Human Genome Research Institute and the National Institute of General Medical Sciences at the National Institutes of Health.

REFERENCES

Ambros, V., Lee, R.C., Lavanway, A., Williams, P.T., and Jewell, D. 2003. MicroRNAs and other tiny endogenous RNAs in *C. elegans*. *Curr. Biol.* **13:** 807–818.

Baugh, L.R., Hill, A.A., Slonim, D.K., Brown, E.L., and Hunter, C.P. 2003. Composition and dynamics of the *Caenorhabditis elegans* early embryonic transcriptome. *Development* **130:** 889–900.

Baugh, L.R., Wen, J.C., Hill, A.A., Slonim, D.K., Brown, E.L., and Hunter, C.P. 2005. Synthetic lethal analysis of *Caenorhabditis elegans* posterior embryonic patterning genes identifies conserved genetic interactions. *Genome Biol.* **6:** R45.

Blacque, O.E., Perens, E.A., Boroevich, K.A., Inglis, P.N., Li, C., Warner, A., Khattra, J., Holt, R.A., Ou, G., Mah, A.K., et al. 2005. Functional genomics of the cilium, a sensory organelle. *Curr. Biol.* **15:** 935–941.

Blumenthal, T. 2005. *Trans*-splicing and operons. In *WormBook* (ed. The *C. elegans* Research Community). doi/10.1895/wormbook.1.5.1, http://www.wormbook.org.

Blumenthal, T. and Gleason, K.S. 2003. *Caenorhabditis elegans* operons: Form and function. *Nat. Rev. Genet.* **4:** 112–120.

Burge, C. and Karlin, S. 1997. Prediction of complete gene structures in human genomic DNA. *J. Mol. Biol.* **268:** 78–94.

Castillo-Davis, C.I., Hartl, D.L., and Achaz, G. 2004. *cis*-Regulatory and protein evolution in orthologous and duplicate genes. *Genome Res.* **14:** 1530–1536.

The *C. elegans* Genome Sequencing Consortium. 1998. Genome sequence of the nematode *C. elegans*: A platform for investigating biology. *Science* **282:** 2012–2018.

Chen, N., Harris, T.W., Antoshechkin, I., Bastiani, C., Bieri, T., Blasiar, D., Bradnam, K., Canaran, P., Chan, J., Chen, C.K., et al. 2005a. WormBase: A comprehensive data resource for *Caenorhabditis* biology and genomics. *Nucleic Acids Res.* **33:** D383–D389.

Chen, N., Pai, S., Zhao, Z., Mah, A., Newbury, R., Johnsen, R.C., Altun, Z., Moerman, D.G., Baillie, D.L., and Stein, L.D. 2005b. Identification of a nematode chemosensory gene family. *Proc. Natl. Acad. Sci.* **102:** 146–151.

Christensen, M., Estevez, A., Yin, X., Fox, R., Morrison, R., McDonnell, M., Gleason, C., Miller III, D.M., and Strange, K. 2002. A primary culture system for functional analysis of C. elegans neurons and muscle cells. Neuron 33: 503–514.

Coulson, A., Sulston, J., Brenner, F.R.S., and Karn, J. 1986. Toward a physical map of the genome of the nematode Caenorhabditis elegans. Proc. Natl. Acad. Sci. 83: 7821–7825.

Coulson, A., Waterston, R., Kiff, J., Sulston, J., and Kohara, Y. 1988. Genome linking with yeast artificial chromosomes. Nature 335: 184–186.

Coulson, A., Kozono, Y., Lutterbach, B., Shownkeen, R., Sulston, J., and Waterston, R. 1991. YACs and the C. elegans genome. Bioessays 13: 413–417.

Coulson, A., Huynh, C., Kozono, Y., and Shownkeen, R. 1995. The physical map of the Caenorhabditis elegans genome. Methods Cell Biol. 48: 533–550.

Coventry, A., Kleitman, D.J., and Berger, B. 2004. MSARI: Multiple sequence alignments for statistical detection of RNA secondary structure. Proc. Natl. Acad. Sci. 101: 12102–12107.

Cutter, A.D. and Ward, S. 2005. Sexual and temporal dynamics of molecular evolution in C. elegans development. Mol. Biol. Evol. 22: 178–188.

Deplancke, B., Dupuy, D., Vidal, M., and Walhout, A.J. 2004. A gateway-compatible yeast one-hybrid system. Genome Res. 14: 2093–2101.

Doggett, N.A., Smith, C.L., and Cantor, C.R. 1992. The effect of DNA concentration on mobility in pulsed field gel electrophoresis. Nucleic Acids Res. 20: 859–864.

Dowell, R.D., Jokerst, R.M., Day, A., Eddy, S.R., and Stein, L. 2001. The distributed annotation system. BMC Bioinformatics 2: 7.

Dupuy, D., Li, Q.R., Deplancke, B., Boxem, M., Hao, T., Lamesch, P., Sequerra, R., Bosak, S., Doucette-Stamm, L., Hope, I.A., et al. 2004. A first version of the Caenorhabditis elegans promoterome. Genome Res. 14: 2169–2175.

Fire, A., Xu, S., Montgomery, M.K., Kostas, S.A., Driver, S.E., and Mello, C.C. 1998. Potent and specific genetic interference by double-stranded RNA in Caenorhabditis elegans. Nature 391: 806–811.

Fitch, D.H. 2005. Evolution: An ecological context for C. elegans. Curr. Biol. 15: R655–R658.

Fox, R.M., Von Stetina, S.E., Barlow, S.J., Shaffer, C., Olszewski, K.L., Moore, J.H., Dupuy, D., Vidal, M., and Miller III, D.M. 2005. A gene expression fingerprint of C. elegans embryonic motor neurons. BMC Genomics 6: 42.

Fraser, A.G., Kamath, R.S., Zipperlen, P., Martinez-Campos, M., Sohrmann, M., and Ahringer, J. 2000. Functional genomic analysis of C. elegans chromosome I by systematic RNA interference. Nature 408: 325–330.

Fukushige, T., Schroeder, D.F., Allen, F.L., Goszczynski, B., and McGhee, J.D. 1996. Modulation of gene expression in the embryonic digestive tract of C. elegans. Dev. Biol. 178: 276–288.

Gaudet, J., Muttumu, S., Horner, M., and Mango, S.E. 2004. Whole-genome analysis of temporal gene expression during foregut development. PLoS Biol. 2: e352.

Griffiths-Jones, S. 2004. The microRNA Registry. Nucleic Acids Res. 32: D109–D111.

GuhaThakurta, D., Schriefer, L.A., Waterston, R.H., and Stormo, G.D. 2004. Novel transcription regulatory elements in Caenorhabditis elegans muscle genes. Genome Res. 14: 2457–2468.

Holt, S.J. and Riddle, D.L. 2003. SAGE surveys C. elegans carbohydrate metabolism: Evidence for an anaerobic shift in the long-lived dauer larva. Mech. Ageing Dev. 124: 779–800.

Hwang, B.J., Muller, H.M., and Sternberg, P.W. 2004. Genome annotation by high-throughput 5' RNA end determination. Proc. Natl. Acad. Sci. 101: 1650–1655.

International Human Genome Sequencing Consortium. 2004. Finishing the euchromatic sequence of the human genome. *Nature* **431:** 931–945.
Jones, S.J., Riddle, D.L., Pouzyrev, A.T., Velculescu, V.E., Hillier, L., Eddy, S.R., Stricklin, S.L., Baillie, D.L., Waterston, R., and Marra, M.A. 2001. Changes in gene expression associated with developmental arrest and longevity in *Caenorhabditis elegans*. *Genome Res.* **11:** 1346–1352.
Kamath, R.S. and Ahringer, J. 2003. Genome-wide RNAi screening in *Caenorhabditis elegans*. *Methods* **30:** 313–321.
Kim, S.K., Lund, J., Kiraly, M., Duke, K., Jiang, M., Stuart, J.M., Eizinger, A., Wylie, B.N., and Davidson, G.S. 2001. A gene expression map for *Caenorhabditis elegans*. *Science* **293:** 2087–2092.
Kimble, J. and Hirsh, D. 1979. The postembryonic cell lineages of the hermaphrodite and male gonads in *Caenorhabditis elegans*. *Dev. Biol.* **70:** 396–417.
Kiontke, K., Gavin, N.P., Raynes, Y., Roehrig, C., Piano, F., and Fitch, D.H. 2004. *Caenorhabditis* phylogeny predicts convergence of hermaphroditism and extensive intron loss. *Proc. Natl. Acad. Sci.* **101:** 9003–9008.
Koch, R., van Luenen, H.G., van der Horst, M., Thijssen, K.L., and Plasterk, R.H. 2000. Single nucleotide polymorphisms in wild isolates of *Caenorhabditis elegans*. *Genome Res.* **10:** 1690–1696.
Kohara, Y. 1996. [Large scale analysis of *C. elegans* cDNA]. *Tanpakushitsu Kakusan Koso* **41:** 715–720.
Korf, I., Flicek, P., Duan, D., and Brent, M.R. 2001. Integrating genomic homology into gene structure prediction. *Bioinformatics* **17 Suppl 1:** S140–S148.
Kunitomo, H., Uesugi, H., Kohara, Y., and Iino, Y. 2005. Identification of ciliated sensory neuron-expressed genes in *Caenorhabditis elegans* using targeted pull-down of poly(A) tails. *Genome Biol.* **6:** R17.
Lamesch, P., Milstein, S., Hao, T., Rosenberg, J., Li, N., Sequerra, R., Bosak, S., Doucette-Stamm, L., Vandenhaute, J., Hill, D.E., et al. 2004. *C. elegans* ORFeome version 3.1: Increasing the coverage of ORFeome resources with improved gene predictions. *Genome Res.* **14:** 2064–2069.
Lee, R.C., Feinbaum, R.L., and Ambros, V. 1993. The *C. elegans* heterochronic gene lin-4 encodes small RNAs with antisense complementarity to lin-14. *Cell* **75:** 843–854.
Li, S., Armstrong, C.M., Bertin, N., Ge, H., Milstein, S., Boxem, M., Vidalain, P.O., Han, J.D., Chesneau, A., Hao, T., et al. 2004. A map of the interactome network of the metazoan *C. elegans*. *Science* **303:** 540–543.
Lim, L.P., Lau, N.C., Weinstein, E.G., Abdelhakim, A., Yekta, S., Rhoades, M.W., Burge, C.B., and Bartel, D.P. 2003. The microRNAs of *Caenorhabditis elegans*. *Genes & Dev.* **17:** 991–1008.
Liu, Y., Liu, N., and Zhao, H. 2005a. Inferring protein–protein interactions through high-throughput interaction data from diverse organisms. *Bioinformatics* **21:** 3279–3285.
Liu, Z.J., Tempel, W., Ng, J.D., Lin, D., Shah, A.K., Chen, L., Horanyi, P.S., Habel, J.E., Kataeva, I.A., Xu, H., et al. 2005b. The high-throughput protein-to-structure pipeline at SECSG. *Acta Crystallogr. D Biol. Crystallogr.* **61:** 679–684.
Lu, S., Symersky, J., Li, S., Carson, M., Chen, L., Meehan, E., and Luo, M. 2004. Structural genomics of *Caenorhabditis elegans*: Crystal structure of the tropomodulin C-terminal domain. *Proteins* **56:** 384–386.
Luersen, K., Eschbach, M.L., Liebau, E., and Walter, R.D. 2004. Functional GATA- and

initiator-like-elements exhibit a similar arrangement in the promoters of *Caenorhabditis elegans* polyamine synthesis enzymes. *Biol. Chem.* **385:** 711–721.

Maeda, I., Kohara, Y., Yamamoto, M., and Sugimoto, A. 2001. Large-scale analysis of gene function in *Caenorhabditis elegans* by high-throughput RNAi. *Curr. Biol.* **11:** 171–176.

McCallum, C.M., Comai, L., Greene, E.A., and Henikoff, S. 2000. Targeted screening for induced mutations. *Nat. Biotechnol.* **18:** 455–457.

McCombie, W.R., Adams, M.D., Kelley, J.M., FitzGerald, M.G., Utterback, T.R., Khan, M., Dubnick, M., Kerlavage, A.R., Venter, J.C., and Fields, C. 1992. *Caenorhabditis elegans* expressed sequence tags identify gene families and potential disease gene homologues. *Nat. Genet.* **1:** 124–131.

McKay, S.J., Johnsen, R., Khattra, J., Asano, J., Baillie, D.L., Chan, S., Dube, N., Fang, L., Goszczynski, B., Ha, E., et al. 2003. Gene expression profiling of cells, tissues, and developmental stages of the nematode *C. elegans*. *Cold Spring Harb. Symp. Quant. Biol.* **68:** 159–169.

Miller, M.A., Cutter, A.D., Yamamoto, I., Ward, S., and Greenstein, D. 2004. Clustered organization of reproductive genes in the *C. elegans* genome. *Curr. Biol.* **14:** 1284–1290.

Mitreva, M., Blaxter, M.L., Bird, D.M., and McCarter, J.P. 2005. Comparative genomics of nematodes. *Trends Genet.* **21:** 573–581.

Nayak, S., Goree, J., and Schedl, T. 2005. fog-2 and the evolution of self-fertile hermaphroditism in *Caenorhabditis*. *PLoS Biol.* **3:** e6.

Nelson, D.W. and Honda, B.M. 1985. Genes coding for 5S ribosomal RNA of the nematode *Caenorhabditis elegans*. *Gene* **38:** 245–251.

Nikolaidis, N. and Nei, M. 2004. Concerted and nonconcerted evolution of the Hsp70 gene superfamily in two sibling species of nematodes. *Mol. Biol. Evol.* **21:** 498–505.

Okkema, P.G. and Fire, A. 1994. The *Caenorhabditis elegans* NK-2 class homeoprotein CEH-22 is involved in combinatorial activation of gene expression in pharyngeal muscle. *Development* **120:** 2175–2186.

Okkema, P.G., Ha, E., Haun, C., Chen, W., and Fire, A. 1997. The *Caenorhabditis elegans* NK-2 homeobox gene ceh-22 activates pharyngeal muscle gene expression in combination with pha-1 and is required for normal pharyngeal development. *Development* **124:** 3965–3973.

Piano, F., Schetter, A.J., Mangone, M., Stein, L., and Kemphues, K.J. 2000. RNAi analysis of genes expressed in the ovary of *Caenorhabditis elegans*. *Curr. Biol.* **10:** 1619–1622.

Reinhart, B.J., Slack, F.J., Basson, M., Pasquinelli, A.E., Bettinger, J.C., Rougvie, A.E., Horvitz, H.R., and Ruvkun, G. 2000. The 21-nucleotide let-7 RNA regulates developmental timing in *Caenorhabditis elegans*. *Nature* **403:** 901–906.

Rivas, E. and Eddy, S.R. 2001. Noncoding RNA gene detection using comparative sequence analysis. *BMC Bioinformatics* **2:** 8.

Roulet, E., Busso, S., Camargo, A.A., Simpson, A.J., Mermod, N., and Bucher, P. 2002. High-throughput SELEX SAGE method for quantitative modeling of transcription-factor binding sites. *Nat. Biotechnol.* **20:** 831–835.

Roy, P.J., Stuart, J.M., Lund, J., and Kim, S.K. 2002. Chromosomal clustering of muscle-expressed genes in *Caenorhabditis elegans*. *Nature* **418:** 975–979.

Sanchez, C., Lachaize, C., Janody, F., Bellon, B., Roder, L., Euzenat, J., Rechenmann, F., and Jacq, B. 1999. Grasping at molecular interactions and genetic networks in *Drosophila melanogaster* using FlyNets, an Internet database. *Nucleic Acids Res.* **27:** 89–94.

Sharan, R., Suthram, S., Kelley, R.M., Kuhn, T., McCuine, S., Uetz, P., Sittler, T., Karp, R.M., and Ideker, T. 2005. Conserved patterns of protein interaction in multiple species. *Proc. Natl. Acad. Sci.* **102:** 1974–1979.

Sonnichsen, B., Koski, L.B., Walsh, A., Marschall, P., Neumann, B., Brehm, M., Alleaume, A.M., Artelt, J., Bettencourt, P., Cassin, E., et al. 2005. Full-genome RNAi profiling of early embryogenesis in *Caenorhabditis elegans*. *Nature* **434:** 462–469.

Spieth, J. and Lawson, D. 2005. Overview of gene structure. In *WormBook* (ed. The *C. elegans* Research Community). doi/10.1895/wormbook.1.5.1, http://www.wormbook.org.

Stein, L.D., Bao, Z., Blasiar, D., Blumenthal, T., Brent, M.R., Chen, N., Chinwalla, A., Clarke, L., Clee, C., Coghlan, A., et al. 2003. The genome sequence of *Caenorhabditis briggsae*: A platform for comparative genomics. *PLoS Biol.* **1:** E45.

Stricklin, S.L., Griffiths-Jones, S., and Eddy, S.R. 2005. *C. elegans* noncoding RNA genes. In *WormBook* (eds. J. Hodgkin and P. Anderson). doi/10.1895/wormbook.1.1.1, http://www.wormbook.org.

Sulston, J.E. and Brenner, S. 1974. The DNA of *Caenorhabditis elegans*. *Genetics* **77:** 95–104.

Sulston, J. and Ferry, G. 2002. *The common thread: A story of science, politics, ethics and the human genome*. Bantam Press, London.

Sulston, J.E. and Horvitz, H.R. 1977. Post-embryonic cell lineages of the nematode, *Caenorhabditis elegans*. *Dev. Biol.* **56:** 110–156.

Sulston, J.E., Schierenberg, E., White, J.G., and Thomson, J.N. 1983. The embryonic cell lineage of the nematode *Caenorhabditis elegans*. *Dev. Biol.* **100:** 64–119.

Sulston, J., Mallett, F., Staden, R., Durbin, R., Horsnell, T., and Coulson, A. 1988. Software for genome mapping by fingerprinting techniques. *Comput. Appl. Biosci.* **4:** 125–132.

Swan, K.A., Curtis, D.E., McKusick, K.B., Voinov, A.V., Mapa, F.A., and Cancilla, M.R. 2002. High-throughput gene mapping in *Caenorhabditis elegans*. *Genome Res.* **12:** 1100–1105.

Symersky, J., Lin, G., Li, S., Qiu, S., Carson, M., Schormann, N., and Luo, M. 2003. Structural genomics of *Caenorhabditis elegans*: Crystal structure of calmodulin. *Proteins* **53:** 947–949.

Teng, Y., Girard, L., Ferreira, H.B., Sternberg, P.W., and Emmons, S.W. 2004. Dissection of *cis*-regulatory elements in the *C. elegans* Hox gene egl-5 promoter. *Dev. Biol.* **276:** 476–492.

Thomas, J.H., Kelley, J.L., Robertson, H.M., Ly, K., and Swanson, W.J. 2005. Adaptive evolution in the SRZ chemoreceptor families of *Caenorhabditis elegans* and *Caenorhabditis briggsae*. *Proc. Natl. Acad. Sci.* **102:** 4476–4481.

Timmons, L. and Fire, A. 1998. Specific interference by ingested dsRNA. *Nature* **395:** 854.

Velculescu, V.E., Zhang, L., Vogelstein, B., and Kinzler, K.W. 1995. Serial analysis of gene expression. *Science* **270:** 484–487.

Venable, J.D., Dong, M.Q., Wohlschlegel, J., Dillin, A., and Yates, J.R. 2004. Automated approach for quantitative analysis of complex peptide mixtures from tandem mass spectra. *Nat. Methods* **1:** 39–45.

Vidalain, P.O., Boxem, M., Ge, H., Li, S., and Vidal, M. 2004. Increasing specificity in high-throughput yeast two-hybrid experiments. *Methods* **32:** 363–370.

Wang, D., Kennedy, S., Conte Jr., D., Kim, J.K., Gabel, H.W., Kamath, R.S., Mello, C.C., and Ruvkun, G. 2005. Somatic misexpression of germline P granules and enhanced RNA interference in retinoblastoma pathway mutants. *Nature* **436:** 593–597.

Washietl, S., Hofacker, I.L., and Stadler, P.F. 2005. Fast and reliable prediction of noncoding RNAs. *Proc. Natl. Acad. Sci.* **102:** 2454–2459.

Waterston, R., Martin, C., Craxton, M., Huynh, C., Coulson, A., Hillier, L., Durbin, R., Green, P., Shownkeen, R., Halloran, N., et al. 1992. A survey of expressed genes in *Caenorhabditis elegans*. *Nat. Genet.* **1:** 114–123.

Wei, C., Lamesch, P., Arumugam, M., Rosenberg, J., Hu, P., Vidal, M., and Brent, M.R. 2005. Closing in on the *C. elegans* ORFeome by cloning TWINSCAN predictions. *Genome Res.* **15:** 577–582.

White, J.G., Southgate, E., Thomson, J.N., and Brenner, F.R.S. 1986. The structure of the nervous system of the nematode *Caenorhabditis elegans*. *Phil. Trans. Roy. Soc. London Ser. B Biol. Sci.* **314:** 1–340.

Wicks, S.R., Yeh, R.T., Gish, W.R., Waterston, R.H., and Plasterk, R.H. 2001. Rapid gene mapping in *Caenorhabditis elegans* using a high density polymorphism map. *Nat. Genet.* **28:** 160–164.

Wicky, C., Villeneuve, A.M., Lauper, N., Codourey, L., Tobler, H., and Muller, F. 1996. Telomeric repeats (TTAGGC)n are sufficient for chromosome capping function in *Caenorhabditis elegans*. *Proc. Natl. Acad. Sci.* **93:** 8983–8988.

Wightman, B., Ha, I., and Ruvkun, G. 1993. Posttranscriptional regulation of the heterochronic gene lin-14 by lin-4 mediates temporal pattern formation in *C. elegans*. *Cell* **75:** 855–862.

Wunderlich, Z., Acton, T.B., Liu, J., Kornhaber, G., Everett, J., Carter, P., Lan, N., Echols, N., Gerstein, M., Rost, B., et al. 2004. The protein target list of the Northeast Structural Genomics Consortium. *Proteins* **56:** 181–187.

Zhang, S., Ma, C., and Chalfie, M. 2004. Combinatorial marking of cells and organelles with reconstituted fluorescent proteins. *Cell* **119:** 137–144.

7

Drosophila melanogaster: A Case Study of a Model Genomic Sequence and Its Consequences

Michael Ashburner and Casey M. Bergman[1]
Department of Genetics
University of Cambridge
Cambridge CB2 3EH, United Kingdom

IT IS ALMOST 100 YEARS SINCE WILLIAM CASTLE INTRODUCED *Drosophila melanogaster* to the pleasures and rigors of biological research (Castle 1906). Four major phases of *Drosophila* research can, perhaps, be distinguished. The period ~1910–1940 of classical genetic analysis was one of rapid development in which most of the major principles of classical genetics were established: the chromosome theory of heredity, the nature of genetic linkage and genetic maps, the genetic behavior of chromosome aberrations, the induction of gene and chromosome mutations by radiation, the discovery of mitotic recombination, and so on. This was followed by a long period, ~1940–1968, of growth but relative sterility, a period in which many of the best minds in genetics turned their attention to microbes and phage. The period from, roughly, ~1968–2000 was a renaissance, witnessed by many molecular biologists moving into the field, creating an analytical, rather than descriptive, study of development and behavior. This metamorphosis was fueled by many major technical advances within the field, for example, the invention of in situ hybridization, of the *P*-element-based transformation technology, of powerful methods for clonal analysis, the discovery of potent chemical mutagens, and by the extraordinary external advances in molecular biology. New generations of researchers selected

---

[1]Present address: Faculty of Life Sciences, University of Manchester, Michael Smith Building, Oxford Road, Manchester, M13 9PT, United Kingdom.

Drosophila as a model organism for the study of fundamental problems in biology. From 2000, fly research has matured into its fourth period: the genome era, for, on March 24, 2000 the first release of the "complete" genomic sequence of *Drosophila melanogaster* was published, timed to coincide with that year's annual fly meeting in Pittsburgh. Five years into the post-genomic era we can begin to ask: What have we learned and what may lie ahead?

THE GENOME

Prior to 1998, two groups, the Berkley *Drosophila* Genome Project and the European *Drosophila* Genome Project, were beginning to sequence the genome of *D. melanogaster* by the tried and tested way of sequencing a minimal tiling path of clones (cosmids, P1 clones, and BACs) chosen from physical maps of the genome (Hartl et al. 1992; Madueno et al. 1995; Kimmerly et al. 1996; Hoskins et al. 2000). That changed on May 12, 1998, when Craig Venter invited Gerry Rubin to participate in an attempt to sequence the genome of this fly by whole-genome shotgun sequencing (WGS), a method untried and untested for anything larger than a bacterial genome of one or a few megabases in size. There was considerable skepticism in the community that WGS would succeed for a large and complex genome with much repetitive DNA (see Green's 1997 riposte to Weber and Myers' 1997 paper in *Genome Research*). It was a leap of faith that the combination of the new capillary sequencing machines, of very careful construction of clone libraries, and of software (then not yet written) would allow the 120-Mb euchromatic genome of *D. melanogaster* to be assembled. By September 1999 this faith had been justified: A WGS assembly of the euchromatic portion of the fly genome had been achieved. This proof-of-principle for a metazoan WGS was the first landmark contribution of the fly genome project (see Ashburner [2006]).

At that time, only one metazoan genome, that of *Caenorhabditis elegans*, had been sequenced and annotated (The *C. elegans* Sequencing Consortium 1998). Experience in genome annotation, both as a technical problem and as a community problem, was sparse. The "annotation jamboree," hosted by Celera in November 1999, was important, not only for what it did, but for how it did it—an intimate and intensive collaboration of software engineers and of biologists drawn from the community working together both to build gene models and to annotate these with functional information using the then fledgling Gene Ontology (http://www.geneontology.org/) (Ashburner et al. 2000; Lewis 2005). This is a model that has been followed by many other communities. Also unusual (although not unique, vide *C. elegans*) to this day was the very close association between the sequencing and annotation groups and FlyBase (http://www.flybase.org/), the community database for Drosophilists. The act of analyzing the fly genome sequence, therefore, was novel in a second respect: It introduced new community methods for genome annotation and curation.

The "complete" sequence of the genome of *D. melanogaster* we have today is not that released in March 2000 (Adams et al. 2000; Myers et al. 2000; Rubin et al. 2000). Since Release 1, there have been three subsequent genome releases (Celniker et al. 2002) (Release 4 was in April 2004; Release 5 is planned to be the final release of the genome sequence); each release is improved in quality, with the correction of errors, both of sequence and assembly, the closure of physical and sequence gaps (only 23 now remain in the Release 4 euchromatin), and the correct assembly of repetitive sequences (this has been the responsibility of the Berkeley *Drosophila* Genome Project; http://www.fruitfly.org/). In addition, a separate project, the *Drosophila* Heterochromatin Genome Project, has been funded to sequence the complex heterochromatic sequences of the telomeres and pericentromeric chromosome regions (http://www.dhgp.org/). FlyBase has been responsible for keeping the annotation of the genome up to date. A major effort by about 10 FlyBase annotators resulted in a complete revision of all gene models and other genome features, based on the first "finished" (Release 3) sequence; this was published in a series of papers in a special issue of *Genome Biology* in December 2002 (http://www.genomebiology.com/Drosophila). Revision of gene models and other features subsequently is an ongoing, reiterative, task being done by FlyBase (Drysdale et al. 2005).

IMMEDIATE LESSONS

Before November 1999 there had been decades of debate as to the number of protein-coding genes in *D. melanogaster*. That debate then stopped: it is ~14,000. Some, for example, Hild et al. (2003), have argued that the number of protein-coding genes had been seriously underestimated (perhaps by as many as 2000 protein-coding genes) by the original annotation. A careful experimental evaluation of these "missed" gene models shows few of them to be real; many are simply new exons of genes already known or predicted (see Yandell et al. 2005). Before December 2002, the abundance and diversity of the transposable elements in the genome of *D. melanogaster* was unknown: The first attempt at their annotation (Kaminker et al. 2002) gave numbers of 1572 elements in 93 families; a more recent analysis using improved methods and including additional families (such as the enigmatic *INE-1* element) (Locke et al. 1999), indicates that the Release 4 "euchromatin" (an operational definition for the assembled chromosome arms including the first few megabases of the pericentromeric heterochromatin) has 6013 elements in 127 families (Quesneville et al. 2005; http://dynagen.ijm.jussieu.fr/repet/dmel4/index.html).

ADDED BENEFITS

In addition to revealing the parts list of the *Drosophila* genome, the completed sequence of *D. melanogaster* has changed the practice of *Drosophila* genetics and led to many unexpected discoveries. Having the genome has enormously accelerated—by a factor of at least 10—the time required to clone a particular gene of interest; this tedious task is no longer rate limiting or essential for biological discovery. The large, and growing, collection of inserted transposons used for gene disruption (mostly P-elements, but also *hobo*, *Minos*, and *piggyBac*) can now be mapped precisely to the genome sequence, rather than to a 50–100-kb interval by in situ hybridization to polytene chromosomes. About 65% of the genes of *D. melanogaster* have been disrupted by at least one transposon insertion (Bellen et al. 2004; Thibault et al. 2004; Venken and Bellen 2005). With advances in P-element technology, this has led to methods for the construction of deletions whose limits are known with base-pair accuracy, and to attempts to cover the entire genome with a minimal tiling path of deletions (Parks et al. 2004; Ryder et al. 2004). Single nucleotide polymorphisms between the sequenced strain and others have led to the construction of several SNP maps, which enormously help the mapping of, for example, EMS-induced point mutations (Berger et al. 2001; Hoskins et al. 2001; Martin et al. 2001). The genome sequence has also greatly facilitated the recovery of EMS-induced mutations in selected gene regions using the method of tilling (Winkler et al. 2005).

The completion of the fly genome in 2000 coincided with great advances in genomic technology that have revolutionized our abilities to study transcription, protein binding to specific DNA sequences, and genetic variation at the molecular level. We can now make microarrays for expression profiling, either targeted to all known or predicted coding regions or against whole-genome tiling paths of high resolution (e.g., the INDAC resource; see http://www.indac.net/); we can now map the binding sites of chromatin-associated proteins to the genome at high resolution, using either DamID (Orian et al. 2003; Sun et al. 2003;

Bianchi-Frias et al. 2004) or chromatin immunoprecipitation (chIP) (MacAlpine et al. 2004; Birch-Machin et al. 2005); we can now conduct genome-scale surveys for polymorphisms using high-throughput PCR strategies (Glinka et al. 2003; Orengo and Aguade 2004), and effectively re-sequence other genomes of the same species, using tiling paths of oligonucleotides (http://www.dpgp.org/). Genome resources have also revolutionized the genetic studies of complex traits in *Drosophila* (Pletcher et al. 2002; Harbison et al. 2005).

The task of obtaining one full-length cDNA from each fly gene is not only facilitated by the genomic sequence (Stapleton et al. 2002a,b), it helps enormously in refining gene models (Misra et al. 2002). We can now look forward to the day when each gene is represented by one full-length cDNA (in a versatile vector that will allow it to be shuttled to a variety of useful constructs) (http://www.fruitfly.org/EST/; S. Celniker, pers. comm.), and perhaps even to the availability of full-length cDNAs from every alternatively spliced transcript. Likewise, the genomic sequence has enabled the design of antisense RNA reagents that are now allowing large-scale, systematic RNAi screens for gene function in tissue culture cells (Bettencourt-Dias et al. 2004; Boutros et al. 2004).

The proper study of the genome is the genome itself. Quite unexpected properties of genomes have come from following this edict. Many individual examples of tandemly repeated genes had been known from work prior to the genome. But it was only the analysis of a 2.9-Mb trial sequence (Ashburner et al. 1999) and of the genome itself in 2000 that showed just how common this is, and the extent to which some protein families (e.g., of serine proteases) had expanded by duplication. Similarly, nested genes were first discovered in flies (Henikoff et al. 1986), but these were thought to be rare exceptions: They are not. More than 7% of the genes in the *D. melanogaster* genome are nested (Ashburner et al. 1999; Misra et. al 2002), and flies have at least a dozen examples of nests within nests. mRNAs that do not encode proteins also appear to be more common than previously thought (Tupy et al. 2005), an observation that may help to explain the phenomenon of "intergenic" transcription (Hild et al. 2003; Stolc et al. 2004).

The analysis of the genome of *D. melanogaster* has led to the insight that this genome is far more complex than we had imagined. In flies, as in other species (Cohen et al. 2000; Caron et al. 2001; Roy et al. 2002), the sequence has allowed us to observe that the genome is organized into large gene-expression neighborhoods, within which even unrelated genes tend to be coexpressed (Boutanaev et al. 2002; Spellman and Rubin 2002; Parisi et al. 2004; Stolc et al. 2004; Belyakin et al. 2005; Thygesen and

Zwinderman 2005). The existence of gene-expression neighborhoods suggests coadapted genomic regions that may be related to chromatin domains that may be preserved as syntenic regions during evolution (see below). The strongest evidence for gene-expression neighborhoods appears to come from genes expressed in the male testis (Boutanaev et al. 2002; Parisi et al. 2004), a genomic organization that may be necessary to facilitate proper gene expression during the final stages of sperm development in a highly condensed chromatin environment. The distribution of intergenic lengths in the compact *D. melanogaster* genome has also been shown to be nonrandom: Genes with complex regulation have long intergenic regions (Nelson et al. 2004). These observations suggest that relationships between genome structure and gene regulation are encoded in the fly genome sequence.

It is no coincidence that perhaps the greatest recent breakthrough in our understanding of gene regulation has come after the completion of genomic sequences of key eukaryotes like *Drosophila*: the discovery of the vast array of microRNAs (miRNAs) and their functions. In fact, the genome sequence of *D. melanogaster* helped reveal the fundamental hairpin structure of pre-miRNAs from mature miRNA expressed sequences (Lagos-Quintana et al. 2001). In turn, this detailed understanding of miRNA structure has allowed their genome-wide prediction in *Drosophila* (Lai et al. 2003). The genome sequence has also been critical for the prediction of miRNA targets (Enright et al. 2003; Stark et al. 2003; Rajewsky and Socci 2004; Brennecke et al. 2005; Burgler and Macdonald 2005).

THERE IS STILL MUCH TO DO

We hope not too many scientists will think that all the fun is over with *Drosophila*, and turn to the study of the *Trichoplax* or *Loxodonta* genomes. There remains much to discover, and many resources are now available to catalyze discovery by individual research groups (Matthews et al. 2005), who will remain the bedrock of the *Drosophila* community in the post-genomic era (Gilbert 1991). Large-scale projects to catalog functional elements in the genome sequence could be integrated and distributed through a *Drosophila* ENCODE project (http://rana.lbl.gov/drosophila/dencode.html), which would capitalize on the tradition of resource-sharing among Drosophilists, and serve as a model for community-driven, comprehensive genome annotation in higher eukaryotes.

The genomic sequences of a further 11 species of *Drosophila* (http://species.flybase.net/) will provide a rich source of data for expanding on lessons learned from the *D. melanogaster* genome. *Drosophila* genome sequences may, in fact, continue to push advances in WGS and comparative assembly techniques by providing in *D. melanogaster* a "finished" reference genome. The impact of finishing on genome assembly, annotation, and biological inference can now be evaluated to direct future strategies for genome sequencing projects (Myers et al. 2000; Benos et al. 2001; Celniker et al. 2002).

Heterochromatin has long been recognized as a major, yet mysterious, component of most metazoan genomes. We have already learned much about its molecular nature from studies with *Drosophila* (Dimitri et al. 2005). We know, for example, that much of the complex heterochromatin of *D. melanogaster* is composed of a graveyard of decaying, often nested, transposable elements with a sprinkling of protein-coding genes (Hoskins et al. 2002; Dimitri et al. 2003). We know that its chromatin differs in the spectrum of its proteins (Elgin and Grewal 2003). But, it would be an exaggeration for even the most zealous "heterochromatist" to claim that we have anything approaching a full understanding of either the structure or function of this important genome component. The completion of the sequence of *D. melanogaster* now requires the sequence of the complex heterochromatin (we except the 36 Mb of simple sequence satellite sequences) and new methods for its analysis.

Straightforward in principle, but demanding in practice, is the challenge to discover "functions" for all of the genes. The Gene Ontology has provided not only a structured language to describe gene "function," but also tools for the prediction of gene function. Yet no scientist should be satisfied for long with only predicted function. Of the 14,461 predicted protein-coding genes of *D. melanogaster*, only 5402 have known mutant alleles; on the other hand, there are 9875 genes in *D. melanogaster* whose existence is reasonably well attested by classical methods but that have yet to be identified on the sequence (data computed from FlyBase) (A. de Grey, pers. comm.). Linking the wealth of results published in the literature to the genome is absolutely necessary if we are to leverage the depth of our understanding of development, behavior, and evolution in *Drosophila* using the genome sequences. Continued progress toward completion of the gene disruption projects and expression profiling (see above) will prove essential for finding functions for the remaining as-yet-uncharacterized genes.

Progress, both experimental and computational, in the understanding of regulatory networks in *Drosophila* is dramatic: Indeed, it can be argued that the regulation of A-P and D-V axes formation in early fly development is one of the best (if not the best) understood complex biological system (http://bdtnp.lbl.gov/). The syncytial embryonic environment is also optimal for the decoding of networks based on transcriptional control. Indeed, these networks can even be emulated in vitro (Isalan et al. 2005). Yet, from experimental analyses, we understand in any detail the structure of the regulatory regions of relatively few genes, and the annotation of even this limited set is regrettably incomplete (http://www.flybase.org/annot/dmel_release4.1.txt). Nevertheless, the ability to reconstruct core features of the *Drosophila* segmentation network automatically in silico from annotated regulatory sequences (Fig. 1) suggests that a complete genomic inventory of regulatory elements will

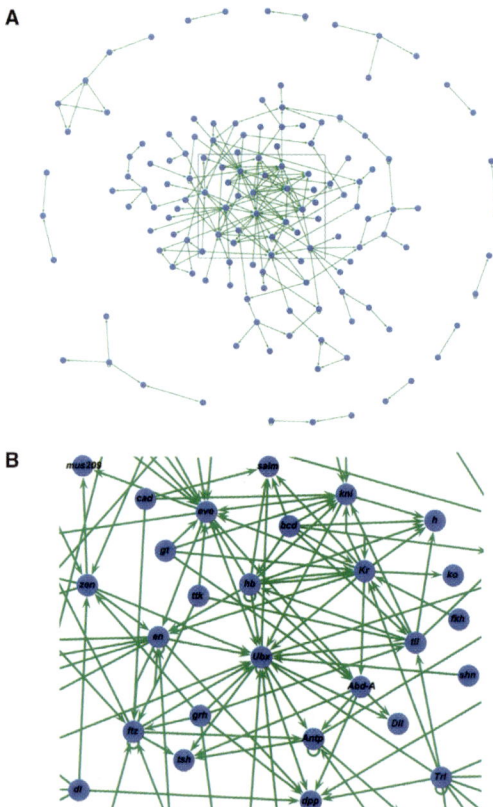

Figure 1. Partial gene regulatory network (GRN) for *Drosophila melanogaster* automatically generated from genome annotations. (*A*) GRN determined from entire set of protein–DNA interactions in the *Drosophila* DNase I footprint database (http://www.flyreg.org/) using Cytoscape (Shannon et al. 2003). Note that the majority of currently annotated protein–DNA interactions coalesce into a single interconnected network. (*B*) Sector of GRN shown in panel *A* demonstrating that known transcriptional regulatory interactions can be computed directly from text-based annotation of genomic DNA.

have direct impact on gene regulatory network analysis in flies. The genomic sequence will enable great advances here, as computational methods for the prediction of regulatory regions (Berman et al. 2002; Ohler et al. 2002; Rajewsky et al. 2002; Ochoa-Espinosa et al. 2005), particularly using comparative data (Bergman et al. 2002; Berman et al. 2004; Grad et al. 2004; Sinha et al. 2004), improve and as the size of the database of functionally characterized *cis*-regulatory sequences mapped to the genome increases (Lifanov et al. 2003; Bergman et al. 2005). The development of a genome tile microarray in *Drosophila* will also be essential for experimentally enumerating the targets and binding specificities of the vast majority of the >700 predicted transcription factors in the fly genome, for which no regulatory information is currently available. Integrating results of gene networks inferred from multiple genomic and proteomic (e.g., two-hybrid screens) (Giot et al. 2003; Stanyon et al. 2004; Formstecher et al. 2005) data sources will hopefully expand and link together functional regulatory modules into coherent systems that specify fly biology.

We doubt that the discovery of gene-expression neighborhoods is the last surprise for our understanding of genome structure at a large scale. Here, we believe that comparative data will have much to say. It was about 70 years ago that Sturtevant and Dobzhansky discovered that overlapping inversions can be used for phylogenetic reconstruction (Sturtevant and Dobzhansky 1936), a fact most remarkably used to study the relationships for the >120 species of endemic Hawaiian species (Carson et al. 1992). Now, the high rate of genome restructuring by inversions in *Drosophila* (Ranz et al. 2001) can be used to reveal groups of genes that maintain positional order, perhaps for functional reasons such as coexpression. The relationship between syntenic regions and gene expression neighborhoods is only beginning to emerge (Stolc et al. 2004); however, the genomic material to reveal the scope of functional constraints on chromosome restructuring is now available. The rigorous definition of syntenic regions and the development of software to achieve this end are an important proximate goal, but the vagaries of incomplete genome sequences, overlapping inversions, and the possibility of breakpoint reuse will present challenges to reconstructing the complete history of inversions in *Drosophila*.

One of the great lessons of the post-genomic era is the added value of comparative sequence data for the functional annotation of model systems such as *Drosophila*. The second genome in the genus, that of *Drosophila pseudoobscura*, was published in January 2005 (Richards et al. 2005) and represents the first stage in the explosion of comparative genomics data being generated currently in flies (http://rana.lbl.gov/drosophila/multipleflies.html), an inevitability heralded by the *D. melanogaster* WGS. As with interpreting all biological systems, the *D. melanogaster* genome must be viewed as the product of evolution; thus, the decoding of functional information in this model genome will be intimately intertwined with knowing the evolutionary history and forces that have produced it. Much, however, remains to be learned about the mechanisms of genome evolution in *Drosophila*: the relative contributions of gene transposition and inversions to gene movement (Ranz et al. 2003b); the gain of lineage-specific genes by retrotransposition and tandem duplication (Betran et al. 2002; Wang et al. 2004); the divergence and proliferation of transposable elements (Kaminker et al. 2002; Lerat et al. 2003; Sanchez-Gracia et al. 2005); the evolution of *cis*-regulatory sequences (Emberly et al. 2003; Phinchongsakuldit et al. 2004; Ludwig et al. 2005; Negre et al. 2005; Sinha and Siggia 2005); the relationship between *cis*-regulatory and transcriptome evolution (Ranz et al. 2003a; Rifkin et al. 2003); and the resolution of the roles that mutation, recombination, genetic drift, population history, and natural selection have jointly played in shaping the genomic landscape we observe today (Andolfatto 2001; Aquadro et al. 2001; Glinka et al. 2003; Orengo and Aguade 2004; Haddrill et al. 2005). By identifying both highly conserved and positively selected sequences to predict and inform function, the integration of evolutionary genomics with classical and forward genetics will continue to propel biological discovery long into the second century of fly research.

ACKNOWLEDGMENTS

We thank Brian Oliver, Nipam Patel, Gerry Rubin, and two anonymous reviewers for comments on the manuscript of this review. We thank Hamid Bolouri for showing us the potential of Cytoscape. We apologize for any oversight in attribution resulting from space limitations. C.M.B. is supported by a USA Research Fellowship from the Royal Society. Work in M.A.'s laboratory is supported by an MRC Programme Grant to M.A. and Steve Russell.

REFERENCES

Adams, M.D., Celniker, S.E., Holt, R.A., Evans, C.A., Gocayne, J.D., Amanatides, P.G., Scherer, S.E., Li, P.W., Hoskins, R.A., Galle, R.F., et al. 2000. The genome sequence of *Drosophila melanogaster*. *Science* **287**: 2185–2195.

Andolfatto, P. 2001. Adaptive hitchhiking effects on genome variability. *Curr. Opin. Genet. Dev.* **11**: 635–641.

Aquadro, C.F., Bauer DuMont, V., and Reed, F.A. 2001. Genome-wide variation in the human and fruitfly: A comparison. *Curr. Opin. Genet. Dev.* **11:** 627–634.

Ashburner, M. 2006. *Won for all: How the* Drosophila *genome was sequenced.* Cold Spring Harbor Laboratory Press, Cold Spring Harbor, New York.

Ashburner, M., Misra, S., Roote, J., Lewis, S.E., Blazej, R., Davis, T., Doyle, C., Galle, R., George, R., Harris, N., et al. 1999. An exploration of the sequence of a 2.9-Mb region of the genome of *Drosophila melanogaster*: The *Adh* region. *Genetics* **153:** 179–219.

Ashburner, M., Ball, C.A., Blake, J.A., Botstein, D., Butler, H., Cherry, J.M., Davis, A.P., Dolinski, K., Dwight, S.S., Eppig, J.T., et al. 2000. Gene Ontology: Tool for the unification of biology. *Nat. Genet.* **25:** 25–29.

Bellen, H.J., Levis, R.W., Liao, G., He, Y., Carlson, J.W., Tsang, G., Evans-Holm, M., Hiesinger, P.R., Schulze, K.L., Rubin, G.M., et al. 2004. The BDGP gene disruption project: Single transposon insertions associated with 40% of *Drosophila* genes. *Genetics* **167:** 761–781.

Belyakin, S.N., Christophides, G.K., Alekseyenko, A.A., Kriventseva, E.V., Belyaeva, E.S., Nanayev, R.A., Makunin, I.V., Kafatos, F.C., and Zhimulev, I.F. 2005. Genomic analysis of *Drosophila* chromosome underreplication reveals a link between replication control and transcriptional territories. *Proc. Natl. Acad. Sci.* **102:** 8269–8274.

Benos, P.V., Gatt, M.K., Murphy, L., Harris, D., Barrell, B., Ferraz, C., Vidal, S., Brun, C., Demaille, J., Cadieu, E., et al. 2001. From first base: The sequence of the tip of the X chromosome of *Drosophila melanogaster*, a comparison of two sequencing strategies. *Genome Res.* **11:** 710–730.

Berger, J., Suzuki, T., Senti, K.A., Stubbs, J., Schaffner, G., and Dickson, B.J. 2001. Genetic mapping with SNP markers in *Drosophila*. *Nat. Genet.* **29:** 475–481.

Bergman, C.M., Pfeiffer, B.D., Rincon-Limas, D.E., Hoskins, R.A., Gnirke, A., Mungall, C.J., Wang, A.M., Kronmiller, B., Pacleb, J., Park, S., et al. 2002. Assessing the impact of comparative genomic sequence data on the functional annotation of the *Drosophila* genome. *Genome Biol.* **3:** research0086.

Bergman, C.M., Carlson, J.W., and Celniker, S.E. 2005. *Drosophila* DNase I footprint database: A systematic genome annotation of transcription factor binding sites in the fruitfly, *Drosophila melanogaster*. *Bioinformatics* **21:** 1747–1749.

Berman, B.P., Nibu, Y., Pfeiffer, B.D., Tomancak, P., Celniker, S.E., Levine, M., Rubin, G.M., and Eisen, M.B. 2002. Exploiting transcription factor binding site clustering to identify *cis*-regulatory modules involved in pattern formation in the *Drosophila* genome. *Proc. Natl. Acad. Sci.* **99:** 757–762.

Berman, B.P., Pfeiffer, B.D., Laverty, T.R., Salzberg, S.L., Rubin, G.M., Eisen, M.B., and Celniker, S.E. 2004. Computational identification of developmental enhancers: Conservation and function of transcription factor binding-site clusters in *Drosophila melanogaster* and *Drosophila pseudoobscura*. *Genome Biol.* **5:** R61.

Betran, E., Thornton, K., and Long, M. 2002. Retroposed new genes out of the X in *Drosophila*. *Genome Res.* **12:** 1854–1859.

Bettencourt-Dias, M., Giet, R., Sinka, R., Mazumdar, A., Lock, W.G., Balloux, F., Zafiropoulos, P.J., Yamaguchi, S., Winter, S., Carthew, R.W., et al. 2004. Genome-wide survey of protein kinases required for cell cycle progression. *Nature* **432:** 980–987.

Bianchi-Frias, D., Orian, A., Delrow, J.J., Vazquez, J., Rosales-Nieves, A.E., and Parkhurst, S.M. 2004. Hairy transcriptional repression targets and cofactor recruitment in *Drosophila*. *PLoS Biol.* **2:** E178.

Birch-Machin, I., Gao, S., Huen, D., McGirr, R., White, R.A., and Russell, S. 2005. Genomic analysis of heat-shock factor targets in *Drosophila*. *Genome Biol.* **6:** R63.

Boutanaev, A.M., Kalmykova, A.I., Shevelyov, Y.Y., and Nurminsky, D.I. 2002. Large clusters of co-expressed genes in the *Drosophila* genome. *Nature* **420:** 666–669.

Boutros, M., Kiger, A.A., Armknecht, S., Kerr, K., Hild, M., Koch, B., Haas, S.A., Heidelberg Fly Array Consortium, Paro, R., and Perrimon, N. 2004. Genome-wide RNAi analysis of growth and viability in *Drosophila* cells. *Science* **303:** 832–835.

Brennecke, J., Stark, A., Russell, R.B., and Cohen, S.M. 2005. Principles of microRNA-target recognition. *PLoS Biol.* **3:** e85.

Burgler, C. and Macdonald, P.M. 2005. Prediction and verification of microRNA targets by MovingTargets, a highly adaptable prediction method. *BMC Genomics* **6:** 88.

Caron, H., van Schaik, B., van der Mee, M., Baas, F., Riggins, G., van Sluis, P., Hermus, M.C., van Asperen, R., Boon, K., Voute, P.A., et al. 2001. The human transcriptome map: Clustering of highly expressed genes in chromosomal domains. *Science* **291:** 1289–1292.

Carson, H.L., Tonzetich, J., and Dopescher, L.T. 1992. Polytene chromosome maps for Hawaiian *Drosophila*. In *Drosophila inversion polymorphism*. (eds. C.B. Krimbas and J.R. Powell), pp. 441–453. CRC Press, Boca Raton, FL.

Castle, W.E. 1906. Inbreeding, cross-breeding and sterility in *Drosophila*. *Science* **23:** 153.

The *C. elegans* Sequencing Consortium. 1998. Genome sequence of the nematode *C. elegans*: A platform for investigating biology. *Science* **282:** 2012–2018.

Celniker, S.E., Wheeler, D.A., Kronmiller, B., Carlson, J.W., Halpern, A., Patel, S., Adams, M., Champe, M., Dugan, S.P., Frise, E., et al. 2002. Finishing a whole-genome shotgun: Release 3 of the *Drosophila melanogaster* euchromatic genome sequence. *Genome Biol.* **3:** research0079.

Cohen, B.A., Mitra, R.D., Hughes, J.D., and Church, G.M. 2000. A computational analysis of whole-genome expression data reveals chromosomal domains of gene expression. *Nat. Genet.* **26:** 183–186.

Dimitri, P., Junakovic, N., and Arca, B. 2003. Colonization of heterochromatic genes by transposable elements in *Drosophila*. *Mol. Biol. Evol.* **20:** 503–512.

Dimitri, P., Corradini, N., Rossi, F., and Verni, F. 2005. The paradox of functional heterochromatin. *Bioessays* **27:** 29–41.

Drysdale, R.A., Crosby, M.A., Gelbart, W., Campbell, K., Emmert, D., Matthews, B., Russo, S., Schroeder, A., Smutniak, F., Zhang, P., et al. 2005. FlyBase: Genes and gene models. *Nucleic Acids Res.* **33:** D390–D395.

Elgin, S.C. and Grewal, S.I. 2003. Heterochromatin: Silence is golden. *Curr. Biol.* **13:** R895–R898.

Emberly, E., Rajewsky, N., and Siggia, E.D. 2003. Conservation of regulatory elements between two species of *Drosophila*. *BMC Bioinformatics* **4:** 57.

Enright, A.J., John, B., Gaul, U., Tuschl, T., Sander, C., and Marks, D.S. 2003. MicroRNA targets in *Drosophila*. *Genome Biol.* **5:** R1.

Formstecher, E., Aresta, S., Collura, V., Hamburger, A., Meil, A., Trehin, A., Reverdy, C., Betin, V., Maire, S., Brun, C., et al. 2005. Protein interaction mapping: A *Drosophila* case study. *Genome Res.* **15:** 376–384.

Gilbert, W. 1991. Towards a paradigm shift in biology. *Nature* **349:** 99.

Giot, L., Bader, J.S., Brouwer, C., Chaudhuri, A., Kuang, B., Li, Y., Hao, Y.L., Ooi, C.E., Godwin, B., Vitols, E., et al. 2003. A protein interaction map of *Drosophila melanogaster*. *Science* **302:** 1727–1736.

Glinka, S., Ometto, L., Mousset, S., Stephan, W., and De Lorenzo, D. 2003. Demography and natural selection have shaped genetic variation in *Drosophila melanogaster*: A multi-locus approach. *Genetics* **165**: 1269–1278.

Grad, Y.H., Roth, F.P., Halfon, M.S., and Church, G.M. 2004. Prediction of similarly-acting *cis*-regulatory modules by subsequence profiling and comparative genomics in *D. melanogaster* and *D. pseudoobscura*. *Bioinformatics* **20**: 2738–2750.

Green, P. 1997. Against a whole-genome shotgun. *Genome Res.* **7**: 410–417.

Haddrill, P.R., Thornton, K.R., Charlesworth, B., and Andolfatto, P. 2005. Multilocus patterns of nucleotide variability and the demographic and selection history of *Drosophila melanogaster* populations. *Genome Res.* **15**: 790–799.

Harbison, S.T., Chang, S., Kamdar, K.P., and Mackay, T.F. 2005. Quantitative genomics of starvation stress resistance in *Drosophila*. *Genome Biol.* **6**: R36.

Hartl, D.L., Ajioka, J.W., Cai, H., Lohe, A.R., Lozovskaya, E.R., Smoller, D.A., and Duncan, I.W. 1992. Towards a *Drosophila* genome map. *Trends Genet.* **8**: 70–75.

Henikoff, S., Keene, M.A., Fechtel, K., and Fristrom, J.W. 1986. Gene within a gene: Nested *Drosophila* genes encode unrelated proteins on opposite DNA strands. *Cell* **44**: 33–42.

Hild, M., Beckmann, B., Haas, S.A., Koch, B., Solovyev, V., Busold, C., Fellenberg, K., Boutros, M., Vingron, M., Sauer, F., et al. 2003. An integrated gene annotation and transcriptional profiling approach towards the full gene content of the *Drosophila* genome. *Genome Biol.* **5**: R3.

Hoskins, R.A., Nelson, C.R., Berman, B.P., Laverty, T.R., George, R.A., Ciesiolka, L., Naeemuddin, M., Arenson, A.D., Durbin, J., David, R.G., et al. 2000. A BAC-based physical map of the major autosomes of *Drosophila melanogaster*. *Science* **287**: 2271–2274.

Hoskins, R.A., Phan, A.C., Naeemuddin, M., Mapa, F.A., Ruddy, D.A., Ryan, J.J., Young, L.M., Wells, T., Kopczynski, C., and Ellis, M.C. 2001. Single nucleotide polymorphism markers for genetic mapping in *Drosophila melanogaster*. *Genome Res.* **11**: 1100–1113.

Hoskins, R.A., Smith, C.D., Carlson, J.W., Carvalho, A.B., Halpern, A., Kaminker, J.S., Kennedy, C., Mungall, C.J., Sullivan, B.A., Sutton, G.G., et al. 2002. Heterochromatic sequences in a *Drosophila* whole-genome shotgun assembly. *Genome Biol.* **3**: research0085.

Isalan, M., Lemerle, C., and Serrano, L. 2005. Engineering gene networks to emulate *Drosophila* embryonic pattern formation. *PLoS Biol.* **3**: e64.

Kaminker, J.S., Bergman, C.M., Kronmiller, B., Carlson, J., Svirskas, R., Patel, S., Frise, E., Wheeler, D.A., Lewis, S.E., Rubin, G.M., et al. 2002. The transposable elements of the *Drosophila melanogaster* euchromatin: A genomics perspective. *Genome Biol.* **3**: research0084.

Kimmerly, W., Stultz, K., Lewis, S., Lewis, K., Lustre, V., Romero, R., Benke, J., Sun, D., Shirley, G., Martin, C., et al. 1996. A P1-based physical map of the *Drosophila* euchromatic genome. *Genome Res.* **6**: 414–430.

Lagos-Quintana, M., Rauhut, R., Lendeckel, W., and Tuschl, T. 2001. Identification of novel genes coding for small expressed RNAs. *Science* **294**: 853–858.

Lai, E.C., Tomancak, P., Williams, R.W., and Rubin, G.M. 2003. Computational identification of *Drosophila* microRNA genes. *Genome Biol.* **4**: R42.

Lerat, E., Rizzon, C., and Biemont, C. 2003. Sequence divergence within transposable element families in the *Drosophila melanogaster* genome. *Genome Res.* **13**: 1889–1896.

Lewis, S.E. 2005. Gene Ontology: Looking backwards and forwards. *Genome Biol.* **6**: 103.

Lifanov, A.P., Makeev, V.J., Nazina, A.G., and Papatsenko, D.A. 2003. Homotypic regulatory clusters in *Drosophila*. *Genome Res.* **13:** 579–588.

Locke, J., Howard, L.T., Aippersbach, N., Podemski, L., and Hodgetts, R.B. 1999. The characterization of *DINE-1*, a short, interspersed repetitive element present on chromosome and in the centric heterochromatin of *Drosophila melanogaster*. *Chromosoma* **108:** 356–366.

Ludwig, M.Z., Palsson, A., Alekseeva, E., Bergman, C.M., Nathan, J., and Kreitman, M. 2005. Functional evolution of a *cis*-regulatory module. *PLoS Biol.* **3:** e93.

MacAlpine, D.M., Rodriguez, H.K., and Bell, S.P. 2004. Coordination of replication and transcription along a *Drosophila* chromosome. *Genes & Dev.* **18:** 3094–3105.

Madueno, E., Papagiannakis, G., Rimmington, G., Saunders, R.D.C., Savakis, C., Siden-Kiamos, I., Skavdis, G., Spanos, L., Trenear, J., Adam, P., et al. 1995. A physical map of the X chromosome of *Drosophila melanogaster*: Cosmid contigs and sequence tagged sites. *Genetics* **139:** 1631–1647.

Martin, S.G., Dobi, K.C., and St Johnston, D. 2001. A rapid method to map mutations in *Drosophila*. *Genome Biol.* **2:** research0036.

Matthews, K.A., Kaufman, T.C., and Gelbart, W.M. 2005. Research resources for *Drosophila*: The expanding universe. *Nat. Rev. Genet.* **6:** 179–193.

Misra, S., Crosby, M.A., Mungall, C.J., Matthews, B.B., Campbell, K.S., Hradecky, P., Huang, Y., Kaminker, J.S., Millburn, G.H., Prochnik, S.E., et al. 2002. Annotation of the *Drosophila melanogaster* euchromatic genome: A systematic review. *Genome Biol.* **3:** research0083.

Myers, E.W., Sutton, G.G., Delcher, A.L., Dew, I.M., Fasulo, D.P., Flanigan, M.J., Kravitz, S.A., Mobarry, C.M., Reinert, K.H., Remington, K.A., et al. 2000. A whole-genome assembly of *Drosophila*. *Science* **287:** 2196–2204.

Negre, B., Casillas, S., Suzanne, M., Sanchez-Herrero, E., Akam, M., Nefedov, M., Barbadilla, A., De Jong, P.J., and Ruiz, A. 2005. Conservation of regulatory sequences and gene expression patterns in the disintegrating *Drosophila* Hox gene complex. *Genome Res.* **15:** 692–700.

Nelson, C.E., Hersh, B.M., and Carroll, S.B. 2004. The regulatory content of intergenic DNA shapes genome architecture. *Genome Biol.* **5:** R25.

Ochoa-Espinosa, A., Yucel, G., Kaplan, L., Pare, A., Pura, N., Oberstein, A., Papatsenko, D., and Small, S. 2005. The role of binding site cluster strength in Bicoid-dependent patterning in *Drosophila*. *Proc. Natl. Acad. Sci.* **102:** 4960–4965.

Ohler, U., Liao, G.C., Niemann, H., and Rubin, G.M. 2002. Computational analysis of core promoters in the *Drosophila* genome. *Genome Biol.* **3:** research0087.

Orengo, D.J. and Aguade, M. 2004. Detecting the footprint of positive selection in a European population of *Drosophila melanogaster*: Multilocus pattern of variation and distance to coding regions. *Genetics* **167:** 1759–1766.

Orian, A., van Steensel, B., Delrow, J., Bussemaker, H.J., Li, L., Sawado, T., Williams, E., Loo, L.W., Cowley, S.M., Yost, C., et al. 2003. Genomic binding by the *Drosophila* Myc, Max, Mad/Mnt transcription factor network. *Genes & Dev.* **17:** 1101–1114.

Parisi, M., Nuttall, R., Edwards, P., Minor, J., Naiman, D., Lu, J., Doctolero, M., Vainer, M., Chan, C., Malley, J., et al. 2004. A survey of ovary-, testis-, and soma-biased gene expression in *Drosophila melanogaster* adults. *Genome Biol.* **5:** R40.

Parks, A.L., Cook, K.R., Belvin, M., Dompe, N.A., Fawcett, R., Huppert, K., Tan, L.R., Winter, C.G., Bogart, K.P., Deal, J.E., et al. 2004. Systematic generation of high-resolution deletion coverage of the *Drosophila melanogaster* genome. *Nat. Genet.* **36:** 288–292.

Phinchongsakuldit, J., MacArthur, S., and Brookfield, J.F. 2004. Evolution of developmental genes: Molecular microevolution of enhancer sequences at the *Ubx* locus in *Drosophila* and its impact on developmental phenotypes. *Mol. Biol. Evol.* **21:** 348–363.

Pletcher, S.D., Macdonald, S.J., Marguerie, R., Certa, U., Stearns, S.C., Goldstein, D.B., and Partridge, L. 2002. Genome-wide transcript profiles in aging and calorically restricted *Drosophila melanogaster*. *Curr. Biol.* **12:** 712–723.

Quesneville, H., Bergman, C.M., Andrieu, O., Autard, D., Nouaud, D., Ashburner, M., and Anxolabehere, D. 2005. Combined evidence annotation of transposable elements in genome sequences. *PLoS Comp. Biol.* **1:** 166–165.

Rajewsky, N. and Socci, N.D. 2004. Computational identification of microRNA targets. *Dev. Biol.* **267:** 529–535.

Rajewsky, N., Vergassola, M., Gaul, U., and Siggia, E.D. 2002. Computational detection of genomic *cis*-regulatory modules applied to body patterning in the early *Drosophila* embryo. *BMC Bioinformatics* **3:** 30.

Ranz, J.M., Casals, F., and Ruiz, A. 2001. How malleable is the eukaryotic genome? Extreme rate of chromosomal rearrangement in the genus *Drosophila*. *Genome Res.* **11:** 230–239.

Ranz, J.M., Castillo-Davis, C.I., Meiklejohn, C.D., and Hartl, D.L. 2003a. Sex-dependent gene expression and evolution of the *Drosophila* transcriptome. *Science* **300:** 1742–1745.

Ranz, J.M., Gonzalez, J., Casals, F., and Ruiz, A. 2003b. Low occurrence of gene transposition events during the evolution of the genus *Drosophila*. *Evol. Int. J. Org. Evol.* **57:** 1325–1335.

Richards, S., Liu, Y., Bettencourt, B.R., Hradecky, P., Letovsky, S., Nielsen, R., Thornton, K., Hubisz, M.J., Chen, R., Meisel, R.P., et al. 2005. Comparative genome sequencing of *Drosophila pseudoobscura*: Chromosomal, gene, and *cis*-element evolution. *Genome Res.* **15:** 1–18.

Rifkin, S.A., Kim, J., and White, K.P. 2003. Evolution of gene expression in the *Drosophila melanogaster* subgroup. *Nat. Genet.* **33:** 138–144.

Roy, P.J., Stuart, J.M., Lund, J., and Kim, S.K. 2002. Chromosomal clustering of muscle-expressed genes in *Caenorhabditis elegans*. *Nature* **418:** 975–979.

Rubin, G.M., Yandell, M.D., Wortman, J.R., Gabor Miklos, G.L., Nelson, C.R., Hariharan, I.K., Fortini, M.E., Li, P.W., Apweiler, R., Fleischmann, W., et al. 2000. Comparative genomics of the eukaryotes. *Science* **287:** 2204–2215.

Ryder, E., Blows, F., Ashburner, M., Bautista-Llacer, R., Coulson, D., Drummond, J., Webster, J., Gubb, D., Gunton, N., Johnson, G., et al. 2004. The DrosDel collection: A set of P-element insertions for generating custom chromosomal aberrations in *Drosophila melanogaster*. *Genetics* **167:** 797–813.

Sanchez-Gracia, A., Maside, X., and Charlesworth, B. 2005. High rate of horizontal transfer of transposable elements in *Drosophila*. *Trends Genet.* **21:** 200–203.

Shannon, P., Markiel, A., Ozier, O., Baliga, N.S., Wang, J.T., Ramage, D., Amin, N., Schwikowski, B., and Ideker, T. 2003. Cytoscape: A software environment for integrated models of biomolecular interaction networks. *Genome Res.* **13:** 2498–2504.

Sinha, S. and Siggia, E.D. 2005. Sequence turnover and tandem repeats in *cis*-regulatory modules in *Drosophila*. *Mol. Biol. Evol.* **22:** 874–885.

Sinha, S., Schroeder, M.D., Unnerstall, U., Gaul, U., and Siggia, E.D. 2004. Cross-species comparison significantly improves genome-wide prediction of *cis*-regulatory modules in *Drosophila*. *BMC Bioinformatics* **5:** 129.

Spellman, P.T. and Rubin, G.M. 2002. Evidence for large domains of similarly expressed genes in the *Drosophila* genome. *J. Biol.* **1:** 5.

Stanyon, C.A., Liu, G., Mangiola, B.A., Patel, N., Giot, L., Kuang, B., Zhang, H., Zhong, J., and Finley Jr., R.L. 2004. A *Drosophila* protein-interaction map centered on cell-cycle regulators. *Genome Biol.* **5:** R96.

Stapleton, M., Carlson, J., Brokstein, P., Yu, C., Champe, M., George, R., Guarin, H., Kronmiller, B., Pacleb, J., Park, S., et al. 2002a. A *Drosophila* full-length cDNA resource. *Genome Biol.* **3:** research0080.

Stapleton, M., Liao, G., Brokstein, P., Hong, L., Carninci, P., Shiraki, T., Hayashizaki, Y., Champe, M., Pacleb, J., Wan, K., et al. 2002b. The *Drosophila* Gene Collection: Identification of putative full-length cDNAs for 70% of *D. melanogaster* genes. *Genome Res.* **12:** 1294–1300.

Stark, A., Brennecke, J., Russell, R.B., and Cohen, S.M. 2003. Identification of *Drosophila* MicroRNA targets. *PLoS Biol.* **1:** e60.

Stolc, V., Gauhar, Z., Mason, C., Halasz, G., van Batenburg, M.F., Rifkin, S.A., Hua, S., Herreman, T., Tongprasit, W., Barbano, P.E., et al. 2004. A gene expression map for the euchromatic genome of *Drosophila melanogaster*. *Science* **306:** 655–660.

Sturtevant, A.H. and Dobzhansky, T. 1936. Inversions in the third chromosome of wild races of *Drosophila pseudoobscura*, and their use in the study of the history of the species. *Proc. Natl. Acad. Sci.* **22:** 448–450.

Sun, L.V., Chen, L., Greil, F., Negre, N., Li, T.R., Cavalli, G., Zhao, H., Van Steensel, B., and White, K.P. 2003. Protein–DNA interaction mapping using genomic tiling path microarrays in *Drosophila*. *Proc. Natl. Acad. Sci.* **100:** 9428–9433.

Thibault, S.T., Singer, M.A., Miyazaki, W.Y., Milash, B., Dompe, N.A., Singh, C.M., Buchholz, R., Demsky, M., Fawcett, R., Francis-Lang, H.L., et al. 2004. A complementary transposon tool kit for *Drosophila melanogaster* using *P* and *piggyBac*. *Nat. Genet.* **36:** 283–287.

Thygesen, H.H. and Zwinderman, A.H. 2005. Modelling the correlation between the activities of adjacent genes in *Drosophila*. *BMC Bioinformatics* **6:** 10.

Tupy, J.L., Bailey, A.M., Dailey, D., Evans-Holm, M., Siebel, C.W., Misra, S., Celniker, S.E., and Rubin, G.M. 2005. Identification of putative noncoding polyadenylated transcripts in *Drosophila melanogaster*. *Proc. Natl. Acad. Sci.* **102:** 5495–5500.

Venken, K.J. and Bellen, H.J. 2005. Emerging technologies for gene manipulation in *Drosophila melanogaster*. *Nat. Rev. Genet.* **6:** 167–178.

Wang, W., Yu, H., and Long, M. 2004. Duplication-degeneration as a mechanism of gene fission and the origin of new genes in *Drosophila* species. *Nat. Genet.* **36:** 523–527.

Weber, J.L. and Myers, E.W. 1997. Human whole-genome shotgun sequencing. *Genome Res.* **7:** 401–409.

Winkler, S., Schwabedissen, A., Backasch, D., Bokel, C., Seidel, C., Bonisch, S., Furthauer, M., Kuhrs, A., Cobreros, L., Brand, M., et al. 2005. Target-selected mutant screen by TILLING in *Drosophila*. *Genome Res.* **15:** 718–723.

Yandell, M., Bailey, A.M., Misra, S., Shu, S., Wiel, C., Evans-Holm, M., Celniker, S.E., and Rubin, G.M. 2005. A computational and experimental approach to validating annotations and gene predictions in the *Drosophila melanogaster* genome. *Proc. Natl. Acad. Sci.* **102:** 1566–1571.

8

Unraveling Genomic Regulatory Networks in the Simple Chordate, *Ciona intestinalis*

Weiyang Shi, Michael Levine, and Brad Davidson
Department of Molecular and Cell Biology
Division of Genetics and Development
Center for Integrative Genomics
University of California
Berkeley, California 94720

THE ANCESTRAL CHORDATE GAVE RISE TO THREE GROUPS, i.e., the vertebrates, cephalochordates, and tunicates (which include the ascidians). While the vertebrates diversified into many familiar aquatic and terrestrial species, the cephalochordates and tunicates remained in the ocean and evolved into highly specialized filter feeders. In the tunicates, acquisition of an endoglucanase gene and cellulose tunic (possibly derived by lateral gene transfer from bacteria) (see Dehal et al. 2002; Matthysse et al. 2004; Nakashima et al. 2004) led to the evolution of a highly divergent adult body plan lacking almost all coelomic cavities. However, the morphogenesis of chordate structures (e.g., notochord and dorsal nerve cord) in tunicate tadpoles has been well conserved despite the "retrograde" evolution of adult anatomy.

The recent sequencing of two closely related ascidian species, *Ciona intestinalis* and *Ciona savignyi*, has revived the status of the ascidians as a major developmental model system (Dehal et al. 2002; Satoh 2003; Satoh et al. 2003). Research has confirmed that many aspects of ascidian and vertebrate embryogenesis rely on a conserved set of orthologous

Supplemental material is available online at www.genome.org.

genes and cellular processes (Satoh 2003; Passamaneck and Di Gregorio 2005). This is clearly the case for development of the nervous system (Meinertzhagen et al. 2004), notochord (Passamaneck and Di Gregorio 2005), and heart (Satou et al. 2004) and may also be true for other tissues and organs, including the blood cells, pharyngeal gill slits, endostyle, and neural crest (Jeffery et al. 2004). However, ascidian embryos are extraordinarily simple, with low cell numbers, rapid development, and well-defined lineages (Fig. 1A,B) (Satou et al. 2004). The compact genomes of both *Ciona* species also facilitate the identification of regulatory DNAs (e.g., enhancers) that direct tissue-specific and lineage-specific patterns of gene expression (Johnson et al. 2004; Kusakabe 2005). In addition, genetic simplicity of the *Ciona* genomes results from a general restriction in gene duplication events (Dehal et al. 2002). For example, there are approximately four vertebrate homologs of the heart homeobox gene *Tinman*, whereas *Ciona* possesses only one such gene (Wada et al. 2003). Thus, the paucity of redundant paralogs for key patterning genes (encoding regulatory proteins or cell-signaling molecules) simplifies the assessment of their function. Additionally, the ability to manipulate isolated blastomeres and to create transgenic embryos through electroporation of fertilized eggs greatly increases the potential to conduct in-depth analyses of *Ciona* development. Thus, research in *Ciona* presents an extraordinarily powerful avenue for deciphering fundamental chordate developmental networks.

There have been a number of recent reviews assessing different aspects of ascidian research in the post-genomic era (Canestro et al. 2003; Satoh 2003; Satoh et al. 2003; Du Pasquier 2004; Kasahara et al. 2004; Meinertzhagen et al. 2004; Kusakabe 2005; Passamaneck and Di Gregorio 2005). In order to avoid redundancy, this review is focused primarily on how the latest research can be compiled in the form of genomic regulatory networks. We also describe research on the cell biological output of these networks. We have structured the review around discussion of three studies that serve as entry points into the state of *Ciona* research following the sequencing of the genome. These studies span three distinct approaches (and continents). The first study is representative of a "systems approach" to unravel complex developmental processes, specifically the early specification of various endomesodermal lineages (Imai et al. 2004). The second study exemplifies the ability of *Ciona* researchers to make rapid progress in dissecting complex regulatory networks underlying the development of conserved chordate tissues, such as the dorsal nerve cord (Bertrand et al. 2003). The third study illustrates the forward genetics approach and how exploitation of *Ciona*'s morphological

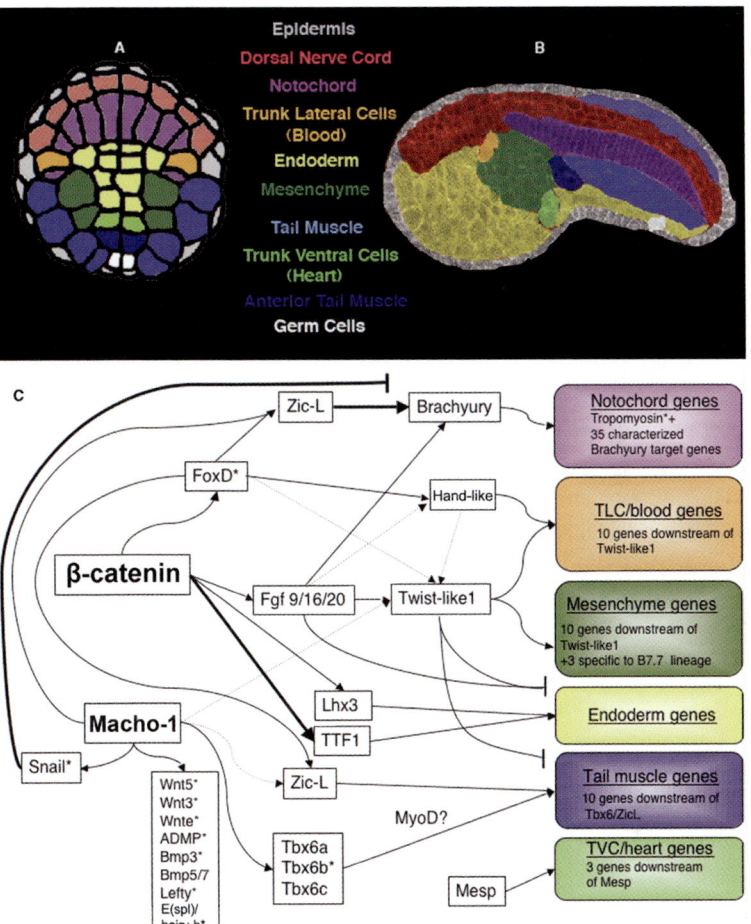

Figure 1. Network map for specification of endomesodermal lineages. (*A*) Diagram of the vegetal hemisphere of a *Ciona* gastrula. Each cell is color coded according to its lineage as indicated by the color labels in the *center*. (*B*) Confocal image of an early tailbud embryo stained with phalloidin. Embryonic tissues have been artificially colored in accordance with the color scheme in *A* to display the fate of each of the lineages. Domains of some tissues are approximate and may not be wholly accurate. (*C*) Summary of data on interactions between genes involved in specification of endomesodermal lineages. Boxes for each lineage are colored according to the scheme in *A* and *B*. Bold lines indicate that this interaction has been verified through the manipulation of binding sites within an enhancer. Asterisks indicate evidence for appropriate binding sites in the noncoding DNA upstream of the target genes. Dotted lines indicate tangential interaction indicated by embryological manipulations. The notochord determinant *Brachyury* encodes a T-box homeodomain protein. The genes required for specifying different mesodermal lineages (*NoTric* [previously known as *Handlike*], *Twist-like1*, *Mesp*, *MyoD*) all encode bHLH transcriptional factors. Further details regarding endomesodermal differentiation can be found in Supplement 1.

simplicity can provide deep insights into genetic regulation of cellular processes (Jiang et al. 2005). Along with a detailed overview of these studies, each section will also use these studies as launching points to discuss other related work and the advantages and challenges inherent to each of these approaches.

COMPREHENSIVE STUDIES OF GENE EXPRESSION AND FUNCTION

Since the sequencing of the genome, the potential of *Ciona* as a model system has been greatly enhanced by a number of systematic annotation studies (Imai 2003; Satou and Satoh 2003; Yagi et al. 2004). Of particular importance was the recent publication of gene-expression patterns for nearly every identified regulatory gene and signaling factor in the *C. intestinalis* genome (Imai et al. 2004). This survey focused mostly on early stages from the zygote through gastrulation, but also included representative stages (neurula, early and late tailbud) from older embryos. This work supplements the sequencing of an extensive set of tissue and stage-specific EST libraries and expression surveys of 1043 transcripts from these libraries (Satou et al. 2002b) along with thorough annotation of predicted genes (Satou and Satoh 2003). Together, these data constitute one of the most comprehensive sets of expression patterns for any model organism and have enormous utility for studies in *Ciona* and other chordates.

An overview of this new expression data generated two important observations regarding early *Ciona* development (Imai et al. 2004). First, over 70% of the genes encoding transcription factors and signaling molecules are maternally expressed. The substantial presence of maternally loaded factors may underlie rapid determination of cell fates in early *Ciona* embryos. Second, only 65 transcription factors and 25 signaling factors are zygotically expressed by the early gastrula stage. As almost all embryonic tissues are specified by this time, it is suggested that "comprehensive transcriptional networks" can be constructed based on the interactions between these 90 genes. Based on this prediction, a quantitative real time PCR-based assay was conducted, measuring the response of these 90 genes to suppression of the characterized tissue-specification factors β-catenin, FoxD, and FGF9/16/20 (Imai et al. 2004).

The comprehensive expression screen and associated RT–PCR assays complement an extensive body of research focused on early specification of *Ciona* endomesoderm lineages. We have attempted to draw together much of this work into a summary network (Fig. 1C). On the right are the regulatory outputs that subdivide the vegetal hemisphere of the 110-cell

Ciona embryo into endoderm and five distinct mesodermal lineages. Details of the genes that control the formation of these lineages and the regulatory interactions can be found in Supplement 1. The subdivision of distinct endomesoderm lineages depends on two crucial maternal factors, Macho-1 and β-catenin (Fig. 1C). β-catenin is translocated into the nuclei of vegetal blastomeres, where it plays an essential and apparently highly conserved role in endomesoderm specification (Imai et al. 2000). Macho-1 functions as the maternal determinant for the tail-muscle lineage, a role that probably evolved within the tunicates in association with the extremely rapid embryogenesis of a functional tail (Nishida and Sawada 2001; Sawada et al. 2005).

To date, the majority of research on these early networks has involved loss-of-function studies to identify potential targets. There are relatively few studies in which direct interactions between regulatory genes and their targets have been verified by the manipulation of specific binding sites within defined enhancers (Fig. 1C, shown by dark lines). Characterization of cell-specific enhancers for the core patterning genes, *ZicL*, *FoxD*, *FGF9/16/20*, *Tbx6*, *Twist-like1*, *Mesp*, *MyoD*, and *NoTric*, would provide the basis for establishing causal interconnections among these genes. A recent study on the *Ciona Otx* enhancer demonstrates the potential of such an approach (Bertrand et al. 2003).

REGULATORY INTERACTIONS UNDERLYING *CIONA* NEURAL SPECIFICATION

Despite the simplicity of the *Ciona* larval central nervous system (~330 cells), the overall organization is comparable to the vertebrate CNS (Fig. 2B) (Meinertzhagen et al. 2004). Many aspects of early neural specification and later neuronal differentiation appear to rely on conserved chordate gene networks that have been expanded during the evolution of more complex vertebrate neural structures (Meinertzhagen et al. 2004). Thus, dissection of gene networks in *Ciona* neurogenesis will provide crucial insights into conserved aspects of these networks in vertebrates. This potential is exemplified by a recent study on the regulation of the early neural specification gene *Otx* (Bertrand et al. 2003), which encodes a homeodomain protein expressed in both *Ciona* and vertebrate anterior neural structures.

The methodology of this study represents an optimal use of *Ciona*'s potential to dissect *cis*-regulation of gene networks. A 3.5-kb enhancer was identified for the *Ci-Otx* gene, and subsequent manipulations characterized an ~120-bp activation element (a-element). This analysis

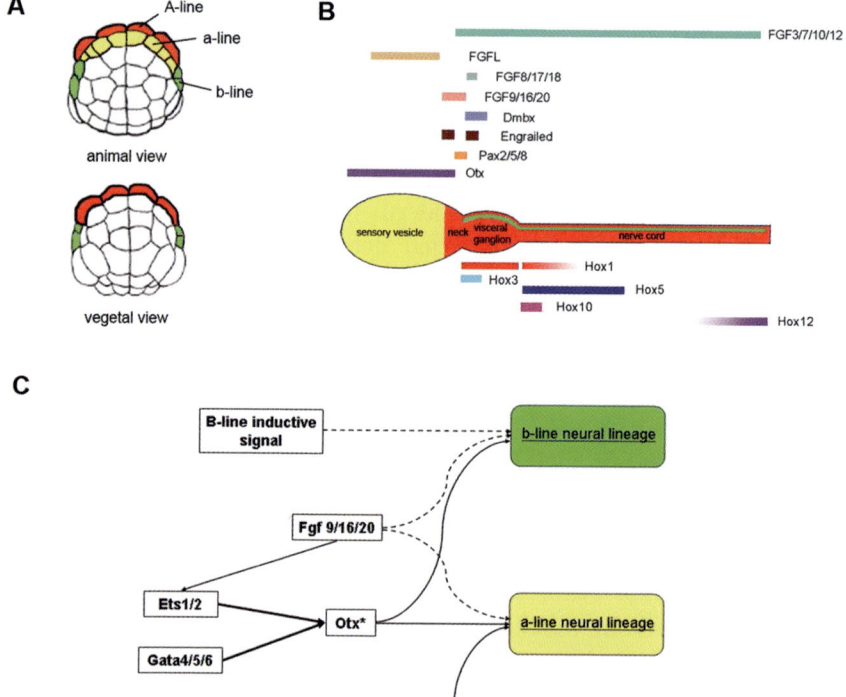

Figure 2. Regulatory network for *Ciona* neural development. (*A*) Animal and vegetal views of 64-cell stage embryo showing the position of A (red), a (yellow), and b-line (green) neural lineages. (*B*) The *Ciona* CNS is subdivided into sensory vesicle, neck, visceral ganglion, and nerve cord along the A-P axis. The progeny of three neural lineages are colored according to the scheme in *A*. Five *Hox* genes exhibit partial spatial colinearity along the visceral ganglion and nerve cord (Ikuta et al. 2004). Additional homeodomain transcriptional factors like *Otx*, *Pax2/5/8*, and *Dmbx* are expressed in the anterior CNS (Takahashi and Holland 2004). Four *FGF* genes are expressed in partially overlapping domains along the CNS (Imai et al. 2002; Satou et al. 2002a). (*C*) Regulatory network for neural specification. The three *Ciona* neural lineages are specified by a combination of maternal and zygotic factors. Bold lines indicate that this interaction has been verified through the manipulation of binding sites within an enhancer. Asterisks indicate evidence for appropriate binding sites in the noncoding DNA upstream of the target genes. Dotted lines indicate tangential interaction indicated by embryological manipulations. Further details of the differentiation of the neural lineages can be found in Supplement 2.

exploited two key features of the *Ciona* system as follows: (1) the *Ciona* genome is very compact, with regulatory regions often located immediately upstream of the genes they control; (2) in *Ciona*, regulatory regions can be rapidly assessed and dissected in great detail by high-throughput electroporation of reporter constructs into fertilized eggs (Corbo et al. 1997; Harafuji et al. 2002; see Fig. 3A,B). The authors utilized the 120-bp *Otx* regulatory element to identify the endogenous activator as *Ci-FGF9/16/20*. Based on previous work, it was clear that *Otx* induction

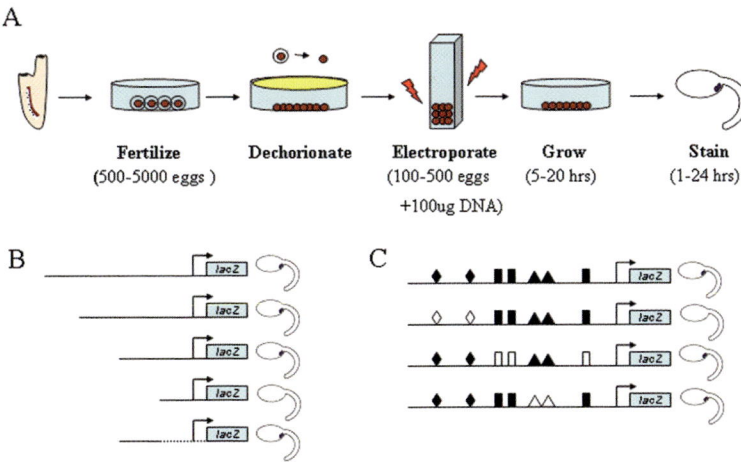

Figure 3. Using electroporation to transiently express plasmid DNA in *Ciona* embryo. The *Ciona* system allows rapid delivery and expression of reporter gene constructs in the developing embryo. It is also convenient to use tissue-specific enhancers for targeted misexpression in specific embryonic lineages. (A) Eggs and sperm are isolated from the adult for in vitro fertilization. The fertilized eggs are dechorionated, mixed with plasmid DNA and electroporated at the 1-cell stage. The electroporated embryos are then allowed to develop to appropriate developmental stages for reporter gene analysis (X-gal staining) or fixed for in situ hybridization. (B) Schematic of serial deletion analysis. The last construct depicts the minimal enhancer element (solid line) activating reporter gene expression with a basal promoter. It often requires testing of 20–40 deletion constructs to identify a minimal enhancer. (C) Schematic of site-directed mutagenesis of a minimal regulatory element. Shapes represent different putative transcription factor binding sites, filled for wild-type and unfilled for mutated. In this case, the binding sites represented by diamonds were required for reporter expression. Thorough characterization of a minimal enhancer through this method would also require testing of at least 20–40 constructs. Because one can test 10–16 constructs in a single round of electroporations, a single researcher can test over 100 constructs/week. Thus, in-depth serial deletion (B) and subsequent site-directed mutagenesis (C) can be conducted over a few weeks rather than many months.

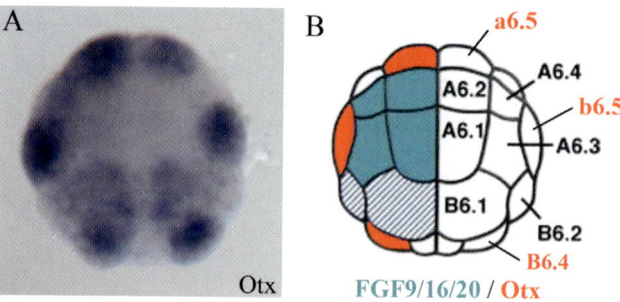

Figure 4. Use of in situ expression patterns to identify candidate genes. (*A*) In 32-cell stage *Ciona* embryos, *Otx* is expressed in the neural lineage a6.5 and b6.5 blastomeres (photo taken from the Ghost database). (*B*) Diagram of a 32-cell stage embryo illustrating that at this stage *FGF9/16/20* is the only *Ciona FGF* zygotically expressed in cells (blue) neighboring *Otx* expressing cells (red) (Bertrand et al. 2003). Multiplex in situ hybridization techniques (Davidson et al. 2005) that permit visualization of multiple genes simultaneously will greatly enhance the ability to identify potential regulatory cascades based on comparative expression patterns.

relied on MEK signaling (Hudson and Lemaire 2001; Hudson et al. 2003). However, identification of the endogenous factor relied on two further advantages of *Ciona*: (1) The limited number of paralogs in the *Ciona* genome reduces the number of candidate factors (6 *Ciona FGFs* vs. 22 vertebrate *FGFs*) and (2) the low cell number of early *Ciona* embryos permits stringent evaluation of candidate factors through in situ hybridization. (The subsequent publication of the comprehensive gene expression database discussed above greatly magnifies this advantage, see Fig. 4.)

At this point, the authors successfully addressed a pivotal obstacle in unraveling signaling networks, i.e., how cells of various lineages differentially interpret a broad signal. In *Ciona*, *FGF9/16/20* is known to mediate induction of the three following lineages: notochord, mesenchyme, and neural cells (Imai et al. 2002; Hudson et al. 2003; see Figs. 1,2). To reveal the mechanistic basis for a neural-specific response to this signal, the authors used phylogenetic footprinting. The availability of both *C. intestinalis* and *C. savignyi* genome assemblies provides an opportunity to identify potential regulatory DNAs by simply aligning orthologous sequences from the two species (Johnson et al. 2004). Alternatively, the high level of polymorphism between individuals from different populations of *C. intestinalis* can also be used to identify regulatory sequences (Boffelli et al. 2004). The authors used the cross-species comparison to reveal a concentration of potential transcriptional factor binding motifs for Ets and Gata factors. They were then able to rapidly confirm the importance of these sites through mutational analysis and by creating versions of the a-element containing only Ets or Gata sites. This analysis showed that the FGF signal causes general activation of *Otx*, while the requirement for Gata coactivation limits this response to the emerging neural lineage. Once again, the ability to perform such in-depth analysis relied on the ability to test reporter constructs by high-throughput electroporation (Fig. 3C). We wish to emphasize that reporter gene activity is monitored in the "zero" generation, just hours after electroporation of 1-cell embryos as compared with the lengthy process of creating germ-line transformation in vertebrate systems. Mosaic expression is not a problem, due to the efficient incorporation of the fusion genes and the small number of cell divisions leading to the formation of the tadpole. Additionally, it is now possible to establish stable transgenic lines through injection or electroporation of recombinant genes flanked by the Minos transposon (Sasakura et al. 2003; Matsuoka et al. 2005). This technique has been used for the identification of regulatory DNAs via enhancer trapping (Awazu et al. 2004).

The results discussed above provide a definitive regulatory network for activation of *Otx* in the anterior neural lineage (the a-line lineage). The posterior CNS in *Ciona* is derived from two distinct lineages, the A-line and b-line (Fig. 2A,B). The regulatory networks specifying these two neural lineages are less well understood and may involve cell-autonomous mechanisms and cell–cell interactions, respectively (for this and other additional details concerning *Ciona* neural development, see Supplement 2). The comprehensive analysis of *Otx* regulation demonstrates the power of the *Ciona* system for the detailed dissection of *cis*-regulatory elements. The challenge is to extend this methodology to help decipher other specification networks underlying early development. Such approaches are now being applied to the early development of the heart (Davidson et al. 2005), endoderm (Satou et al. 2001; Fanelli et al. 2003; Imai 2003; Oda-Ishii et al. 2005), tail muscles (Yagi et al. 2005), and stomodaeum (Christiaen et al. 2005). In combination with the comprehensive gene-expression studies and functional assays discussed previously, these enhancer studies will set the stage for deciphering the basic gene networks in the *Ciona* embryo. However, such networks are only the first step in understanding *Ciona* development. The next step is to investigate the roles of downstream genes in influencing the cellular components underlying tissue morphogenesis. In other words, how do interconnected networks of regulatory factors and cell-signaling molecules control the detailed behavior of individual cells? As discussed in the next section,

Ciona is also an excellent model system for exploring these downstream cellular processes.

NOTOCHORD MORPHOGENESIS

During *Ciona* notochord development, many of the genes downstream of *Brachyury* appear to control cellular processes such as adhesion, intercalation, and cell shape (Hotta et al. 2000). Studies on notochord differentiation and morphogenesis highlight the broad range of tools that are available for the dissection of complex morphogenetic processes (Di Gregorio et al. 2002; Keys et al. 2002). A recent study used forward genetics (Fig. 5) combined with detailed analysis of mutant cell behavior to characterize the role of *Prickle*, a component of the planar polarity pathway, in the morphogenesis of the notochord (Jiang et al. 2005).

The authors began with a screen for spontaneous mutations, focusing on defects in notochord formation. Because the larval body is largely

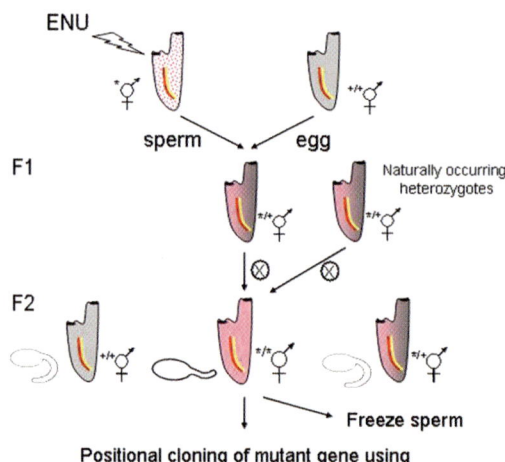

Figure 5. Forward genetic approach to identify recessive mutations in *Ciona*. *Ciona* mutagenesis involves a classical F_2 screen (Moody et al. 1999). The F_0 animals are treated with the chemical mutagen ENU and maintained for several weeks to reduce mosaicism. The mutagenized sperm is then crossed to wild-type animals to obtain heterozygous F_1 progenies. The F_1 animals are individually self-fertilized to produce F_2 progeny that are screened for phenotypic abnormalities. The sperm of mutant lines can then be cryo-preserved for long-term storage. Alternatively, naturally occurring recessive mutations can be isolated by self-crossing single animals from the wild population. The point mutation can be determined by various mapping methods, including linkage analysis of amplified length polymorphism (Jiang et al. 2005).

disposable, *Ciona* embryos with severe defects in larval structures (including the notochord and dorsal nerve cord) can still metamorphose into reproductive juveniles. The authors isolated the *aimless* (*aim*) mutant based on severe defects in larval tail formation. A variety of genomic tools identified *aim* as a mutation in the conserved cell polarity gene, *Prickle*. This gene was previously identified in a comprehensive subtractive hybridization screen for notochord-specific genes downstream of *Brachyury* (Takahashi et al. 1999).

To explore the precise role of *Prickle* in notochord formation, the authors compared the behavior of isolated notochord precursor cells from mutant and wild-type embryos. Early lineage determination in *Ciona* embryos permits the isolation of individual blastomeres from defined lineages. Confocal analysis led to the observation that mutant cells exhibit loss of localized bipolar protrusions. The transparency and low cell number of *Ciona* embryos make it feasible to observe cell behavior in vivo. The authors exploited high-throughput *Ciona* transgenesis to target tagged transcripts of *Prickle* and the conserved cell polarity gene *Dishevelled* to the developing notochord. Careful observation of wild-type and mutant transgenic embryos provided evidence that Prickle mediates the localization of Dishevelled along the medio-lateral axis of intercalating notochord cells.

The power of *Ciona* for determining the genetic basis of cellular processes is only now being exploited. There is great potential in applying similar techniques to other aspects of *Ciona* development, including directed cell migration of heart and blood cells, cell movements underlying gastrulation, and morphogenesis of the neural tube.

EMERGING RESOURCES AND TECHNIQUES

There are a number of significant resources and techniques that will promote further research. Some of these tools are currently being developed, while others are still under consideration.

WEB RESOURCES

The *Ciona intestinalis* genome assembly and annotation is provided by the Joint Genome Institute at http://genome.jgi-psf.org/ciona4/ciona4.home.html. The newly released assembly version 2.0 contains 141Mbase pairs of non-N sequences (A. Sidow, pers. comm.) and is a significant improvement over the original version 1.0, which covers only about 2/3 of the predicted genome (Dehal el al. 2002). The Ghost database (http://ghost.zool.kyoto-u.ac.jp/indexr1.html) provides the transcriptional factor/signaling molecule expression profiles and the cDNA and EST database information. In addition to these two most widely accessed databases, a number of computer resources are under development.

1. An integrated and searchable database of in situ expression profiles is currently under development (http://aniseed-ibdm.univ-mrs.fr). This database will include detailed three-dimensional representations of expression patterns using precisely staged embryos. This will be particularly useful, as there is currently no systematic, consistent staging criteria for *Ciona* embryos following the onset of gastrulation.
2. A searchable database of the *Ciona savignyi* genome was recently placed in Ensembl (http://www2.bioinformatics.tll.org.sg:8082/Ciona_savignyi/) and the *Ciona intestinalis* genome has been incorporated into the Vista browser (http://pipeline.lbl.gov/cgi-bin/gateway2). The Vista browser provides easy acquisition of orthologous *Ciona savignyi* sequences needed for the phylogenetic identification of regulatory DNAs.
3. A reference assembly of the *Ciona savignyi* genome was produced by joint efforts of the Broad Institute (Vinson el al. 2005) and Stanford University. The annotated version 2.0 covering ~170 Mb will soon be available for download (A. Sidow, pers. comm.).
4. The current version of the genome did not incorporate the extensive EST database in its gene predictions. More accurate annotations of many genes are available on the Ghost database (see above). Efforts are also underway at Ensembl to annotate both new *Ciona* genome assemblies based on gene predication programs, the EST database, and the comparisons between the two *Ciona* genomes (A. Sidow, A. Ureta-Vidal, pers. comm.).
5. Efforts are underway to produce improved genome assembly and scaffolds for each of the 14 chromosomes (Shoguchi et al. 2004). Due to the high sequence polymorphism of the *C. intestinalis* genome, repeat sequences were not assembled in the initial assembly, resulting in shorter genomic scaffolds (Dehal et al. 2002). Comparisons to the *C. savignyi* genome should also greatly assist in a more complete assembly.
6. In *Ciona* mRNA 5′-leader, *trans*-splicing may be a common occurrence (Vandenberghe et al. 2001) and thus may represent a challenge for accurate annotation of full-length transcripts. Efforts

to systematically obtain full 5′ sequences for annotated genes are currently underway.

OTHER RESOURCES

1. DNA oligo microarrays covering the whole transcriptome are under development to replace the existing cDNA microarray, which has only about 85% coverage (Azumi et al. 2003). Although the gene expression survey discussed above (Imai et al. 2004) represents a powerful resource, it suffers from inherent problems associated with in situ hybridization, namely, that the quality of probes vary, leading to background noise that may obscure the true expression pattern. For example, in comparing the gene expression survey (Imai et al. 2004) with the study on *Macho-1* targets (Yagi et al. 2004), it becomes clear that a large number of transcription/signaling factors expressed by the 32-cell stage (and subsequently down-regulated by *Macho-1* suppression) were not detected by in situ hybridization until after the 110-cell stage. Extensive microarray screens using staged embryos and isolated blastomeres will provide a more complete assessment of the gene networks underlying embryogenesis.
2. Although the *Forkhead* promoter has been used to visualize the activities of a variety of tissue-specific enhancers, it is far from ideal, since it sometimes causes spurious expression patterns (Di Gregorio et al. 2001). The *Brachyury* promoter may be better suited for some enhancer analysis (Bertrand et al. 2003), but there is still a great need for additional core promoters.
3. Recently completed EST libraries of early juvenile stages should help launch studies into the differentiation of critical post larval structures such as the endostyle, branchial gill slits, and heart. Many of the putative regulatory genes that are not expressed in the embryo might be essential for post-metamorphic events.
4. Establishment of ascidian stock centers will overcome the main obstacle to a robust *Ciona* research community, which is year-round access to gravid adults. Most research takes place in "land-locked" laboratories, and it is not always easy to find animals (particularly in winter months) that provide embryos for electroporation assays. To overcome this, centers where wild-type and stable transgenic lines are developed and maintained are under development in both Japan (Shimoda) and the USA (UC Santa Barbara). Such centers

will be critical for providing gravid animals year round and for distribution of stable transgenic lines. Particularly useful lines might include Gal4 UAS transgenics for expression of constructs in particular lineages and lines with GFP-tagged lineages for easy assessment of mutant phenotypes.

CONCLUSIONS

The *Ciona* system possesses virtually every modern analytical tool for the comprehensive determination of the genomic regulatory networks underlying development. We anticipate that some of the current deficiencies, such as access to gravid animals and the availability of inexpensive microarrays, will be remedied in the near future. It is easy to envision complete networks governing the specification, differentiation, and morphogenesis of key chordate tissues and organs such as the blood, heart, brain, nerve cord, and notochord. These networks should provide the foundation for unraveling the more elaborate genetic interactions used for the construction of homologous structures in vertebrates. In principle, it should be possible to integrate gene expression profiles, interlocking networks of regulatory genes and cell-signaling pathways to identify the on/off state of every gene in the *Ciona* genome in every blastomere at each stage in embryogenesis.

REFERENCES

Awazu, S., Sasaki, A., Matsuoka, T., Satoh, N., and Sasakura, Y. 2004. An enhancer trap in the ascidian *Ciona intestinalis* identifies enhancers of its Musashi orthologous gene. *Dev. Biol.* **275:** 459–472.

Azumi, K., Takahashi, H., Miki, Y., Fujie, M., Usami, T., Ishikawa, H., Kitayama, A., Satou, Y., Ueno, N., and Satoh, N. 2003. Construction of a cDNA microarray derived from the ascidian *Ciona intestinalis*. *Zool. Sci.* **20:** 1223–1229.

Bertrand, V., Hudson, C., Caillol, D., Popovici, C., and Lemaire, P. 2003. Neural tissue in ascidian embryos is induced by FGF9/16/20, acting via a combination of maternal GATA and Ets transcription factors. *Cell* **115:** 615–627.

Boffelli, D., Weer, C.V., Weng, L., Lewis, K.D., Shoukry, M.I., Pachter, L., Keys, D.N., and Rubin, E.M. 2004. Intraspecies sequence comparisons for annotating genomes. *Genome Res.* **14:** 2406–2411.

Canestro, C., Bassham, S., and Postlethwait, J.H. 2003. Seeing chordate evolution through the *Ciona* genome sequence. *Genome Biol.* **4:** 208.

Christiaen, L., Bourrat, F., and Joly, J.S. 2005. A modular *cis*-regulatory system controls isoform-specific pitx expression in ascidian stomodaeum. *Dev. Biol.* **277:** 557–566.

Corbo, J.C., Levine, M., and Zeller, R.W. 1997. Characterization of a notochord-specific enhancer from the Brachyury promoter region of the ascidian, *Ciona intestinalis*. *Development* **124:** 589–602.

Davidson, B., Shi, W., and Levine, M. 2005. Uncoupling heart cell specification and migration in the simple chordate *Ciona intestinalis*. *Development* **132:** 4811–4818.

Dehal, P., Satou, Y., Campbell, R.K., Chapman, J., Degnan, B., De Tomaso, A., Davidson, B., Di Gregorio, A., Gelpke, M., Goodstein, D.M., et al. 2002. The draft genome of *Ciona intestinalis*: Insights into chordate and vertebrate origins. *Science* **298:** 2157–2167.

Di Gregorio, A., Corbo, J.C., and Levine, M. 2001. The regulation of forkhead/HNF-3β expression in the *Ciona* embryo. *Dev. Biol.* **229:** 31–43.

Di Gregorio, A., Harland, R.M., Levine, M., and Casey, E.S. 2002. Tail morphogenesis in the ascidian, *Ciona intestinalis*, requires cooperation between notochord and muscle. *Dev. Biol.* **244:** 385–395.

Du Pasquier, L. 2004. Innate immunity in early chordates and the appearance of adaptive immunity. *CR Biol.* **327:** 591–601.

Fanelli, A., Lania, G., Spagnuolo, A., and Di Lauro, R. 2003. Interplay of negative and positive signals controls endoderm-specific expression of the ascidian Cititf1 gene promoter. *Dev. Biol.* **263:** 12–23.

Harafuji, N., Keys, D.N., and Levine, M. 2002. Genome-wide identification of tissue-specific enhancers in the *Ciona* tadpole. *Proc. Natl. Acad. Sci.* **99:** 6802–6805.

Hotta, K., Takahashi, H., Asakura, T., Saitoh, B., Takatori, N., Satou, Y., and Satoh, N. 2000. Characterization of Brachyury-downstream notochord genes in the *Ciona intestinalis* embryo. *Dev. Biol.* **224:** 69–80.

Hudson, C. and Lemaire, P. 2001. Induction of anterior neural fates in the ascidian *Ciona intestinalis*. *Mech. Dev.* **100:** 189–203.

Hudson, C., Darras, S., Caillol, D., Yasuo, H., and Lemaire, P. 2003. A conserved role for the MEK signalling pathway in neural tissue specification and posteriorisation in the invertebrate chordate, the ascidian *Ciona intestinalis*. *Development* **130:** 147–159.

Ikuta, T., Yoshida, N., Satoh, N., and Saiga, H. 2004. *Ciona* intestinalis *Hox* gene cluster: Its dispersed structure and residual colinear expression in development. *Proc. Natl. Acad. Sci.* **101:** 15118–15123.

Imai, K.S. 2003. Isolation and characterization of β-catenin downstream genes in early embryos of the ascidian *Ciona savignyi*. *Differentiation* **71:** 346–360.

Imai, K., Takada, N., Satoh, N., and Satou, Y. 2000. β-catenin mediates the specification of endoderm cells in ascidian embryos. *Development* **127:** 3009–3020.

Imai, K.S., Satoh, N., and Satou, Y. 2002. Early embryonic expression of FGF4/6/9 gene and its role in the induction of mesenchyme and notochord in *Ciona savignyi* embryos. *Development* **129:** 1729–1738.

Imai, K.S., Hino, K., Yagi, K., Satoh, N., and Satou, Y. 2004. Gene expression profiles of transcription factors and signaling molecules in the ascidian embryo: Towards a comprehensive understanding of gene networks. *Development* **131:** 4047–4058.

Jeffery, W.R., Strickler, A.G., and Yamamoto, Y. 2004. Migratory neural crest-like cells form body pigmentation in a urochordate embryo. *Nature* **431:** 696–699.

Jiang, D., Munro, E.M., and Smith, W.C. 2005. Ascidian prickle regulates both mediolateral and anterior-posterior cell polarity of notochord cells. *Curr. Biol.* **15:** 79–85.

Johnson, D.S., Davidson, B., Brown, C.D., Smith, W.C., and Sidow, A. 2004. Noncoding regulatory sequences of *Ciona* exhibit strong correspondence between evolutionary constraint and functional importance. *Genome Res.* **14:** 2448–2456.

Kasahara, M., Suzuki, T., and Pasquier, L.D. 2004. On the origins of the adaptive immune system: Novel insights from invertebrates and cold-blooded vertebrates. *Trends Immunol.* **25:** 105–111.

Keys, D.N., Levine, M., Harland, R.M., and Wallingford, J.B. 2002. Control of intercalation is cell-autonomous in the notochord of *Ciona intestinalis*. *Dev. Biol.* **246:** 329–340.

Kusakabe, T. 2005. Decoding *cis*-regulatory systems in ascidians. *Zool. Sci.* **22:** 129–146.

Matsuoka, T., Awazu, S., Shoguchi, E., Satoh, N., and Sasakura, Y. 2005. Germline transgenesis of the ascidian *Ciona intestinalis* by electroporation. *Genesis* **41:** 67–72.

Matthysse, A.G., Deschet, K., Williams, M., Marry, M., White, A.R., and Smith, W.C. 2004. A functional cellulose synthase from ascidian epidermis. *Proc. Natl. Acad. Sci.* **101:** 986–991.

Meinertzhagen, I.A., Lemaire, P., and Okamura, Y. 2004. The neurobiology of the ascidian tadpole larva: Recent developments in an ancient chordate. *Annu. Rev. Neurosci.* **27:** 453–485.

Moody, R., Davis, S.W., Cubas, F., and Smith, W.C. 1999. Isolation of developmental mutants of the ascidian *Ciona savignyi*. *Mol. Gen. Genet.* **262:** 199–206.

Nakashima, K., Yamada, L., Satou, Y., Azuma, J., and Satoh, N. 2004. The evolutionary origin of animal cellulose synthase. *Dev. Genes Evol.* **214:** 81–88.

Nishida, H. and Sawada, K. 2001. macho-1 encodes a localized mRNA in ascidian eggs that specifies muscle fate during embryogenesis. *Nature* **409:** 724–729.

Oda-Ishii, I., Bertrand, V., Matsuo, I., Lemaire, P., and Saiga, H. 2005. Making very similar embryos with divergent genomes: Conservation of regulatory mechanisms of Otx between the ascidians *Halocynthia roretzi* and *Ciona intestinalis*. *Development* **132:** 1663–1674.

Passamaneck, Y.J. and Di Gregorio, A. 2005. *Ciona intestinalis*: Chordate development made simple. *Dev. Dyn.* **233:** 1–19.

Sasakura, Y., Awazu, S., Chiba, S., and Satoh, N. 2003. Germ-line transgenesis of the Tc1/mariner superfamily transposon Minos in *Ciona intestinalis*. *Proc. Natl. Acad. Sci.* **100:** 7726–7730.

Satoh, N. 2003. The ascidian tadpole larva: Comparative molecular development and genomics. *Nat. Rev. Genet.* **4:** 285–295.

Satoh, N., Satou, Y., Davidson, B., and Levine, M. 2003. *Ciona intestinalis*: An emerging model for whole-genome analyses. *Trends Genet.* **19:** 376–381.

Satou, Y. and Satoh, N. 2003. Genomewide surveys of developmentally relevant genes in *Ciona intestinalis*. *Dev. Genes Evol.* **213:** 211–212.

Satou, Y., Imai, K.S., and Satoh, N. 2001. Early embryonic expression of a LIM-homeobox gene Cs-lhx3 is downstream of β-catenin and responsible for the endoderm differentiation in *Ciona savignyi* embryos. *Development* **128:** 3559–3570.

Satou, Y., Imai, K.S., and Satoh, N. 2002a. Fgf genes in the basal chordate *Ciona intestinalis*. *Dev. Genes Evol.* **212:** 432–438.

Satou, Y., Takatori, N., Fujiwara, S., Nishikata, T., Saiga, H., Kusakabe, T., Shin-i, T., Kohara, Y., and Satoh, N. 2002b. *Ciona intestinalis* cDNA projects: Expressed sequence tag analyses and gene expression profiles during embryogenesis. *Gene* **287:** 83–96.

Satou, Y., Imai, K.S., and Satoh, N. 2004. The ascidian Mesp gene specifies heart precursor cells. *Development* **131:** 2533–2541.

Sawada, K., Fukushima, Y., and Nishida, H. 2005. Macho-1 functions as transcriptional activator for muscle formation in embryos of the ascidian *Halocynthia roretzi*. *Gene Expr. Patterns* **5:** 429–437.

Shoguchi, E., Ikuta, T., Yoshizaki, F., Satou, Y., Satoh, N., Asano, K., Saiga, H., and Nishikata, T. 2004. Fluorescent in situ hybridization to ascidian chromosomes. *Zool. Sci.* **21:** 153–157.

Takahashi, T. and Holland, P.W. 2004. Amphioxus and ascidian Dmbx homeobox genes give clues to the vertebrate origins of midbrain development. *Development* **131:** 3285–3294.

Takahashi, H., Hotta, K., Erives, A., Di Gregorio, A., Zeller, R.W., Levine, M., and Satoh, N. 1999. Brachyury downstream notochord differentiation in the ascidian embryo. *Genes & Dev.* **13:** 1519–1523.

Vandenberghe, A.E., Meedel, T.H., and Hastings, K.E. 2001. mRNA 5′-leader *trans*-splicing in the chordates. *Genes & Dev.* **15:** 294–303.

Vinson, J.P., Jaffe, D.B., O'Neill, K., Karlsson, E.K., Stange-Thomann, N., Anderson, S., Mesirov, J.P., Satoh, N., Satou, Y., Nusbaum, C., et al. 2005. Assembly of polymorphic genomes: Algorithms and application to *Ciona savignyi*. *Genome Res.* **15:** 1127–1135.

Wada, S., Tokuoka, M., Shoguchi, E., Kobayashi, K., Di Gregorio, A., Spagnuolo, A., Branno, M., Kohara, Y., Rokhsar, D., Levine, M., et al. 2003. A genomewide survey of developmentally relevant genes in *Ciona intestinalis*. II. Genes for homeobox transcription factors. *Dev. Genes Evol.* **213:** 222–234.

Yagi, K., Satoh, N. and Satou, Y. 2004. Identification of downstream genes of the ascidian muscle determinant gene Ci-macho1. *Dev. Biol.* **274:** 478–489.

Yagi, K., Takatori, N., Satou, Y., and Satoh, N. 2005. Ci-Tbx6b and Ci-Tbx6c are key mediators of the maternal effect genes Ci-macho-1 in muscle cell differentiation in *Ciona intestinalis* embryos. *Dev. Biol.* **282:** 535–549.

9

Fish Genomics and Biology

Hugues Roest Crollius
Dyogen Lab
Centre National de la Recherche Scientifique UMR8541
Ecole Normale Supérieure
75005 Paris, France

Jean Weissenbach
Genoscope and Centre National de la
  Recherche Scientifique UMR8030
91057 Evry Cedex, France

"THERE THE NETS BROUGHT UP BEAUTIFUL SPECIMENS OF FISH: Some with azure fins and tails like gold, the flesh of which is unrivalled; some nearly destitute of scales, but of exquisite flavour; others, with bony jaws, and yellow-tinged gills, as good as bonitos; all fish that would be of use to us." While the gastronomic qualities of fish did not escape Jules Verne in his 1870 *20,000 Leagues Under the Sea*, fish are no less put to good use in twenty-first century biology. In this new context, one could easily replace fin color and flesh quality by genome size and embryo transparency in a similar enumeration of the advantages of these animals for biology in general and molecular genetics in particular. If Captain Nemo was in a position to offer such variety on his menu, it is partly because fish comprise more than 25,000 species, by far the most successful vertebrate group. Indeed few aquatic ecosystems have eluded colonization by at least some fish species, from Tibetan streams to the abyss of the oceans via sub-zero Antarctic seas (Nelson 1994). Of these species, many have long been used as models in different disciplines of biology (Fig. 1) because of this very diversity: The atrophy or exaggeration of important anatomical or physiological functions occur with sufficient frequency to

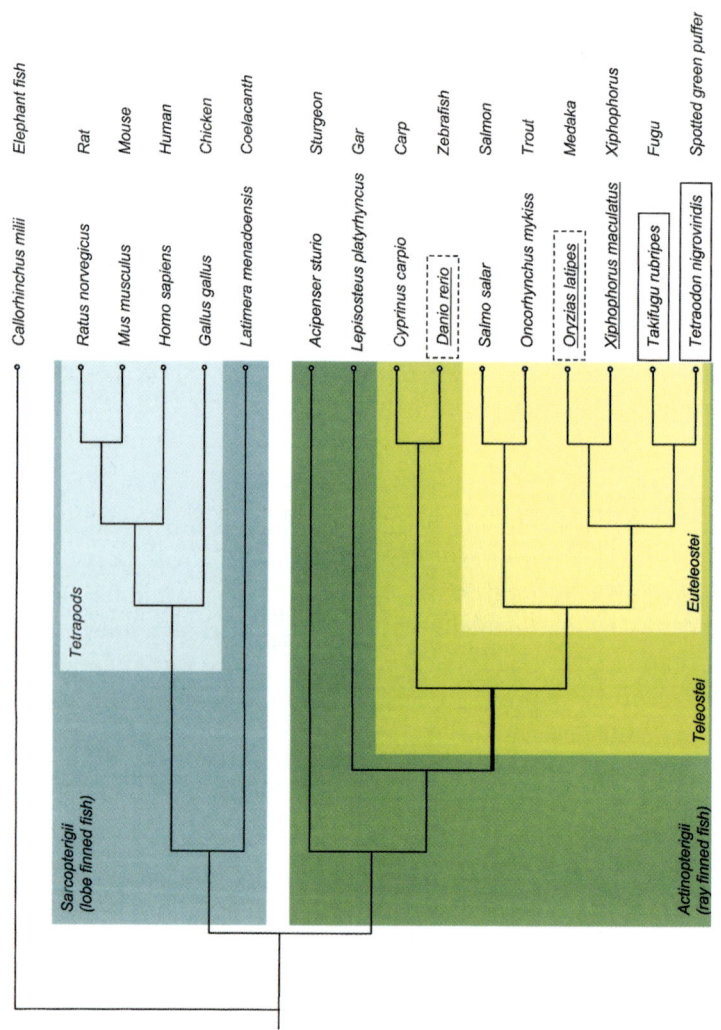

Figure 1. Consensus phylogenetic relationships between fish (simplified from Nelson 1994; Inoue et al. 2003) and tetrapods, including fish considered as biological models (underlined), and those for which a sequence has been published (boxed, continuous line) or is underway (boxed, dashed line). The thick branch indicates the most likely position of the whole-genome duplication at the root of the teleosts, based on Hoegg et al. (2004).

have attracted biologists to fish models (Epstein and Epstein 2005). This includes molecular genetics and genome research, for which fish also possess interesting and outstanding features, if not all-time records, among vertebrates.

About 30 years ago, a popular tropical aquarium fish named *Danio rerio* (zebrafish) was already seen as endowed with many advantages for genetic analysis: a short generation time (about 3 mo), large egg clutches all year round, easy maintenance, and external development of a transparent embryo (Streisinger et al. 1981). Combined with large-scale mutagenesis screens initiated in the early 1990s (Haffter et al. 1996; Stainier et al. 1996), zebrafish filled a gaping hole in vertebrate developmental biology: the ability to study genes via their mutant phenotypes on a large scale as in *Drosophila melanogaster* or *Caenorhabditis elegans*. Since then, capital discoveries for our understanding of vertebrate development and human disease have already emerged from zebrafish studies. However if one considers genome analysis a question of DNA sequence acquisition and "mining," then fishes really became a major player in 1993 when Sydney Brenner suggested a new species as a genome model, the marine pufferfish *Takifugu rubripes* (*fugu*) (Brenner et al. 1993). Aside from its gastronomic delicacy status in Japan and China, *fugu* possess one of the smallest vertebrate genomes. This feature, already recorded for its freshwater relative *Tetraodon nigroviridis* in 1968 (Hinegardner 1968), is a major advantage to rapidly gain access to a large catalog of genes in a vertebrate at a cost comparatively smaller than for the much larger genome of a mammalian species. However, both pufferfish are species for which we know little in terms of physiology, reproduction, or life cycle. Since biology is still a science largely driven by the quality and depth of the experimental data and our ability to extract meaning from it, the outcome of both pufferfish genome programs has until recently been confined to discoveries on the structure and evolution of genes and genomes, with few connections to development, cell biology, or physiology.

However, this is about to change, with the emergence of networks of scientists and cross-disciplinary platforms where precise biological questions are examined with an array of tools and resources that include genome sequences and genomic techniques. Here we review the impact of the genome sequence for those fish species for which it is already available, and we examine how the combination of genomics with more traditional disciplines might pave the way for a much wider impact on biology.

ANGLING IN THE GENOMIC AQUARIUM

The sequence of the human genome was still a distant goal in 1993, but the project of sequencing the entire genome of a multicellular eukaryote, that of the nematode worm *C. elegans*, was well on its track (Sulston et al. 1992). As it happens, many important eukaryotic model organisms that were already being studied with molecular biology approaches had genome sizes within the reach of sequencing technologies of the time: *Saccharomyces cerevisiae* (14 Mb), *D. melanogaster* (180 Mb), *C. elegans* (100 Mb), and *Arabidopsis thaliana* (125 Mb). But none of these were vertebrates, a situation which motivated a pilot project to evaluate the usefulness of *fugu* as a model vertebrate genome (Brenner et al. 1993). This influential analysis showed that the genome was indeed about 400 Mb, or eight times smaller than the human or mouse genomes. Just as important, exon–intron boundaries seemed conserved, suggesting that gene structures were likely to be very similar, albeit in a much more compact sequence. Rapidly, other studies strengthened the notion that pufferfish DNA could help identify and better understand the structure and sometimes the function of mammalian sequences.

In one such early attempt, a *fugu* sequence next to the *Hoxb4a* gene and conserved with mouse, showed enhancer activity in transgenic mice (Aparicio et al. 1995). The power of comparative genomics in vertebrates was also soon illustrated when the complete Huntington disease (HD) gene from *fugu* was sequenced and compared to its human ortholog (Baxendale et al. 1995). The analysis showed that the *fugu* HD gene possesses a four-glutamine repeat, whereas mouse has seven and healthy humans a minimal of eight. Because a tract of four glutamines is unlikely to form a functional site in itself (such as a polar zipper, Perutz et al. 1994), the *fugu* protein supported the view that it is only after expanding to over 37 residues in HD patients that it somehow gained a new pathogenic function. This initial sequencing project in *fugu* was followed by many more that generally showed limited long-range conservation of gene order, rarely more than four genes per conserved synteny block, thus dampening down the initial hope that this compact genome was a valuable tool to accelerate the mapping of human genes, then a priority in the Human Genome Project (Gilley et al. 1997). However there were still major reasons to establish complete sequences of fish genomes: (1) these early studies had shown the usefulness of comparative sequencing when constructing gene models on the human genome, (2) the initiation of large scale mutagenesis projects on zebrafish was calling for a global genome effort already in preparation with the construction of genetic maps (Shimoda et al. 1999; Kelly et al. 2000) and radiation hybrid maps (Geisler et al. 1999; Hukriede et al. 2001).

TRAWLING FOR WHOLE GENOMES

Reevaluating the Number of Human Genes

Fugu is a relatively large marine fish that contains elevated doses of tetrodotoxin causing live specimen or frozen samples to be the subject of restrictive importation laws in most countries outside of Asia and thus posing practical problems for genomic analyses. A different pufferfish, *Tetraodon nigroviridis* (Green spotted puffer, sometimes confused in the aquarium fish market with *Tetraodon fluviatilis*, which is a different species) was proposed (Crnogorac-Jurcevic et al. 1997) that alleviates this restriction: *Tetraodon* also possesses a small compact genome (Hinegardner 1968) but it is a popular aquarium fish that can live in freshwater (Ebert 2001). In contrast to *fugu*, few specific *Tetraodon* loci were sequenced and studied in comparison to their homologs in other species. From the beginning instead, *Tetraodon* genomic DNA was exploited in large-scale comparisons between different vertebrate genomes. The initial rationale behind this second pufferfish project was to assist in the annotation of human genes, a slow and fastidious task when performed by humans, often unreliable when performed by automatic approaches, and yet one of the primary goals of the Human Genome Project. On the basis of an initial sampling of random sequences from the *Tetraodon* genome (about 30%), a tool named Exofish that was based on BLAST sequence alignments was developed to identify conserved regions in human genomic DNA that correspond to coding exons, rapidly and with high specificity (Roest Crollius et al. 2000). A surprising outcome of this first example of a global sequence comparison between large samples of two vertebrate genomes is that the number of conserved sequences identified by Exofish in the human genome was not compatible with the 60,000 to 150,000 genes that it was thought to possess at the time. Indeed comparisons with the *Tetraodon* sequence sample indicated that the entire human genome would contain about 88,000 evolutionary conserved regions (termed ecores) corresponding to human exons, while known human genes possessed on average between 2.6 and 3.2 ecores. A simple ratio between these figures yields a total of about 30,000 human genes. This new estimate, confirmed later by the initial analysis of the human genome sequence (Lander et al. 2001), challenged the notion that the complexity of genetic information contained in a genome is a function of the number of protein-coding genes.

Puffer Fish Genome Features in Draft Sequences

The *fugu* genome, the first vertebrate genome to be sequenced after human, was obtained using the whole-genome shotgun method (Aparicio et al. 2002). This sequence draft enabled a number of interesting observations, such as differences in specific protein families between human and *fugu*. The *Tetraodon* genome sequence was subsequently produced (Jaillon et al. 2004), also with the whole-genome shotgun method albeit with a higher redundancy in sequence reads (8.3 vs. 5.6). Both pufferfish possess about 70 different families of transposable elements against only 20 for human or mouse, but in pufferfish they comprise two to three orders of magnitude fewer copies. Interestingly in *Tetraodon*, SINE and LINE families are distributed in opposite regions of the genome compared to human or mouse: SINEs are more abundant in G + C-rich sequences in mammals, and in A + T-rich regions in *Tetraodon*, and vice versa for LINE elements. More surprisingly, these initial studies of *Tetraodon* and *fugu* showed a number of differences in their genomes. For instance a G + C-rich region present in both *Tetraodon* and mammal genomes is absent in *fugu*. Also some gene families such as type I cytokines and their receptors, present in all vertebrates studied so far, were notably difficult to find in *fugu*, while over 30 members of the family could be identified in *Tetraodon*. These discrepancies are most likely attributable to biases in clone libraries or differences in methodologies, and hopefully should be resolved as the genomes reach completion. When comparing fish and mammal gene catalogs, surprisingly few major differences could be documented when using the Gene Ontology (Harris et al. 2004) classification system. More striking differences could be seen using protein domain comparisons: Proteins involved in sodium transport are more abundant in fish, which also contain an allantoin pathway for purine degradation that is absent in humans. Neutral nucleotidic sequence evolution per year was found to be twice as fast in pufferfish as between human and mouse, and protein evolution also appears to proceed at a faster rate in fish, although the reasons for this are still unclear. It should be noted that these results depend on the dating of the divergence between *Tetraodon* and *fugu* (18–30 Mya) (Crnogorac-Jurcevic et al. 1997).

Insights into Vertebrate Genome Evolution

Perhaps one of the major differences in the two pufferfish draft sequences resides in the fact that the *fugu* genome sequence was assembled purely by the whole-genome shotgun method with no physical mapping, whereas most (64%) of the *Tetraodon* genomic sequence is anchored on each of the chromosomes, providing a long-range view of gene organization in the genome. This added information primarily resolved a long-standing issue on the occurrence of a whole-genome duplication in the fish lineage.

Remarkably, the idea that an increase in chromosome numbers may be a source of phenotypic novelty is nearly a century old. In 1911, Kuwada already observed that some varieties of maize were tetraploid and suggested that this may be the source of "innumerable races" (Taylor and Raes 2004). During the following 60 years, the occurrences and consequences of gene and genome duplications continued to be discussed (Taylor and Raes 2004), with for instance the proposal by Stephens in 1951 that increasing the number of genetic loci was the only path to "evolutionary progress" and his suggestion that genome duplication could be one way of achieving this (Stephens 1951). These theories reached a high point in 1970 with the publication of Susumo Ohno's book (Ohno 1970) that stated several landmark notions: (1) without duplicated genes, the emergence of metazoans, vertebrates, and mammals from unicellular organisms would have been impossible, (2) this process required the creation of new loci with previously nonexistent functions, and (3) he postulated that at least one whole-genome duplication facilitated the evolution of vertebrates. While these ideas were met with mild enthusiasm at the time (Lewin 1971; Spofford 1972), it is today a widely accepted notion that to create functional novelty, gene duplications are at least as important as point mutations in individual loci. Moreover when the duplication affects the entire genome at once, this potential for novelty is theoretically amplified by allowing the duplication and retention of partial or complete metabolic pathways.

However, the fate of most duplicate copies of genes, over tens or hundreds of million years, is to be eliminated from the genome in a global process called diploidization (Wolfe 2001; Jaillon et al. 2004; Kellis et al. 2004) (Fig. 2). The fact that most duplicate copies of genes have long disappeared from anciently duplicated genomes is one reason why providing proof of the duplication is often difficult. In the case of fish, a strong indication in support of the duplication came from the revelation that zebrafish possess seven HOX clusters on seven different chromosomes, instead of the four clusters found in mammals (Amores et al. 1998). This observation immediately suggested that these four chromosomes at least—but most likely the entire genome—duplicated once in an ancient teleost, followed by the loss of one HOX cluster in the zebrafish lineage. This conclusion was sustained by examples in other fish

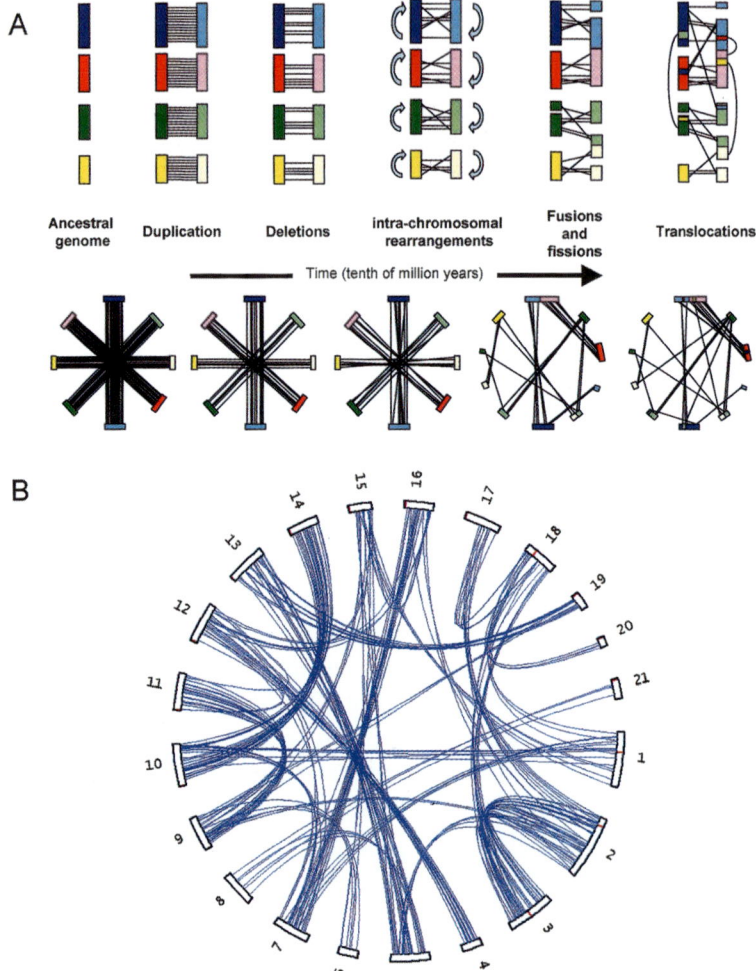

Figure 2. (*A*) A schematic model of whole-genome duplication with four chromosomes, followed by massive gene loss, chromosome fusions and fissions, inter- and intrachromosomal rearrangements. Each colored rectangle is a chromosome, and lines are drawn between duplicate copies of genes present on sister chromosomes or chromosome segments. The *top* panel starts with the sister chromosomes facing each other and illustrates the changes induced by chromosome rearrangements, while the *lower* panel shows the situation in a circular representation, which assumes one does not know the relationships between chromosomes a priori. After several million years of evolution, the distribution of the few duplicate genes that remain do not bear a trace of the ancient duplication event. (*B*) The same representation as in the *lower* panel in (*A*), but with real data from the *Tetraodon* genome: Despite more than 300 million years since the duplication, the distribution of about 2% of *Tetraodon* genes that remain strictly in two copies (joined by blue lines) in the genome shows a striking pattern where chromosomes are associated in pairs (e.g., chromosome 9 and 11, or 10 and 14), or sometimes in triplets (e.g., chromosomes 5, 13, and 19). The former suggests that no interchromosomal rearrangements have occurred on these chromosome pairs since the duplication, while the latter is reminiscent of a chromosome fusion or fission.

species (Meyer and Schartl 1999) but stronger support came from comparative analyses in zebrafish using many more gene loci (Postlethwait et al. 2000) as well as results from Expressed Sequence Tags (ESTs) positioned on the zebrafish genetic linkage map (Woods et al. 2000), where it was clear that this genome contained large duplicated segments that did not exist in human or mouse. However, the possibility that the frequent occurrences of duplicated genes in fish originated from a high level of segmental or local duplications could not be entirely ruled out (Robinson-Rechavi et al. 2001). These studies were followed by several attempts at dating the emergence of duplicate gene copies in *fugu* (Christoffels et al. 2004; Vandepoele et al. 2004) but with a relatively large uncertainty.

The ultimate demonstration rested on the long-range continuity of the *Tetraodon* sequence assembly covering parts of all chromosomes (Jaillon et al. 2004). Although the paralogs (pairs of genes that appear by duplication of an ancestral gene) that were identified using very conservative criteria represent less than 2% of the current set of genes, their distribution in the genome clearly associates chromosomes in pairs or in triplets, a situation expected if a single tetraploidization event took place at some point in the past, followed by a few chromosome fissions or fusions (Fig. 2). A second and even more telling signature of the duplication was found in the alternating pattern of *Tetraodon* syntenic groups along human chromosomes. Indeed, the return to a diploid state after the whole-genome duplication mostly affects genes, not necessarily chromosomes. While the two duplicated sister chromosomes remain an integral part of the genome, they each gradually loose about 50% of their genes, in small clusters that alternate between the two (Fig. 3A). To observe the result of this process, one needs to compare the position of genes on a chromosome of a species that did not duplicate during the same period of time, with orthologous genes of the species that did duplicate. The characteristic pattern that results seems quite universal, from single-cell eukaryotes (Kellis et al. 2004) to vertebrates (Jaillon et al. 2004), and is illustrated in Figure 3B. The deletion process takes place because the supernumerary gene copies are not under selection and they thus rapidly acquire deleterious mutations. The choice of which copy will be deleted is driven by the first mutation to occur in one of the two copies, a process that must be random since ultimately each sister chromosome inherits about half of the initial gene complement.

However, some duplicate copies are not deleted, in this case implying either that both copies are immediately placed under selection, that a deleterious mutation partially obliterates the function of one copy

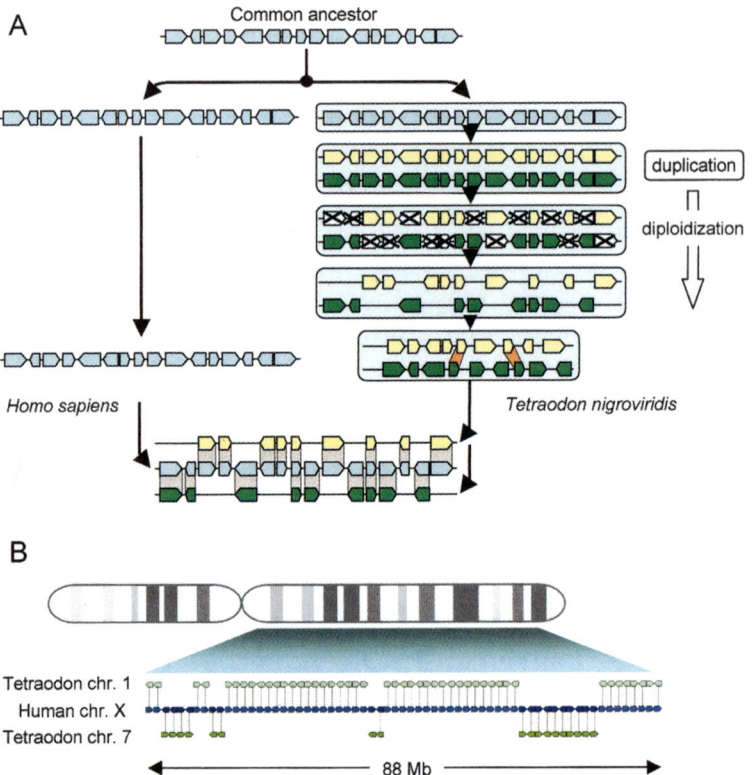

Figure 3. (A) Duplication leads to double-conserved synteny. After speciation, a chromosome or chromosome segment from an ancestral species is duplicated in one lineage but not in the other. In the former (e.g., Tetraodon), supernumerary copies of genes are progressively deleted from each of the duplicated segments in approximately equal proportion (diploidization). Ultimately the two duplicated chromosomes only contain 50% of the initial gene complement and are thus very different from each other. The difficulty of finding the original pair of sister chromosomes can be alleviated by a comparison with a genome that originates from the same pre-duplication ancestor, but that did not duplicate (e.g., human). The nonduplicated chromosome segment should contain genes with orthologs alternating between the two duplicated chromosomes (adapted from Kellis et al. 2004). (B) Example of double-conserved synteny. An 88-Mb region covering the majority of the long arm of human chromosome X contains 65 genes with orthologs in the Tetraodon genome, alternating between chromosome 1 and chromosome 7. Genes are represented by small arrows that indicate the orientation of transcription.

making the other gene essential, or that the first mutation is not deleterious but favorable. Because it is a genome-wide process of gene selection affecting all functional classes at the same time, it is of interest to investigate, in the case of fish, which classes have emerged as advantageous compared to genomes that did not duplicate. Ideally, the question would be best addressed using one of the living fish species that diverged just before the whole-genome duplication, such as sturgeons (acipenseriformes) or gar (semionotiformes) (Hoegg et al. 2004) (Fig. 1). Until the genome sequence of such a species becomes available, a comparison with the gene catalogs of much more distant species such as mammals using Gene Ontology annotations (Harris et al. 2004) and phylogenetic classifications reveals significant differences: Development, cell differentiation and cell communication classes are enriched in gene duplicates (F.G. Brunet, H. Roest Crollius, M. Paris, J.M. Aury, P. Gibert, O. Jaillon, V. Laudet, and M. Robinson-Rechavi, in prep.). This result is interesting in the light of the teleost radiation that took place approximately at the time of the genome duplication. Also strikingly and as previously shown in yeast and nematode (Davis and Petrov 2004), fish genes that evolve slowly prior to the duplication seem to be preferentially retained in two copies after the duplication: but what of the rate of evolution between the two retained copies? Whereas the subfunctionalization model (Force et al. 1999) predicts that the two copies will share the ancestral function and thus evolve under the same constraints, the neofunctionalization model (Ohno 1970) proposes that the emergence of a new function in one copy occurs under positive selection, i.e., one copy will evolve faster than the other. Interestingly, in the above-mentioned study, *Tetraodon* gene duplicates recurrently show a markedly accelerated evolution of one of the two copies, in agreement with the latter model and supporting the view that genome (and hence gene) duplication is a driving force behind the emergence of functional novelty. The ancestral teleost genome duplication provided its owner with a powerful toolkit to adapt and diversify: twice as many genes as any other emerging vertebrate. It is tempting to propose that a consequence of this genome doubling is to be found in the rich diversity of extant fish species, unparalleled today among vertebrates (Amores et al. 1998; Meyer and Schartl 1999). Ultimately, the availability of more fish genome sequences will help distinguish between teleost-wide and lineage-specific strategies for the retention of beneficial duplicate functional classes. But for now, the teleost genome duplication provides a direct entry point into another exciting theme: the reconstruction of the ancestral genome prior to the duplication, which would closely resemble that of the ancestral bony vertebrate genome. Studies in zebrafish

(Postlethwait et al. 2000), medaka (Naruse et al. 2004), and Tetraodon (Jaillon et al. 2004) have already delineated the probable 12 protochromosomes with increasing precision.

Zebrafish and Medaka: Biological Models Come Back

The zebrafish as a model system has accompanied the development of molecular biology from the 1960s to the genomic revolution of the 1990s (Grunwald and Eisen 2002). Following the publication of the results from two large-scale genetic screens in 1996 in a special issue of *Development* (1996;123:1–461), zebrafish was propelled at the forefront of developmental biology research. While such screens are currently being expanded to tackle a wider range of mutations in embryonic and adult stages (see below), many researchers are also studying zebrafish models of human diseases (Dooley and Zon 2000) including blood diseases (de Jong and Zon 2005), heart disorders (MacRae and Fishman 2002), and cancer (Amatruda et al. 2002). As mentioned above, genome research on zebrafish started in parallel to these developments. Early efforts to map genes and genetic markers served to identify the loci of specific mutant genes, and to compare the organization of genes with other vertebrates (Barbazuk et al. 2000; Woods et al. 2000). Microarray technology has also been applied, in combination with single nucleotide polymorphism (SNP) identification, to accelerate the linking of mutant phenotypes to their genotypes (Stickney et al. 2002). The zebrafish genome is about 1700 Mb, and that of the Tübingen strain is now being sequenced at the Sanger Institute using a mixed strategy where both large insert clones and the entire genome are submitted to the whole-genome shotgun method. Today the zebrafish community worldwide has expanded tremendously with over 3500 scientists registered in the ZFIN community database (zfin.org). The availability of the genome sequence will provide a unifying reference to integrate the wealth of functional data accumulated so far.

The Medaka (*Oryza latipes*) has also long been used as a model in genetics, dating back to the beginning of last century (Wittbrodt et al. 2002). While in many respects it matches zebrafish in its advantages as a laboratory model, Medaka possesses several characteristics of its own as a useful biological model such as the existence of several fertile inbred strains and embryonic stem cells that can be stably cultured long enough to enable genetic manipulation (Hong et al. 1996), although their use is currently limited by the fact that they do not contribute to the germ line. The complementarity between medaka and zebrafish is obvious when considering the numerous mutations of the same locus that affect each fish differently or even uniquely, thus enabling the deciphering of species-specific patterns for these mutations (Furutani-Seiki and Wittbrodt 2004). Interestingly, Medaka is the only fish so far for which a single genetic locus, *DMY*, has been found to be responsible for sex determination, as in mammals (Matsuda et al. 2002; Nanda et al. 2002). While *DMY* is a recent invention in the Medaka lineage (Lutfalla et al. 2003; Volff et al. 2003) and thus was not found in other model fish species (Kondo et al. 2003), it provides the first opportunity to study the molecular basis of sex determination in fish. Many resources for genome research have been developed for Medaka, including genetic maps (Naruse et al. 2004), EST sequences (Kimura et al. 2004), and physical maps (Khorasani et al. 2004). The 800-Mb genome of Medaka has now been sequenced and assembled, and the analysis that is under way should reveal exciting insights into vertebrate genome evolution.

THE FUTURE OF FISH GENOMICS

The future of fish genomics is bright, and this prediction is sustained below by three examples where specific characteristics of fish species were successfully exploited to gain penetrating insights in a broad range of subjects.

Fish, as a highly diversified group of the vertebrate family, experience an astonishing range of environmental conditions to which their physiologies, body shapes, and lifestyles have adapted. However, a common denominator of all fish species is their aquatic habitat, meaning that water is in direct contact with several tissues and internal compartments of the animal, potentially inducing a high sensitivity to water-borne parameters such as temperature, oxygen levels, salinity, and sometimes toxic chemicals. This intimate relationship between an organism and a wide range of different environments has recently prompted the view that fish could be used as models for "environmental genomics" (Cossins and Crawford 2005), in other words the study of the interface between an organism and its environment using genomic approaches. This concept had recently been illustrated by an elegant study involving the exposure of carp to increasing levels of cold, from 30°C to 10°C and measuring the change in the level of expression of several thousand genes using microarray techniques (Gracey et al. 2004). Carp, as most fish, is poikilothermic (cold-blooded), which implies that its body temperature follows that of the water it is immersed in. In the seven tissues that were monitored, the decrease in temperature was correlated with a graded increase in expression of a core set of 252 genes predominantly from transcriptional regulation, RNA splicing, and translation systems, while very few genes common to all tissues showed a decrease in expression. Conversely, tissue specific responses showed that brain modifies its glycolytic activity while a switch to lipid metabolism is observed in liver. Interestingly this study was performed in a nonmodel species, exemplifying the use of a specific species of fish with a specific feature of interest, in this case the carp which tolerates a wide range of temperature, to investigate a whole body physiological adaptation to a change in environment. While the results are useful to better understand the molecular basis of the response to cold in carp, fish share so many aspects of their developmental pathways, physiological mechanisms, and organ systems with mammals that these results are also relevant to human physiology.

The second illustration of the strength of future genomic research using fish species is its recognition by international funding agencies. One such recent example is the major impetus to zebrafish genomics provided by the European Union Framework Programme for Research. By funding the ZF-MODEL consortium (www.zf-models.org), an Integrated Project devoted to the study of zebrafish models for human development and disease, the project links about 30 groups from seven European countries (Bradbury 2004). The research plan is to take advantage of the zebrafish model in combination with an array of large-scale techniques to identify zebrafish genes that replicate to some degree human pathologies, or are involved in developmental pathways similar to ours. The scale of this endeavor can be measured by the thousands of mutants already generated by the consortium members using either forward genetics (for instance with chemical mutagenesis screens), or reverse genetics (for instance using the TILLING technique; McCallum et al. 2000) and the scope of the screens used to analyze them. Designed to identify mutants that resemble human disorders, these screens are also carried out on adult fish and encompass for instance bone malformation, skin development, eye movement, or addiction behavior. Combined with enhancer detection techniques using green fluorescent protein (GFP) transgenic lines and expression profiling with microarray analysis, the consortium will integrate a broad range of expertise to tackle the function of the mutated genes, which hopefully will lead to a better understanding of corresponding human disorders.

In addition to counting several member species elevated to the status of genomic models such as zebrafish or pufferfish, fish also represent a major source of food for humans. The European Union thus also recognized the need for improving aquaculture research by funding the AQUAFIRST consortium to identify genes associated with stress and disease resistance in sea bream, sea bass, and rainbow trout in order to provide a physiological and genetic basis for marker-assisted selective breeding. Genomic techniques and sequence comparisons with genomic models are cornerstones of this project, which illustrates how fish genomics may grow in coming years outside of fundamental research labs toward more applied objectives nonetheless essential to human welfare.

A final example of the use of fish genomes that may expand in the near future rests on the phylogenetic position of fish species in comparison to mammals (Fig. 1). Today a great deal of attention is focused on the part of eukaryote genomes that do not code for proteins but are nevertheless functional. Elements that may carry out specific functions in these regions include noncoding RNA genes and regulatory control regions. One of the most powerful techniques to identify such elements is to align genomic sequences of distantly related organisms and look for regions that have remained similar during evolution, thus suggesting that a functional constraint is acting to preserve the sequence from mutations. The advantage of fish in this context is the long evolutionary distance, approximately 450 million years, since their last ancestor with mammals. Neutral mutations have since saturated the genome to a point where any conserved region between for instance human and pufferfish is indicative of a functional constraint. This comparative approach was first applied on a genome scale to identify coding exons (Roest Crollius et al. 2000), and more recently to identify ultra-conserved regions (UCRs) of unknown function (Sandelin et al. 2004; Woolfe et al. 2004). However an additional assumption that can be deduced from the discovery of UCRs conserved between such distant organisms is their fundamental importance across vertebrates. In line with this, UCRs found in this way lie in clusters around genes involved in the regulation of development. Indeed when the orthologous regions were assayed using GFP reporter constructs in zebrafish embryos, most showed significant enhancer activity in one or more tissues (Woolfe et al. 2004). So fish–mammal sequence alignments not only provide the means to identify functional elements, they also act as a screen to select those elements essential to vertebrates. With the production of new fish genome sequences as well as new mammalian sequences, such comparative studies are likely to play an important role in guiding the identification of functional noncoding elements, and in deciphering the subtle sequence variations that might lead to phenotypic changes (Ahituv et al. 2004; Boffelli et al. 2004).

Many fish species are routinely being studied at the molecular level and even at the genomic level and have not been cited here. For instance thousands of EST sequences are available for carp, catfish, salmon, trout, killifish, stickleback, or tilapia and large insert BAC libraries are also available for several of these species, further illustrating the widespread interests in using fish for genomic research. Four fish genome sequences are or will be available soon: *fugu*, *Tetraodon*, medaka, and zebrafish. The stickleback genome (*Gasterosteus aculeatus*) is also well advanced (Table 1), but to our knowledge no other is currently ongoing beyond these, although obviously several interest groups are actively working towards promoting certain species. Research on salmon and trout for instance would greatly benefit from the availability of the genome sequence. Indeed more is known about the physiology and biology of rainbow trout than any other fish species, although genomic sequencing could be complicated by an additional genome duplication in the salmonid lineage some 25–100 Mya (Allendorf and Thorgaard 1984). The sarcopterygian fish coelacanth is the nearest living relative of tetrapods (Gorr et al. 1991) (Fig. 1) and thus escaped the whole-genome duplication that affected the teleosts. This may be a serious advantage since it might provide access to a genome that resembles the early tetrapod genome, unaffected by the consequence of massive gene duplications such as gene conversion. Altogether the coelacanth would indeed be an excellent candidate for genome sequencing as it would provide a reference genome for tetrapods while allowing the identification of genomic features that differentiate them from teleosts (Noonan et al. 2004). For similar reasons but across a wider evolutionary scale, the elephant fish (*Callorhinchus milii*, a cartilaginous fish; Fig. 1) has recently been proposed as a good model to study the genome structure and gene content of a basal jawed vertebrate, and provide a common reference for tetrapods and ray-finned fishes (Venkatesh et al. 2005).

Up to now fish genomics has been able to draw on the similarities and the differences between mammalian and fish genomes to gain profound insights into the evolution of vertebrate genomes in general, and into the function of individual genes often associated with human disorders in particular. Captain Nemo's fishing exploits with the *Nautilus* may be hard to match but the net cast by genome scientists has also reeled in some unexpected surprises, and the end of this story is certainly a long way off.

Table 1. Summary of current genomic resources on fish species

	Genome projects	Genome size (Mb)	Assembled genome sequence coverage	Shotgun reads	ESTs or cDNAs
Zebrafish	http://www.sanger.ac.uk/Projects/D_rerio/	1700	6.5–7	24,535,919	673078
Tetraodon	http//www.genoscope.cns.fr/tetraodon	350	8.3	2,975,798	99204[a]
Fugu	http://genome.jgi-psf.org/Takru4/Takru4.home.html	380	8.7	3,638,510	25860
Medaka	http://dolphin.lab.nig.ac.jp/medaka/	800	9.0	15,171,833	—
Salmon	http://www.salmongenome.no	3100	—	22,677	114911
Stickleback	http://www.genome.gov/12512292	675	—	6,959,213	171001
Trout	NA	2700	—	47,051	231820

[a]*Tetraodon* cDNA sequences are full-length insert sequences; sequences from other species are ESTs.
NA: Not available.

REFERENCES

Ahituv, N., Rubin, E.M., and Nobrega, M.A. 2004. Exploiting human—fish genome comparisons for deciphering gene regulation. *Hum. Mol. Genet.* **13 Spec No 2**: R261–R266.

Allendorf, F.W. and Thorgaard, G.H. 1984. Tetraploidy and the evolution of Salmonid fishes. In *Evolutionary genetics of fishes* (ed. B.J. Turner), pp. 1–53. Plenum Publishing Corporation, New York.

Amatruda, J.F., Shepard, J.L., Stern, H.M., and Zon, L.I. 2002. Zebrafish as a cancer model system. *Cancer Cell* **1**: 229–231.

Amores, A., Force, A., Yan, Y.L., Joly, L., Amemiya, C., Fritz, A., Ho, R.K., Langeland, J., Prince, V., Wang, Y.L., et al. 1998. Zebrafish hox clusters and vertebrate genome evolution. *Science* **282**: 1711–1714.

Aparicio, S., Morrison, A., Gould, A., Gilthorpe, J., Chaudhuri, C., Rigby, P., Krumlauf, R., and Brenner, S. 1995. Detecting conserved regulatory elements with the model genome of the Japanese puffer fish, *fugu* rubripes. *Proc. Natl. Acad. Sci.* **92**: 1684–1688.

Aparicio, S., Chapman, J., Stupka, E., Putnam, N., Chia, J.M., Dehal, P., Christoffels, A., Rash, S., Hoon, S., Smit, A.F., et al. 2002. Whole-genome shotgun assembly and analysis of the genome of *Fugu rubripes*. *Science* **25**: 25.

Barbazuk, W.B., Korf, I., Kadavi, C., Heyen, J., Tate, S., Wun, E., Bedell, J.A., McPherson, J.D., and Johnson, S.L. 2000. The syntenic relationship of the zebrafish and human genomes. *Genome Res.* **10**: 1351–1358.

Baxendale, S., Abdulla, S., Elgar, G., Buck, D., Berks, M., Micklem, G., Durbin, R., Bates, G., Brenner, S., and Beck, S. 1995. Comparative sequence analysis of the human and pufferfish Huntington's disease genes [see comments]. *Nat. Genet.* **10**: 67–76.

Boffelli, D., Nobrega, M.A., and Rubin, E.M. 2004. Comparative genomics at the vertebrate extremes. *Nat. Rev. Genet.* **5**: 456–465.

Bradbury, J. 2004. Small fish, big science. *PLoS Biol.* **2**: E148.

Brenner, S., Elgar, G., Sandford, R., Macrae, A., Venkatesh, B., and Aparicio, S. 1993. Characterization of the pufferfish (*Fugu*) genome as a compact model vertebrate genome. *Nature* **366**: 265–268.

Christoffels, A., Koh, E.G., Chia, J.M., Brenner, S., Aparicio, S., and Venkatesh, B. 2004. *Fugu* genome analysis provides evidence for a whole-genome duplication early during the evolution of ray-finned fishes. *Mol. Biol. Evol.* **21**: 1146–1151.

Cossins, A.R. and Crawford, D.L. 2005. Fish as models for environmental genomics. *Nat. Rev. Genet.* **6**: 324–333.

Crnogorac-Jurcevic, T., Brown, J.R., Lehrach, H., and Schalkwyk, L.C. 1997. *Tetraodon fluviatilis*, a new puffer fish model for genome studies. *Genomics* **41**: 177–184.

Davis, J.C. and Petrov, D.A. 2004. Preferential duplication of conserved proteins in eukaryotic genomes. *PLoS Biol.* **2**: E55.

de Jong, J.L. and Zon, L.I. 2005. Use of the zebrafish to study primitive and definitive hematopoiesis. *Annu. Rev. Genet.* **39**: 481–501.

Dooley, K. and Zon, L.I. 2000. Zebrafish: A model system for the study of human disease. *Curr. Opin. Genet. Dev.* **10**: 252–256.

Ebert, E. 2001. *Aqualog: The puffers of fresh and brackish waters*. Aqualog verlag GmbH, Rodgau, Germany.

Epstein, F.H. and Epstein, J.A. 2005. A perspective on the value of aquatic models in biomedical research. *Exp. Biol. Med. (Maywood)* **230**: 1–7.

Force, A., Lynch, M., Pickett, F.B., Amores, A., Yan, Y.L., and Postlethwait, J. 1999. Preservation of duplicate genes by complementary, degenerative mutations. *Genetics* **151**: 1531–1545.

Furutani-Seiki, M. and Wittbrodt, J. 2004. Medaka and zebrafish, an evolutionary twin study. *Mech. Dev.* **121**: 629–637.

Geisler, R., Rauch, G.J., Baier, H., van Bebber, F., Bross, L., Dekens, M.P., Finger, K., Fricke, C., Gates, M.A., Geiger, H., et al. 1999. A radiation hybrid map of the zebrafish genome. *Nat. Genet.* **23**: 86–89.

Gilley, J., Armes, N., and Fried, M. 1997. *Fugu* genome is not a good mammalian model. *Nature* **385**: 305–306.

Gorr, T., Kleinschmidt, T., and Fricke, H. 1991. Close tetrapod relationships of the coelacanth *Latimeria* indicated by haemoglobin sequences. *Nature* **351**: 394–397.

Gracey, A.Y., Fraser, E.J., Li, W., Fang, Y., Taylor, R.R., Rogers, J., Brass, A., and Cossins, A.R. 2004. Coping with cold: An integrative, multitissue analysis of the transcriptome of a poikilothermic vertebrate. *Proc. Natl. Acad. Sci.* **101**: 16970–16975.

Grunwald, D.J. and Eisen, J.S. 2002. Headwaters of the zebrafish—Emergence of a new model vertebrate. *Nat. Rev. Genet.* **3**: 717–724.

Haffter, P., Granato, M., Brand, M., Mullins, M.C., Hammerschmidt, M., Kane, D.A., Odenthal, J., van Eeden, F.J., Jiang, Y.J., Heisenberg, C.P., et al. 1996. The identification of genes with unique and essential functions in the development of the zebrafish, *Danio rerio*. *Development* **123**: 1–36.

Harris, M.A., Clark, J., Ireland, A., Lomax, J., Ashburner, M., Foulger, R., Eilbeck, K., Lewis, S., Marshall, B., Mungall, C., et al. 2004. The Gene Ontology (GO) database and informatics resource. *Nucleic Acids Res.* **32**: D258–D261.

Hinegardner, R. 1968. Evolution of celullar DNA content in Teleost fishes. *Am. Nat.* **102**: 517–523.

Hoegg, S., Brinkmann, H., Taylor, J.S., and Meyer, A. 2004. Phylogenetic timing of the fish-specific genome duplication correlates with the diversification of teleost fish. *J. Mol. Evol.* **59**: 190–203.

Hong, Y., Winkler, C., and Schartl, M. 1996. Pluripotency and differentiation of embryonic stem cell lines from the medakafish (*Oryzias latipes*). *Mech. Dev.* **60**: 33–44.

Hukriede, N., Fisher, D., Epstein, J., Joly, L., Tellis, P., Zhou, Y., Barbazuk, B., Cox, K., Fenton-Noriega, L., Hersey, C., et al. 2001. The LN54 radiation hybrid map of zebrafish expressed sequences. *Genome Res.* **11**: 2127–2132.

Inoue, J.G., Miya, M., Tsukamoto, K., and Nishida, M. 2003. Basal actinopterygian relationships: A mitogenomic perspective on the phylogeny of the "ancient fish." *Mol. Phylogenet. Evol.* **26**: 110–120.

Jaillon, O., Aury, J.M., Brunet, F., Petit, J.L., Stange-Thomann, N., Mauceli, E. Bouneau, L., Fischer, C., Ozouf-Costaz, C., Bernot, A., et al. 2004. Genome duplication in the teleost fish *Tetraodon nigroviridis* reveals the early vertebrate proto-karyotype. *Nature* **431**: 946–957.

Kellis, M., Birren, B.W., and Lander, E.S. 2004. Proof and evolutionary analysis of ancient genome duplication in the yeast *Saccharomyces cerevisiae*. *Nature* **428**: 617–624.

Kelly, P.D., Chu, F., Woods, I.G., Ngo-Hazelett, P., Cardozo, T., Huang, H., Kimm, F., Liao, L., Yan, Y.L., Zhou, Y., et al. 2000. Genetic linkage mapping of zebrafish genes and ESTs. *Genome Res.* **10**: 558–567.

Khorasani, M.Z., Hennig, S., Imre, G., Asakawa, S., Palczewski, S., Berger, A., Hori, H., Naruse, K., Mitani, H., Shima, A., et al. 2004. A first generation physical map of the medaka genome in BACs essential for positional cloning and clone-by-clone based genomic sequencing. *Mech. Dev.* **121:** 903–913.

Kimura, T., Jindo, T., Narita, T., Naruse, K., Kobayashi, D., Shin, I.T., Kitagawa, T., Sakaguchi, T., Mitani, H., Shima, A., et al. 2004. Large-scale isolation of ESTs from medaka embryos and its application to medaka developmental genetics. *Mech. Dev.* **121:** 915–932.

Kondo, M., Nanda, I., Hornung, U., Asakawa, S., Shimizu, N., Mitani, H., Schmid, M., Shima, A., and Schartl, M. 2003. Absence of the candidate male sex-determining gene dmrt1b(Y) of medaka from other fish species. *Curr. Biol.* **13:** 416–420.

Lander, E.S., Linton, L.M., Birren, B., Nusbaum, C., Zody, M.C., Baldwin, J., Devon, K., Dewar, K., Doyle, M., FitzHugh, W., et al. 2001. Initial sequencing and analysis of the human genome. *Nature* **409:** 860–921.

Lewin, B. 1971. Genes in tandem. *Nature* **230:** 314.

Lutfalla, G., Roest Crollius, H., Brunet, F., Laudet, V., and Robinson-Rechavi, M. 2003. Inventing a sex-specific gene: A conserved role of DMRT1 in teleost fishes plus a recent duplication in the medaka *Oryzias latipes* resulted in DMY. *J. Mol. Evol.* **57 Suppl 1:** S148–S153.

MacRae, C.A. and Fishman, M.C. 2002. Zebrafish: The complete cardiovascular compendium. *Cold Spring Harb. Symp. Quant. Biol.* **67:** 301–308.

Matsuda, M., Nagahama, Y., Shinomiya, A., Sato, T., Matsuda, C., Kobayashi, T., Morrey, C.E., Shibata, N., Asakawa, S., Shimizu, N., et al. 2002. DMY is a Y-specific DM-domain gene required for male development in the medaka fish. *Nature* **417:** 559–563.

McCallum, C.M., Comai, L., Greene, E.A., and Henikoff, S. 2000. Targeted screening for induced mutations. *Nat. Biotechnol.* **18:** 455–457.

Meyer, A. and Schartl, M. 1999. Gene and genome duplications in vertebrates: The one-to-four (-to-eight in fish) rule and the evolution of novel gene functions. *Curr. Opin. Cell Biol.* **11:** 699–704.

Nanda, I., Kondo, M., Hornung, U., Asakawa, S., Winkler, C., Shimizu, A., Shan, Z., Haaf, T., Shimizu, N., Shima, A., et al. 2002. A duplicated copy of DMRT1 in the sex-determining region of the Y chromosome of the medaka, *Oryzias latipes*. *Proc. Natl. Acad. Sci.* **99:** 11778–11783.

Naruse, K., Tanaka, M., Mita, K., Shima, A., Postlethwait, J., and Mitani, H. 2004. A medaka gene map: The trace of ancestral vertebrate proto-chromosomes revealed by comparative gene mapping. *Genome Res.* **14:** 820–828.

Nelson, J.S. 1994. *Fishes of the world*. Wiley, New York.

Noonan, J.P., Grimwood, J., Danke, J., Schmutz, J., Dickson, M., Amemiya, C.T., and Myers, R.M. 2004. Coelacanth genome sequence reveals the evolutionary history of vertebrate genes. *Genome Res.* **14:** 2397–2405.

Ohno, S. 1970. *Evolution by gene duplication*. Allen and Unwin, London.

Perutz, M.F., Johnson, T., Suzuki, M., and Finch, J.T. 1994. Glutamine repeats as polar zippers: Their possible role in inherited neurodegenerative diseases. *Proc. Natl. Acad. Sci.* **91:** 5355–5358.

Postlethwait, J.H., Woods, I.G., Ngo-Hazelett, P., Yan, Y.L., Kelly, P.D., Chu, F., Huang, H., Hill-Force, A., and Talbot, W.S. 2000. Zebrafish comparative genomics and the origins of vertebrate chromosomes *Genome Res.* **10:** 1890–1902.

Robinson-Rechavi, M., Marchand, O., Escriva, H., and Laudet, V. 2001. An ancestral whole-genome duplication may not have been responsible for the abundance of duplicated fish genes. *Curr. Biol.* **11:** R458–R459.

Roest Crollius, H., Jaillon, O., Bernot, A., Dasilva, C., Bouneau, L., Fizames, C., Wincker, P., Brottier, P., Quetier, F., Saurin, W., et al. 2000. Human gene number estimate provided by genome wide analysis using *Tetraodon nigroviridis* genomic DNA. *Nat. Genet.* **25:** 235–238.

Sandelin, A., Bailey, P., Bruce, S., Engstrom, P.G., Klos, J.M., Wasserman, W.W., Ericson, J., and Lenhard, B. 2004. Arrays of ultraconserved non-coding regions span the loci of key developmental genes in vertebrate genomes. *BMC Genomics* **5:** 99.

Shimoda, N., Knapik, E.W., Ziniti, J., Sim, C., Yamada, E., Kaplan, S., Jackson, D., de Sauvage, F., Jacob, H., and Fishman, M.C. 1999. Zebrafish genetic map with 2000 microsatellite markers. *Genomics* **58:** 219–232.

Spofford, J.B. 1972. Phylogenetic mechanism. *Science* **175:** 617–618.

Stainier, D.Y., Fouquet, B., Chen, J.N., Warren, K.S., Weinstein, B.M., Meiler, S.E., Mohideen, M.A., Neuhauss, S.C., Solnica-Krezel, L., Schier, A.F., et al. 1996. Mutations affecting the formation and function of the cardiovascular system in the zebrafish embryo. *Development* **123:** 285–292.

Stephens, S.G. 1951. Possible significance of duplications in evolution. *Adv. Genet.* **4:** 247–256.

Stickney, H.L., Schmutz, J., Woods, I.G., Holtzer, C.C., Dickson, M.C., Kelly, P.D., Myers, R.M., and Talbot, W.S. 2002. Rapid mapping of zebrafish mutations with SNPs and oligonucleotide microarrays. *Genome Res.* **12:** 1929–1934.

Streisinger, G., Walker, C., Dower, N., Knauber, D., and Singer, F. 1981. Production of clones of homozygous diploid zebra fish (*Brachydanio rerio*). *Nature* **291:** 293–296.

Sulston, J., Du, Z., Thomas, K., Wilson, R., Hillier, L., Staden, R., Halloran, N., Green, P., Thierry-Mieg, J., Qiu, L., et al. 1992. The *C. elegans* genome sequencing project: A beginning. *Nature* **356:** 37–41.

Taylor, J.S. and Raes, J. 2004. Duplication and divergence: The evolution of new genes and old ideas. *Annu. Rev. Genet.* **38:** 615–643.

Vandepoele, K., De Vos, W., Taylor, J.S., Meyer, A., and Van de Peer, Y. 2004. Major events in the genome evolution of vertebrates: Paranome age and size differ considerably between ray-finned fishes and land vertebrates. *Proc. Natl. Acad. Sci.* **101:** 1638–1643.

Venkatesh, B., Tay, A., Dandona, N., Patil, J.G., and Brenner, S. 2005. A compact cartilaginous fish model genome. *Curr. Biol.* **15:** R82–R83.

Volff, J.N., Kondo, M., and Schartl, M. 2003. Medaka dmY/dmrt1Y is not the universal primary sex-determining gene in fish. *Trends Genet.* **19:** 196–199.

Wittbrodt, J., Shima, A., and Schartl, M. 2002. Medaka—A model organism from the far East. *Nat. Rev. Genet.* **3:** 53–64.

Wolfe, K.H. 2001. Yesterday's polyploids and the mystery of diploidization. *Nat. Rev. Genet.* **2:** 333–341.

Woods, I.G., Kelly, P.D., Chu, F., Ngo-Hazelett, P., Yan, Y.L., Huang, H., Postlethwait, J.H., and Talbot, W.S. 2000. A comparative map of the zebrafish genome. *Genome Res.* **10:** 1903–1914.

Woolfe, A., Goodson, M., Goode, D.K., Snell, P., McEwen, G.K., Vavouri, T., Smith, S.F., North, P., Callaway, H., Kelly, K., et al. 2004. Highly conserved non-coding sequences are associated with vertebrate development. *PLoS Biol.* **3:** e7.

10

Xenomics

Enrique Amaya
The Wellcome Trust/Cancer Research UK Gurdon Institute
University of Cambridge
Cambridge CB2 1QN, United Kingdom, and
Department of Zoology
University of Cambridge
Cambridge CB2 3EJ, United Kingdom

XENOPUS GENOMICS IS VERY MUCH IN ITS INFANCY. Although large-scale sequencing efforts were slow to be initiated in this system, in the past 3–4 yr there has been an explosion of genomic information accumulating in *Xenopus laevis* and its diploid relative, *Xenopus tropicalis* (Fig. 1). Since the beginning of 2003, >320,000 sequences have been deposited in public repositories for *X. laevis* and >1,100,000 for *X. tropicalis*, mostly in the form of expressed sequence tags (ESTs). With this expansive amount of new sequence information, *X. tropicalis* recently jumped into third place on the list of organisms with the most EST's, behind human and mouse. During the same period of time, the Joint Genome Institute (JGI) has been sequencing the *X. tropicalis* genome, using a shotgun approach, and it has recently completed 8× coverage. The JGI is currently in the process of assembling the *X. tropicalis* genome, and it is expected that the JGI will announce its results by the end of 2005.

Now that such an extensive amount of genomic information is becoming available in *Xenopus*, how will this be useful in our scientific pursuits? This question is best answered by further asking, "What is the ultimate value of obtaining sequence information?" If the ultimate aim is not simply to catalog genes but to understand their function, then it would be very advantageous to find the ideal organisms to study gene function. It is here where the marriage of *Xenopus* and genomics will reap its full benefits, as *Xenopus* is perhaps the best vertebrate model organism for functional genomics.

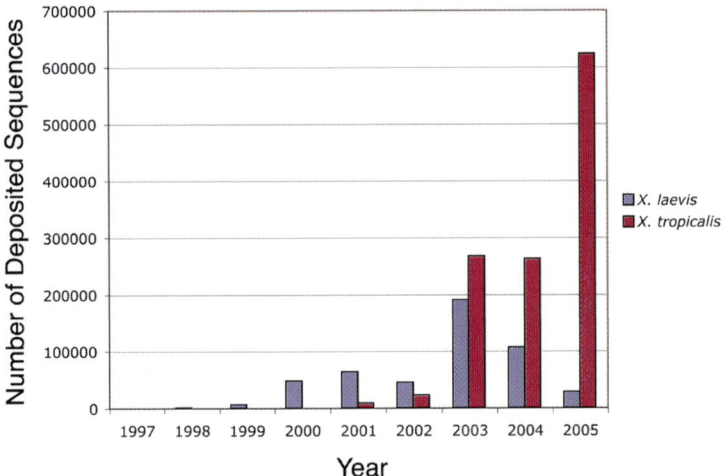

Figure 1. Graph of number of deposited ESTs for *Xenopus laevis* (blue) and *Xenopus tropicalis* (red) for each year since 1997. In the past 2–3 yr, there has been an explosion of sequence information in the public repositories, heralding the arrival of these two systems into the genomics era.

FUNCTIONAL SCREENS

Xenopus is arguably the best available vertebrate model for systematic large-scale in vivo gene function analysis. In this section I review the different types of functional screens that can be performed in *Xenopus*.

Gain-of-Function Screens

Oocyte Expression Screens

It has been recognized since the early 1970s that the *Xenopus* oocyte can be used as an in vivo test tube to study the function of biological macromolecules (Gurdon 1975). First, *Xenopus* oocytes can be cultured in vitro for many days, and second, microinjection of exogenous macromolecules, such as DNA or mRNA, will behave appropriately in the oocytes, resulting in their transcription and/or translation, respectively. However, before *Xenopus* oocytes could be used in large-scale expression screens, it was important that purified mRNAs could be generated in large quantities for any gene of interest. This next critical step was made in the mid-1980s, when a method was found for producing large quantities of synthetic mRNA in vitro by using bacteriophage promoters, and it was shown that these synthetic mRNAs would be efficiently translated when injected into *Xenopus* oocytes (Krieg and Melton 1984; Melton et al. 1984).

Xenopus oocytes were first exploited in large-scale functional screens by groups interested in identifying receptors for neuropeptides and neurotransmitters (Masu et al. 1987; Julius et al. 1988). The first group to successfully use oocytes to clone a novel neuropeptide receptor was Masu and colleagues (1987). This group was interested in identifying the receptor for the tachykinin neuropeptide, substance K. They knew from previous work that *Xenopus* oocytes injected with total mRNA isolated from bovine brain and stomach expressed functional mammalian substance K receptors on their membranes, as assayed by electrophysiological measurements. Now that a sensitive assay for receptor function and a source of mRNA encoding the receptor were available, all that was required was to develop a large-scale functional screen to clone the receptor. To do this, they constructed a cDNA library from bovine stomach in a vector containing a bacteriophage promoter (SP6), so that in vitro transcribed RNA could be made from this library. Then, they injected in vitro transcribed RNA from this library in pools and assayed for substance K activity, based on electrophysiological measurements. Once an active pool was identified, subpools were tested for activity until an individual clone encoding the receptor for tachykinin neuropeptide, substance K, was identified (Masu et al. 1987). A similar strategy was used a year later by Julius and colleagues (1988) to clone the 5HT1c serotonin receptor in *Xenopus* oocytes. Since these two pioneering screens, many other neuropeptide receptors and ligand-gated ion channels have been identified by using large-scale functional screens in *Xenopus* oocytes. A screen in oocytes has also been developed aimed at identifying secreted molecules that contain mesoderm and/or neural-inducing activities (Lustig and Kirschner 1995). In this assay, pools of in vitro transcribed RNA made from a cDNA library from tissue with inducing activity are injected into oocytes. Then an explant of competent tissue is placed on top of the oocytes. If the oocyte secretes factors that change the fate of the explant, then that pool is sib-selected until a single active clone is identified.

Gain-of-Function Screens in Embryos

In 1991 Smith and Harland (1991) took a similar approach to the one taken to clone neuropeptide receptors but, in their case, with the aim of identifying genes that could mimic the vegetal dorsal inducer in the early frog embryo. For their assay, they irradiated the vegetal hemisphere of one-cell-stage embryos with UV, in order to remove the endogenous vegetal dorsal inducing signal, resulting in ventralized embryos. Then they prepared a cDNA library from gastrula embryos and injected in vitro transcribed RNA from this library in pools and assayed for rescue of dorsal axis in the UV-irradiated/ventralized embryos. By using this approach, they identified *Xwnt8* as a potent vegetal dorsalizing factor, which could rescue a complete dorsal axis in UV-irradiated embryos (Smith and Harland 1991). A year later, Smith and Harland (1992) used the same approach to clone *noggin*, a novel gene that was later shown to encode a potent extracellular antagonist of Bone Morphogenetic Protein 4 (BMP4) (Zimmerman et al. 1996). Since the pioneering work of Smith and Harland, many important molecules have been identified in *Xenopus*, using large-scale gain-of-function screens (also known as expression cloning screens); see Table 1 for examples. Interestingly, most of these are

Table 1. Genes identified in large-scale gain-of-function screens in *Xenopus*

Gene name	Gene product function	Reference
Xwnt8	Signaling molecule	Smith and Harland 1991
noggin	Antagonist of signaling molecule	Smith and Harland 1992
siamois	Transcription factor	Lemaire et al. 1995
Xnr3	Signaling molecule	Glinka et al. 1996; Smith et al. 1995
Xnr1	Signaling molecule	Lustig et al. 1996a
Mix.1	Transcription factor	Mead et al. 1996
Xombi	Transcription factor	Lustig et al. 1996b
Mard2	Transcription factor	Baker and Harland 1996
twin	Transcription factor	Laurent et al. 1997
sizzled	Antagonist of signaling molecule	Salic et al. 1997
dickkopf-1	Antagonist of signaling molecule	Glinka et al. 1998
geminin	Regulates DNA replication/transcription	Kroll et al. 1998
gremlin	Antagonist of signaling molecule	Hsu et al. 1998
XBF-2	Transcription factor	Mariani and Harland 1998
laloo	Signaling molecule	Weinstein et al. 1998
XSox17	Transcription factor	Zorn et al. 1999
E2F	Cell cycle control	Suzuki and Hemmati-Brivanlou 2000
XOs4	Unknown	Zohn and Brivanlou 2001
Xath2	Transcription factor	Taelman et al. 2001
β-catenin	Transcription factor	Domingos et al. 2001
Baf57	Chromatin remodeling	Domingos et al. 2002
wise	Signaling molecule	Itasaki et al. 2003
coco	Signaling molecule	Bell et al. 2003
R-Spondin2	Signaling molecule	Kazanskaya et al. 2004
Ectodermin	Antagonist of signaling molecule	Dupont et al. 2005

signaling molecules, antagonists of signaling molecules, or transcription factors downstream of signaling molecules.

The large-scale gain-of-function screens used to identify these genes were done with redundant, non-normalized libraries, with pool sizes ranging from 96 clones per pool to several thousand clones per pool (Smith and Harland 1991; Grammer et al. 2000). Although this approach has given rise to the identification of a large number of important genes during early development, the approach has been inherently inefficient for several reasons. First, most of the clones that are screened do not contain the full-length coding sequence of the protein, as any given library contains mostly truncated clones. Second, given that the libraries used for these screens were non-normalized, clones encoding the same protein product were screened multiple times, especially those genes transcribed in the embryo at high levels, while genes present at more modest levels in the embryo were screened only rarely. However, by using sequence information and bioinformatics tools, one can select a non-redundant full-length clone set, thus allowing the functional screens to be performed much more efficiently (Fig. 2; Gilchrist et al. 2004; Chen et al. 2005; Voigt et al. 2005; http://www.gurdon.cam.ac.uk/informatics/Xenopus.html). By using such a streamlined full-length clone set, it has been possible to decrease the clone size per pool during the functional screens to eight clones per pool (Voigt et al. 2005). This dramatically increases the sensitivity and efficiency of the functional screens that are performed (Chen et al. 2005; Voigt et al. 2005). To date, a wealth of genes have been uncovered by using large-scale gain-of-function screens in *Xenopus*. Now that a non-redundant full-length clone set has been developed as a physical resource to the community, *Xenopus* is very likely to remain one of the ideal systems for large-scale functional genomic efforts.

Expression Screens in the Test Tube

Xenopus eggs and embryos have also been exploited with much success for investigating many basic cell biological and biochemical principles, due to the ready availability of very large numbers of eggs and embryos from these frogs. In particular, it is possible to generate extracts from eggs and embryos, which recapitulate many cell biological processes, such as nuclear disassembly and reassembly, nuclear import and export, DNA replication, chromosome assembly and disassembly, mitotic spindle assembly and function, protein synthesis and degradation, cell cycle control, apoptosis, and microtubule and microfilament assembly and disassembly (Murray and Kirschner 1989a,b; Murray et al. 1989; Glotzer

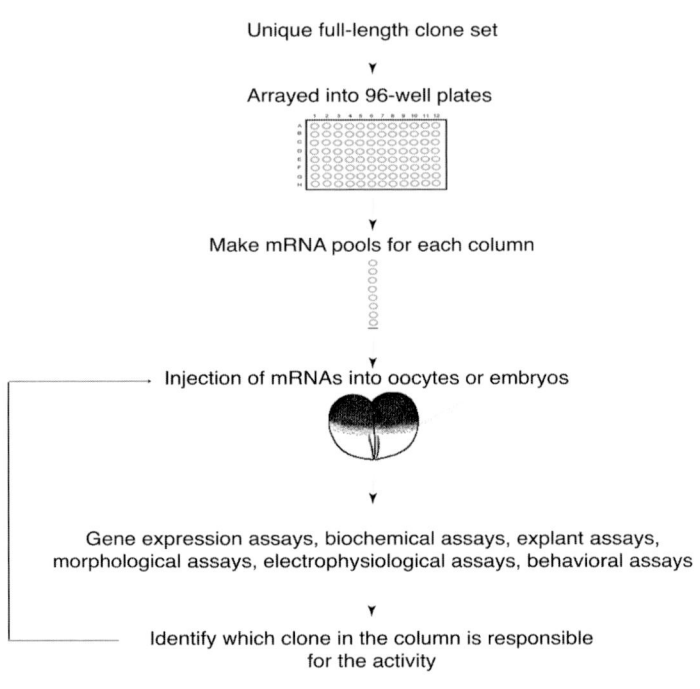

Figure 2. Large-scale gain-of-function screen strategy in *Xenopus*. A unique full-length clone set is established and arrayed into 96-well plates. Miniplasmid preps are made and pools of plasmids made for each column. Each pooled set of plasmids is transcribed into mRNA in vitro. Pooled mRNA is injected into oocytes or embryos, and then a variety of functional screens are performed, depending on the type of molecules that are being sought. Once an active pool is identified, the pool is broken down to individual clones and assayed again to identify the active clone.

et al. 1991; Newmeyer and Wilson 1991; Pfaller et al. 1991; Smythe and Newport 1991; Dasso et al. 1992; Allan 1993; Holloway et al. 1993; Newmeyer et al. 1994; Hengartner 1995; King et al. 1995, 1996; Yu et al. 1996; Evans et al. 1997; Kornbluth 1997; Thommes and Blow 1997; Lohka 1998; Pain et al. 1998; de la Barre et al. 1999; Desai et al. 1999; Shirasu et al. 1999; von Ahsen and Newmeyer 2000; Mandato et al. 2001; Arias and Walter 2004). Since these cell-free extracts can be manipulated in many different ways, it has been possible to assess the role of specific proteins in distinct processes. For example, one can immuno-deplete the extracts of particular proteins or protein complexes and address the effect in different cellular processes. It is also possible to use *Xenopus* extracts in combination with expression-based screens to identify proteins that are substrates for particular biochemical pathways (King et al. 1997; Lustig et al. 1997). In this case a cDNA library is split into pools, as is done for the functional screens described above, but instead of injecting in vitro transcribed RNAs from pooled clones into oocytes or embryos, the pooled clones are transcribed and translated together in vitro, in the presence of radioactively labeled amino acids. In this way, the behavior of the labeled translated proteins can be monitored after being added to the egg extracts. Such screens have been used to identify proteins that are phosphorylated in a cell cycle–dependent manner (Stukenberg et al. 1997). They have also been used to identify proteins that are specifically degraded during mitosis (McGarry and Kirschner 1998). This general strategy has been modified in a variety of ways, as, for example, in order to identify new substrates for caspase 3 (Kothakota et al. 1997) or to identify novel uracil-DNA glycosylases (Haushalter et al. 1999). In the future, these screens will benefit from the availability of non-redundant full-length clone sets, as has been used for functional screens in embryos (Gilchrist et al. 2004; Chen et al. 2005; Voigt et al. 2005). Combining the tractability and manipulability of cell-free extracts from eggs and embryos with large-scale functional genomic approaches, such as those described above, will continue to provide valuable approaches for identifying new components involved in cell biological and biochemical processes in the future.

Loss-of-Function Screens and Genetic Approaches

Although for many years gain-of-function screens have been very fruitful in *Xenopus* for identifying genes involved in cell and developmental biology processes, loss-of-function screens or genetic approaches have not been traditionally used in this system. However, such approaches have recently begun to be employed as well. One of the primary reasons that such approaches have not traditionally been adopted is that *X. laevis*, the more commonly used species, is allotetraploid. This means that it essentially has two nonidentical copies for each gene, which complicates loss of function experiments due to potential genetic redundancy. In addition, *X. laevis* has a long generation time of around a year, further complicating its use as a straightforward genetic system. However, in recent years *X. tropicalis*, a diploid relative of *X. laevis*, has begun to be used more and more due to its simpler genome and shorter generation time, making it a more ideal system for both genomic studies and loss-of-function experiments (Amaya et al. 1998).

For >10 yr, loss-of-function experiments in *Xenopus* have traditionally been based on overexpression of dominant-negative variants of genes (Amaya et al. 1991). The disadvantage of this method is that it requires previous knowledge of protein domains and function for designing appropriate dominant-negative variants of genes. In addition, dominant-negative constructs are seldom specific to only one gene product. Several years ago Janet Heasman and colleagues (2000, 2002) reported that antisense morpholino oligonucleotides (aMOs) complementary to the start of translation of the mRNA could specifically and efficiently inhibit translation in *X. laevis* embryos. It was also shown that aMOs can efficiently and specifically inhibit translation of mRNAs in *X. tropicalis* into the tadpole stages (Nutt et al. 2001). In addition, aMOs complementary to splice junction inhibit splicing of pre-mRNA in vivo, providing another means of inhibiting gene function in the embryo (Kenwrick et al. 2004; Sivak et al. 2005). Targeting aMOs to splice junctions has the additional benefit that its effect can be monitored by RT-PCR (Kenwrick et al. 2004; Sivak et al. 2005). The disadvantage in targeting aMOs to splice junctions is that it requires knowledge of the genomic structure of the gene in question. However, given the extent of genome information now available in *X. tropicalis* (http://genome.jgi-psf.org/; http://www.ensembl.org/Xenopus_tropicalis/index.html), it is now possible to identify splice junctions to most genes. Given that only sequence information around the start of translation and/or around splice junctions is needed to design aMOs, it has become possible to consider performing mid- to large-scale loss-of-function screens in *X. tropicalis* (Kenwrick et al. 2004). Such an approach is likely to be very fruitful in the future for addressing the function of genes during early development on a large scale. Even a small-scale aMO screen uncovered a novel gene, pinhead, essential for head formation (Kenwrick et al. 2004).

Another approach likely to be valuable for identifying mutation in genes of interest is combining chemical mutagenesis screens with direct sequencing (Hurlstone et al. 2003; Stemple 2004). To this end, Lyle Zimmerman and Derek Stemple (pers. comm.) are currently using this approach to identify mutations in hundreds of genes in *X. tropicalis*. But what about forward genetic screens in *X. tropicalis*? Several groups are currently performing chemical, irradiation, and insertional mutagenesis screens, but such approaches remain in their infancy due to the lack of easily scored mutations for optimizing mutagen dose levels. However, recently two groups published the first set of naturally occurring mutations in

X. tropicalis, which should help facilitate such optimization experiments (Grammer et al. 2005; Noramly et al. 2005).

LARGE-SCALE WHOLE-MOUNT IN SITU HYBRIDIZATION SCREENS

One of the primary means of specifying cell fate during development is through differential gene expression. With this in mind, Christof Niehrs and colleagues began a systematic large-scale whole-mount in situ hybridization screen using *Xenopus* embryos in the early 1990s with the aim of identifying genes with localized expression patterns during development (Gawantka et al. 1995, 1998; Pollet et al. 2005). They started this screen several years before any large-scale sequencing projects in *Xenopus* had been initiated, so their approach was simply to take random clones from several embryonic staged cDNA libraries and to perform whole-mount in situ hybridizations systematically on thousands of clones on embryos at the gastrula, neurula, and tailbud stage. In the initial publication of the screen, they performed whole-mount in situ hybridization on 1765 randomly picked clones (Gawantka et al. 1998). From these, they identified 273 genes that were differentially expressed. More recently, the Niehrs group published the second phase of their large-scale gene expression screen, in this case presenting the results after performing whole-mount in situ hybridizations on 8369 cDNA clones. From these, they identified 773 genes with restricted expression (Pollet et al. 2005). Given that there was an overlap of 39 genes identified in both screens, up to now, this large-scale gene expression screen has resulted in the identification of 957 unique genes in *Xenopus* with restricted expression patterns. No doubt there will be more genes found in phase 3, but as it stands, the cataloging of the restricted expression patterns of nearly a thousand genes during early *Xenopus* development is an ambitious and exceedingly important project. A database (AxelDB) containing this information is available at http://indigene.ibaic.u-psud.fr/article.php3?id_article=48 (Pollet et al. 2000).

Several important insights have been gained from this gargantuan project. One is that a significant number of genes have restricted expression patterns. Overall, one-fourth to one-fifth of genes tested had restricted expression patterns during early development. Surprisingly, many of these genes encode proteins thought to be involved in basic cellular housekeeping functions. About one-third of the genes with restricted expression patterns are cell type and/or organ specific. These genes are immediately useful as markers, but many of these are also likely to have important roles in specifying cell fate. Although identifying cell type–specific genes was one of the aims of the large-scale gene expression screen, the more interesting insights have been gained from genes with more complex expression patterns. One is the concept of tissue relatedness, gained from correlating the cell types and/or tissue types that share common expression of genes; and the other is the concept of synexpression groups, which comprise genes with common complex expression patterns (Gawantka et al. 1998; Niehrs and Pollet 1999; Pollet et al. 2005).

Tissue Relatedness

By using hierarchical clustering, the tissue types that shared common expression of genes were correlated, thus resulting in a tissue-relatedness tree (Gawantka et al. 1998; Pollet et al. 2005). The purpose of this analysis was to arrive at a relatively unbiased measure of tissue relatedness. Two types of tissue relatedness were found by using this analysis. The first is tissues that share a common lineage relationship, termed monofatic relatedness (Pollet et al. 2005). Examples of these include such tissues as the diencephalon and telencephalon, and lateral plate mesoderm and ventral blood islands. The monofatic relationships are not surprising, and indeed, most of the tissues are monofatic. However, there was another type of relatedness that came out of this analysis, termed parafatic relationships, which was less expected (Pollet et al. 2005). These are tissues that share the expression of many genes, although they do not share a common lineage relationship. In these cases it is believed that these tissues share common physiological processes. Examples of parafatic relationships include the cement gland and notochord, both being highly secretory tissues. In other cases, the underlying common physiological process underlying other parafatic relationships, such as the hypophysis (pineal gland) and several mesodermal derivatives, is less obvious. Another example of a parafatic relationship is the endocrine pancreas and the nervous system, which share the expression of many genes in common although they are not related by lineage (Edlund 2001).

Synexpression Groups

Perhaps the most interesting feature to come out of the large-scale gene expression screen is the identification of groups of genes that share identical or very similar complex expression patterns, termed synexpression groups (Gawantka et al. 1998; Niehrs and Pollet 1999; Pollet et al. 2005).

These are groups of genes that encode proteins likely to be involved in common functional pathways but are used at multiple embryonic stages and/or tissues, and are thus under common transcriptional control. Synexpression groups include the *BMP4* group, the delta group, the chromatin group, the endoplasmic reticulum group, the karyopherin group, and the *FGF8* group (Gawantka et al. 1998; Wischnewski et al. 2000; Furthauer et al. 2002; Pera et al. 2002; Tsang et al. 2002; Pollet et al. 2005). One of the most valuable aspects of the concept of synexpression groups is that one is able to predict the likely function of a gene product based on its expression, which can then form the basis of further functional experiments. In many cases this has been borne to be true. For example, the Xvent2 protein, which is a homeodomain-containing protein in the *BMP4* synexpression group, is indeed a target and downstream mediator of BMP signaling (Onichtchouk et al. 1996, 1998). *BAMBI*, another gene in the *BMP4* group, has likewise been shown to be involved in the regulation of BMP signaling (Onichtchouk et al. 1999). Similarly, the *Delta* synexpression group contains several enhancers of *split* (*ESR*)–like genes, known to be targets of Delta-Notch signaling (Jen et al. 1999; Li et al. 2003). However the more interesting results have arisen from experiments in other genes within this group, which had not previously been implicated in Delta-Notch signaling. One such gene, *Nrarp* (also know as *NAP*), is a novel ankyrin-containing protein, which is both a target and modulator of Delta-Notch signaling (Lamar et al. 2001; Lahaye et al. 2002). The concept of synexpression has been adopted by other systems, and one synexpression group that has arisen from a large-scale in situ screen in zebrafish is the *FGF* synexpression group (Furthauer et al. 2002; Pera et al. 2002; Tsang et al. 2002). This group includes *FGF8* and *FGF3*, as well as several genes known to be involved in the regulation of FGF signaling, such as the *sprouty2* and *sprouty4* (Furthauer et al. 2001, 2002). However it also includes other genes, previously not implicated in FGF signaling, such as sef, which stands for similar expression to *fgf* genes (Furthauer et al. 2002; Tsang et al. 2002). Sef interacts with the intracellular domain of the FGF receptor and functions as a feedback inhibitor of FGF dependent ras/MAPK signaling (Furthauer et al. 2002; Tsang et al. 2002). Another gene with an expression pattern very similar to *FGF8* is *FLRT3*, which stands for fibronectin-leucine-rich transmembrane protein 3. This gene also has been shown to be a transmembrane modulator of FGF-mediated MAPK signaling (Bottcher et al. 2004). In summary, members of synexpression groups have often been shown subsequently to be involved in common pathways, thus confirming the value of these groupings in providing good predictive value for functional analysis.

The Future of Large-Scale Whole-Mount in Situ Hybridization Screens

The gene expression screen carried out in the Niehrs group has been fantastically useful in uncovering new marker genes as well as novel protein products in established molecular pathways. What is the future of such screens? Given that a relatively small percentage of total genes (~4000 of the predicted 30,000 genes in the vertebrate genome) have been included in this screen, there is certainly scope for the incorporation of further genes into the whole-mount in situ hybridization screen. Indeed, most of the genes selected for the published screens originated from early-stage libraries. Simply expanding the selection of clones to later-staged libraries is likely to result in the identification of many more genes with differential gene expression patterns. In addition, a large amount of genomic information is now available, in the form of ESTs in both *X. laevis* and *X. tropicalis*, as well as genome sequence information in the case of *X. tropicalis*. Therefore, in the next phase of screens, it would seem prudent to use a unique set of genes, rather than randomly selected clones for the screen. The Ueno group in Japan is taking essentially this approach in their large-scale in situ hybridization screen, which they make available at http://xenopus.nibb.ac.jp/. To date, they have made available the expression patterns of 801 genes in their database. A large-scale in situ screen using the unique set of full-length clones identified in *X. tropicalis* is similarly underway (E. Amaya and N. Papalopulu, unpubl.). The ultimate aim is to determine the expression patterns of all genes in *X. laevis* and *X. tropicalis*, but this highly ambitious goal will take many years to complete. Another essential aim for the future will be to coordinate the results of the large-scale gene expression screens in *Xenopus* with those underway in other chordate models systems, such as the mouse, chick, ascidians, zebrafish, and medaka (Bettenhausen and Gossler 1995; Neidhardt et al. 2000; Crosier et al. 2001; Kudoh et al. 2001; Makabe et al. 2001; Nishikata et al. 2001; Satou et al. 2001; Thut et al. 2001; Gammill and Bronner-Fraser 2002; Bell et al. 2004; Quiring et al. 2004; Thisse et al. 2004). Ideally one should be able to query the expression pattern of a given gene in a model organism, find the ortholog in another model system, and compare the expression patterns between systems. Such sort of analysis will be invaluable in correlating expression and functional data between one organism and others.

GENE EXPRESSION PROFILING AND MICROARRAYS

While assaying the expression pattern of one gene at a time, as has been done in the large-scale whole-mount in situ screen described above, is enormously useful due to the detailed information that can be gained from this sort of analysis, it is exceedingly time consuming to do this on the tens of thousands of genes present in the genome. However one way that thousands of genes can be assayed simultaneously is through the use of microarrays (Brown and Botstein 1999). Given that *Xenopus* embryos can be generated in the thousands and it is relatively easy to manipulate them, it can be argued that this system is ideally suited for microarray experiments. The Brivanlou laboratory was the first to recognize this potential, and in 1999 Curtis Altmann and Ali H. Brivanlou announced the first *Xenopus* cDNA microarray. It contained 864 gastrula cDNAs, including 768 randomly selected clones from an expression library and the rest made up of known markers (Altmann et al. 2001). This early prototype microarray heralded the first steps toward the development of genomic resources in *Xenopus*, 2 yr before any large concerted sequencing effort in *Xenopus* had come off the ground. This prototype array was used to identify new genes with regionalized expression pattern, as well as novel activin inducible genes (Altmann et al. 2001).

The Brivanlou then followed this prototype array with a second-generation cDNA microarray, containing, in addition to the clones in the prototype array, a further 4000 randomly picked gastrula clones (Munoz-Sanjuan et al. 2002). The second array again predated the availability of concerted sequencing efforts, but nevertheless, they were able to use this array to monitor the transcription changes that occur in ectodermal explants that had been neuralized by the inhibitory Smad, Smad7. By using this 5000 cDNA microarray, they identified 142 genes whose transcriptional profile changed in the neuralized explants. Surprisingly, many of the genes identified in their screen were predicted to encode proteins involved in post-transcriptional rather than transcriptional control, suggesting an important role for post-transcriptional control during the initial stages of neural development (Munoz-Sanjuan et al. 2002). They followed up the screen by performing whole-mount in situ hybridization on the candidate clones, and finally, on those that were full-length, they tested them in functional assays of neuralization, either on their own or in combination with other neuralizing factors (Munoz-Sanjuan et al. 2002). In this way, they were able to carry the experiment from profiling, to confirmation by RT-PCR, to determination of expression patterns of the candidate genes by whole-mount in situ hybridization, and finally to function analysis, confirming the value of this system to quickly go from profiling experiments using microarray to studies of gene function in vivo.

Since this early work, other groups have begun to generate more *Xenopus* cDNA and oligonucleotide microarrays, in some cases using sequence information to try to limit the amount of redundancy in the arrays (Tran et al. 2002; Chung et al. 2004; Konig et al. 2004; Chalmers et al. 2005; Shin et al. 2005). However, the most comprehensive use of microarray analysis in *Xenopus* has been done by Christof Niehrs and colleagues (Baldessari et al. 2005). In their study they performed gene profiling experiments with 37 different samples and followed this with cluster analysis to determine coregulated gene clusters. The goal was to determine whether they could identify additional synexpression groups to the ones they had identified previously in their large-scale whole-mount in situ hybridization screens (Gawantka et al. 1998; Pollet et al. 2005), described previously. To be as comprehensive as possible, they took samples from different stages of development; they dissected embryos into different domains at four stages; they also generated ventralized and dorsalized embryos; and finally they took samples from adult organs. In their analysis, they identified 15 coregulated gene clusters, ranging in size from 1328 clones to 23 clones per cluster (Baldessari et al. 2005). These gene clusters were the following: a protein biosynthesis group, a chromatin group, an RNA processing group, a respiratory chain and Krebs cycle group, a cell cycle group, an endoplasmic reticulum group, a vesicle transport group, a synaptic vesicle group, a microtubule group, an intermediate filament group, an epithelial protein group, a collagen group, a cement gland/hatching gland group, a muscle group, and a muscle and heart group. In addition, they filtered the data in order to extract new region and tissue specific genes. Undoubtedly these data will add to the wealth of information already present in *Xenopus* and will provide essential information for further functional experiments on these and other genes in the future.

In addition to the comprehensive analysis by the Niehrs group described above, many other gene profiling studies using microarrays have been published in the past year. These include the use of microarrays to identify gene targets of retinoid signaling (Arima et al. 2005), targets of FGF signaling (Chung et al. 2004), targets of BMP signaling (Peiffer et al. 2005), and genes activated when BMP signaling is inhibited, thus identifying neural specific genes (Shin et al. 2005). Thus in a short period of time, it has been possible to elucidate the gene targets of some of the major signaling pathways in early development in the frog, using gene profiling experiments on microarrays. In addition, microarray experiments will allow the identification of direct targets of transcription factors, as has been done already for VegT (Taverner et al. 2005). To this end, *Xenopus* is an excellent system for identifying direct targets of transcription factors, since it is relatively easy to combine the use of hormone-inducible transcription factors with the protein translation inhibitors, such as cycloheximide (Smith et al. 1991; Kolm and Sive 1995; Taverner et al. 2005). Therefore it will not be long before such experiments are performed on a large number of transcription factors using microarrays in *Xenopus*.

Eventually it will be possible to begin to build complex gene regulatory networks in *Xenopus*, as is being done for *Drosophila* and the sea urchin (Davidson et al. 2002; Levine and Davidson 2005). Indeed, such gene networks have already been initiated in *Xenopus* (Loose and Patient 2004; Koide et al. 2005). Although such efforts are only just beginning, there is little doubt that with the recent acquisition of extensive genomic information, together with the facility to perform elegant large-scale functional screens, large-scale whole-mount in situ hybridization screens, microarray experiments, and transgenic experiments (Kroll and Amaya 1996; Bronchain et al. 1999; Sparrow et al. 2000; Karaulanov et al. 2004), a complete understanding of the genomic program of development will soon be available in *Xenopus*.

REFERENCES

Allan, V.J. 1993. Assay of membrane motility in interphase and metaphase *Xenopus* extracts. *Methods Cell Biol.* **39:** 203–226.

Altmann, C.R., Bell, E., Sczyrba, A., Pun, J., Bekiranov, S., Gaasterland, T., and Brivanlou, A.H. 2001. Microarray-based analysis of early development in *Xenopus laevis*. *Dev. Biol.* **236:** 64–75.

Amaya, E., Musci, T.J., and Kirschner, M.W. 1991. Expression of a dominant negative mutant of the FGF receptor disrupts mesoderm formation in *Xenopus* embryos. *Cell* **66:** 257–270.

Amaya, E., Offield, M.F., and Grainger, R.M. 1998. Frog genetics: *Xenopus* tropicalis jumps into the future. *Trends Genet.* **14:** 253–255.

Arias, E.E. and Walter, J.C. 2004. Initiation of DNA replication in *Xenopus* egg extracts. *Front. Biosci.* **9:** 3029–3045.

Arima, K., Shiotsugu, J., Niu, R., Khandpur, R., Martinez, M., Shin, Y., Koide, T., Cho, K.W., Kitayama, A., Ueno, N., et al. 2005. Global analysis of RAR-responsive genes in the *Xenopus* neurula using cDNA microarrays. *Dev. Dyn.* **232:** 414–431.

Baker, J.C. and Harland, R.M. 1996. A novel mesoderm inducer, Madr2, functions in the activin signal transduction pathway. *Genes & Dev.* **10:** 1880–1889.

Baldessari, D., Shin, Y., Krebs, O., Konig, R., Koide, T., Vinayagam, A., Fenger, U., Mochii, M., Terasaka, C., Kitayama, A., et al. 2005. Global gene expression profiling and cluster analysis in *Xenopus laevis*. *Mech. Dev.* **122:** 441–475.

Bell, E., Munoz-Sanjuan, I., Altmann, C.R., Vonica, A., and Brivanlou, A.H. 2003. Cell fate specification and competence by Coco, a maternal BMP, TGFβ and Wnt inhibitor. *Development* **130:** 1381–1389.

Bell, G.W., Yatskievych, T.A., and Antin, P.B. 2004. GEISHA, a whole-mount in situ hybridization gene expression screen in chicken embryos. *Dev. Dyn.* **229:** 677–687.

Bettenhausen, B. and Gossler, A. 1995. Efficient isolation of novel mouse genes differentially expressed in early postimplantation embryos. *Genomics* **28:** 436–441.

Bottcher, R.T., Pollet, N., Delius, H., and Niehrs, C. 2004. The transmembrane protein XFLRT3 forms a complex with FGF receptors and promotes FGF signalling. *Nat. Cell Biol.* **6:** 38–44.

Bronchain, O.J., Hartley, K.O., and Amaya, E. 1999. A gene trap approach in *Xenopus*. *Curr. Biol.* **9:** 1195–1198.

Brown, P.O. and Botstein, D. 1999. Exploring the new world of the genome with DNA microarrays. *Nat. Genet.* **21:** 33–37.

Chalmers, A.D., Goldstone, K., Smith, J.C., Gilchrist, M., Amaya, E., and Papalopulu, N. 2005. A *Xenopus tropicalis* oligonucleotide microarray works across species using RNA from *Xenopus laevis*. *Mech. Dev.* **122:** 355–363.

Chen, J.A., Voigt, J., Gilchrist, M., Papalopulu, N., and Amaya, E. 2005. Identification of novel genes affecting mesoderm formation and morphogenesis through an enhanced large scale functional screen in *Xenopus*. *Mech. Dev.* **122:** 307–331.

Chung, H.A., Hyodo-Miura, J., Kitayama, A., Terasaka, C., Nagamune, T., and Ueno, N. 2004. Screening of FGF target genes in *Xenopus* by microarray: Temporal dissection of the signalling pathway using a chemical inhibitor. *Genes Cells* **9:** 749–761.

Crosier, P.S., Bardsley, A., Horsfield, J.A., Krassowska, A.K., Lavallie, E.R., Collins-Racie, L.A., Postlethwait, J.H., Yan, Y.L., McCoy, J.M., and Crosier, K. 2001. In situ hybridization screen in zebrafish for the selection of genes encoding secreted proteins. *Dev. Dyn.* **222:** 637–644.

Dasso, M., Smythe, C., Milarski, K., Kornbluth, S., and Newport, J.W. 1992. DNA replication and progression through the cell cycle. *Ciba Found. Symp.* **170:** 161–180; discussion 180–166.

Davidson, E.H., Rast, J.P., Oliveri, P., Ransick, A., Calestani, C., Yuh, C.H., Minokawa, T., Amore, G., Hinman, V., Arenas-Mena, C., et al. 2002. A genomic regulatory network for development. *Science* **295:** 1669–1678.

de la Barre, A.E., Robert-Nicoud, M., and Dimitrov, S. 1999. Assembly of mitotic chromosomes in *Xenopus* egg extract. *Methods Mol. Biol.* **119:** 219–229.

Desai, A., Murray, A., Mitchison, T.J., and Walczak, C.E. 1999. The use of *Xenopus* egg extracts to study mitotic spindle assembly and function in vitro. *Methods Cell Biol.* **61:** 385–412.

Domingos, P.M., Itasaki, N., Jones, C.M., Mercurio, S., Sargent, M.G., Smith, J.C., and Krumlauf, R. 2001. The Wnt/β-catenin pathway posteriorizes neural tissue in *Xenopus* by an indirect mechanism requiring FGF signalling. *Dev. Biol.* **239:** 148–160.

Domingos, P.M., Obukhanych, T.V., Altmann, C.R., and Hemmati-Brivanlou, A. 2002. Cloning and developmental expression of Baf57 in *Xenopus laevis*. *Mech. Dev.* **116:** 177–181.

Dupont, S., Zacchigna, L., Cordenonsi, M., Soligo, S., Adorno, M., Rugge, M., and Piccolo, S. 2005. Germ-layer specification and control of cell growth by Ectodermin, a Smad4 ubiquitin ligase. *Cell* **121:** 87–99.

Edlund, H. 2001. Developmental biology of the pancreas. *Diabetes* **50 (Suppl 1):** S5–S9.

Evans, E.K., Lu, W., Strum, S.L., Mayer, B.J., and Kornbluth, S. 1997. Crk is required for apoptosis in *Xenopus* egg extracts. *EMBO J.* **16:** 230–241.

Furthauer, M., Reifers, F., Brand, M., Thisse, B., and Thisse, C. 2001. sprouty4 acts in vivo as a feedback-induced antagonist of FGF signaling in zebrafish. *Development* **128:** 2175–2186.

Furthauer, M., Lin, W., Ang, S.L., Thisse, B., and Thisse, C. 2002. Sef is a feedback-induced antagonist of Ras/MAPK-mediated FGF signalling. *Nat. Cell Biol.* **4:** 170–174.

Gammill, L.S. and Bronner-Fraser, M. 2002. Genomic analysis of neural crest induction. *Development* **129:** 5731–5741.

Gawantka, V., Delius, H., Hirschfeld, K., Blumenstock, C., and Niehrs, C. 1995. Antagonizing the Spemann organizer: Role of the homeobox gene Xvent-1. *EMBO J.* **14:** 6268–6279.

Gawantka, V., Pollet, N., Delius, H., Vingron, M., Pfister, R., Nitsch, R., Blumenstock, C., and Niehrs, C. 1998. Gene expression screening in *Xenopus* identifies molecular pathways, predicts gene function and provides a global view of embryonic patterning. *Mech. Dev.* **77:** 95–141.

Gilchrist, M.J., Zorn, A.M., Voigt, J., Smith, J.C., Papalopulu, N., and Amaya, E. 2004. Defining a large set of full-length clones from a *Xenopus tropicalis* EST project. *Dev. Biol.* **271:** 498–516.

Glinka, A., Delius, H., Blumenstock, C., and Niehrs, C. 1996. Combinatorial signalling by Xwnt-11 and Xnr3 in the organizer epithelium. *Mech. Dev.* **60:** 221–231.

Glinka, A., Wu, W., Delius, H., Monaghan, A.P., Blumenstock, C., and Niehrs, C. 1998. Dickkopf-1 is a member of a new family of secreted proteins and functions in head induction. *Nature* **391:** 357–362.

Glotzer, M., Murray, A.W., and Kirschner, M.W. 1991. Cyclin is degraded by the ubiquitin pathway. *Nature* **349:** 132–138.

Grammer, T.C., Liu, K.J., Mariani, F.V., and Harland, R.M. 2000. Use of large-scale expression cloning screens in the *Xenopus laevis* tadpole to identify gene function. *Dev. Biol.* **228:** 197–210.

Grammer, T.C., Khokha, M.K., Lane, M.A., Lam, K., and Harland, R.M. 2005. Identification of mutants in inbred *Xenopus tropicalis*. *Mech. Dev.* **122:** 263–272.

Gurdon, J.B. 1975. Attempts to analyse the biochemical basis of regional differences in animal eggs. *Ciba Found. Symp.* 223–239.

Haushalter, K.A., Todd Stukenberg, M.W., Kirschner, M.W., and Verdine, G.L. 1999. Identification of a new uracil-DNA glycosylase family by expression cloning using synthetic inhibitors. *Curr. Biol.* **9:** 174–185.

Heasman, J. 2002. Morpholino oligos: Making sense of antisense? *Dev. Biol.* **243:** 209–214.

Heasman, J., Kofron, M., and Wylie, C. 2000. β-Catenin signaling activity dissected in the early *Xenopus* embryo: A novel antisense approach. *Dev. Biol.* **222:** 124–134.

Hengartner, M.O. 1995. Out-of body experiences: Cell-free cell death. *Bioessays* **17:** 549–552.

Holloway, S.L., Glotzer, M., King, R.W., and Murray, A.W. 1993. Anaphase is initiated by proteolysis rather than by the inactivation of maturation-promoting factor. *Cell* **73:** 1393–1402.

Hsu, D.R., Economides, A.N., Wang, X., Eimon, P.M., and Harland, R.M. 1998. The *Xenopus* dorsalizing factor Gremlin identifies a novel family of secreted proteins that antagonize BMP activities. *Mol. Cell* **1:** 673–683.

Hurlstone, A.F., Haramis, A.P., Wienholds, E., Begthel, H., Korving, J., Van Eeden, F., Cuppen, E., Zivkovic, D., Plasterk, R.H., and Clevers, H. 2003. The Wnt/β-catenin pathway regulates cardiac valve formation. *Nature* **425:** 633–637.

Itasaki, N., Jones, C.M., Mercurio, S., Rowe, A., Domingos, P.M., Smith, J.C., and Krumlauf, R. 2003. Wise, a context-dependent activator and inhibitor of Wnt signalling. *Development* **130:** 4295–4305.

Jen, W.C., Gawantka, V., Pollet, N., Niehrs, C., and Kintner, C. 1999. Periodic repression of Notch pathway genes governs the segmentation of *Xenopus* embryos. *Genes & Dev.* **13:** 1486–1499.

Julius, D., MacDermott, A.B., Axel, R., and Jessell, T.M. 1988. Molecular characterization of a functional cDNA encoding the serotonin 1c receptor. *Science* **241:** 558–564.

Karaulanov, E., Knochel, W., and Niehrs, C. 2004. Transcriptional regulation of BMP4 synexpression in transgenic *Xenopus*. *EMBO J.* **23:** 844–856.

Kazanskaya, O., Glinka, A., del Barco Barrantes, I., Stannek, P., Niehrs, C., and Wu, W. 2004. R-Spondin2 is a secreted activator of Wnt/β-catenin signaling and is required for *Xenopus* myogenesis. *Dev. Cell* **7:** 525–534.

Kenwrick, S., Amaya, E., and Papalopulu, N. 2004. Pilot morpholino screen in *Xenopus tropicalis* identifies a novel gene involved in head development. *Dev. Dyn.* **229:** 289–299.

King, R.W., Peters, J.M., Tugendreich, S., Rolfe, M., Hieter, P., and Kirschner, M.W. 1995. A 20S complex containing CDC27 and CDC16 catalyzes the mitosis-specific conjugation of ubiquitin to cyclin B. *Cell* **81:** 279–288.

King, R.W., Glotzer, M., and Kirschner, M.W. 1996. Mutagenic analysis of the destruction signal of mitotic cyclins and structural characterization of ubiquitinated intermediates. *Mol. Biol. Cell* **7:** 1343–1357.

King, R.W., Lustig, K.D., Stukenberg, P.T., McGarry, T.J., and Kirschner, M.W. 1997. Expression cloning in the test tube. *Science* **277:** 973–974.

Koide, T., Hayata, T., and Cho, K.W. 2005. *Xenopus* as a model system to study transcriptional regulatory networks. *Proc. Natl. Acad. Sci.* **102:** 4943–4948.

Kolm, P.J. and Sive, H.L. 1995. Efficient hormone-inducible protein function in *Xenopus laevis*. *Dev. Biol.* **171:** 267–272.

Konig, R., Baldessari, D., Pollet, N., Niehrs, C., and Eils, R. 2004. Reliability of gene expression ratios for cDNA microarrays in multiconditional experiments with a reference design. *Nucleic Acids Res.* **32:** e29.

Kornbluth, S. 1997. Apoptosis in *Xenopus* egg extracts. *Methods Enzymol.* **283:** 600–614.

Kothakota, S., Azuma, T., Reinhard, C., Klippel, A., Tang, J., Chu, K., McGarry, T.J., Kirschner, M.W., Koths, K., Kwiatkowski, D.J., et al. 1997. Caspase-3–generated fragment of gelsolin: Effector of morphological change in apoptosis. *Science* **278:** 294–298.

Krieg, P.A. and Melton, D.A. 1984. Functional messenger RNAs are produced by SP6 in vitro transcription of cloned cDNAs. *Nucleic Acids Res.* **12:** 7057–7070.

Kroll, K.L. and Amaya, E. 1996. Transgenic *Xenopus* embryos from sperm nuclear transplantations reveal FGF signaling requirements during gastrulation. *Development* **122:** 3173–3183.

Kroll, K.L., Salic, A.N., Evans, L.M., and Kirschner, M.W. 1998. Geminin, a neuralizing molecule that demarcates the future neural plate at the onset of gastrulation. *Development* **125:** 3247–3258.

Kudoh, T., Tsang, M., Hukriede, N.A., Chen, X., Dedekian, M., Clarke, C.J., Kiang, A., Schultz, S., Epstein, J.A., Toyama, R., et al. 2001. A gene expression screen in zebrafish embryogenesis. *Genome Res.* **11:** 1979–1987.

Lahaye, K., Kricha, S., and Bellefroid, E.J. 2002. XNAP, a conserved ankyrin repeat-containing protein with a role in the Notch pathway during *Xenopus* primary neurogenesis. *Mech. Dev.* **110:** 113–124.

Lamar, E., Deblandre, G., Wettstein, D., Gawantka, V., Pollet, N., Niehrs, C., and Kintner, C. 2001. Nrarp is a novel intracellular component of the Notch signaling pathway. *Genes & Dev.* **15:** 1885–1899.

Laurent, M.N., Blitz, I.L., Hashimoto, C., Rothbacher, U., and Cho, K.W. 1997. The *Xenopus* homeobox gene twin mediates Wnt induction of goosecoid in establishment of Spemann's organizer. *Development* **124:** 4905–4916.

Lemaire, P., Garrett, N., and Gurdon, J.B. 1995. Expression cloning of Siamois, a *Xenopus* homeobox gene expressed in dorsal-vegetal cells of blastulae and able to induce a complete secondary axis. *Cell* **81:** 85–94.

Levine, M. and Davidson, E.H. 2005. Gene regulatory networks for development. *Proc. Natl. Acad. Sci.* **102:** 4936–4942.

Li, Y., Fenger, U., Niehrs, C., and Pollet, N. 2003. Cyclic expression of esr9 gene in *Xenopus* presomitic mesoderm. *Differentiation* **71:** 83–89.

Lohka, M.J. 1998. Analysis of nuclear envelope assembly using extracts of *Xenopus* eggs. *Methods Cell Biol.* **53:** 367–395.

Loose, M. and Patient, R. 2004. A genetic regulatory network for *Xenopus* mesendoderm formation. *Dev. Biol.* **271:** 467–478.

Lustig, K.D. and Kirschner, M.W. 1995. Use of an oocyte expression assay to reconstitute inductive signaling. *Proc. Natl. Acad. Sci.* **92:** 6234–6238.

Lustig, K.D., Kroll, K., Sun, E., Ramos, R.R., Elmendorf, H., and Kirschner, M.W. 1996a. A *Xenopus* nodal-related gene that acts in synergy with noggin to induce complete secondary axis and notochord formation. *Development* **122:** 3275–3282.

Lustig, K.D., Kroll, K.L., Sun, E.E., and Kirschner, M.W. 1996b. Expression cloning of a *Xenopus* T-related gene (Xombi) involved in mesodermal patterning and blastopore lip formation. *Development* **122:** 4001–4012.

Lustig, K.D., Stukenberg, P.T., McGarry, T.J., King, R.W., Cryns, V.L., Mead, P.E., Zon, L.I., Yuan, J., and Kirschner, M.W. 1997. Small pool expression screening: Identification of genes involved in cell cycle control, apoptosis, and early development. *Methods Enzymol.* **283:** 83–99.

Makabe, K.W., Kawashima, T., Kawashima, S., Minokawa, T., Adachi, A., Kawamura, H., Ishikawa, H., Yasuda, R., Yamamoto, H., Kondoh, K., et al. 2001. Large-scale cDNA analysis of the maternal genetic information in the egg of *Halocynthia roretzi* for a gene expression catalog of ascidian development. *Development* **128:** 2555–2567.

Mandato, C.A., Weber, K.L., Zandy, A.J., Keating, T.J., and Bement, W.M. 2001. *Xenopus* egg extracts as a model system for analysis of microtubule, actin filament, and intermediate filament interactions. *Methods Mol. Biol.* **161:** 229–239.

Mariani, F.V. and Harland, R.M. 1998. XBF-2 is a transcriptional repressor that converts ectoderm into neural tissue. *Development* **125:** 5019–5031.

Masu, Y., Nakayama, K., Tamaki, H., Harada, Y., Kuno, M., and Nakanishi, S. 1987. cDNA cloning of bovine substance-K receptor through oocyte expression system. *Nature* **329:** 836–838.

McGarry, T.J. and Kirschner, M.W. 1998. Geminin, an inhibitor of DNA replication, is degraded during mitosis. *Cell* **93:** 1043–1053.

Mead, P.E., Brivanlou, I.H., Kelley, C.M., and Zon, L.I. 1996. BMP-4–responsive regulation of dorsal-ventral patterning by the homeobox protein Mix.1. *Nature* **382:** 357–360.

Melton, D.A., Krieg, P.A., Rebagliati, M.R., Maniatis, T., Zinn, K., and Green, M.R. 1984. Efficient in vitro synthesis of biologically active RNA and RNA hybridization probes from plasmids containing a bacteriophage SP6 promoter. *Nucleic Acids Res.* **12:** 7035–7056.

Munoz-Sanjuan, I., Bell, E., Altmann, C.R., Vonica, A., and Brivanlou, A.H. 2002. Gene profiling during neural induction in *Xenopus laevis*: Regulation of BMP signaling by post-transcriptional mechanisms and TAB3, a novel TAK1-binding protein. *Development* **129:** 5529–5540.

Murray, A.W. and Kirschner, M.W. 1989a. Cyclin synthesis drives the early embryonic cell cycle. *Nature* **339:** 275–280.

———. 1989b. Dominoes and clocks: the union of two views of the cell cycle. *Science* **246:** 614–621.

Murray, A.W., Solomon, M.J., and Kirschner, M.W. 1989. The role of cyclin synthesis and degradation in the control of maturation promoting factor activity. *Nature* **339:** 280–286.

Neidhardt, L., Gasca, S., Wertz, K., Obermayr, F., Worpenberg, S., Lehrach, H., and Herrmann, B.G. 2000. Large-scale screen for genes controlling mammalian embryogenesis, using high-throughput gene expression analysis in mouse embryos. *Mech. Dev.* **98:** 77–94.

Newmeyer, D.D. and Wilson, K.L. 1991. Egg extracts for nuclear import and nuclear assembly reactions. *Methods Cell Biol.* **36:** 607–634.

Newmeyer, D.D., Farschon, D.M., and Reed, J.C. 1994. Cell-free apoptosis in *Xenopus* egg extracts: Inhibition by Bcl-2 and requirement for an organelle fraction enriched in mitochondria. *Cell* **79:** 353–364.

Niehrs, C. and Pollet, N. 1999. Synexpression groups in eukaryotes. *Nature* **402:** 483–487.

Nishikata, T., Yamada, L., Mochizuki, Y., Satou, Y., Shin-i, T., Kohara, Y., and Satoh, N. 2001. Profiles of maternally expressed genes in fertilized eggs of *Ciona intestinalis*. *Dev. Biol.* **238:** 315–331.

Noramly, S., Zimmerman, L., Cox, A., Aloise, R., Fisher, M., and Grainger, R.M. 2005. A gynogenetic screen to isolate naturally occurring recessive mutations in *Xenopus tropicalis*. *Mech. Dev.* **122:** 273–287.

Nutt, S.L., Bronchain, O.J., Hartley, K.O., and Amaya, E. 2001. Comparison of morpholino based translational inhibition during the development of *Xenopus laevis* and *Xenopus tropicalis*. *Genesis* **30:** 110–113.

Onichtchouk, D., Gawantka, V., Dosch, R., Delius, H., Hirschfeld, K., Blumenstock, C., and Niehrs, C. 1996. The Xvent-2 homeobox gene is part of the BMP-4 signalling pathway controlling [correction of controlling] dorsoventral patterning of *Xenopus* mesoderm. *Development* **122:** 3045–3053.

Onichtchouk, D., Glinka, A., and Niehrs, C. 1998. Requirement for Xvent-1 and Xvent-2 gene function in dorsoventral patterning of *Xenopus* mesoderm. *Development* **125:** 1447–1456.

Onichtchouk, D., Chen, Y.G., Dosch, R., Gawantka, V., Delius, H., Massague, J., and Niehrs, C. 1999. Silencing of TGF-β signalling by the pseudoreceptor BAMBI. *Nature* **401:** 480–485.

Pain, V.M., Patrick, T.D., Cox, R., and Morley, S.J. 1998. Analysis of translational activity of extracts derived from oocytes and eggs of *Xenopus laevis*. *Methods Mol. Biol.* **77:** 195–209.

Peiffer, D.A., Von Bubnoff, A., Shin, Y., Kitayama, A., Mochii, M., Ueno, N., and Cho, K.W. 2005. A *Xenopus* DNA microarray approach to identify novel direct BMP target genes involved in early embryonic development. *Dev. Dyn.* **232:** 445–456.

Pera, E.M., Kim, J.I., Martinez, S.L., Brechner, M., Li, S.Y., Wessely, O., and De Robertis, E.M. 2002. Isthmin is a novel secreted protein expressed as part of the Fgf-8 synexpression group in the *Xenopus* midbrain-hindbrain organizer. *Mech. Dev.* **116:** 169–172.

Pfaller, R., Smythe, C., and Newport, J.W. 1991. Assembly/disassembly of the nuclear envelope membrane: Cell cycle–dependent binding of nuclear membrane vesicles to chromatin in vitro. *Cell* **65:** 209–217.

Pollet, N., Schmidt, H.A., Gawantka, V., Vingron, M., and Niehrs, C. 2000. Axeldb: A *Xenopus laevis* database focusing on gene expression. *Nucleic Acids Res.* **28:** 139–140.

Pollet, N., Muncke, N., Verbeek, B., Li, Y., Fenger, U., Delius, H., and Niehrs, C. 2005. An atlas of differential gene expression during early *Xenopus* embryogenesis. *Mech. Dev.* **122:** 365–439.

Quiring, R., Wittbrodt, B., Henrich, T., Ramialison, M., Burgtorf, C., Lehrach, H., and Wittbrodt, J. 2004. Large-scale expression screening by automated whole-mount in situ hybridization. *Mech. Dev.* **121:** 971–976.

Salic, A.N., Kroll, K.L., Evans, L.M., and Kirschner, M.W. 1997. Sizzled: a secreted Xwnt8 antagonist expressed in the ventral marginal zone of *Xenopus* embryos. *Development* **124:** 4739–4748.

Satou, Y., Takatori, N., Yamada, L., Mochizuki, Y., Hamaguchi, M., Ishikawa, H., Chiba, S., Imai, K., Kano, S., Murakami, S.D., et al. 2001. Gene expression profiles in *Ciona intestinalis* tailbud embryos. *Development* **128:** 2893–2904.

Shin, Y., Kitayama, A., Koide, T., Peiffer, D.A., Mochii, M., Liao, A., Ueno, N., and Cho, K.W. 2005. Identification of neural genes using *Xenopus* DNA microarrays. *Dev. Dyn.* **232:** 432–444.

Shirasu, M., Yonetani, A., and Walczak, C.E. 1999. Microtubule dynamics in *Xenopus* egg extracts. *Microsc. Res. Tech.* **44:** 435–445.

Sivak, J.M., Petersen, L.F., and Amaya, E. 2005. FGF signal interpretation is directed by sprouty and spred proteins during mesoderm formation. *Dev. Cell* **8:** 689–701.

Smith, W.C. and Harland, R.M. 1991. Injected Xwnt-8 RNA acts early in *Xenopus* embryos to promote formation of a vegetal dorsalizing center. *Cell* **67:** 753–765.

———. 1992. Expression cloning of noggin, a new dorsalizing factor localized to the Spemann organizer in *Xenopus* embryos. *Cell* **70:** 829–840.

Smith, J.C., Price, B.M., Green, J.B., Weigel, D., and Herrmann, B.G. 1991. Expression of a *Xenopus* homolog of Brachyury (T) is an immediate-early response to mesoderm induction. *Cell* **67:** 79–87.

Smith, W.C., McKendry, R., Ribisi Jr., S., and Harland, R.M. 1995. A nodal-related gene defines a physical and functional domain within the Spemann organizer. *Cell* **82:** 37–46.

Smythe, C. and Newport, J.W. 1991. Systems for the study of nuclear assembly, DNA replication, and nuclear breakdown in *Xenopus laevis* egg extracts. *Methods Cell Biol.* **35:** 449–468.

Sparrow, D.B., Latinkic, B., and Mohun, T.J. 2000. A simplified method of generating transgenic *Xenopus*. *Nucleic Acids Res.* **28:** e12.

Stemple, D.L. 2004. TILLING: A high-throughput harvest for functional genomics. *Nat. Rev. Genet.* **5:** 145–150.

Stukenberg, P.T., Lustig, K.D., McGarry, T.J., King, R.W., Kuang, J., and Kirschner, M.W. 1997. Systematic identification of mitotic phosphoproteins. *Curr. Biol.* **7:** 338–348.

Suzuki, A. and Hemmati-Brivanlou, A. 2000. *Xenopus* embryonic E2F is required for the formation of ventral and posterior cell fates during early embryogenesis. *Mol. Cell* **5:** 217–229.

Taelman, V., Opdecamp, K., Avalosse, B., Ryan, K., and Bellefroid, E.J. 2001. Xath2, a bHLH gene expressed during a late transition stage of neurogenesis in the forebrain of *Xenopus* embryos. *Mech. Dev.* **101:** 199–202.

Taverner, N.V., Kofron, M., Shin, Y., Kabitschke, C., Gilchrist, M.J., Wylie, C., Cho, K.W., Heasman, J., and Smith, J.C. 2005. Microarray based identification of VegT targets in *Xenopus*. *Mech. Dev.* **122:** 333–354.

Thisse, B., Heyer, V., Lux, A., Alunni, V., Degrave, A., Seiliez, I., Kirchner, J., Parkhill, J.P., and Thisse, C. 2004. Spatial and temporal expression of the zebrafish genome by large-scale in situ hybridization screening. *Methods Cell Biol.* **77:** 505–519.

Thommes, P. and Blow, J.J. 1997. The DNA replication licensing system. *Cancer Surv.* **29:** 75–90.

Thut, C.J., Rountree, R.B., Hwa, M., and Kingsley, D.M. 2001. A large-scale in situ screen provides molecular evidence for the induction of eye anterior segment structures by the developing lens. *Dev. Biol.* **231:** 63–76.

Tran, P.H., Peiffer, D.A., Shin, Y., Meek, L.M., Brody, J.P., and Cho, K.W. 2002. Microarray optimizations: Increasing spot accuracy and automated identification of true microarray signals. *Nucleic Acids Res.* **30:** e54.

Tsang, M., Friesel, R., Kudoh, T., and Dawid, I.B. 2002. Identification of Sef, a novel modulator of FGF signalling. *Nat. Cell Biol.* **4:** 165–169.

Voigt, J., Chen, J.A., Gilchrist, M., Amaya, E., and Papalopulu, N. 2005. Expression cloning screening of a unique and full-length set of cDNA clones is an efficient method for identifying genes involved in *Xenopus* neurogenesis. *Mech. Dev.* **122:** 289–306.

von Ahsen, O. and Newmeyer, D.D. 2000. Cell-free apoptosis in *Xenopus laevis* egg extracts. *Methods Enzymol.* **322:** 183–198.

Weinstein, D.C., Marden, J., Carnevali, F., and Hemmati-Brivanlou, A. 1998. FGF-mediated mesoderm induction involves the Src-family kinase Laloo. *Nature* **394:** 904–908.

Wischnewski, J., Solter, M., Chen, Y., Hollemann, T., and Pieler, T. 2000. Structure and expression of *Xenopus* karyopherin-β3: Definition of a novel synexpression group related to ribosome biogenesis. *Mech. Dev.* **95:** 245–248.

Yu, H., King, R.W., Peters, J.M., and Kirschner, M.W. 1996. Identification of a novel ubiquitin-conjugating enzyme involved in mitotic cyclin degradation. *Curr. Biol.* **6:** 455–466.

Zimmerman, L.B., De Jesus-Escobar, J.M., and Harland, R.M. 1996. The Spemann organizer signal noggin binds and inactivates bone morphogenetic protein 4. *Cell* **86:** 599–606.

Zohn, I.E. and Brivanlou, A.H. 2001. Expression cloning of *Xenopus* Os4, an evolutionarily conserved gene, which induces mesoderm and dorsal axis. *Dev. Biol.* **239:** 118–131.

Zorn, A.M., Barish, G.D., Williams, B.O., Lavender, P., Klymkowsky, M.W., and Varmus, H.E. 1999. Regulation of Wnt signaling by Sox proteins: XSox17 α/β and XSox3 physically interact with β-catenin. *Mol. Cell* **4:** 487–498.

11

Chicken Genome: Current Status and Future Opportunities

David W. Burt
Department of Genomics and Genetics
Roslin Institute (Edinburgh)
Midlothian EH25 9PS, United Kingdom

AVIAN GENOMICS HAS ITS ORIGINS IN GENETIC LINKAGE MAPPING (Burt and Cheng 1998), but our knowledge of the chicken genome has been transformed in recent years, mostly through the analysis of large numbers of partial cDNA sequences (Abdrakhmanov et al. 2000; Tirunagaru et al. 2000; Boardman et al. 2002) and culminating with the chicken genome sequence (Hillier et al. 2004). These were landmark events in our understanding of avian biology, developmental biology, and the evolution of vertebrates and will facilitate applications in agriculture and medicine.

Chicken research has had a significant impact on fundamental biology and the chicken has been a popular model organism for at least 100 years, for example, with the discovery of B cells and tumor viruses (Brown et al. 2003). Ready access to the chicken embryo using incubated eggs and the ease of manipulation make this system ideal for studies of vertebrate development (Stern 2004, 2005). The chicken has been used in many of the classical studies on the molecular basis of patterning in the vertebrate embryo, in particular, the limb bud. In recent times, other model organisms, such as the mouse and zebrafish, have been in greater demand because of increased genetic resources and the ability to manipulate their genomes. The chicken EST and genome programs have removed many of these limitations in the chicken. In addition, new tools such as the electroporation of chicken embryos and the use of RNAi to knock down gene expression are likely to make the chicken embryo a powerful model for the molecular study of development in vertebrates (Stern 2004, 2005).

During the past 80 years, modern selective breeding has made spectacular progress in both egg and meat production traits (Burt 2002). World egg production has increased to 795 billion/year in 2002 (Commodity Research Bureau [CRB]) and broiler meat to 6.5 million tons/year (USDA foreign agricultural service [FAS]), during this period. Associated with these successes have been a number of undesirable traits. In meat-type chickens, there has been an increase in the incidence of congenital disorders, such as ascites and lameness, reduced fertility, and reduced resistance to infectious disease. In egg-type chickens, there has been an increase in the incidence of osteoporosis associated with increased egg production. Given the possibility that genetic progress in egg and meat production will reach its limit within the next 20 years (Burt 2002), priorities in the poultry industry will be to reduce these costs and develop new products. The consumer wants high-quality products (e.g., increased egg shell strength), which requires greater uniformity and predictability in production. With an increased requirement for food safety, there will be a need to reduce the use of chemicals and antibiotics and increase genetic resistance to pathogens. These new traits are difficult and costly to measure by conventional genetic selection, and the developments in poultry genomics in the last few years promises new solutions to these problems.

In this report, key features and limitations of the draft chicken genome sequence will be discussed. More detailed reviews have been presented elsewhere from the viewpoint of genomics (Burt 2004a; Dequeant and Pourquié 2005), developmental biology (Burt 2004b; Stern 2005), evolution (Ellegren 2005), and genomic tools (Antin and Konieczka 2005).

CHICKEN GENOME

Genome Sequence

The current draft of the chicken genome (Hillier et al. 2004) was assembled using a whole-genome sequencing strategy, including BAC, fosmid, and plasmid paired-end reads (WASHU). This approach produced a high-quality assembly, in part because of the relatively small size of the chicken genome, one third that of a typical mammal. However, it was the low repetitive DNA content, only 11% compared with 40%–50% found in mammals, that was a key contributing factor to the quality of the final assembly. This sequence employed DNA from a single inbred female

Jungle Fowl (*Gallus gallus gallus*, the ancestor of domesticated chickens; Fumihito et al. 1994) and represented a 6.6-fold coverage of the genome. Together with genetic and BAC maps, almost 100,000 contigs were assembled into a scaffold of 907 Mb, or 86% of a 1050-Mb genome. In birds, it is the female that is the heterogametic sex, with single copies of the Z and W chromosomes. Therefore, these chromosomes were poorly represented in the final assembly. In addition, unlike the rest of the genome, the W chromosome has a high repeat content and so very little sequence was assembled. Targeted sequencing of the sex chromosomes will be necessary to complete their assemblies. For autosomes, sequence coverage was 98% based on overlaps with an independent set of BAC clones sequenced to high quality. Overlaps with cDNA clones suggested 5%–10% of genes were missing from the final assembly; gene duplications and GC-rich sequences were a particular problem. The MHC region on chromosome 16, a rich source of duplicated genes, was very poorly represented. Further work to complete the chicken genome sequence to a high quality for comparative genomics and gene discovery is required.

Genome Organization

A unique characteristic of avian genomes is the large variability in chromosome size. In addition to a pair of sex chromosomes, chickens have 38 pairs of autosomes: 5 macro-, 5 intermediate, and 28 microchromosomes. Since each chromosome arm must have at least one obligate crossover, it follows that the microchromosomes will have the highest rate of recombination. Comparison of genetic maps (Schmid et al. 2005) and genome sequences confirms this expectation, with crossover rates of 2.8 cM/Mb for macrochromosomes and 6.4 cM/Mb for microchromosomes. This is in contrast to 1–2 cM/Mb for most human chromosomes, making the chicken ideal for genetic linkage studies. High-resolution genetic maps will be necessary to define variation in recombination rates within chromosomes.

Many sequence characteristics, such as %GC content, CpG island density, and gene density, show clear relationships with chromosome size and therefore recombination rate (Table 1). However, we must be cautious about making any conclusions on cause and effect with these correlations (Fazzari and Greally 2004). The density of genes is highest on the microchromosomes, confirming earlier conclusions based on mapping genes (Smith et al. 2000) and CpG islands (McQueen et al. 1996). The estimated number of CpG islands based on bioinformatics approaches depends on the definition in use. In this case (Hillier et al. 2004), ~70,000

Table 1. General characteristics of macro- and microchromosomes

Feature	Macrochromosomes	Microchromosomes	Reference
Cytogenetic band type	G-band	R-band	Ponce de Leon et al. 1992
Gene density (per Mb)	9.0 to 15.4	13.8 to 41.2	Hillier et al. 2004
Intron length (bp)	4066 to 5742	1867 to 4128	Hillier et al. 2004
Exon length (bp)	164 to 171	157 to 172	Hillier et al. 2004
Intergenic gap length (kb)	18 to 31	8 to 24	Hillier et al. 2004
Intron length/ exon length	24.6 to 34.6	11.7 to 23.2	Hillier et al. 2004
%G + C content	38.4 to 40.1	40.9 to 50.1	Hillier et al. 2004
CpG island density (per Mb)	29 to 49	73 to 266	Hillier et al. 2004
LINEs (%)	6.0 to 11.9	2.5 to 10.0	Hillier et al. 2004
SINEs (%)	~none	~none	Hillier et al. 2004
Synonymous rate	0.090 to 0.125	0.111 to 0.156	Axelsson et al. 2005
Nonsynonymous rate	0.011 to 0.021	0.007 to 0.016	Axelsson et al. 2005
Ka/Ks ratio	0.128 to 0.360	0.066 to 0.177	Axelsson et al. 2005
GC3 (%)	49 to 53	56 to 65	Hillier et al. 2004
DNA replication	Late	Early	Ponce de Leon et al. 1992
Recombination rate (cM/Mb)	2.5 to 3.2	2.5 to 17.1	Hillier et al. 2004

CpG islands were predicted in the chicken, with 38% of these located in regions of conserved synteny with mammalian genomes. Since 48% are associated with a gene, CpG island density mimics gene density and is highest on microchromosomes. Conversely, sizes of introns and intergenic regions and density of repetitive elements correlate negatively with gene density and are reduced on microchromosomes. If we assume that genomes balance selective constraints favoring DNA loss over those that favor expansion and that selection will be most efficient in regions of high recombination where linkage of alleles are more readily broken (Hill and Robertson 1966), then the correlation of the densities of genes, CpG islands, repeats, etc. with chromosome size (and therefore recombination rate) is to be expected.

Comparison of orthologous chicken and turkey sequences revealed that different chromosome size classes are subject to different evolutionary forces

(Axelsson et al. 2005). Microchromosomes show 18% higher sequence divergence in introns and a 26% higher rate of synonymous substitution in coding sequences than macrochromosomes, indicating that the smaller chromosomes are more susceptible to germline mutations. A possible cause for the differences in mutation rate is "biased-gene-conversion" (Meunier and Duret 2004), a recombination-induced mutation mechanism.

Ever since the first gene maps were created (Haldane 1927), comparative maps have been used to examine the evolution of the vertebrate genome. Comparisons between the early gene maps of human and chicken (Burt et al. 1999) suggested extensive conservation of synteny, possibly more than found between mouse and human. The comparison of chicken with mammalian and fish genomes has confirmed and extended this view (Bourque et al. 2005). The estimated number of interchromosomal rearrangements between the mammalian ancestor and chicken, during an estimated period of 500 million years (Myr), is almost the same as the number found in the mouse lineage, over the course of ~87 Myr.

Genes and Proteins

A major benefit of the chicken genome sequence has been the set of gene predictions. The most conservative evidence-based approach of Ensembl generated 17,709 predictions (Table 2). The comparative ab initio methods, TWINSCAN (Korf et al. 2001) and SGP-2 (Syntenic Gene Prediction-2) (Parra et al. 2003), predict larger gene sets but likely include false positives. In total, there may be 20,000–23,000 genes; suggesting we still have more to learn about gene prediction (Eyras et al. 2005). When used to identify novel genes missed in the current human gene set (Ensembl 22,287 genes), only an additional 37 were predicted (Castelo et al. 2005), which suggests we have identified most of the "conserved" genes found in birds and mammals. Only 75 processed (or retrotransposed) pseudogenes were found in the chicken genome (Hillier et al. 2004), compared with 15,000 in mammals. The reason for this low number may be the sequence specificity of reverse transcription by avian LINES (long interspersed elements). Mammalian LINES are more promiscuous and able to retrotranspose most mRNAs. It was hoped that the lack of pseudogenes in the chicken would help to identify functional noncoding RNA genes in mammalian genomes via conservation of chromosomal gene location. (Because of their noncoding character, it is difficult to distinguish functional RNA genes from the large excess of RNA pseudogenes in mammals by ab initio methods.) In chicken, 571 RNA genes in 20 distinct families were predicted and only the miRNA and snoRNA families (that usually lie within introns of coding genes) show conserved synteny to the

Table 2. Frequency and class of gene/protein predictions (Ensembl, June 2004)

Gene family	Number	Function
tRNA	280	Transfer RNA, adaptor in translation
5S rRNA	12	Ribosomal RNA, component of ribosome
5.8S rRNA[a]	~145	Ribosomal RNA, component of ribosome
18S rRNA[a]	~145	Ribosomal RNA, component of ribosome
28S rRNA[a]	~145	Ribosomal RNA, component of ribosome
snRNP U1	18	Major spliceosome
snRNP U2	6	Major spliceosome
snRNP U4	4	Major spliceosome
snRNP U5	9	Major and minor spliceosomes
snRNP U6	15	Major spliceosome
snRNP U4atac	1	Minor spliceosome
snRNP U6atac	4	Minor spliceosome
snRNP U11	1	Minor spliceosome
snRNP U12	1	Minor spliceosome
miRNA	121	Translation repression
snoRNA	83	Small nucleolar RNA, takes part in processing of rRNA
RNaseP	1	Ribozyme, processes tRNA
snRNP U7	1	3'-end processing of replication-dependent histone pre-mRNAs
SRP	3	RNA component of signal recognition particle
7SK	4	Binds P-TEFb, which activates transcription by phosphorylating C-terminal domain of RNA Pol II. This process is negatively regulated by the 7SK RNP.
Y RNA	2	Component of the Ro RNP, in association with Ro60 and La, function of Ro RNP not known
Telomerase RNA	1	Provides the template for telomeric DNA addition
BIC	1	Cooperates with c-myc in B lymphomagenesis and erythroleukemogenesis
Total RNA genes	571 (1441 incl. rRNA genes)	
Total pseudogenes	75	
Total protein-coding genes	17,709	

[a]Ribosomal DNA 40-kb repeat (18S, 5.8S, 28S); Muscarella et al. (1985).

extent that protein coding genes do. That the other noncoding RNA families did not suggests that they may transpose throughout the genome in ways that differ from coding genes.

Comparisons between mammals and birds can also start to address questions about gene gains/losses (Hillier et al. 2004). Comparisons between human, chicken, and *Fugu* suggest a core set of almost one third of all genes (7606) is conserved in all vertebrates. These comparisons also suggest that the rates of gene loss were higher in the avian lineage and fewer gene duplications were found in birds. Careful comparisons detected some genes lost from the chicken lineage, including vomeronasal receptors, caseins, and some genes of the immune system. Similarly, birds have more keratins specific to feathers and mammals have lost the avidin egg proteins. The discovery that all enzymes in the urea cycle were present but apparently not used for this function in birds was perplexing.

NEW TOOLS FOR GENOME ANALYSIS

Important by-products of any genome project are the resources (cDNA and BAC clones, genetic markers, etc.) and information it provides for future research (Antin and Konieczka 2005). Together with chromosome paints, BAC clones (BPRC) have been used to define cytogenetically all chicken chromosomes (Masabanda et al. 2004). Because of the nearly identical sizes of the microchromosomes in mitotic chromosome spreads, this was not previously feasible. A BAC map with 20-fold redundancy or 91% coverage of the chicken genome has been assembled into 260 contigs (Wallis et al. 2004; ChickFPC). BAC contig maps are under construction for other birds; including turkey, California condor, and zebra finch (Edwards et al. 2005). These clones can be used to target specific genomic regions and to create whole-genome BAC arrays for comparative surveys of avian genomes. These arrays may be able to classify many avian species into unique clades, a notoriously difficult task (Edwards et al. 2005). From the very start, ESTs and cDNA clones have been important (Boardman et al. 2002; ChickEST), in particular for the prediction of chicken genes. ESTs have been used to create cDNA microarrays (Burnside et al. 2005) and design DNA chips (Affymetrix) for high-throughput gene expression assays. A total of 4532 full length cDNA clones (Caldwell et al. 2004; Hubbard et al. 2005), representing ~25% of known gene predictions in chicken, can now be used in evolutionary and functional studies (available from ARK-Genomics). RNAi and transgenic technologies are now available in the chicken, which when combined with the accessible chicken embryo, makes this a powerful system for functional studies in vivo (Brown et al. 2003;

Table 3. Online resources for avian genomics

Web site	Description
http://bacpac.chori.org/	BPRC: BAC resources center
http://chick.umist.ac.uk/	ChickEST: BBSRC chicken EST database
http://chicken.genomics.org.cn/index.jsp	ChickVD: chicken variation database
http://geisha.biosci.arizona.edu/	GEISHA: *Gallus gallus* EST and in situ hybridization analysis
http://genex.hgu.mrc.ac.uk/intro.html	EMAP: Edinburgh mouse atlas project
http://genome.ucsc.edu/cgi-bin/hgGateway	UCSC genome browser
http://genomeold.wustl.edu/	WASHU: Washington University genome sequencing center
http://www.affymetrix.com/	Affymetrix
http://www.animalsciences.nl/ChickFPC/	ChickFPC: Chicken FPC BAC map
http://www.ark-genomics.org/	ARK-Genomics: Center for functional genomics in farm animals
http://www.chicken-genome.org/	AvianNET: the avian genome information network
http://www.crbtrader.com/fund/articles/eggs.asp	CRB: commodity research bureau
http://www.ensembl.org/Gallus_gallus/index.html	Ensembl genome browser
http://www.fas.usda.gov/currwmt.asp	FAS: USDA foreign agricultural service
http://www.gmod.org/	GMOD: Generic Model Organism Database
http://www.nature.com/nature/focus/avianflu/index.html	Avian flu: Web Focus
http://www.ncbi.nlm.nih.gov/genome/guide/chicken/	NCBI genome browser
http://www.thearkdb.org/browser?species=chicken	ARKdb: genetic mapping databases
http://acedb.asg.wur.nl/	ChickACE: Wageningen Animal Sciences Group ACEbrowser

Nakamura et al 2004; Sang 2004; Stern 2004). The application of these tools and access to the biological information they generate is a huge and complex task. There are a number of databases distributed throughout the world (Table 3), including genome browsers (Ensembl, NCBI, and UCSC), genetic maps (ARKdb and ChickACE), gene expression (GEISHA), and others, but there is a need to integrate these views into a single Model Organism Database (GMOD).

APPLICATIONS OF THE CHICKEN GENOME SEQUENCE

Birds and mammals shared a common ancestor ~310 million years ago (Mya) (Hedges 2002). Sequence comparisons between these groups are characterized with a high signal-to-noise ratio for the detection of functional elements. Taken together with the ready access to chicken embryos and as a major food source, chicken genomics is likely to have major applications and benefits in comparative genomics, evolutionary biology and systematics, models of development and human disease, and agriculture.

Comparative Genomics

A major reason for sequencing the chicken genome was to increase our understanding of the human genome through comparative genomics, for example, to define regions under selection such as coding and regulatory elements (Hillier et al. 2004). Comparisons with known functional sequences suggested that 75% of coding regions and 30%–40% of regulatory elements are conserved. Only 2.5% of the chicken sequence could be aligned with that of the human (44% coding, 25% intronic, and 31% intergenic) and, given that 5% of the mammalian genome is under selection, almost all of this is likely to be of functional significance.

Comparative genomics has identified ~400 ultra-conserved regions (UCR) greater than 200 bp sharing at least 95% sequence identity between human and chicken (Sandelin et al. 2004). Surprisingly, highly conserved, noncoding regions like the UCR often exist far from any predicted gene within so-called "gene deserts" that are apparently free of any known protein-coding genes and are often clustered (Ovcharenko et al. 2005). Genes with a role in transcriptional regulation and development flank many of these UCR and gene deserts. These regions are often far from genes and may represent distant regulatory signals.

Parent-specific gene expression by genomic imprinting is only found in mammals and not birds or lower vertebrates. Therefore, comparison of imprinted genes in mammals with orthologs in the chicken may uncover features about the origins of imprinting. Comparative mapping suggests these genes cluster on macrochromosomes in regions that preferentially undergo asynchronous DNA replication (Dunzinger et al. 2005). Analysis of the chicken region orthologous to the imprinted mammalian *ASCL2–H19* region (Yokomine et al. 2005) revealed extensive conservation of gene organization, except *H19*, a critical noncoding imprinted gene. This gene and its regulatory elements were absent from the chicken genome. These studies suggest that imprinted genes were clustered before the evolution of imprinting, an event that occurred after the divergence of birds and mammals ~310 Mya. Subsequently, imprinting control elements, such as the *H19* gene region, must have evolved by duplication and/or transposition into these gene clusters.

A long-standing question in genome evolution has been the question of genome size. The chicken genome is 35% the size of the human and 45% of the mouse. In part, this can be explained in terms of the low frequency of repeats, pseudogenes, segmental duplication, and gene duplications (Hillier et al. 2004). However, these factors only account for 20%–25% of the variation in genome size, so other factors are at work, possibly a dearth of ancient repeats (that are no longer detectably repetitive) or reduction in cell size and energy conservation (Hughes and Piontkivska 2005).

Developmental Biology

Applications in developmental biology are likely to be another major beneficiary of the genome sequence (Burt 2004b; Stern 2005). The chicken has always been a favorite among developmental biologists (Brown et al. 2003; Stern 2005) because of easy access to the chick embryo and ease of manipulation. These features, when combined with the new tools of genomics, are ideal for testing gene function and predicted regulatory sequences in vivo. For example, studies on the conservation of the avian *SOX2* genes have identified neural specific enhancers, confirmed in vivo by electroporation of chick embryo neural tubes (Uchikawa et al. 2004).

In the mouse and other model systems, whole-mount in situ hybridization screens have been useful in identifying patterns of expression that may suggest developmental functions of novel genes (EMAP). A similar effort has started in the chicken using the large collection of sequenced chicken ESTs (Boardman et al. 2002; ARK-Genomics; ChickEST). Data can be accessed at GEISHA and standard three-dimensional embryo reconstructions are under development (EMAP).

Genetic Variation and Complex Trait Analysis

In parallel with the chicken genome sequencing project, a consortium (Wong et al. 2004; Wang et al. 2005b; ChickVD) generated 2.8 million SNPs from a comparison of the Red Jungle Fowl reference sequence and partial genome scans of Silkie, Broiler, and Layer lines. Nucleotide diversity (5×10^{-3} per nucleotide) was six times the rate found in humans (Ellegren 2005). Resequencing confirmed 94% of the total and 83% of the nonsynonymous SNPs. An initial surprise was that ~70% of SNPs were common to all breeds, suggesting an origin prior to domestication 5,000–10,000 years ago. Another possibility is that their ancestry has been lost because of extensive cross breeding between Asian and western poultry populations. The next steps are to verify a larger sample of SNPs and create high-resolution genetic and linkage disequilibrium maps of chicken populations. These assays will be used to map and identify genes controlling traits of economic and biological interest at quantitative trait loci (QTL). Currently, more than 600 QTL have been mapped using microsatellites (Andersson and Georges 2004; Hocking 2005; Wang et al. 2005b). The availability of a standard set of 10,000 or more SNPs combined with the ease of building structured large resource populations hold much promise toward the identification of genes controlling these traits.

Animal Health and the Avian Immune System

One area that has benefited most from genomic approaches has been the characterization of the genes and proteins in the avian immune system. The MHC was the first major chicken genome sequence to be assembled (Kaufman et al. 1999) and was a surprise, being relatively compact and simpler than those of mammals. Since then, there has been slow progress in the isolation of avian cytokines and other signaling molecules. The main problem has been their high rate of evolution, limiting their detection using homology to mammalian sequences (Staeheli et al. 2001). Even now, one must be careful in concluding that avian homologs to mammalian immune genes do not exist, as several examples known from ESTs or directed sequencing were not found in the genome assembly. This started to change when analysis of large EST data sets identified 185 immune-related sequences (Lynn et al. 2003; Smith et al. 2004). This compared with the 80 genes identified by Tirunagaru et al. (2000) and the 28 genes listed in the review by Staeheli et al. (2001). Sequences included interleukins, transcription factors, chemokines, differentiation antigens, receptors, genes involved in the Toll pathway, and MHC-associated genes. The discovery of *IL4* and other cytokines involved in the Th2 response (Smith et al. 2004) was a surprise, since it had previously been speculated that the chicken does not elicit a typical Th2 response (Staeheli et al. 2001). The receptors for *IL10* and *IL13* were also identified, indicating that the chicken probably also contained these genes, which are typical Tr1 and Th2 cytokines. This was confirmed by sequencing specific BAC clones identified assuming conservation of synteny between chicken and mammalian genomes (Avery et al. 2004; Rothwell et al. 2004).

A comprehensive analysis of the chicken genome sequence has identified many cytokines, chemokines, and their receptors (Hillier et al. 2004; Kaiser et al. 2004, 2005; Wang et al. 2005a). Even genes once thought to be mammalian-specific, including *IL3, IL7, IL9, IL26, CSMF, LIF,* and Cathelicidin, were found (Hillier et al. 2004). These are proteins that evolve rapidly and require more effort to detect. A number of orthologs to human chemokines are absent from the chicken genome, including *CCL2, 7, 8, 11, 15, 18, 23, 24,* and *26; CXCL1–7, 9, 10,* and *11,* possibly products of independent gene duplications in mammals. Similarly, missing chemokine receptors included *CCR1, CCR3, CCR10, CXCR3,* and *CXCR6.* The lack of functional eosinophils correlates with the absence of the eotaxin genes (*CCL22, CCL24, CCL26*) and their receptor (*CCR3*). Chickens lack lymph nodes and also the genes for the lymphotoxins (*LT-α* and *-β*) and their receptors. *TNF* is also absent, but its receptor, *TNFRSF1A* (*ENSGALG00000014890*) is present, suggesting that further sequencing will reveal this gene in the chicken. Similar analyses have been performed on the leukocyte receptor complex (Nikolaidis et al. 2005) that regulates the activity of T- and B-lymphocytes and NK cells. A model of evolution by repeated birth and death of these Ig-like receptors' genes was proposed.

CONCLUSIONS

When the first issue of *Genome Research* appeared 10 years ago, avian genomics was still in a mapping phase (Burt and Cheng 1998). The idea of sequencing the chicken genome was only a dim possibility and comparative maps were hailed as an alternative mapping resource. As the first livestock species to be fully sequenced, the chicken genome sequence is a landmark in both avian biology and agriculture. The avian community was small but has grown rapidly in the last two years thanks to the EST and genome sequencing programs. The challenge now is to keep the momentum going and to exploit these resources. The creation of AvianNET, an organization to encourage the exchange of tools and resources in avian biology, is a start but only a beginning. The chicken genome was determined to inform us about the nature and function of the human genome. It has also informed us about the nature of birds and other vertebrates. With 9600 extant avian species, there is still a lot to learn. Birds, in particular, poultry and ducks are a source of many infectious diseases (Avian Flu: Web Focus 2005) and genomics is going to tell us a lot about host responses to these pathogens. There is therefore a need to sequence and characterize other avian genomes. This time these sequences will be used to inform us about responses to pathogens that infect both humans and birds.

ACKNOWLEDGMENTS

I would like to thank many colleagues and collaborators for their continued support and enthusiasm on issues related to avian genomics and acknowledge financial support from the Biotechnology and Biological Science Research Council (UK). In addition, I would like to thank the useful comments and suggestions from the anonymous reviewers.

REFERENCES

Abdrakhmanov, I., Lodygin, D., Geroth, P., Arakawa, H., Law, A., Plachy, J., Korn, B., and Buerstedde, J.M. 2000. A large database of chicken bursal ESTs as a resource for the analysis of vertebrate gene function. *Genome Res.* **10:** 2062–2069.

Andersson, L. and Georges, M. 2004. Domestic-animal genomics: Deciphering the genetics of complex traits. *Nat. Rev. Genet.* **5:** 202–212.

Antin, P.B. and Konieczka, J.H. 2005. Genomic resources for chicken. *Dev. Dyn.* **232:** 877–882.

Avery, S., Rothwell, L., Degen, W.D., Schijns, V.E., Young, J., Kaufman, J., and Kaiser, P. 2004. Characterization of the first non-mammalian T2 cytokine gene cluster: The cluster contains functional single-copy genes for IL-3, IL-4, IL-13, and GM-CSF, a gene for IL-5 that appears to be a pseudogene, and a gene encoding another cytokine-like transcript, KK34. *J. Interferon Cytokine Res.* **24:** 600–610.

Axelsson, E., Webster, M.T., Smith, N.G.C., Burt, D.W., and Ellegren, H. 2005. Comparison of the chicken and turkey genomes reveals a higher rate of nucleotide divergence on microchromosomes than macrochromosomes. *Genome Res.* **15:** 120–125.

Boardman, P.E., Sanz-Ezquerro, J., Overton, I.M., Burt, D.W., Bosch, E., Fong, W.T., Tickle, C., Brown, W.R., Wilson, S.A., and Hubbard, S.J. 2002. A comprehensive collection of chicken cDNAs. *Curr. Biol.* **12:** 1965–1969.

Bourque, G., Zdobnov, E.M., Bork, P., Pevzner, P.A., and Tesler, G. 2005. Comparative architectures of mammalian and chicken genomes reveal highly variable rates of genomic rearrangements across different lineages. *Genome Res.* **15:** 98–110.

Brown, W.R., Hubbard, S.J., Tickle, C., and Wilson, S.A. 2003. The chicken as a model for large-scale analysis of vertebrate gene function. *Nat. Rev. Genet.* **4:** 87–98.

Burnside, J., Neiman, P., Tang, J., Basom, R., Talbot, R., Aronszajn, M., Burt, D.W., and Delrow, J. 2005. Development of a cDNA array for chicken gene expression analysis. *BMC Genomics* **6:** 13.

Burt, D.W. 2002. Applications of biotechnology in the poultry industry. *Worlds Poult. Sci. J.* **58:** 5–13.

———. 2004a. Chicken genomics charts a path to the genome sequence. *Brief. Funct. Genomic Proteomic* **3:** 60–67.

———. 2004b. The chicken genome and the developmental biologist. *Mech. Dev.* **121:** 1129–1135.

Burt, D.W. and Cheng, H.H. 1998. Chicken gene maps. *ILAR J.* **39:** 229–236.

Burt, D.W., Bruley, C.K., Dunn, I., Jones, C.T., Ramage, A., Law, A.S., Morrice, D.R., Paton, I.R., Smith, J., Windsor, D., et al. 1999. Dynamics of chromosome evolution: Clues from comparative gene mapping in birds and mammals. *Nature* **402:** 411–413.

Caldwell, R.B., Kierzek, A.M., Arakawa, H., Bezzubov, Y., Zaim, J., Fiedler, P., Kutter, S., Blagodatski, A., Kostovska, D., Koter, M., et al. 2004. Full-length cDNAs from chicken bursal lymphocytes to facilitate gene function analysis. *Genome Biol.* **6:** R6

Castelo, R., Reymond, A., Wyss, C., Camara, F., Parra, G., Antonarakis, S.E., Guigó, R., and Eyras, E. 2005. Comparative gene finding in chicken indicates that we are closing in on the set of multi-exonic widely expressed human genes. *Nucleic Acids Res.* **33:** 1935–1939.

Dequeant, M.L. and Pourquié, O. 2005. Chicken genome: New tools and concepts. *Dev. Dyn.* **232:** 883–886.

Dunzinger, U., Nanda, I., Schmid, M., Haaf, T., and Zechner, U. 2005. Chicken orthologues of mammalian imprinted genes are clustered on macrochromosomes and replicate asynchronously. *Trends Genet.* **21:** 488–492.

Edwards, S.V., Jennings, B.W., and Shedlock, A.M. 2005. Phylogenetics of modern birds in the era of genomics. *Proc. Royal Sci. B.* **272:** 979–992.

Ellegren, H. 2005. The avian genome uncovered. *Trends Ecol. Evol.* **20:** 180–186.

Eyras, E., Reymond, A., Castelo, R., Bye, J.M., Camara, F., Flicek, P., Huckle, E.J., Parra, G., Shteynberg, D.D., Wyss, C., et al. 2005. Gene finding in the chicken genome. *BMC Bioinformatics* **6:** 131.

Fazzari, M.J. and Greally, J.M. 2004. Epigenomics: Beyond CpG islands. *Nat. Rev. Genet.* **5:** 446–455.

Fumihito, A., Miyake, T., Sumi, S., Takada, M., Ohno, S., and Kondo, N. 1994. One subspecies of the red junglefowl (*Gallus gallus gallus*) suffices as the matriarchic ancestor of all domestic breeds. *Proc. Natl. Acad. Sci.* **91:** 12505–12509.

Haldane, J.B.S. 1927. The comparative genetics of color in rodents and carnivora. *Biol. Rev. Camb. Philos. Soc.* **2:** 199–212.

Hedges, S.B. 2002. The origin and evolution of model organisms. *Nat. Rev. Genet.* **3:** 838–849.

Hill, W.G. and Robertson, A. 1966. The effect of linkage on limits to artificial selection. *Genet. Res.* **8:** 269–294.

Hillier, L.W., Miller, W., Birney, E., Warren, W., Hardison, R.C., Ponting, C.P., Bork, P., Burt, D.W., Groenen, M.A., Delany, M.E., et al. 2004. Sequence and comparative analysis of the chicken genome provide unique perspectives on vertebrate evolution. *Nature* **432:** 695–716.

Hocking, P. 2005. Review on QTL mapping results in chickens. *Worlds Poult. Sci. J.* **61:** 215–226.

Hubbard, S.J., Grafham, D.V., Beattie, K.J., Overton, I.M., McLaren, S.R., Croning, M.D., Boardman, P.E., Bonfield, J.K., Burnside, J., Davies, R.M., et al. 2005. Transcriptome analysis for the chicken based on 19,626 finished cDNA sequences and 485,337 expressed sequence tags. *Genome Res.* **15:** 174–183.

Hughes, A.L. and Piontkivska, H. 2005. DNA repeat arrays in chicken and human genomes and the adaptive evolution of avian genome size. *BMC Evol. Biol.* **5:** 12.

Kaiser, P., Rothwell, L., Avery, S., and Balu, S. 2004. Evolution of the interleukins. *Dev. Comp. Immunol.* **28:** 375–394.

Kaiser, P., Poh, T.Y., Rothwell, L., Avery, S., Balu, S., Pathania, U.S., Hughes, S., Goodchild, M., Morrell, S., Watson, M., et al. 2005. A genomic analysis of chicken cytokines and chemokines. *J. Interferon Cytokine Res.* **25:** 467–484.

Kaufman, J., Milne, S., Gobel, T.W., Walker, B.A., Jacob, J.P., Auffray, C., Zoorob, R., and Beck, S. 1999. The chicken B locus is a minimal essential major histocompatibility complex. *Nature* **401:** 923–925.

Korf, I., Flicek, P., Duan, D., and Brent, M.R. 2001. Integrating genomic homology into gene structure prediction. *Bioinformatics* **17 Suppl 1:** 140–148.

Lynn, D.J., Lloyd, A.T., and O'Farrelly, C. 2003. In silico identification of components of the Toll-like receptor (TLR) signaling pathway in clustered chicken expressed sequence tags (ESTs). *Vet. Immunol. Immunopathol.* **93:** 177–184.

Masabanda, J.S., Burt, D.W., O'Brien, P.C., Vignal, A., Fillon, V., Walsh, P.S., Cox, H., Tempest, H.G., Smith, J., Habermann, F., et al. 2004. Molecular cytogenetic definition of the chicken genome: The first complete avian karyotype. *Genetics* **166:** 1367–1373.

McQueen, H.A., Fantes, J., Cross, S.H., Clark, V.H., and Archibald, A.L. 1996. CpG islands of chicken are concentrated on microchromosomes. *Nat. Genet.* **12:** 321–324.

Meunier, J. and Duret, L. 2004. Recombination drives the evolution of GC-content in the human genome. *Mol. Biol. Evol.* **21:** 984–990.

Muscarella, D.E., Vogt, V.M., and Bloom, S.E. 1985. The ribosomal RNA gene cluster in aneuploid chickens: Evidence for increased gene dosage and regulation of gene expression. *J. Cell Biol.* **101:** 1749–1756.

Nakamura, H., Katahira, T., Sato, T., Watanabe, Y., and Funahashi, J.-I. 2004. Gain- and loss of-function in chick embryos by electroporation. *Mech. Dev.* **121:** 1137–1143.

Nikolaidis, N., Makalowska, I., Chalkia, D., Makalowski, W., Klein, J., and Nei, M. 2005. Origin and evolution of the chicken leukocyte receptor complex. *Proc. Natl. Acad. Sci.* **102:** 4057–4062.

Ovcharenko, I., Loots, G.G., Nobrega, M.A., Hardison, R.C., Miller, W., and Stubbs, L. 2005. Evolution and functional classification of vertebrate gene deserts. *Genome Res.* **15:** 137–145.

Parra, G., Agarwal, P., Abril, J.F., Wiehe, T., Fickett, J.W., and Guigó, R. 2003. Comparative gene prediction in human and mouse. *Genome Res.* **13:** 108–117.

Ponce de Leon, F.A., Li, Y., and Weng, Z. 1992. Early and late replicative chromosomal banding patterns of *Gallus domesticus*. *J. Hered.* **83:** 36–42.

Rothwell, L., Young, J.R., Zoorob, R., Whittaker, C.A., Hesketh, P., Archer, A., Smith, A.L., and Kaiser, P. 2004. Cloning and characterization of chicken IL-10 and its role in the immune response to *Eimeria maxima*. *J. Immunol.* **173:** 2675–2682.

Sandelin, A., Bailey P., Bruce, S., Engström, P.G., Klos, J.M., Wasserman, W.W., Ericson, J., and Lenhard, B. 2004. Arrays of ultraconserved non-coding regions span the loci of key developmental genes in vertebrate genomes. *BMC Genomics* **5:** 99.

Sang, H. 2004. Prospects for transgenesis in the chick. *Mech. Dev.* **121:** 1179–1186.

Schmid, M., Nanda, I., Hoehn, H., Schartl, M., Haaf, T., Buerstedde, J.M., Arakawa, H., Caldwell, R.B., Weigend, S., Burt, D.W., et al. 2005. Second report on chicken genes and chromosomes 2005. *Cytogenet. Genome Res.* **109:** 415–479.

Smith, J., Bruley, C.K., Paton, I.R., Dunn, I., Jone, C.T., Windsor, D., Morrice, D.R., Law, A.S., Masabanda, J., Sazanov, A., et al. 2000. Differences in gene density on the Chicken macrochromosomes and microchromosomes: A tool for gene discovery in vertebrate genomes. *Anim. Genet.* **31:** 96–103.

Smith, J., Speed, D., Law, A.S., Glass, E.J., and Burt, D.W. 2004. In-silico identification of chicken immune-related genes. *Immunogenetics* **56:** 122–133.

Staeheli, P., Puehler, F., Schneider, K., Göbel, T.W., and Kaspers, B. 2001. Cytokines of birds: Conserved functions—a largely different look. *J. Interferon Cytokine Res.* **21:** 993–1010.

Stern, C.D. 2004. The chick embryo—Past, present and future as a model system in developmental biology. *Mech. Dev.* **121:** 1011–1013.

———. 2005. The chick: A great model system becomes even greater. *Dev. Cell* **8:** 9–17.

Tirunagaru, V.G., Sofer, L., Cui, J., and Burnside, J. 2000. An expressed sequence tag database of T-cell enriched activated chicken splenocytes: Sequence analysis of 5251 clones. *Genomics* **66:** 144–151.

Uchikawa, M., Takemoto, T., Kamachi, Y., and Kondoh, H. 2004. Efficient identification of regulatory sequences in the chicken genome by a powerful combination of embryo electroporation and genome comparison. *Mech. Dev.* **121:** 1145–1158.

Wallis, J.W., Aerts, J., Groenen, M.A., Crooijmans, R.P., Layman, D., Graves, T.A., Scheer, D.E., Kremitzki, C., Fedele, M.J., Mudd, N.K., et al. 2004. A physical map of the chicken genome. *Nature* **432:** 761–764.

Wang, J., Adelson, D.L., Yilmaz, A., Sze, S.H., Jin, Y., and Zhu, J.J. 2005a. Genomic organization, annotation, and ligand-receptor inferences of chicken chemokines and chemokine receptor genes based on comparative genomics. *BMC Genomics* **6:** 45.

Wang, J., He, X., Dai, M., Ruan, J., Chen, J., Zhang, Y., Hu, Y., Ye, C., Li, S., Cong, L., et al. 2005b. ChickVD: A sequence variation database in the chicken genome. *Nucleic Acids Res.* **33:** D438–D441.

Wong, G.K., Liu, B., Wang, J., Zhang, Y., Yang, X., Zhang, Z., Meng, Q., Zhou, J., Li, D., Zhang, J., et al. 2004. A genetic variation map for chicken with 2.8 million single-nucleotide polymorphisms. *Nature* **432:** 717–722.

Yokomine, T., Shirohzu, H., Purbowasito, W., Toyoda, A., Iwama, H., Ikeo, K., Hori, T., Mizuno, S., Tsudzuki, M., Matsuda, Y., et al. 2005. Structural and functional analysis of a 0.5-Mb chicken region orthologous to the imprinted mammalian Ascl2/Mash2-Igf2-H19 region. *Genome Res.* **15:** 154–165.

12

Advances in Livestock Genomics: Opening the Barn Door

James E. Womack
Department of Veterinary Pathobiology
Center for Animal Biotechnology and Genomics
Texas A&M University
College Station, Texas 77843-4467

THE HUMAN GENOME PROJECT IS PROPERLY CREDITED with accelerating the discovery of disease genes and providing a totally new paradigm for medical research. It has sharpened our approach to the study of development, behavior, cancer, and infectious diseases, and provided a toolbox for deciphering genetic diversity and human origins. It probably does not get enough credit, however, for opening the door for genetic analysis of other animals, particularly those species used in agriculture. Livestock genomics has followed in the footsteps of the human genome initiative, adopting both its successful strategies and technologies to advance our understanding of livestock genomes with shoestring budgets relative to the resources available for human and medical research. In turn, livestock genomics contributes to informing the human genome. Mapping and sequencing species from clades other than primates and rodents contribute to our understanding of evolutionary history and its underlying mechanisms. As demonstrated by Thomas et al. (2003) and others, sequences from these diverse clades contribute to the identification of functional elements in the human genome outside the more easily annotated coding regions.

In addition to informing the human genome, agricultural science has a unique responsibility to human health and social stability, and that is feeding an expanding world population while minimizing environmental and ecological risks. Clearly, the identification of variation in livestock genomes that predisposes health and productivity with less reliance on hormones, antibiotics, and pesticides will be a major step in meeting this global challenge. A review of recent advances in livestock genomics and their realized and potential contributions to both human biology and agricultural science is the goal of this paper.

One of the world's most important agricultural animals and one in which genomics research advanced rapidly to the sequencing phase, is the domestic chicken. That genome is the subject of Chapter 11, however, and this review focuses on the mammals, principally cattle, pigs, sheep, and horses, and, to a lesser extent, river buffalo and goats.

STATUS OF LIVESTOCK GENOMICS

Early attempts to construct whole-genome maps of livestock species were based on the two technologies underlying the first human genome maps, somatic cell genetics and in situ hybridization (Womack and Moll 1986; Yerle et al. 1995). These early maps defined synteny (genes on the same chromosome but not necessarily linked) and cytogenetic locations of sequences hybridizing specific DNA probes. These strategies proved extremely important to early comparative mapping because the mapped markers were generally genes or gene products, highly conserved across mammalian genomes. The synteny and cytogenetic maps they produced gave us our first insights into the relative stability of the mammalian genome throughout its evolutionary history and pointed to certain groups of animals with a much higher degree of genome similarity than others, suggesting some lineages of animals with highly conserved genomes and others in which genomic evolution was much more dynamic. Linkage mapping, however, lagged behind, awaiting the development of highly polymorphic markers with sufficient density in the genomes of outbred animal populations to efficiently map traits with whole-genome approaches. Beckmann and Soller (1983), inspired by advances in human genetics, were early proponents of the use of DNA level markers for building maps and mapping traits in livestock species. Modern genomics in livestock followed the lead of the human genome initiative at all levels and had its formal origins in a series of conferences in which strategies were distilled, and more importantly, collaborations were established to maximize the relatively meager resources available to animal genetics in the early 1990s. Two international conferences in 1990—a Banbury Conference on "Mapping the Genomes of Agriculturally Important Animals" at Cold Spring Harbor and the first Allerton Conference in Illinois on "Gene Mapping of Domestic

Table 1. Map status and genomic resource for livestock species

	Cow	Pig	Sheep	Horse	Buffalo	Goat
Chromosome ($2n$)	60	38	54	64	50	60
Mapped markers						
Synteny	1800	1000	250	500	60	—
Linkage	2500	2700	1500	480	—	350
Cytogenetic	300	300	850	450	300	350
RH	3200	3000	300	800	30	—
Resources						
EST sequence	450k	300k	20k	30k	—	—
cDNA libraries	80	100	12	15	—	—
BAC libraries	3	5	3	3	—	—
Microarrays	3	2	—	—	—	—
Whole-genome sequence	6×	a	—	—	—	—

[a] Whole-genome sequencing of pig genome to begin in 2005.

Animal Genomes: Needs and Opportunities"—were springboards for collaborative use of available genomic resources and the prioritization and development of resources unavailable at that time. From these and other similar gatherings, international groups of animal geneticists launched both formal and informal genome projects for some of the most widely used livestock species, resulting in our current inventory of genomic resources (Table 1). Following is a brief discussion of the development and current status of genomic information for cattle, pigs, sheep, horses, and the lesser developed genomes of river buffalo and goats. Recent comprehensive reviews focused on individual species are available for cattle (Lewin 2003; Sonstegard and Van Tassell 2004), pig (Rothschild 2003), sheep (Cockett 2003), horse (Chowdhary and Bailey 2003), buffalo (Iannuzzi et al. 2003), and goat (Schibler et al. 1998). Moreover, Andersson and Georges (2004) reviewed livestock genomics in the context of grand challenges of human genomics (Collins et al. 2003), providing both a model for this paper and a standard for measuring the development of livestock genomics in its role of helping to define the human genome through the genetic analysis of complex traits.

Cattle

Cattle genomics had its origins in somatic cell genetics (Heuertz and Hors-Cayla 1981; Womack and Moll 1986; Womack 1987). The first "genome maps" for cattle were synteny groups, genes on the same chromosome, defined by protein gene products segregating in hybrid somatic cell lines. These synteny groups were assigned to specific chromosomes by integrating somatic cell genetics with in situ hybridization (Fries et al. 1986, 1993; Gallagher Jr. et al. 1993) and were greatly expanded with the advent of molecular markers, initially defined by probed Southern blots and later by PCR-based markers. An international consortium organized at the 1988 meeting of the International Society for Animal Genetics (ISAG) assembled a set of families for linkage mapping, and the development of microsatellite markers in the early 1990s resulted in an international linkage map (Barendse et al. 1994), almost concurrently with a linkage map developed at USDA-MARC (Bishop et al. 1994). These maps were quickly expanded (Barendse et al. 1997; Kappes et al. 1997) into tools that have proved effective for mapping loci underlying both monogenic and quantitative traits. The next significant advance in cattle genomics was the development of radiation hybrid (RH) maps (Womack et al. 1997; Williams et al. 2002) and the use of these maps for high-resolution comparative mapping (Band et al. 2000; Everts-van der Wind et al. 2004; Itoh et al. 2005). A consortium to generate a bacterial artificial chromosome (BAC) map of the bovine genome has generated a 294,651 whole clone HindIII fingerprint map that is currently being refined by BAC end sequencing and is scheduled for completion in 2005. Highly developed linkage and RH maps and the progress of the BAC consortium were instrumental in the success of a White Paper proposal to the NHGRI for whole-genome sequencing in cattle (http://www.genome.gov/Pages/Research/Sequencing/SeqProposals/BovineSEQ.pdf). A single partially inbred Hereford female was selected to contribute 6× whole-genome shotgun (WGS) reads and another 1.5× will come from individual animals of the Holstein, Angus, Jersey, Limousin, Brahman, and Norwegian Red breeds for SNP detection. In addition, an ~1× BAC skim sequence will aid assembly. Sixfold WGS has been achieved and breed skims for SNP detection are under way. Approximately 10,000 nonredundant full-length cDNAs are being sequenced at the Michael Smith Genome Science Center in Vancouver. The University of Illinois has generated a 3800-element bovine cDNA microarray (Band et al. 2002), recently upgraded to 7872 elements (Everts et al. 2005), and The Bovine Functional Genomics Consortium (Suchyta et al. 2003) has produced an 18,000-element array.

Pigs

The history of pig genomics does not parallel that of cattle in that the linkage map was the first whole-genome map produced. As a result of the European PiGMaP initiative (Archibald et al. 1995) and USDA-MARC (Rohrer et al. 1996), linkage maps were developed and subsequently expanded to a high level of resolving power for trait mapping. An estimated 2700 markers are represented on these two maps. Somatic cell genetics (Yerle et al. 1996) and RH mapping (Yerle et al. 1998; Hawken et al. 1999) have facilitated comparative mapping in pigs. As in cattle, RH mapping of ESTs with human orthologs has provided the power for definitive comparative mapping of the pig and human genomes (Rink et al. 2002; Tuggle et al. 2003). An international group has begun construction of a BAC map (http://www.genomic.iastate.edu/newsletter/PigWhitePaper.html, NHGRI), and large cDNA sequencing and EST projects are advancing rapidly (Fahrenkrug et al. 2002; Tuggle et al. 2003). Multinational funding has recently been secured for whole-genome sequencing to begin in 2005. Meanwhile, the "Sino-Danish Pig Genome Project" has published pig genome sequence with $<1\times$ coverage (Wernersson et al. 2005). Functional genomics in pigs is now the beneficiary of a 3468-element microarray (Niewold et al. 2005) and a 3867-element microarray (Dvorak et al. 2005) both from intestinal mucosa.

Sheep

Burkin et al. (1993) developed a somatic cell hybrid panel segregating sheep chromosomes and assigned a few dozen markers to syntenic groups before somatic cell genetics in sheep gave way to linkage mapping. Building on linkage maps developed to find the Booroola fecundity gene, Montgomery et al. (1993) and Crawford et al. (1995) produced a map with 246 markers. Interestingly, approximately one-half of these markers were bovine-derived microsatellites. Second-generation (de Gortari et al. 1998) and third-generation (Maddox et al. 2001) linkage maps have been expanded to include well over 1000 markers. Radiation hybrid maps were developed at INRA (France) and Utah State University (Cockett 2003) and are currently being populated with both EST and microsatellite markers, a strategy that integrates genetic and physical maps. BAC libraries have been produced and contigs assembled around several regions of interest to individual laboratories.

Horse

Horse genomics was slow out of the gate but has picked up its pace significantly in the last five years. A workshop in 1995 (http://www.uky.edu/AG/Horsemap/) launched an international initiative that has produced three linkage maps (Lindgren et al. 1998; Guérin et al. 1999, 2003; Swinburne et al. 2000) totaling in excess of 450 markers, more than 400 of which are microsatellites. A comprehensive RH map was produced by Chowdhary et al. (2002) that contains 730 markers and integrates somatic cell, linkage, and cytogenetic maps into a valuable tool for comparative mapping (Chowdhary and Bailey 2003). Radiation hybrid mapping of the horse X-chromosome revealed perfectly conserved gene order relative to the human X-chromosome at a moderate level of RH resolution (Raudsepp et al. 2002).

Buffalo

The world population of river buffalo used for meat and milk consists of >130,000,000 animals. Scientific resources are limited in many of the countries where buffalo are economically important livestock, and as a consequence genome research has not been supported at the level of some of the other species. Excellent cytogenetics and fluorescence in situ hybridization, however, principally in the laboratory of Leopoldo Iannuzzi, has established a strong foundation for buffalo genomics. Almost 300 loci are on the cytogenetic map (Iannuzzi et al. 2003), most of them with homologs mapped in other species, and thus have contributed significantly to comparative mapping. The development of a hybrid somatic cell panel (El Nahas et al. 1996) produced synteny maps that were integrated with cytogenetic maps, resulting in the immediate assignment of syntenic groups to chromosomes. Chromosome banding patterns revealed almost identical karyotypes of river buffalo ($2N = 50$) and cattle ($2N = 60$) with five bi-armed buffalo chromosomes appearing to be fusions of five pairs of single-armed cattle chromosomes. Identity of these chromosome arms has been verified by comparative mapping in the two bovid species. A recently developed radiation hybrid panel for buffalo (E. Amaral, J. Elliott, J.E. Womack, pers. comm.) will advance comparative mapping to a higher level of resolution. There are presently no linkage maps for river buffalo. The work of E. Amaral, J. Elliott, J.E. Womack (pers. comm.), however, suggests that as in sheep, primers for most cattle-derived microsatellites amplify buffalo sequences in homologous regions of the respective genomes. If sufficient numbers of these microsatellites are polymorphic in buffalo, they will facilitate the development of a linkage map when pedigreed families are properly identified and DNA is made available to the growing buffalo mapping community.

Goat

A linkage map of the goat (Vaiman et al. 1996) was expanded significantly by Schibler et al. (1998) and now contains >300 markers. The latter study also added 202 cytogenetic localizations, many of which were also mapped by linkage, thus integrating the two maps. This map was among the first integrated maps in ruminants and served as a prototypic ruminant map and an excellent tool for comparative mapping prior to the generation of radiation hybrid maps in cattle and sheep.

CONTRIBUTIONS OF LIVESTOCK SPECIES TO COMPARATIVE MAPPING

Along with that of the domestic cat (O'Brien and Nash 1982), genome maps of livestock species were the first to expand the comparative maps of mammalian genomes beyond primates and rodents. Early somatic cell maps were largely driven by comparative mapping interest; thus homologs of genes previously mapped, usually in humans, were genotyped in panels of hybrid somatic cells derived from domestic animals. Markers segregating together in these panels were said to be syntenic or on the same strand (chromosome). The term "synteny" is in and of itself irrelevant to comparative genomics and was coined by Frank Ruddle to describe genes determined to be on the same chromosome by somatic cell genetics as opposed to the term "linkage," which has traditionally been associated with nonrandom assortment of alleles of two or more genes in meiosis. Unfortunately, the term synteny is often adulterated to imply evolutionary conservation of homologous chromosome segments between species. Nonetheless, the first autosomal comparative maps in non-rodent species were comparisons of syntenic groups in humans to syntenic groups in domestic animals (O'Brien and Nash 1982; Womack and Moll 1986). These were subsequently aided by cytogenetic mapping of homologous genes in different species.

Comparison of syntenic groups between two species is a type of comparative chromosomal painting if the markers are ordered in one species. Developing unordered synteny maps in cattle, for example, using ordered homologous markers from the human map results in the equivalent of "cattle on human" painting of the human map. The opposite, human on cow (or pig, horse, etc.) painting, was made possible by the technique of Zoo-FISH (Wienberg et al. 1982; Jauch et al. 1992). The hybridization of cocktails of fluorescence labeled unique sequence probes derived from isolated human chromosomes to chromosomes of another species was first applied to non-human primates but subsequently to other species including cattle (Hayes 1995; Solinas-Toldo et al. 1995; Chowdhary et al. 1996), pigs (Goureau et al. 1996), and horses (Raudsepp et al. 1996). Single chromosome paints are not generally available from the domestic animal species for Zoo-FISH painting with the exception of pig (Goureau et al. 1996).

As mentioned earlier, comparative genomics at the level of DNA sequence is particularly instructive in the identification of highly conserved genomic elements other than coding sequences (Margulies et al. 2003; Thomas et al. 2003). The power of this approach has been enhanced by the availability of whole-genome sequence from the livestock species. An exciting harvest from comparative genomics will come when multiple species from different evolutionary clades have been sequenced or mapped at a high level of resolution. Highly conserved elements in cattle, sheep, goat, and buffalo, for example, might point toward what makes a ruminant a ruminant if conserved homologs are not obvious in other clades. Similar comparisons could point to unique functional elements in primates, rodents, and other groups with multiple sequenced species.

A significant contribution to biology from farm animal genomes is evident in the recent discovery of extensive reuse of chromosome breakpoints during mammalian evolution (Murphy et al. 2005). These sites of evolutionary activity are marked by high gene density, accumulation of segmental duplications in humans, and footprints of telomeres and centromeres. Similar studies will undoubtedly continue to uncover biologically significant sites in animal genomes that are not revealed by concentrated focus on a single genome.

INFORMING HUMAN MEDICINE

The livestock species have some distinct advantages over other animals for studying the underlying mechanisms of phenotypic variation between and within species. Yet, the farm animals have not been fully appreciated or exploited in biomedical research. All the livestock species were domesticated from wild ancestors in the last 10,000 years, and highly differentiated phenotypes have resulted from intensive selective breeding, much of it in the last century. In addition, large numbers of offspring can be produced from a single mating in most species, and excellent phenotypic records accompany many of the pedigrees of animals with extreme phenotypes. And finally, highly developed genomic tools and rapidly developing databases are now available for the study of the major domestic animal genomes.

Biological insights gained from animal genomics include aiding the discovery of genes for human diseases. An excellent example is the discovery of a mutation in the *MC4R* gene in pigs (Kim et al. 2000) that results in obesity similar to that in humans. A good example of gene discovery in animals leading directly to gene discovery in humans is a mutation in the limbin gene responsible for chondrodysplastic dwarfism in Japanese Brown cattle (Takeda et al. 2002). Limbin had not previously been associated with any of the inherited dwarfisms in humans but was subsequently determined to be the homolog of *EVC2*, a gene responsible for Ellis-van Creneld syndrome, an autosomal recessive chondrodyplastic dwarfism in humans (Galdzicka et al. 2002; Ruiz-Perez et al. 2003). Discovery of the cattle gene clearly aided the discovery of the gene underlying the human disease.

Double muscling, a generalized muscular hypertrophy in cattle, has been recognized for almost 200 years and has been positively selected in some breeds such as the Belgian Blue, where the recessive *mh* allele is practically fixed. Conversely, it is an undesirable phenotype in many breeding programs because of the high incidence of dystocia. The gene was mapped to bovine chromosome 2 by Charlier et al. (1995) and fine-mapped by "identity by descent" (IBD) by Dunner et al. (1997). Comparative mapping (Dunner et al. 1997; Sonstegard et al. 1997) placed the mutation in a region of cattle chromosome 2 that is conserved relative to human chromosome 2. Comparative candidate positional cloning (Womack 1996) suggested several candidate genes from the human map, including myostatin (Grobet et al. 1997; Smith et al. 1997). Meanwhile, McPherron et al. (1997) demonstrated pronounced muscular hypertrophy in mice homozygous for a knockout deletion of *Gdf8*, the locus encoding myostatin. Based on this strong comparative genomic information, Grobet et al. (1997) sought and found a deletion in the bovine myostatin gene responsible for the *ms/ms* phenotype. Other loss-of-function mutations in the bovine gene have been identified in Belgian Blue and other breeds (Kambadur et al. 1997). Schuelke et al. (2004) recently described a child with gross muscular hypertrophy. The wealth of literature on double muscling in cattle and the mouse knockout experiments immediately led these investigators to the human myostatin gene, where, indeed, a splice-site mutation was discovered. Thus, loss of function of the myostatin gene produces similar phenotypes across the three mammalian species (Fig. 1), and the identification of mutations in double-muscled cattle was instrumental in discovery of the homologous gene underlying

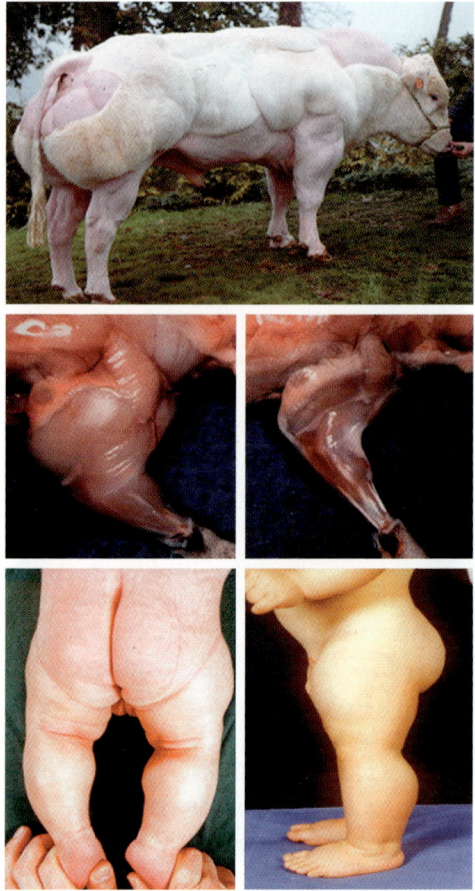

Figure 1. Phenotypic comparisons of: (*top*) "double muscled" bull homozygous for loss-of-function allele at the myostatin (*MSTN*) locus. Photograph courtesy of Michel Georges, University of Liège, Belgium. (*Middle*) Forelimb of GDF-8 (myostatin) knockout mouse (*left*) versus wild-type littermate (*right*). Photograph courtesy of Se-Jin Lee, Johns Hopkins University School of Medicine, Baltimore. Reprinted with copyright permission from Nature Publishing Group © 1997, from McPherron et al. (1997) (http://www.nature.com/). (*Lower*) Child at age of 6 d (*left*) and 7 mo (*right*) homozygous for loss-of-function allele in the myostatin gene. Photograph courtesy of Markus Schuelke, University Medical Center, Berlin. Reprinted with copyright permission from the Massachusetts Medical Society and the *New England Journal of Medicine* © 2004, from Schuelke et al. (2004).

the similar human phenotype. Various methods of blocking myostatin function are being considered as therapies for muscular-degenerative disorders in humans (Bogdanovich et al. 2002).

IMPROVING ANIMAL HEALTH AND PRODUCTION

Selection for desirable traits, or conversely, selection against undesirable traits, has been practiced since the domestication of animals almost 10,000 years ago. The promise of more accuracy, efficiency, and economy in selecting animals that will produce offspring with desirable phenotypes, however, underpins a substantial portion of the funding for livestock genome projects over the past two decades. The early linkage maps for most livestock species were constructed as tools for mapping traits and developing molecular markers for use in marker-assisted selection (MAS). The ultimate marker for MAS is, of course, the mutation underlying the selected phenotype. Although mapped QTLs in livestock species now number in the hundreds, very few mutations underlying quantitative trait variation have been identified. For obvious reasons, success has been better with several monogenic traits of economic and biological interest. In an extremely valuable database called "Online Mendelian Inheritance in Animals (OMIA)" (http://www.angis.org.au/oma/), Frank Nicholas lists 56 single-locus traits in cattle, 27 of which have had the causative mutation identified. The same listing provides equivalent numbers of 33 and 11 for pig, 59 and 9 for sheep, 26 and 9 for horse, and 8 and 5 for goat. Thus, the mutation has been identified in only about one-third of the single-gene traits cataloged in these species. Of the hundreds of QTLs now mapped in livestock, we have only two examples of the elucidation of mutations underpinning the QTL, both in dairy cattle. The first discovery of a quantitative trait nucleotide (QTN) was provided by comparative candidate positional cloning of the *DGAT1* locus as a gene contributing to fat composition in milk on chromosome 14 (Grisart et al. 2002), followed by functional confirmation of the effect of a mis-sense mutation (Grisart et al. 2004). Another QTL for milk fat and protein concentration was identified on chromosome 6 (Ron et al. 2001), and recently localized to a mis-sense mutation in the *ABCG2* gene, again with the aid of comparative and functional analysis (Cohen-Zinder et al. 2005).

Although QTN discovery has been slow in all mammals, including humans, a promising future was predicted by Korstanje and Paigen (2002) with their charting of exponential growth of genes and mutations identified in mammalian QTL studies beginning in 1999. These discoveries parallel the development of genome sequencing initiatives. The emergence of sequence from livestock species over the next few years bodes well for the discovery of genes underlying health and production traits in economically important species.

THE NOT TOO DISTANT FUTURE

My public statements in the late 1990s that domestic animal genomes would likely never be fully sequenced disqualify me as a prognosticator in the dynamic discipline of livestock genomics. Nonetheless, it seems inappropriate to conclude a review of our discipline without a brief guess, albeit a conservative one, as to where it might be headed. Genome sequencing, database development, expression arrays, and SNP maps with automated genotyping will obviously become staples of our genomic toolbox, probably before our current generation of graduate students leaves our laboratories. RNA interference may soon find its way into animal improvement, likely in conjunction with cloning from modified somatic cells. It is interesting to speculate on the early applications of these genomic resources. While the bottleneck between mapped QTL and gene discovery will not be cleared immediately, we should expect a rapidly accelerated harvest of causative mutations for biologically interesting and economically important phenotypes. The next wave of livestock QTLs will likely lead us to the discovery of new genes for disease resistance. Selection of animals with innate resistance to pathogens is, of course, important to sustainable agriculture and one potential defense against agricultural bioterrorism. With few exceptions such as the discovery of the role of *BoLA-DBR3* in bovine leucosis (Xu et al. 1993) and mapping of the QTLs for trypanotolerance in cattle (Hanotte et al. 2003), variation in disease resistance has been recalcitrant to either candidate gene studies or QTL mapping in livestock species. This generally reflects our inability to safely contain pathogens in challenge experiments requiring large numbers of cows, pigs, sheep, and the like. Gene sequencing and SNP discovery in our domestic animal species will soon give us information about linkage disequilibrium over large genomic regions and the identification of haplotype blocks in various populations and breeds of livestock. It is likely that these blocks will be large enough in many populations for retroactive association studies in herds subjected to pathogen exposure. The list of molecules involved in host recognition of pathogens and associated cell signaling grows almost daily, suggesting ample opportunity for mutations throughout the genome to differentiate the response of individual animals to pathogen contact. The identification of QTLs for disease resistance in livestock may be the next big frontier for the contribution of domestic animal genomics to the understanding of host–pathogen interaction and the subsequent improvement of both animal and human health.

REFERENCES

Andersson, L. and Georges, M. 2004. Domestic animal genomics: Deciphering the genetics of complex traits. *Nat. Rev. Genet.* **5**: 202–212.

Archibald, A.L., Haley, C.S., Brown, J.F., Couperwhite, S., McQueen, H.A., Nicholson, D., Coppieters, W., Van de Weghe, A., Stratil, A., Wintero, A.K., et al. 1995. The PiGMaP consortium linkage map of the pig (*Sus scrofa*). *Mamm. Genome* **6**: 157–175.

Band, M.R., Larson, J.H., Rebeiz, M., Green, C.A., Heyen, D.W., Donovan, J., Windish, R., Steining, C., Mahyuddin, P., Womack, J.E., et al. 2000. An ordered comparative map of the cattle and human genomes. *Genome Res.* **10**: 1359–1368.

Band, M.R., Olmstead, C., Everts, R.E., Liu, Z.L., and Lewin, H.A. 2002. A 3800 gene microarray for cattle functional genomics: Comparison of gene expression in spleen, placenta and brain. *Anim. Biotechnol.* **13**: 163–172.

Barendse, W., Armitage, S.M., Kossarek, L.M., Shalom, A., Kirkpatrick, B.W., Ryan, A.M., Clayton, D., Li, L., Neibergs, H.L., Zhang, N., et al. 1994. A genetic linkage map of the bovine genome. *Nat. Genet.* **6**: 227–235.

Barendse, W., Vaiman, D., Kemp, S.J., Sugimoto, Y., Armitage, S.M., Williams, J.L., Sun, H.S., Eggen, A., Agaba, M., Aleyasin, S.A., et al. 1997. A medium-density genetic linkage map of the bovine genome. *Mamm. Genome* **8**: 21–28.

Beckmann, J.S. and Soller, M. 1983. Restriction fragment length polymorphisms in genetic improvement: Methodologies, mapping and costs. *Theo. Appl. Genet.* **67**: 35–43.

Bishop, M.D., Kappes, S.M., Keele, J.W., Stone, R.T., Sunden, S.L.F., Hawkins, G.A., Toldo, S.S., Fries, R., Grosz, M.D., Yoo, J.Y., et al. 1994. A genetic linkage map for cattle. *Genetics* **136**: 619–639.

Bogdanovich, S., Krag, T.O., Barton, E.R., Morris, L.D., Whittemore, L.A., Ahima, R.S., and Khurana, T.S. 2002. Functional improvement of dystrophic muscle by myostatin blockade. *Nature* **420**: 418–421.

Burkin, D.J., Morse, H.G., Broad, T.E., Pearce, P.D., Ansari, H.A., Lewis, P.E., and Jones, C. 2003. Mapping the sheep genome: Production of characterized sheep × hamster cell hybrids. *Genomics* **16**: 466–472.

Charlier, C., Coppieters, W., Farnir, F., Grobert, L., Leroy, P.L., Michaux, C., Min, M., Schwers, A., Vanmanshoven, P., Hanset, R., et al. 1995. The *mh* gene causing double-muscling in cattle maps to bovine Chromosome 2. *Mamm. Genome* **6**: 788–792.

Chowdhary, B.P. and Bailey, E. 2003. Equine genomics: Galloping to new frontiers. *Cytogenet. Genome Res.* **102**: 184–188.

Chowdhary, B.P., Frönicke, L., Gustavsson, I., and Scherthan, H. 1996. Comparative analysis of the cattle and human genomes: Detection of ZOO-FISH and gene mapping-based chromosomal homologies. *Mamm. Genome* **7**: 297–302.

Chowdhary, B.P., Raudsepp, T., Honeycutt, D., Owens, E.K., Piumi, F., Guerin, G., Matise, T.C., Kata, S.R., Womack, J.E., and Skow, L.C. 2002. Construction of a 5000 (rad) whole-genome radiation hybrid panel in the horse and generation of a comprehensive and comparative map for ECA11. *Mamm. Genome* **13**: 89–94.

Cockett, N. 2003. Current status of the ovine genome map. *Cytogenet. Genome Res.* **102**: 76–78.
Cohen-Zinder, M., Seroussi, E., Larkin, D.M., Loor, J.J., Everts-van der Wind, A., Lee, J.H., Drackley, J.K., Band, M.R., Hernandez, A.G., Shani, M., et al. 2005. Identification of a missense mutation in the bovine ABCG2 gene with a major effect on the QTL on chromosome 6 affecting milk yield and composition in Holstein cattle. *Genome Res.* **15**: 936–944.
Collins, F.S., Green, E.D., Guttmacher, A.E., and Guyer, M.S. 2003. US National Human Genome Research Institute, a vision for the future of genomics research. *Nature* **422**: 835–847.
Crawford, A.M., Dodds, K.G., Ede, A.J., Pierson, C.A., Montgomery, G.W., Garmonsway, H.G., Beattie, A.E., Davies, K., Maddox, J.F., Kappes, S.W., et al. 1995. An autosomal genetic linkage map of the sheep genome. *Genetics* **140**: 703–724.
de Gortari, M.J., Freking, B.A., Cuthbertson, R.P., Kappes, S.M., Keele, J.W., Stone, R.T., Leymaster, K.A., Dodds, K.G., Crawford, A.M., and Beattie, C.W. 1998. A second-generation linkage map of the sheep genome. *Mamm. Genome* **9**: 204–209.
Dunner, S., Charlier, C., Farnir, F., Brouwers, B., Canon, J., and Georges, M. 1997. Towards interbreed IBD fine mapping of the *mh* locus: Double-muscling in the *Asturiana de lo Valles* breeds involves the same locus as in the *Belgian Blue* cattle breed. *Mamm. Genome* **8**: 430–435.
Dvorak, C.M.T., Hyland, K.A., Machado, J.G., Zhang, Y., Fahrenkrug, S.C., and Murtaugh, M.P. 2005. Gene discovery and expression profiling in porcine Peyer's patch. *Vet. Immun. Immunopath.* **105**: 301–315.
El Nahas, S.M., Oraby, H.A., de Hondt, H.A., Medhat, A.M., Zahran, M.M., Mahfouz, E.R., and Karim, A.M. 1996. Synteny mapping in river buffalo. *Mamm. Genome* **7**: 831–834.
Everts, R.E., Band, M.R., Liu, Z.L., Kumar, C.G., Liu, L., Loor, J.J., Olivera, R., and Lewin, H.A. 2005. A 7872 cDNA microarray and its use in bovine functional genomics. *Vet. Immun. Immunopath.* **105**: 235–245.
Everts-van der Wind, A., Kata, S.R., Band, M.R., Rebeiz, M., Larkin, D.M., Everts, R., Green, C.A., Liu, L., Natarajan, S., Goldammer, T., et al. 2004. A 1463 gene cattle–human comparative map with anchor points defined by human genome sequence coordinates. *Genome Res.* **14**: 1424–1437.
Fahrenkrug, S.C., Smith, T.P., Freking, B.A., Cho, J., White, J., Vallet, J., Wise, T., Rohrer, G., Pertea, G., Sultana, R., et al. 2002. Porcine gene discovery by normalized cDNA-library sequencing and EST cluster assembly. *Mamm. Genome* **13**: 475–478.
Fries, R., Hediger, R., and Stranzinger, G. 1986. Tentative chromosomal localization of the bovine major histocompatibililty complex by in situ hybridization. *Anim. Genet.* **17**: 287–294.
Fries, R., Eggen, A., and Womack, J.E. 1993. The bovine genome map. *Mamm. Genome* **4**: 405–428.
Galdzicka, M., Patnala, S., Hirshman, M.G., Cai, J.-F., Nitowsky, H., Egeland, J.A., and Ginns, E.I. 2002. A new gene, EVC2, is mutated in Ellis-van Creveld syndrome. *Mol. Gene Metabo.* **77**: 291–295.
Gallagher Jr., D.S., Threadgill, D., Ryan, A.M., Womack, J.E., and Irwin, D.M. 1993. Physical mapping of the lysozyme gene family in cattle. *Mamm. Genome* **4**: 386–373.
Goureau, A., Yerle, M., Schmitz, A., Riquet, J., Milan, D., Pinton, P., Frelat, G., and Gellin, J. 1996. Human and porcine correspondence of chromosome segments using bidirectional chromosome painting. *Genomics* **36**: 252–262.

Grisart, B., Coppieters, W., Farnir, F., Karim, L., Ford, C., Berzi, P., Cambisano, N., Mni, M., Reid, S., Simmon, P., et al. 2002. Positional candidate cloning of a QTL in dairy cattle: Identification of a missense mutation in the bovine DGAT1 gene with major effect on milk yield and composition. *Genome Res.* **12**: 222–231.

Grisart, B., Farnir, F., Karim, L., Cambisano, N., Kim, J.J., Kvasv, A., Mini, M., Simon, P., Frere, J.M., Copieters, W., et al. 2004. Genetic and functional confirmation of the causality of the DGAT1 K232A quantitative trait nucleotide in affecting milk yield and composition. *Proc. Natl. Acad. Sci.* **101**: 2398–2403.

Grobet, L., Martin, L.J.R., Poncelet, D., Pirottin, D., Brouwers, B., Riquet, J., Schoeberlein, A., Dunner, S., Ménissier, F., Massabanda, J., et al. 1997. A deletion in the bovine myostatin gene causes the double-muscled phenotype in cattle. *Nat. Genet.* **17**: 71–74.

Guérin, G., Bailey, E., Bernoco, D., Anderson, I., Antczak, D.F., Bell, K., Binns, M.M., Bowling, A.T., Brandon, R., Cholewinski, G., et al. 1999. Report of the International Equine Gene Mapping Workshop: Male linkage map. *Anim. Genet.* **30**: 341–354.

Guérin, G., Bailey, E., Bernoco, D., Anderson, I., Antczak, D.F., Bell, K., Biros, I., Bjornstad, G., Bowling, A.T., Brandon, R., et al. 2003. The second generation of the International Equine Gene Mapping Workshop half-sibling linkage map. *Anim. Genet.* **34**: 161–168.

Hanotte, O., Ronin, Y., Agaba, M., Nilsson, P., Gelhaus, A., Horstmann, R., Sugimoto, Y., Kemp, S., Gibson, J., Korol, A., et al. 2003. Mapping of quantitative trait loci controlling trypanotolerance in a cross of tolerant West African N'Dama and susceptible East African Born cattle. *Proc. Natl. Acad. Sci.* **100**: 7443–7448.

Hawken, R.J., Murtaugh, J., Flickinger, G.H., Yerle, M., Robic, A., Milan, D., Gellin, J., Beattie, C.W., Schook, L.B., and Alexander, L.J. 1999. A first-generation porcine whole-genome radiation hybrid map. *Mamm. Genome* **10**: 824–830.

Hayes, H. 1995. Chromosome painting with human chromosome specific DNA libraries reveals the extent and the distribution of conserved segments in bovine chromosomes. *Cytogenet. Cell Genet.* **71**: 168–174.

Heuertz, S. and Hors-Cayla, M.-C. 1981. Cattle gene mapping by somatic cell hybridization study of 17 marker enzymes. *Cytogenet. Cell Genet.* **30**: 137–145.

Iannuzzi, L., Di Meo, G.P., Perucatti, A., Schibler, L., Incarnato, D., Gallagher, D., Eggen, A., Ferretti, L., Cribiu, E.P., and Womack, J. 2003. The river buffalo (*Bubalus bubalis*, 2n = 50) cytogenetic map: Assignment of 64 loci by fluorescence in situ hybridization and R-banding. *Cytogenet. Genome Res.* **102**: 65–75.

Itoh, T., Watanabe, T., Ihara, N., Mariani, P., Beattie, C.W., Sugimoto, Y., and Takasuga, A. 2005. A comprehensive radiation hybrid map of the bovine genome comprising 5593 loci. *Genomics* **85**: 413–424.

Jauch, A., Wienberg, J., Stanyon, R., Arnold, N., Tofanelli, S., Ishida, T., and Cremer, T. 1992. Reconstruction of genomic rearrangements in great apes and gibbons by chromosome painting. *Proc. Natl. Acad. Sci.* **89**: 8611–8615.

Kambadur, R., Sharma, M., Smith, T.P.L., and Bass, J.J. 1997. Mutations in *myostatin* (*GDF8*) double-muscled Belgian Blue and Piedmontese cattle. *Genome Res.* **7**: 910–915.

Kappes, S.M., Keele, J.W., Stone, R.T., McGraw, R.A., Sonstegard, T.S., Smith, T.P., Lopez-Corrales, N.L., and Beattie, C.W. 1997. A second-generation linkage map of the bovine genome. *Genome Res.* **7**: 235–249.

Kim, K.S., Larsen, N., Short, T., Plastow, G., and Rothschild, M.F. 2000. A missense variant of the melanocortin 4 receptor (MC4R) gene is associated with fatness, growth and feed intake traits. *Mamm. Genome* **11**: 131–135.

Korstanje, R. and Paigen, B. 2002. From QTL to gene: The harvest begins. *Nat. Genet.* **31:** 235–236.

Lewin, H.A. 2003. The future of cattle genome research: The beef is here. *Cytogenet. Genome Res.* **102:** 10–15.

Lindgren, G., Sandberg, K., Persson, H., Marklund, S., Breen, M., Sandgren, B., Carlstén, J., and Ellegren, H. 1998. A primary male autosomal linkage map of the horse genome. *Genome Res.* **8:** 951–966.

Maddox, J.F., Davies, K.P., Crawford, A.M., Hulme, D.J., Vaiman, D., Cribiu, E.P., Freking, B.A., Beh, K.J., Cockett, N.E., Kang, N., et al. 2001. An enhanced linkage map of the sheep genome comprising more than 1000 loci. *Genome Res.* **11:** 1275–1289.

Margulies, E.H., Blanchette, M., Haussler, D., and Green, E.D. NISC Comparative Sequencing Program. 2003. Identification and characterization of multi-species conserved sequencing. *Genome Res.* **13:** 2507–2518.

McPherron, A.C., Lawler, A.M., and Lee, S.J. 1997. Regulation of skeletal muscle mass in mice by a new TGF-B superfamily member. *Nature* **387:** 83–90.

Montgomery, C.W., Crawford, A.M., Penty, A.M., Dodds, K.G., Ede, A.J., Henry, H.M., Pierson, C.A., Lord, E.A., Galloway, S.M., Schmack, A.E., et al. 1993. The ovine Booroola fecundity gene (*FecB*) is linked to markers from a region of human chromosome 4q. *Nat. Genet.* **4:** 410–414.

Murphy, W.J., Larkin, D.M., Everts-van der Wind, A., Gurque, G., Tesler, G., Auvil, L., Beever, J.E., Chowdhary, B.P., Galibert, F., Gatzke, L., et al. 2005. Dynamics of mammalian chromosome evolution inferred from multispecies comparative maps. *Science* **309:** 613–617.

Niewold, T.A., Kerstens, H.H.D., van der Meulen, J., Smits, M.A., and Hulst, M.M. 2005. Development of a porcine small intestinal cDNA micro-array: Characterization and functional analysis of the response to enterotoxigenic E. coli. *Vet. Immun. Immunopath.* **105:** 317–329.

O'Brien, S.J. and Nash, W.G. 1982. Genetic mapping in mammals: Chromosome map of domestic cat. *Science* **216:** 257–265.

Raudsepp, T., Fronicke, L., Scherthan, H., Gustavsson, I., and Chowdhary, B.P. 1996. Zoo-FISH delineates conserved chromosomal segments in horse and man. *Chromosome Res.* **4:** 218–225.

Raudsepp, T., Kata, S.R., Piumi, F., Swinburne, J., Womack, J.E., Skow, L.E., and Chowdhary, B.P. 2002. Conservation of gene order between horse and human X chromosomes as evidenced through radiation hybrid mapping. *Genomics* **79:** 451–457.

Rink, A., Santschi, E.M., Eyer, K.M., Roelofs, B., Hess, M., Godfrey, M., Karajusuf, E.K., Yerle, M., Milan, D., and Beattie, C.W. 2002. A first generation EST RH comparative map of the porcine and human genome. *Mamm. Genome* **14:** 578–587.

Rohrer, G.A., Alexander, L.J., Hu, Z., Smith, T.P., Keele, J.W., and Beattie, C.W. 1996. A comprehensive map of the porcine genome. *Genome Res.* **6:** 371–391.

Ron, M., Kliger, D., Feldmesser, E., Seroussi, E., Ezra, E., and Weller, J.I. 2001. Multiple QTL analysis of bovine chromosome 6 in the Israeli Holstein population by a daughter design. *Genetics* **159:** 727–735.

Rothschild, M.F. 2003. From a sow's ear to a silk purse: Real progress in porcine genomics. *Cytogenet. Genome Res.* **102:** 95–99.

Ruiz-Perez, V., Tompson, S.W.J., Blair, H.J., Espinoza-Valdez, C., Lapunzina, P., Silva, E.O., Hamel, B., Gibbs, J.L., Young, I.D., Wright, M.J., et al. 2003. Mutations in two nonhomologous genes in a head-to-head configuration cause Ellis-van Creveld syndrome. *Am. J. Hum. Gene* **72:** 728–732.

Schibler, L., Vaiman, D., Oustry, A., Giarud-Delville, C., and Cribiu, E.P. 1998. Comparative gene mapping: A fine-scale survey of chromosome rearrangements between ruminants and humans. *Genome Res.* **8:** 901–915.

Schuelke, M., Wagner, K.R., Stolz, L.E., Hubner, C., Hübner, C., Riebel, T., Kömen, W., Braun, T., Tobin, J.F., and Lee, S.-J. 2004. Myostatin mutation associated with gross muscle hypertrophy in a child. *N. Engl. J. Med.* **350:** 2682–2688.

Smith, T.P.L., Lopez-Corrales, N.L., Knappes, S.M., and Sonstegard, T.S. 1997. Myostatin maps of the interval containing the above mh locus. *Mamm. Gen.* **8:** 742–744.

Solinas-Toldo, S., Lengauer, C., and Fries, R. 1995. Comparative genome map of human and cattle. *Genomics* **27:** 489–496.

Sonstegard, T.S. and van Tassell, C.P. 2004. Bovine genomics update: Making the cow jump over the moon. *Genet. Res. Camb.* **84:** 3–9.

Sonstegard, T.S., Lopez-Corrales, N.L., Knappes, S.M., Beattie, C.W., and Smith, T.P.L. 1997. Comparative mapping of human Chromosome 2 identifies segments of conserved synteny near the bovine mh locus. *Mamm. Gen.* **8:** 751–755.

Suchyta, S.P., Sipkovsky, S., Kruska, R., Jeffers, A., McNutty, A., Coussens, M.J., Tempelmann, R.J., Halgren, R.G., Saama, P.M., Bauman, D.E., et al. 2003. Development and testing of a high-density cDNA microarray resource for cattle. *Physiol. Genomics* **15:** 158–164.

Swinburne, J., Gersenberg, C., Breen, M., Aldridge, V., Lockhart, L., Marti, E., Antczak, D., Egglestone-Stott, M., Bailey, E., Mickelson, J., et al. 2000. First comprehensive low-density horse linkage map based on two, three-generation, full-sibling, cross-bred horse reference families. *Genomics* **66:** 123–134.

Takeda, H., Takami, M., Oguni, T., Tsuji, T., Yoneda, K., Sato, H., Ihara, N., Itoh, T., Kata, S.R., Mishina, Y., et al. 2002. Positional cloning of the gene LIMBIN responsible for bovine chondrodysplastic dwarfism. *Proc. Natl. Acad. Sci.* **99:** 10549–10554.

Thomas, J.W., Touchman, J.W., Blakesley, R.W., Bouffard, G.G., Beckstrom-Sternberg, S.M., Margulies, E.H., Blanchette, M., Siepel, A.C., Thomas, P.J., McDowell, J.C., et al. 2003. Comparative analyses of multi-species sequences from targeted genomic regions. *Nature* **424:** 788–793.

Tuggle, C.K., Green, J.A., Fitzsimmons, C., Woods, R., Prather, R.S., Malchenko, S., Soares, M.B., Kucaba, T., Crouch, K., Smith, C., et al. 2003. EST-based gene discovery in pig: Virtual expression patterns and comparative mapping to human. *Mamm. Genome* **14:** 565–579.

Vaiman, D., Schibler, L., Bourgeois, F., Oustry, A., Amigues, Y., and Cribiu, E.P. 1996. A genetic linkage map of the male goat genome. *Genetics* **144:** 279–305.

Wernersson, R., Schierup, M.H., Jorgensen, F.G., Gorodkin, J., Panitz, F., Staerfeldt, H.-H., Christensen, O.F., Mailund, T., Hornshoj, H., Klein, A., et al. 2005. Pigs in sequence space: A 0.66 coverage pig genome survey based on shotgun sequencing. *BMC Genomics* **6:** 70.

Wienberg, J., Stanyon, R., Jauch, A., and Cremer, T. 1982. Homologies in human and *Macaca fuscata* chromosome revealed by in situ suppression hybridization with human chromosomes specific DNA libraries. *Chromosoma* **101:** 265–270.

Williams, J.L., Eggen, A., Ferretti, L., Farr, C.J., Gautier, M., Amati, G., Ball, G., Caramorr, T., Critcher, R., Costa, S., et al. 2002. A bovine whole-genome radiation hybrid panel and outline map. *Mamm. Genome* **13:** 469–474.

Womack, J.E. 1987. Gene map of the cow (*Bos taurus*). *Genet. Maps* **4:** 499.

———. 1996. The bovine gene map: A tool for comparative candidate positional cloning. *Arch. Zootec.* **45:** 151–164.

Womack, J.E. and Moll, Y.D. 1986. Gene map of the cow: Conservation of linkage with mouse and man. *J. Hered.* **77:** 2–7.

Womack, J.E., Johnson, J.S., Owens, E.K., Rexroad III, C.E., Schlapfer, J., and Yang, Y.P. 1997. A whole-genome radiation hybrid panel for bovine gene mapping. *Mamm. Genome* **8:** 854–856.

Xu, A., van Eijk, M.J.T., Park, C., and Lewin, H.A. 1993. Polymorphism in BoLA-DRB3 exon 2 correlates with resistance to persistent lymphocytosis caused by bovine leukemia virus. *J. Immun.* **151:** 6977–6985.

Yerle, M., Lahbib-Mansais, Y., Mellink, C., Goureau, A., Pinton, P., Echard, G., Gellin, J., Zijlstra, C., De Haan, N., Bosma, A.A., et al. 1995. The PiGMap Consortium cytogenetic map of the domestic pig (*Sus scrofa domestica*). *Mamm. Genome* **6:** 176–186.

Yerle, M., Echard, G., Robic, A., Mairal, A., Dubut-Fontana, C., Riquet, J., Pinton, P., Milan, D., Lahbib-Mansais, Y., and Gellin, J. 1996. A somatic cell hybrid panel for pig regional gene mapping characterized by molecular cytogenetics. *Cytogenet. Cell Genet.* **73:** 194–202.

Yerle, M., Pinton, P., Robic, A., Alfonso, A., Palvadeau, Y., Delcros, C., Hawken, R., Alexander, L., Beattie, C., Schook, L., et al. 1998. Construction of a whole-genome radiation hybrid panel for high-resolution gene mapping in pigs. *Cytogenet. Cell Genet.* **82:** 182–188.

13

The Canine Genome

Elaine A. Ostrander
Cancer Genetics Branch
National Human Genome Research Institute
National Institutes of Health
Bethesda, Maryland 20892

Robert K. Wayne
Department of Ecology and Evolutionary Biology
University of California at Los Angeles
Los Angeles, California 90095

As one of the premier journals in genome biology celebrates its 10th anniversary, the scientific community studying dogs also enjoys a year of major advances and milestones, particularly with regard to canine genomics and comparative genetics. In July of 2004, the first high-quality draft (7.5×) sequence of the Boxer dog was made publicly available (Lindblad-Toh et al. 2005). This advance followed on the heels of other major milestones in the past several months, including the availability of a 1.5× Poodle sequence (Kirkness et al. 2003), a dense high quality radiation hybrid (RH) map (Breen et al. 2004), a detailed comparative map (Hitte et al. 2005), the localization and cloning of several disease genes, the successful application of dogs for gene therapy studies (Howell et al. 1997; Acland et al. 2001; Mount et al. 2002; Ponder et al. 2002), and new insights into the evolution of dogs and dog breeds (Parker et al. 2004).

As a result, the genome community is well poised to take advantage of the canine system and begin to fulfill some of the expectations advanced nearly 15 yr ago. First, with the development of appropriate molecular resources, the canine system was proposed to hold the power to map and clone disease genes that had proven intractable through studies of human families. Second, the variation in size and skeletal proportions that are segregated into distinct breeds of dog was hypothesized to provide a unique resource for dissecting genetic pathways underlying skeletal development. Finally, the range of behavioral traits that appeared strongly associated with individual breeds suggested a mechanism to decipher the basic genetic vocabulary of behavior (Patterson et al. 1982; Ostrander et al. 1993, 2000; Galibert et al. 1998; Patterson 2000). At the heart of these questions lies a fundamental conundrum. Why has the wolf genome, from which the dog is recently evolved, retained alleles controlling such a large amount of genetic variability, particularly as regards morphology? Is the dog genome somehow unique from other genomes? Or would strong selective pressures applied to any mammalian genome result in a range of species with a level of phenotypic variation that rivals the dog? Research done to date cannot readily answer these questions. However, we are beginning to understand how to localize the genes that regulate morphology (Chase et al. 2002). In so doing, we can begin to understand how genetic variation leads to major phenotypic changes. With the sequencing of the dog genome, it may be within our grasp to localize genes that cause the difference between Giant Mastiffs and Pekingese, Pointer and Terrier, and sight and scent hounds.

In this celebratory review, we first discuss the evolutionary framework and domestication of dogs. We then consider the recent accomplishments of the canine genome community. Finally, we highlight ongoing studies aimed at addressing some of the questions above.

THE EVOLUTIONARY FRAMEWORK

The domestic dog is the most recently evolved species in the dog family Canidae, a group that has a long history spanning the last 50 million years (Myr). This history can be portrayed as a succession of phylogenetic hierarchies defined by DNA sequence information (Fig. 1) and is a necessary structure for understanding molecular data. Of note is that dogs are the earliest divergence in the superfamily Canoidae that includes bears, weasels, skunks, raccoons, and the pinnipeds (seals, sea lions, and walruses) (Fig. 1A). This kinship predicts dogs will share more molecular similarities with these taxa than with cats, mongooses, civets, or hyenas. However, because of the early divergence of dogs from all other carnivores, only slowly evolving regions will show substantial sequence similarities. A second important point is that the 35 species of extant canids are genetically very similar, having radiated from a common ancestor less than about 10 Mya. The recent radiation in a family that

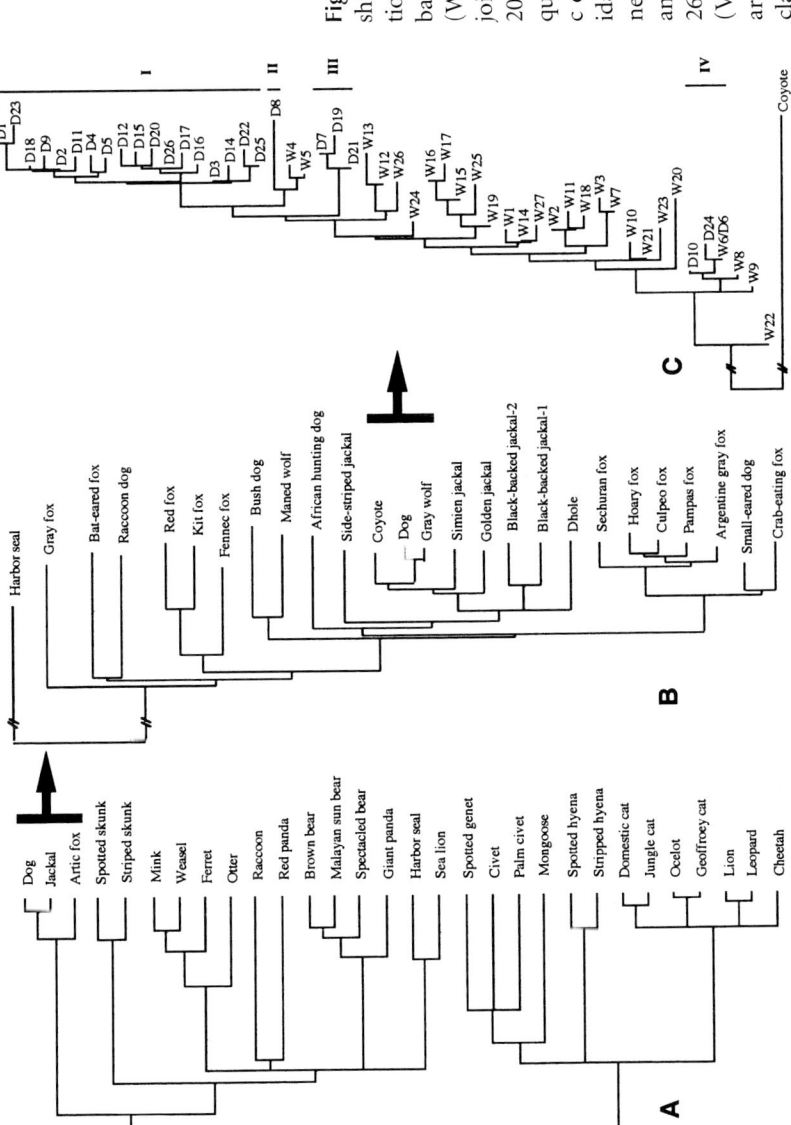

Figure 1. Evolutionary relationships of the dog. (A) The evolutionary relationships of carnivores based on DNA hybridization data (Wayne et al. 1989). (B) A neighbor-joining tree of canids based on 2001 bp of mitochondrial DNA sequence (cytochrome b, cytochrome c oxidase I, and cytochrome c oxidase II) (Wayne et al. 1997). (C) A neighbor-joining tree of wolf (W) and dog (D) haplotypes based on 261 bp of control region I sequences (Vila et al. 1997). Dog haplotypes are grouped in four sequence clades, numbered I to IV.

otherwise has a long evolutionary history suggests that genetic comparisons among extant canids will highlight rapidly evolving sequences and that they all may share uniquely evolved molecular structures such as SINE elements inherited from their recent common ancestor (Fanning et al. 1988; Kirkness et al. 2003) or rapidly evolving genes such as olfactory receptors, immune related genes, or reproductive proteins (e.g., Clark et al. 2003). In fact, although the dog family has a diverse chromosome complement ranging from 36 to 78 chromosomes, they all can be reconstructed through simple chromosome rearrangement from a common ancestral karyotype (Nash et al. 2001).

Within the Canidae, three distinct phylogenetic groupings are apparent (Fig. 1B) (Wayne et al. 1987a,b, 1997) as follows: (1) the fox-like canids, which include species closely related to the red fox (genus *Vulpes*), as well as the arctic and fennec fox (genus *Alopex* and *Fennecus*, respectively); (2) the wolf-like canids including dog, wolf, coyote, Ethiopian wolf or Simien jackal, and three other species of jackals (genus *Canis*), as well as the African hunting dog (genus *Lycaon*) and the dhole (genus *Cuon*); and (3) the South American canids including fox-sized canids such as the pampus fox, crab-eating fox, and small-eared dog (genus *Pseudolopex, Lycolopex, Atelocynus*) and the maned wolf (genus *Chrysocyon*) and bushdog (genus *Speothos*). Additionally, there are several canids that have no close living relatives and define distinct evolutionary lineages such as the gray fox (genus *Urocyon*), the bat-eared fox (genus *Otocyon*), and the raccoon dog (genus *Nyctereutes*).

These phylogenetic relationships imply that the dog has several close relatives within its genus, in fact, all members of *Canis* can produce fertile hybrids and several species may have genomes that reflect hybridization in the wild (Wayne and Jenks 1991; Gottelli et al. 1994; Roy et al. 1996; Wilson et al. 2000; Adams et al. 2003). Furthermore, the wolf-like canids are grouped more closely with the South American canids and the red and gray fox are very distinct groups whose common ancestry with dogs extends to the beginning of the modern radiation. Consequently, molecular tools developed from the dog genome sequencing project are likely to be most applicable to the wolf-like canids. For instance, fewer than half of microsatellite primers developed in the dog amplify DNA in the gray fox (Goldstein et al. 1999).

THE DOMESTICATION OF THE DOG

The essential questions about dog domestication concern the species from which the dog originated and the location, number, and timing of domestication or interbreeding events. Molecular data has shed some light on all of these questions. First, with regard to species origins, Charles Darwin and others such as Konard Lorenz, the renowned behavioral biologist, speculated that given the great diversity in form and behavior of dogs, they might share ancestry with wolves and other canids, such as any one of the three species of jackals. However, extensive genetic analyses of the dog and other wolf-like canids clearly show that the dog is derived from gray wolves only, rather than jackals, coyotes, or Ethiopian wolves (Fig. 1C; Wayne et al. 1987a,b; Vila et al. 1997, 2005; Leonard et al. 2002; Savolainen et al. 2002). Consequently, the immense phenotypic diversity in the dog owes its origin to primarily the standing genetic variation existing in the ancestral population of gray wolves and any subsequent mutations that occurred during the brief history of domestication. At least for structural genes, such mutations are expected to be few since their mutation rate is so low, on the order of 10^{-5} mutations per gene per generation (Hartl and Clark 1997).

Mitochondrial DNA (mtDNA) sequence analysis has shed some light on the location of dog domestication as well as the number of founding matralines. mtDNA analysis offers a unique perspective on evolutionary history because the mitochondrial genome is maternally inherited, and hence, only females leave a genetic legacy. Moreover, because the mitochondrial genome does not recombine, phylogenetic analysis of mtDNA sequence data defines a uniquely bifurcating haplotype tree (Fig. 1A,B,C). Phylogenetic analysis of dog and gray wolf mitochondrial sequences clearly show that dog sequences are found in at least four distinct clades, implying a single origination event and at least three other origination or interbreeding events. The latter are difficult to distinguish once the first domestication had occurred, although extensive marker analysis of the nuclear genome might be able to discriminate the two alternatives. A striking finding of the mtDNA analysis is that one sequence clade (clade I, Fig. 1C) contains the majority of dog sequences and that the nucleotide diversity of this clade is high, implying an origin of the clade from 40 to 135 thousand years ago (Vila et al. 1997; Savolainen et al. 2002). This date exceeds the 15,000-yr-old archeological record of dogs and suggests that dogs may have had a long prehistory when they were not phenotypically distinct from wolf progenitors. These early dogs may not have been recognized as domesticated by study of the archeological record before 15,000 yr ago because of their physical similarity to gray wolves. The initial change to the diagnostic phenotype of domestic dogs beginning about 15,000 yr ago may have instead indicated a change in the selection pressures associated with the transition from hunter gatherer to more sedentary lifestyles (Wayne et al. 2006).

Conceivably, a more recent date can be made consistent with the archeological record if it is assumed that dogs were founded from multiple matralines in clade one (Savolainen et al. 2002). To determine whether such a diverse founding is likely, analysis of nuclear genes sequence data is needed (e.g., Parker et al. 2004). In fact, recent analysis of major histocompatability (MHC) genes in dogs and wolves suggest that the origin of dogs involved several populations and hundreds of individuals (Vila et al. 2005). Consequently, the model emerging from mitochondrial DNA, MHC analysis, and microsatellite loci is that the dogs had a diverse origin in East Asia that likely involved multiple contributions from several populations, and thereafter, there may have been other origins of domestication and backcrossing (Vila et al. 1997, 2005; Leonard et al. 2002; Savolainen et al. 2002; Parker et al. 2004). A multiple and diverse origin model describes domestication in other domestic animals such as cattle, sheep, and goats (Bruford et al. 2003). Furthermore, once domesticated, dogs rapidly spread around the earth and as a result, genetically divergent populations and breeds are found in Africa, Asia, the Arctic, Australia, the Middle East, and historically, the New World (Leonard et al. 2002; Parker et al. 2004; Savolainen et al. 2004).

BREED DIVERSITY AND GENETIC STRUCTURE

The explosion of dog breeds over the past two centuries represents perhaps one of the greatest genetic experiments ever conducted by humans. Distilled from the genome of the wild wolf are animals that differ by more than 40-fold in size with the ability to herd, guard, hunt, and guide (American Kennel Club 1998). Behavioral variation is surpassed by morphologic variation, with individual breeds represented by dogs of every imaginable size and proportion. Coats alone can be described by color, texture, length, thickness, and curl. Tails can be described as plumed, curled, double curled, gay (upright), sickled (arching), otter (down and flat), whipped, ringed, screwed, or snapped (American Kennel Club 1998). The diversity in skeletal size and proportion of dogs is greater than any mammalian species and even exceeds that of the entire canid family (Wayne 1986a,c). Such variation may reflect simple modifications of post-natal development (Wayne 1986a,c), but the specific genetic mechanisms are not well known (see below).

Much of the morphologic variation in dogs is partitioned into over 350 distinct breeds worldwide as a result of the development of breed standards and controlled breeding. In general, in order to register a dog in the American Kennel Club at least both parents must have been registered in the same breed. Consequently, purebred dogs are members of closed breeding populations, which receive little genetic variation beyond that existing in the original founders (Ostrander and Giniger 1997; Galibert et al. 1998; Ostrander et al. 2000; Sutter and Ostrander 2004).

Common to the origin and development of many breeds is a founder event involving only a few dogs and, thereafter, reproductive dominance by popular sires that conform most closely to the breed standard. These restrictive breeding practices reduce effective population size and increase genetic drift, resulting in the loss of genetic diversity within breeds and allele frequency divergence among them. For example, in a genetic study of 85 breeds, Parker et al. (2004) showed that humans and dogs have similar levels of overall nucleotide diversity, 8×10^{-4}, which represent the overall number of nucleotide substitutions per base/pair. However, the variation between dog breeds is much greater than the variation between human populations (27.5% versus 5.4%). Conversely, the degree of genetic homogeneity is much greater within individual dog breeds than within distinct human populations (94.6% versus 72.5%). Furthermore, in some breeds, genetic variation has been additionally reduced by bottlenecks associated with catastrophic events such as war and economic depression, making them analogous to human populations of limited genetic variation used for disease-mapping studies such as the Finns, Icelanders, and Bedouins. As a result, the unique pattern of LD in dogs provides an exceptional opportunity to study complex traits that are relevant to human biology using robust approaches that would not be possible in human populations.

Because many breeds represent closed gene pools, they may define distinct genetic clusters. Analysis of microsatellite loci have strongly supported this notion (Koskinen 2003). For example, in the Parker et al. (2004) study, 96 microsatellite markers were genotyped that spanned all dog autosomes at approximately a 30-Mb resolution (Parker et al. 2004). Excluding data from the highly related Belgian Sheepdog and Belgian Tervuren breeds, they observed that 99% of 414 dogs were correctly assigned to breed. Consequently, a "breed" can be defined at the molecular level and dogs can be correctly assigned to their breed with small amounts of data. These results strongly imply that breeds are distinct genetic units and even closely related breeds do not represent genetic replicates.

BREED ORIGIN AND RELATIONSHIP

Mitochondrial DNA studies have not been useful for the reconstruction of breed origins or relationships because the origin of the vast majority of sequence polymorphisms found in dogs preceded the development of modern breeds. Therefore, phylogenetic hierarchies based on DNA sequences reveal the history of mutations that occurred before dogs were domesticated (e.g., Fig. 1C). However, many breeds contain several mitochondrial DNA haplotypes, suggesting that multiple matralines were involved in the founding of a dog breed. To assess the recent evolution and relationships of breeds, microsatellite loci provide a better tool, as their high variability insures allele frequency divergence through drift. Genetic distance trees based on the microsatellite dataset from Parker et al. (2004) revealed several distinct breed clusters. The most divergent grouping presumably contained the most ancient breeds, but none of these nine ancient breeds were of European origin. The ancient breeds included dogs from a wide geographic area including the Arctic, Asia, Africa, and the Middle East. By comparison, the majority of breeds, including European breeds, appeared to stem from a single node without significant phylogenetic structure, which has been termed a "hedge," indicating a recent origin and extensive hybridization between the breeds (Parker et al. 2004; Fig. 2). The focus on breeds belonging to this hedge in past studies probably explains the observed lack of phylogenetic resolution (Zajc et al. 1997; Koskinen and Bredbacka 2000; Irion et al. 2003).

This evolutionary hierarchy suggests breeds should cluster genetically into groups sharing recent common ancestry. A genetic clustering algorithm, deployed in the computer program "structure" was used to explore the possible groupings within dogs (Pritchard et al. 2000). Structure assigned 335 dogs correctly to 69 unique breed specific clusters that represented either single breeds or sets of very closely related breeds. However, the program could not easily distinguish a half-dozen obviously related pairs such as the Bernese Mountain Dog and Greater Swiss Mountain Dog or Mastiff and Bullmastiff. This lack of resolution in these few breeds is predicted based on breed history. For instance, the Bullmastiff is reported to be 60% Mastiff and 40% Bulldog and was created by crossing the two breeds in the mid-1800s (Rogers and Brace 1995).

Individual breeds represented the smallest definable cluster; however, higher order clusters are expected given the origins of many dogs breeds. Consequently, the number of groups (K) was set to 2, 3, and, finally, 4. The first distinct cluster to be defined at $K = 2$ included nearly all breeds of Asian origin (Akita, Shiba Inu, Shar-Pei, Lhasa Apso, etc.), some sled dogs, and some known ancient hounds such as the Saluki (Fig. 2). When added to the analysis, gray wolves from eight countries all grouped in the first cluster as well. The early divergence of the Asian breeds on the phylogenetic tree and their association with the wolves in clustering analysis (Fig. 2) supports the conclusions of mitochondrial DNA analysis that domestication first took place in East Asia (Savolainen et al. 2002).

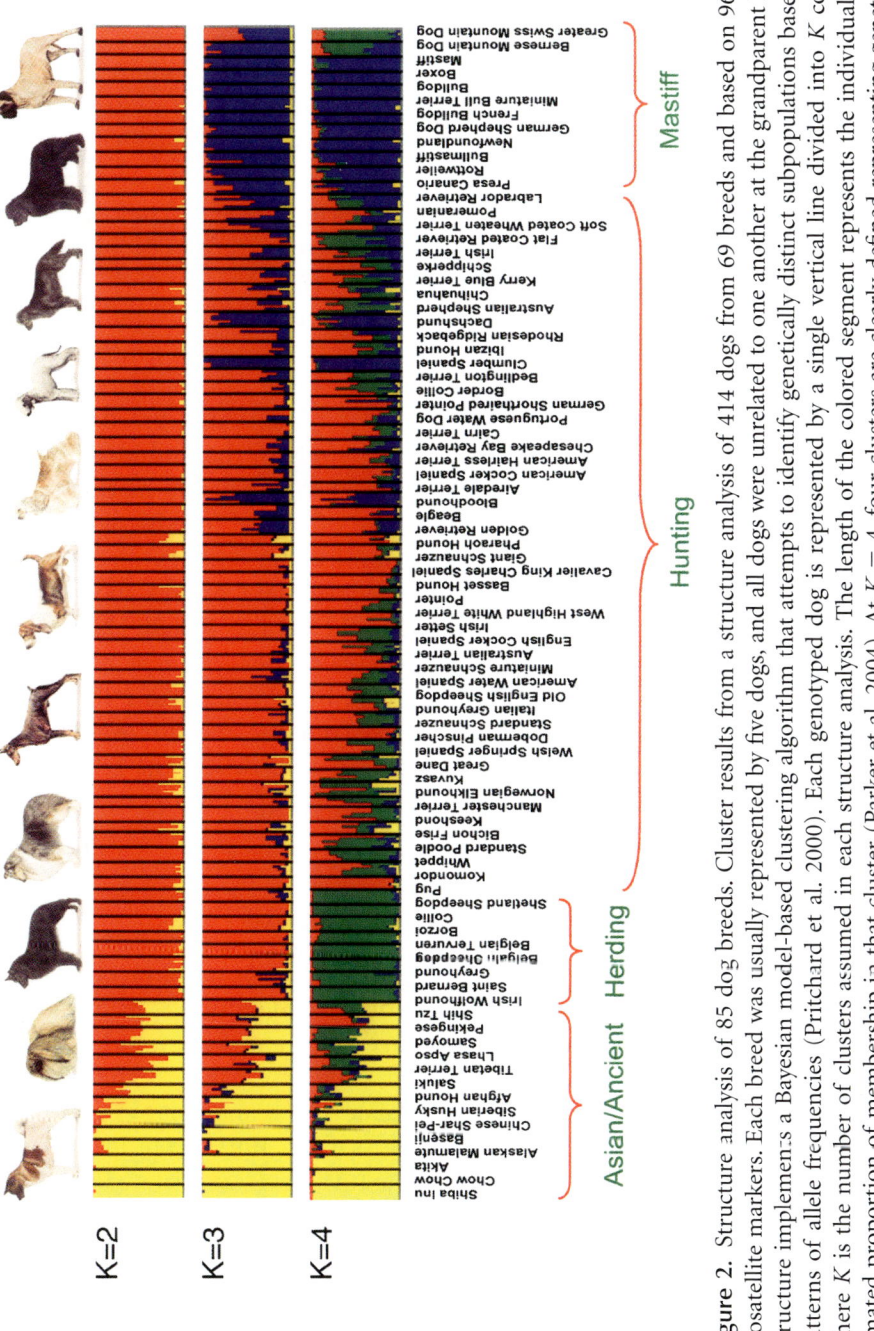

Figure 2. Structure analysis of 85 dog breeds. Cluster results from a structure analysis of 414 dogs from 69 breeds and based on 96 microsatellite markers. Each breed was usually represented by five dogs, and all dogs were unrelated to one another at the grandparent level. Structure implements a Bayesian model-based clustering algorithm that attempts to identify genetically distinct subpopulations based on patterns of allele frequencies (Pritchard et al. 2000). Each genotyped dog is represented by a single vertical line divided into K colors, where K is the number of clusters assumed in each structure analysis. The length of the colored segment represents the individual's estimated proportion of membership in that cluster (Parker et al. 2004). At $K = 4$, four clusters are clearly defined representing genetically distinct breed grouping within the domestic dog (see text).

The next cluster to be defined at $K = 3$ was comprised of mastiff-type dogs including the Mastiff, Bullmastiff, Bulldog, Boxer, etc. Finally, at $K = 4$, the third cluster to be defined included working dogs such as the Collie and Shetland Sheepdog, together with a subset of the sight hounds, such as the Greyhound. The final cluster comprised mostly modern breeds used in hunting and included gun dogs, hounds, and terriers. On-going analysis is focusing on defining clusters within this hedge group, using more highly mutable tetranucleotide-based microsatellite markers (Francisco et al. 1996) and less mutable markers based on single nucleotide polymorphisms (SNPs). However, the structure analysis for the first time defined groups based on common ancestry and genetic similarity rather than function (e.g., hunting or herding breeds) and provides a genetic guide to the design of whole-genomic scans (see below).

Another promising approach toward reconstructing breed history utilizes single gene histories. For example, study of the multidrug resistance gene (*MDR1*) and four closely linked microsatellite markers was used to reconstruct the history of a group of related breeds (Neff et al. 2004). A single *MDR1* mutation was found to segregate in nine breeds that included seven herding breeds and two sight hound subgroups, which were likely related to at least one of the herding breeds. Haplotype analysis confirmed this relationship by revealing that the region around *MDR1* was identical by descent in all nine breeds, suggesting that they inherited this haplotype from an exclusive common ancestor. Additional study of single gene mutations in dogs will help dissect the branching structure of "twigs" in the phylogenetic tree of dogs.

MAPPING AND SEQUENCING THE DOG GENOME

The success of disease-mapping studies and those unraveling the mysteries of canine evolution were clearly dependent on the prior development of key resources. Meiotic linkage maps and RH maps based on family studies (Mellersh et al. 1997) and a 5000 rad panel (Vignaux et al. 1999) were first made available in the late 1990s and were essential to subsequent map-building efforts (Mellersh et al. 1997, 2000; Priat et al. 1998; Neff et al. 1999). The first comparative maps and later dense RH maps that followed allowed researchers to take full advantage of the much more well-developed human and mouse genome mapping resources (Breen et al. 2001; Guyon et al. 2003, 2004). A recent integrated RH map of the dog, including microsatellites, genes, and BAC ends (Breen et al. 2004), has proven invaluable in allowing investigators to do positional cloning experiments following initial findings of linkage. Most recent mapping efforts focused on developing a high-resolution 9000 rad comparative map (Hitte et al. 2004), which includes 10,348 canine markers, 9850 corresponding to canine orthologs of human genes derived from a 1.5× poodle shotgun sequence (Kirkness et al. 2003). For online information, see http://sun-recomgen.med.univ-rennes1.fr/Dogs/ and http://research.nhgri.nih.gov/dog_genome/.

Very recently, the landscape for canine genome studies has been changed by the availability of a 7.5× assembled sequence of the Boxer genome (http://www.genome.ucsc.edu), completed by investigators at the Broad Institute (CanFam1.0 and CanFam2.0) (Lindblad-Toh et al. 2005). These data suggest that the euchromatic portion of the dog genome is ~18% smaller than the human genome and 6% smaller than the mouse genome. The size difference is explained by a lower rate of repeat insertions in the dog genome relative to both human and mouse, while the deletion rate of ancestral bases has been approximately equal between the dog and human lineages. The relatively low level of recent repeats in the dog genome contributes, together with high quality data and improved assembly algorithms, to the high connectivity and quality of the dog genome assembly. This is well supported by the above-mentioned RH gene map of the dog, which shows high concordance with the assembled sequence as well as a set of several hundred BAC ends previously localized by FISH (Hitte et al. 2005).

The assembled sequence demonstrates that ~94% of the dog genome is contained in clear segments of conserved synteny relative to the human and mouse genomes. The gene count of ~19,000 canine genes is slightly lower than that currently considered for human, which is somewhat surprising. The accuracy of these data, however, is high; of the 19,000 reported canine genes, 14,200 represent 1-1-1 orthologs between dog, human, and mouse. Approximately 5.4% of the orthologous nucleotides between human and dog appears to be under purifying selection. The purifying selection acting on conserved orthologous genes appears significantly higher in the lineage leading to dog than in that leading to human, but lower than in the lineage leading to mouse. However, the relative constraints between orthologs with different functions have been highly correlated between the three lineages. Only genes involved in nervous system function have diverged faster in both dog and human relative to mouse, but not relative to each other, consistent with similar selection pressures, and possibly, convergent evolution. Finally, gene family expansions are less common in dog than in human, suggesting that the dog has the most primitive gene content of the currently sequenced placental mammals.

LINKAGE DISEQUILIBRIUM ACROSS AND BETWEEN DOG BREEDS

To fully exploit the unique genetic characteristics of the dog, the architecture of linkage disequilibrium (LD) in the canine genome needs to be understood. This knowledge would facilitate the mapping and cloning of genes important to canine health, as well as the discovery of loci regulating phenotypic traits. The importance of this knowledge is demonstrated in human studies where LD mapping in well-defined populations has simplified locus heterogeneity problems associated with complex traits (Kruglyak 1999a; Sundin et al. 2000; Ophoff et al. 2002; Friedrichsen et al. 2004). Three fundamental questions have been addressed. First, how does the extent of LD compare to that which has been reported in humans? Second, how does LD differ between breeds, and finally, how well does breed history predict the extent of LD?

These issues have been addressed in two major studies (Sutter et al. 2004; Wade et al. 2005). Sutter et al. (2004) examined 189 SNPs from five unlinked loci in five breeds using 20 unrelated dogs from each breed (Fig. 3). They

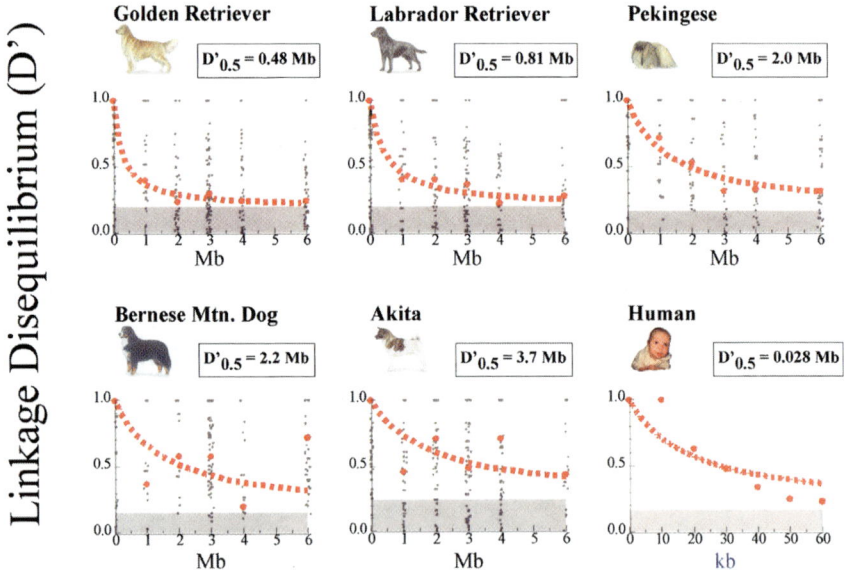

Figure 3. LD in five breeds of dog. LD in 20 unrelated dogs from each of the five breeds scanned for a total of 51 Kb in five unlinked regions on chromosomes 1, 2, 3, 34, and 37. The scan revealed 189 SNPs and those with a minor allele frequency greater then 0.2 in each breed were used on LD calculations. Data were averaged across the five sites and the D' statistic used to indicate the level of linkage disequilibrium. $D'_{0.5}$ indicates the point at which the D' statistic decays by 50%. Data are given in Mb for dog and kb for human.

found that in the Golden Retriever, LD falls to half of its maximum value at about 0.48 Mb. However, in the other breeds, LD is more extensive, increasing to about 0.9 in the Pekingese and Labrador Retriever and to 2.2 Mb in the Bernese Mountain Dog. Finally, at 3.8 Mb, LD in the Akita is nearly 10× greater than that observed in the Golden Retriever. In some cases, these observations agree well with recorded breed history (Fogel 1995; Wilcox and Walkowicz 1995; American Kennel Club 1998;). For instance, the Golden and Labrador Retriever are among the most popular breeds and neither breed has experienced significant population bottlenecks (Fogel 1995; Wilcox and Walkowicz 1995). By comparison, LD is expected to be greater in the Pekingese, as these dogs are derived from a small number of founders that came to the U.S. from China (Fogel 1995; Wilcox and Walkowicz 1995). LD is predicted to be most extreme in the Akita, a relatively rare breed with a restricted gene pool.

These results suggest two important considerations for the design of mapping and cloning studies. First, as there is at least a 10-fold difference in the extent of LD between dog breeds, breed selection deserves careful consideration. Second, LD in dogs is 20–50 times more extensive than that found in humans, where LD is typically reported to be about 0.28 Mb (Reich et al. 2001; Weiss and Clark 2002). More than 500,000 SNPs must be genotyped for whole-genome association studies in humans (Kruglyak 1999b; The International HapMap Consortium 2003). In contrast, only about 10,000 SNPs are hypothesized to be needed for the comparable dog study (Sutter et al. 2004). Thus, the mapping of common and complex diseases such as epilepsy, cancer, autoimmune disease, deafness, and heart disease in dogs may be more economical than similar efforts in humans.

The canine genome sequencing effort has made 2.1 million SNPs publicly available (http://www.broad.mit.edu/mammals/dog/snp/) (Lindblad-Toh et al. 2005). To determine how to best use this resource, Sutter et al. (2004) examined the extent of haplotype sharing for the five breeds described above. For any one breed, 80% of chromosomes examined had, on average, just 2.7 haplotypes. For all 100 dogs examined, 80% of chromosomes carried just 4.5 haplotypes. The overall degree of haplotype sharing, measured as the proportion of a breed's chromosomes carrying haplotypes shared with another breed, ranged from 46% to 84%. These findings of low haplotype diversity and high haplotype sharing, albeit with great variability, suggest that a universal SNP set of modest size will be sufficient to successfully accomplish whole-genome association studies in most breeds.

A more in-depth analysis of the same general questions, as well as issues regarding the overall haplotype structure of the dog were examined using ~1300 SNPs plus resequencing data drawn from 10 random regions covering 6% of the genome. The study was undertaken as part of the canine genome sequencing effort (Lindblad-Toh et al. 2005) and the conclusions largely agree with those of Sutter et al. (2004). In addition to the 7.5× Boxer sequence, the genome sequencing effort generated 100,000 sequence reads from each of nine diverse breeds representing all seven AKC groups, and 20,000 reads from each of five wild canids (four wolves and one coyote). The resulting SNP frequencies of 1/900 bp between breeds, 1/580 bp between dogs and wolves, and 1/420 bp between dogs and coyote, emphasizes that all three species are more closely related than human and chimpanzee. The resulting set of 2.1 million SNPs have a polymorphism rate across breeds of ~72% within any given breed, suggesting that most SNPs discovered as part of the sequencing effort will be useful for mapping in any breed.

Comparison of the two boxer haplotypes, as well as extensive resequencing and genotyping in 10 breeds by the sequencing group has been illustrative for understanding the detailed haplotype structure of the dog. Such analyses demonstrate megabase sized portions of the genome that are alternatively homozygous and heterozygous exist both for the sequenced boxer, as well as for 24 dogs from different breeds and 20 dogs from each of 10 breeds. Thus, megabase-sized haplotypes will be common within virtually any purebred dog.

Lindblad-Toh and collaborators conclude that LD within any breed is actually dependent on the intensity and duration of two bottlenecks. The first is an ancient bottleneck occurring at the time of canine domestication that is common to all dogs. The second likely occurred during breed formation. In combination, these bottlenecks resulted in LD that extends for megabases in most breeds and limited haplotype diversity. Indeed, across the dog population as a whole, ancestral haplotype blocks are roughly 5–10 kb long with approximately five alleles in each block. Thus, when LD is examined carefully across many breeds, typically, five haplotypes are observed across each 10–500-kb window, with one or two being common and the rest rare. The recent ancestry of these haplotypes supports the idea that a modest number of SNPs, perhaps as few as 5000, will be sufficient for genome-wide association mapping. However, the underlying ancestral haplotype block structure implies that the false-positive rate will be high if only single SNP association is used. Consequently, haplotype-based association should be used instead for most mapping studies.

CANINE DISEASE GENE MAPPING

Billions of dollars are spent on canine health in the United States each year (Association 2002) and much of it is focused on a limited number of diseases including cancer, epilepsy, blindness, cataracts, autoimmune disease, and heart disease. Over 360 genetic disorders found in humans have also been described in the dog (Patterson 2000; Sargan 2004), and about 46% of these genetic diseases occur predominantly or exclusively in one or a few breeds. A detailed listing of over 1000 canine diseases, and descriptions of each, appears in the database of inherited diseases in dogs (IDID, http://www.vet.cam.ac.uk/idid) (Sargan 2004).

To date, the location of many canine disease loci has been determined, and in some cases the underlying gene has been cloned (for review, see Patterson et al. 1982; Ostrander and Giniger 1997; Galibert et al. 1998; Ostrander et al. 2000; Sutter and Ostrander 2004; Switonski et al. 2004).

In some cases, identification of canine disease genes has opened new avenues of research for human biologists. For instance, the identification of a mutation in the hypocretin 2 receptor gene (Lin et al. 1999) in Doberman Pinschers with inherited narcolepsy has proven key to understanding the molecular mechanisms which regulate sleep (Nishino et al. 2000; Thannickal et al. 2000). In humans, the disease is associated with a progressive loss of hypocretin-expressing neurons and is a non-Mendelian trait mediated by a unique mechanism different from that causing the disease in Dobermans. However, study of the simpler etiology in dogs provided the requisite tools for understanding the more complex disease in humans.

In other cases, study of canine disease genes has increased our understanding of the interaction between genes and how such interactions affect disease. Such interactions have proven difficult to study in human populations, where the size of even the largest case-control studies is simply too small to identify anything but major effects. The identification of the *MURR1* gene associated with copper toxicosis in Bedlington Terriers (van De Sluis et al. 2002) provides an excellent example. Contrary to expectation, this disease did not map to the portion of the canine genome analogous to the Wilson's disease locus in humans (Yuzbasiyan-Gurkan et al. 1997; van de Sluis et al. 1999). Analysis of the human homolog of *MURR1* in Wilson's disease patients has subsequently proven provocative, as those who carry particular sequence variants appear to present with earlier onset disease (Stuehler et al. 2004), suggesting that the two genes or their products interact to accelerate disease.

Another significant advance concerns the identification of novel disease mechanisms through the study of dog genetics. Lohi et al. (2005) recently identified a gene for progressive myoclonic epilepsy (PME) in a population of purebred miniature wirehaired dachshunds. About 5% of the breed suffers from this autosomal recessive disease, which was shown to be analogous to the human disorder, Lafora disease. As in the human disease, affected individuals carry mutations in the *NHLRC1* gene. However, in contrast to the human disease, the disease in dogs is due exclusively to bi-allelic expansion of a dodecamer repeat found within the 5'end of the genes' single large exon. Affected individuals carry 19 to 26 copies of the repeat sequence rather than the expected two copies. This is the first example of a dodecamer repeat expansion associated with disease in any mammalian system and suggests a potential novel mechanism for human disorders.

Currently, perhaps the greatest concentrated collaborative efforts are focused on the study of canine cancer (Chun and de Lorimier 2003; Ettinger 2003; Fan 2003; London and Seguin 2003; Porrello et al. 2004; Modiano et al. 2005). Dogs develop cancer about twice as frequently as humans and the disease presentation and pathology of canine cancers is similar to analogous human tumors. Genetic studies are ongoing to find susceptibility genes for canine osteosarcoma, lymphoma, mast cell tumors, malignant histiocytosis, and kidney cancer, and a BAC CGH array resource is in development to better understand somatic events leading to tumor growth and metastasis (Thomas et al. 2003a,b). Of primary interest is determining whether different types of tumors have unique or shared origins. If a common origin of a particular canine cancer is established, then considering data from several breeds simultaneously can facilitate the localization of the susceptibility gene. Breeds of similar appearance and sharing common ancestry as suggested by historical record may often share variants for disease phenotypes (e.g., Neff et al. 2004). However, in most cases, rigorous studies such as those described below are needed to address the issue.

GENETICS OF MORPHOLOGY

The genetic basis for differences in size and proportion among dogs has yet to be revealed. However, both candidate gene and association studies are beginning to provide insight into the complexity underlying morphological differentiation. For example, two potential candidate genes, *MSX2* and *TCOF1*, which are expressed during cranial facial development, were sequenced in 10 different dog breeds that varied in cranial and face shape (Haworth et al. 2001a,b). However, only a single amino acid change in the TCOF1 protein showed an association with short and broad skulls. Nonetheless, greatly expanded surveys of candidate genes may prove more fruitful; for example, variation in the production of insulin-like growth factor 1 (IGF-1) was shown to correlate with differences in the body size of poodles, suggesting it may be a candidate gene for size variation in dogs (Eigenmann et al. 1984).

More definitive associations have been demonstrated through quantitative analysis of morphologic measurements combined with genome marker scans. For example, Chase et al. (2002) analyzed data from nearly 700 Portuguese Water Dogs genotyped with ~500 markers and http://www.georgieproject.com/. For 460 dogs, they recorded 91 measurements from a set of five x-rays taken on each dog. The data were analyzed using principal component analysis, which defines independent component axes based on linear combinations of variables. Each axis is ordered by a decreasing fraction of the total variation in the data set. The first four axes explained 61% of the variation in the data set and represented different components of size and shape. For example, the first principal component axis reflected overall size variation of the skeleton, whereas the second reflected the relationship between the pelvis, head, and neck, such that the size and strength of the pelvis and head–neck musculoskeletal systems are inversely related. Quantitative Trait Loci (QTLs) have been localized that are related to variation on each of the above four principal components. Moreover, using a data set of 286 phenotyped dogs, Chase et al. (2004) defined two loci on chromosome one spaced 95 Mb apart that appear to account for a modest percentage of hip dysplasia, as defined by Norberg angle in the Portuguese Water Dog.

Nonclassical genetic variation may also be an important source of phenotypic variation in dogs. Fondon III and Garner (2004) suggested that highly mutable simple tandem repeats imbedded in genes may be the source of new variation in recent developed lines and may explain their high rate of morphologic change. To test this hypothesis, these investigators analyzed three-dimensional models of dog skulls from 20 breeds and seven mongrels. In representatives of 92 different breeds, they also sequenced 37 repeat-containing regions from 17 genes known or thought to be involved in craniofacial development. In general, they found that dogs had more perfect repeats than humans and may be changing faster in length. Additionally, they found that the size and the ratio of lengths of two tandem repeats in the *Runx 2* gene correlated with the degree of dorsoventral nose bend (clinorhyncy) and mid-face length in a variety of breeds. Although this evidence is suggestive, clearly more detailed studies are needed associating repeat change with specific phenotypic traits (Pennisi 2000). If such genetic mechanisms are unique to the dog, they may explain, in part, the apparent phenotypic plasticity of dogs. However, dogs also have a unique skeletal development whose alterations may more readily result in novel phenotypes (Wayne 1986a,b,c; Morey 1992, 1994).

One area of morphology we do not discuss in detail is that of canine coat color, which has been written about extensively in the past. More recently, progress on dissecting coat color genetics in the dog has been done by two groups (Kerns et al. 2003; Berryere et al. 2005). Particular progress has been made in understanding the interactions between the Agouti protein and the Melanocortin 1 receptor, which control the type of pigment synthesized in mammalian hair (Berryere et al. 2005). Additional recent work has focused on black color in dogs, which appears to be independent of the above interactions (Schmutz et al. 2002; Kerns et al. 2003). Very interesting work that is just beginning focuses on the role of polymorphisms in coat color affecting genes, such as the melanophilin gene (Philipp et al. 2005). With the availability of the canine genome sequence, this is an area that will surely expand in the coming years.

GENETICS OF BEHAVIOR

Dog breeds have distinct behaviors, and dogs as a whole have unique behaviors not found in gray wolves (Hare et al. 2002). However, the genetic basis of behavior is less well understood than morphology. In general, the greatest need remains the development of assays to reproducibly score specific behaviors. However, some understanding is likely to come from the study of pedigrees of dogs displaying aberrant behaviors. For example, Moon-Fanelli et al. (1998) have characterized pedigrees of Bull Terriers displaying obsessive compulsive disease (OCD) phenotypes, such as tail chasing, which in other respects is similar to human OCD. As genome scans of affected pedigrees are completed, they may shed light on both the human and canine disease conditions.

Expression patterns may also provide clues to the genetic basis of behavior. Saetre et al. (2004) surveyed the expression pattern of 7762 genes in three different regions in the brains of domestic dogs and in gray wolves and coyotes. They found that the pattern of gene expression in the hypothalamus of domestic dogs was different from that in gray wolves and coyotes, whereas patterns of gene expression in the amygdala and frontal cortex were less differentiated. The hypothalamus controls specific emotional, endocrinological, and autonomic responses of dogs and is highly conserved throughout mammals. The results of Saetre et al. (2004) suggest that behavioral selection in dogs may have affected this central part of the brain, initiating a cascade of effects that result in some of the unique behaviors found in dogs.

CONCLUSIONS

The domestic dog has long fascinated evolutionary biologists and geneticists because of the extreme phenotypic diversity exhibited by the species and the short time frame over which this diversity has evolved. Molecular genetic evidence suggests that dogs are indeed the oldest domesticated species and their origin may have even well preceded their first appearance in the archeological record about 15,000 yr ago. The dog has a diverse genetic origin that likely involved multiple gray wolf populations and subsequently was enriched by backcrossing with wolves throughout their history. This substantial input of variation from wild ancestors has provided the raw material for phenotypic change, but unique development and genetic mechanisms may also have assisted the course of artificial selection. Dogs clearly have behaviors, phenotypes, and diseases that are not evident in their wild progenitors. Finally, in the more recent evolution of dog breeds, limited interbreeding has imposed a remarkable genetic structure such that nearly all breeds represent distinct genetic pools that can be divided into at least four distinct genetic groupings.

Understanding the genetic mechanisms that have given rise to the unique attributes of domestic dogs may finally be within reach. A complete and a partial genome sequence are available from a boxer and a poodle, respectively, and mapping resources are well developed and increasing in sophistication. The dog genome in general has high levels of LD, such that whole-genome association studies will be facilitated and genomic scans of specific breeds segregating traits of interest may readily be found through patterns of LD or reductions in heterozygosity due to selective sweeps (Weiss and Clark 2002; Bamshad and Wooding 2003; Luikart et al. 2003; Pollinger et al. 2005). In this review, we have provided the evolutionary and empirical framework for understanding the molecular diversity of dogs with the aim of taking the first step toward answering the questions posed in the introduction. The primary intent of this article was to help generate the enthusiasm that will lead to realizing the promise of the dog genome for solving significant problems in evolution, genetics, and human health.

ACKNOWLEDGMENTS

We thank two anonymous reviewers, Kerstin Lindblad-Toh, Heidi Parker, Nate Sutter, Ed Giniger, and Francis Galibert for thoughtful comments and helpful suggestions on this manuscript. We also thank Kerstin Lindblad-Toh for sharing data in advance of publication. Finally, we thank the many colleagues, dog owners, and breeders who have generously shared samples and made much of the work reviewed here possible.

REFERENCES

Acland, G.M., Aguirre, G.D., Ray, J., Zhang, Q., Aleman, T.S., Cideciyan, A.V., Pearce-Kelling, S.E., Anand, V., Zeng, Y., Maguire, A.M., et al. 2001. Gene therapy restores vision in a canine model of childhood blindness. *Nat. Genet.* **28**: 92–95.

Adams, J.R., Leonard, J.A., and Waits, L.P. 2003. Widespread occurrence of a domestic dog mitochondrial DNA haplotype in southeastern US coyotes. *Mol. Ecol.* **12**: 541–546.

American Kennel Club. 1998. *The complete dog book.* Howell Book House, New York.

American Veterinary Medical Association. 2002. *U.S. pet ownership and demographics sourcebook.* American Veterinary Medical Association, Schaumburg, IL.

Bamshad, M. and Wooding, S.P. 2003. Signatures of natural selection in the human genome. *Nat. Rev. Genet.* **4**: 99–111.

Berryere, T.G., Kerns, J.A., Barsh, G.S., and Schmutz, S.M. 2005. Association of an Agouti allele with fawn or sable coat color in domestic dogs. *Mamm. Genome* **16**: 262–272.

Breen, M., Jouquand, S., Renier, C., Mellersh, C.S., Hitte, C., Holmes, N.G., Cheron, A., Suter, N., Vignaux, F., Bristow, A.E., et al. 2001. Chromosome-specific single-locus FISH probes allow anchorage of an 1800-marker integrated radiation-hybrid/linkage map of the domestic dog genome to all chromosomes. *Genome Res.* **11**: 1784–1795.

Breen, M., Hitte, C., Lorentzen, T.D., Thomas, R., Cadieu, E., Sabacan, L., Scott, A., Evanno, G., Parker, H.G., Kirkness, E., et al. 2004. An integrated 4249 marker FISH/RH map of the canine genome. *BMC Genomics* **5**: 1–11.

Bruford, M.W., Bradley, D.G., and Luikart, G. 2003. DNA markers reveal the complexity of livestock domestication. *Nat. Rev. Genet.* **4**: 900–910.

Chase, K., Carrier, D.R., Adler, F.R., Jarvik, T., Ostrander, E.A., Lorentzen, T.D., and Lark, K.G. 2002. Genetic basis for systems of skeletal quantitative traits: Principal component analysis of the canid skeleton. *Proc. Natl. Acad. Sci.* **99**: 9930–9935.

Chase, K., Lawler, D.F., Adler, F.R., Ostrander, E.A., and Lark, K.G. 2004. Bilaterally asymmetric effects of quantitative trait loci (QTLs): QTLs that affect laxity in the right versus left coxofemoral (hip) joints of the dog (*Canis familiaris*). *Am. J. Med. Genet. A* **124**: 239–247.

Chun, R. and de Lorimier, L.P. 2003. Update on the biology and management of canine osteosarcoma. *Vet. Clin. North Am. Small Anim. Pract.* **33**: 491–516, vi.

Clark, A.G., Glanowski, S., Nielsen, R., Thomas, P.D., Kejarwal, A., Todd, M.A., Tanenbaum, D.M., Civello, D., Lu, F., Murphy, B,. et al. 2003. Inferring nonneutral evolution from human–chimp–mouse orthologous gene trios. *Science* **302**: 1960–1963.

Eigenmann, J.E., Patterson, D.F., and Froesch, E.R. 1984. Body size parallels insulin-like growth factor I levels but not growth hormone secretory capacity. *Acta Endocrinol.* **106**: 448–453.

Ettinger, S.N. 2003. Principles of treatment for soft-tissue sarcomas in the dog. *Clin. Tech. Small Anim. Pract.* **18**: 118–122.

Fan, T.M. 2003. Lymphoma updates. *Vet. Clin. North Am. Small Anim. Pract.* **33**: 455–471.

Fanning, T.G., Modi, W.S., Wayne, R.K., and O'Brien, S.J. 1988. Evolution of heterochromatin-associated satellite DNA loci in felids and canids (Carnivora). *Cytogenet. Cell Genet.* **48**: 214–219.

Fogel, B. 1995. *The encyclopedia of the dog.* DK Publishing, Inc., New York.
Fondon III, J.W. and Garner, H.R. 2004. Molecular origins of rapid and continuous morphological evolution. *Proc. Natl. Acad. Sci.* **101:** 18058–18063.
Francisco, L.V., Langston, A.A., Mellersh, C.S., Neal, C.L., and Ostrander, E.A. 1996. A class of highly polymorphic tetranucleotide repeats for canine genetic mapping. *Mamm. Genome* **7:** 359–362.
Friedrichsen, D.M., Stanford, J.L., Isaacs, S.D., Janer, M., Chang, B.L., Deutsch, K., Gillanders, E., Kolb, S., Wiley, K.E., Badzioch, M.D., et al. 2004. Identification of a prostate cancer susceptibility locus on chromosome 7q11-21 in Jewish families. *Proc. Natl. Acad. Sci.* **101:** 1939–1944.
Galibert, F., Andre, C., Cheron, A., Chuat, J.C., Hitte, C., Jiang, Z., Jouquand, S., Priat, C., Renier, C., and Vignaux, F. 1998. The importance of the canine model in medical genetics. *Bull. Acad. Natl. Med.* **182:** 811–821.
Goldstein, D.B., Roemer, G.W., Smith, D.A., Reich, D.E., Bergman, A., and Wayne, R.K. 1999. The use of microsatellite variation to infer population structure and demographic history in a natural model system. *Genetics* **151:** 797–801.
Gottelli, D., Sillero-Zubiri, C., Applebaum, G.D., Roy, M.S., Girman, D.J., Garcia-Moreno, J., Ostrander, E.A., and Wayne, R.K. 1994. Molecular genetics of the most endangered canid: The Ethiopian wolf *Canis simensis. Mol. Ecol.* **3:** 301–312.
Guyon, R., Lorentzen, T.D., Hitte, C., Kim, L., Cadieu, E., Parker, H.G., Quignon, P., Lowe, J.K., Renier, C., Gelfenbeyn, B., et al. 2003. A 1-Mb resolution radiation hybrid map of the canine genome. *Proc. Natl. Acad. Sci.* **100:** 5296–5301.
Guyon, R., Kirkness, E.F., Lorentzen, T.D., Hitte, C., Comstock, K.E., Quignon, P., Derrien, T., Andréa, C., Fraser, C.M., Galibert, F., et al. 2004. Comparative mapping of human chromosome 1p and the canine genome. In *The genome of Homo sapiens. The 68th Cold Spring Harbor Symposium*, pp. 171–177. Cold Spring Harbor Press, Cold Spring Harbor, NY.
Hare, B., Brown, M., Williamson, C., and Tomasello, M. 2002. The domestication of social cognition in dogs. *Science* **298:** 1634–1636.
Hartl, D. and Clark, A. 1997. *Principles of population genetics.* Sinauer Associates, Inc., Sunderland, MA.
Haworth, K., Breen, M., Binns, M., Hopkinson, D.A., and Edwards, Y.H. 2001a. The canine homeobox gene MSX2: Sequence, chromosome assignment and genetic analysis in dogs of different breeds. *Anim. Genet.* **32:** 32–36.
Haworth, K.E., Islam, I., Breen, M., Putt, W., Makrinou, E., Binns, M., Hopkinson, D., and Edwards, Y. 2001b. Canine TCOF1; cloning, chromosome assignment and genetic analysis in dogs with different head types. *Mamm. Genome* **12:** 622–629.
Hitte, C., Derrien, T., Andre, C., Ostrander, E.A., and Galibert, F. 2004. CRH_Server: An online comparative and radiation hybrid mapping server for the canine genome. *Bioinformatics* **20:** 3665–3667.
Hitte, C., Madeoy, J., Kirkness, E.F., Priat, C., Lorentzen, T.D., Senger, F., Thomas, D., Derrien, T., Ramirez, C., Scott, C., et al. 2005. Survey sequencing combined with dense radiation hybrid gene mapping facilitates genome navigation. *Nat. Rev. Genet.* **6:** 643–648.
Howell, J.M., Fletcher, S., Kakulas, B.A., O'Hara, M., Lochmuller, H., and Karpati, G. 1997. Use of the dog model for Duchenne muscular dystrophy in gene therapy trials. *Neuromus. Disord.* **7:** 325–328.
The International HapMap Consortium 2003. The International HapMap Project. *Nature* **426:** 789–796.

Irion, D.N., Schaffer, A.L., Famula, T.R., Eggleston, M.L., Hughes, S.S., and Pedersen, N.C. 2003. Analysis of genetic variation in 28 dog breed populations with 100 microsatellite markers. *J. Hered.* **94:** 81–87.

Kerns, J.A., Olivier, M., Lust, G., and Barsh, G.S. 2003. Exclusion of melanocortin-1 receptor (mc1r) and agouti as candidates for dominant black in dogs. *J. Hered.* **94:** 75–79.

Kirkness, E.F., Bafna, V., Halpern, A.L., Levy, S., Remington, K., Rusch, D.B., Delcher, A.L., Pop, M., Wang, W., Fraser, C.M., et al. 2003. The dog genome: Survey sequencing and comparative analysis. *Science* **301:** 1898–1903.

Koskinen, M.T. 2003. Individual assignment using microsatellite DNA reveals unambiguous breed identification in the domestic dog. *Anim. Genet.* **34:** 297–301.

Koskinen, M.T. and Bredbacka, P. 2000. Assessment of the population structure of five Finnish dog breeds with microsatellites. *Anim. Genet.* **31:** 310–317.

Kruglyak, L. 1999a. Genetic isolates: Separate but equal? *Proc. Natl. Acad. Sci.* **96:** 1170–1172.

———. 1999b. Prospects for whole-genome linkage disequilibrium mapping of common disease genes. *Nat. Genet.* **22:** 139–144.

Leonard, J.A., Wayne, R.K., Wheeler, J., Valadez, R., Guillen, S., and Vila, C. 2002. Ancient DNA evidence for Old World origin of New World dogs. *Science* **298:** 1613–1616.

Lin, L., Faraco, J., Li, R., Kadotani, H., Rogers, W., Lin, X., Qiu, X., Jong, P.J.D., Nishino, S., and Mignot, E. 1999. The sleep disorder canine narcolepsy is caused by a mutation in the hypocretin (orexin) receptor 2 gene. *Cell* **98:** 365–376.

Lindblad-Toh, K., Wade, C.M., Mikkelsen, T.S., Karlsson, E.K., Jaffe, D.B., Kamal, M., Clamp, M., Chang, J.L., Kulbokas III, E.J., Zody, M.C., et al. 2005. Genome sequence, comparative analysis and haplotype structure of the domestic dog. *Nature* **438:** 803–819.

Lohi, H., Young, E.J., Fitzmaurice, S.N., Rusbridge, C., Chan, E.M., Vervoort, M., Turnbull, J., Zhao, X.C., Ianzano, L., Paterson, A.D., et al. 2005. Expanded repeat in canine epilepsy. *Science* **307:** 81.

London, C.A. and Seguin, B. 2003. Mast cell tumors in the dog. *Vet. Clin. North Am. Small Anim. Pract.* **33:** 473–489, v.

Luikart, G., England, P.R., Tallmon, D., Jordan, S., and Taberlet, P. 2003. The power and promise of population genomics: From genotyping to genome typing. *Nat. Rev. Genet.* **4:** 981–994.

Mellersh, C.S., Langston, A.A., Acland, G.M., Fleming, M.A., Ray, K., Wiegand, N.A., Francisco, L.V., Gibbs, M., Aguirre, G.D., and Ostrander, E.A. 1997. A linkage map of the canine genome. *Genomics* **46:** 326–336.

Mellersh, C.S., Hitte, C., Richman, M., Vignaux, F., Priat, C., Jouquand, S., Werner, P., Andre, C., DeRose, S., Patterson, D.F., et al. 2000. An integrated linkage-radiation hybrid map of the canine genome. *Mamm. Genome* **11:** 120–130.

Modiano, J.F., Breen, M., Burnett, R.C., Parker, H.G., Inusah, S., Thomas, R., Avery, P.R., Lindblad-Toh, K., Ostrander, E.A., Cutter, G.C., et al. 2005. Distinct B and T cell lymphoproliferative disease prevalence among dog breeds indicates heritable risk. *Cancer Res.* **65:** 5654–5661.

Moon-Fanelli, A.A. and Dodman, N.H. 1998. Description and development of compulsive tail chasing in terriers and response to clomipramine treatment. *J. Am. Vet. Med. Assoc.* **212:** 1252–1257.

Morey, D. 1994. The early evolution of the domestic dog. *Am. Scientist* **82:** 336–347.

Morey, D.F. 1992. Size, shape, and development in the evolution of the domestic dog. *J. Archaeol. Sci.* **19:** 181–204.

Mount, J.D., Herzog, R.W., Tillson, D.M., Goodman, S.A., Robinson, N., McCleland, M.L., Bellinger, D., Nichols, T.C., Arruda, V.R., Lothrop Jr., C.D., et al. 2002. Sustained phenotypic correction of hemophilia B dogs with a factor IX null mutation by liver-directed gene therapy. *Blood* **99:** 2670–2676.

Nash, W.G., Menninger, J.C., Wienberg, J., Padilla-Nash, H.M., and O'Brien, S.J. 2001. The pattern of phylogenomic evolution of the Canidae. *Cytogenet. Cell Genet.* **95:** 210–224.

Neff, M.W., Broman, K.W., Mellersh, C.S., Ray, K., Acland, G.M., Aguirre, G.D., Ziegle, J.S., Ostrander, E.A., and Rine, J. 1999. A second-generation genetic linkage map of the domestic dog, Canis familiaris. *Genetics* **151:** 803–820.

Neff, M.W., Robertson, K.R., Wong, A.K., Safra, N., Broman, K.W., Slatkin, M., Mealey, K.L., and Pedersen, N.C. 2004. Breed distribution and history of canine mdr1-1Delta, a pharmacogenetic mutation that marks the emergence of breeds from the collie lineage. *Proc. Natl. Acad. Sci.* **101:** 11725–11730.

Nishino, S., Ripley, B., Overeem, S., Lammers, G.J., and Mignot, E. 2000. Hypocretin (orexin) deficiency in human narcolepsy. *Lancet* **355:** 39–40.

Ophoff, R.A., Escamilla, M.A., Service, S.K., Spesny, M., Meshi, D.B., Poon, W., Molina, J., Fournier, E., Gallegos, A., Mathews, C., et al. 2002. Genomewide linkage disequilibrium mapping of severe bipolar disorder in a population isolate. *Am. J. Hum. Genet.* **71:** 565–574.

Ostrander, E.A. and Giniger, E. 1997. Semper fidelis: What man's best friend can teach us about human biology and disease. *Am. J. Hum. Genet.* **61:** 475–480.

Ostrander, E.A., Rine, J., Sack Jr., G.H., and Cork, L.C. 1993. What is the role of molecular genetics in modern veterinary practice? *J. Am. Vet. Med. Assoc.* **203:** 1259–1262.

Ostrander, E.A., Galibert, F., and Patterson, D.F. 2000. Canine genetics comes of age. *Trends Genet.* **16:** 117–124.

Parker, H.G., Kim, L.V., Sutter, N.B., Carlson, S., Lorentzen, T.D., Malek, T.B., Johnson, G.S., DeFrance, H.B., Ostrander, E.A., and Kruglyak, L. 2004. Genetic structure of the purebred domestic dog. *Science* **304:** 1160–1164.

Patterson, D. 2000. Companion animal medicine in the age of medical genetics. *J. Vet. Internal. Med.* **14:** 1–9.

Patterson, D.F., Haskins, M.E., and Jezyk, P.F. 1982. Models of human genetic disease in domestic animals. *Adv. Hum. Genet.* **12:** 263–339.

Pennisi, E. 2000. Human genome. Finally, the book of life and instructions for navigating it. *Science* **288:** 2304–2307.

Philipp, U., Hamann, H., Mecklenburg, L., Nishino, S., Mignot, E., Gunzel-Apel, A.R., Schmutz, S.M., and Leeb, T. 2005. Polymorphisms within the canine MLPH gene are associated with dilute coat color in dogs. *BMC Genet.* **6:** 34.

Pollinger, J.P., Bustamente, C.D., Fledel-Alon, A., Schmutz, S., Gray, M.M., and Wayne, R.K. 2005. Selective sweep mapping of genes with large phenotypic effects. *Genome Res.* **15:** 1809–1819.

Ponder, K.P., Melniczek, J.R., Xu, L., Weil, M.A., O'Malley, T.M., O'Donnell, P.A., Knox, V.W., Aguirre, G.D., Mazrier, H., Ellinwood, N.M., et al. 2002. Therapeutic neonatal hepatic gene therapy in mucopolysaccharidosis VII dogs. *Proc. Natl. Acad. Sci.* **99:** 13102–13107.

Porrello, A., Cardelli, P., and Spugnini, E.P. 2004. Pet models in cancer research: General principles. *J. Exp. Clin. Cancer Res.* **23:** 181–193.

Priat, C., Hitte, C., Vignaux., F., Renier, C., Jiang, Z., Jouquand, S., Cheron, A., Andre, C., and Galibert, F. 1998. A whole-genome radiation hybrid map of the dog genome. *Genomics* **54:** 361–378.

Pritchard, J.K., Stephens, M., Rosenberg, N.A., and Donnelly, P. 2000. Association mapping in structured populations. *Am. J. Hum. Genet.* **67:** 170–181.

Reich, D.E., Cargill, M., Bolk, S., Ireland, J., Sabeti, P.C., Richter, D.J., Lavery, T., Kouyoumjian, R., Farhadian, S.F., Ward, R., et al. 2001. Linkage disequilibrium in the human genome. *Nature* **411:** 199–204.

Rogers, C.A. and Brace, A.H. 1995. *The international encyclopedia of dogs.* Howell Book House, New York.

Roy, M.S., Geffen, E., Smith, D., and Wayne, R.K. 1996. Molecular genetics of pre-1940 red wolves. *Conservation Biol.* **10:** 1413–1424.

Saetre, P., Lindberg, J., Leonard, J.A., Olsson, K., Pettersson, U., Ellegren, H., Bergstrom, T.F., Vila, C., and Jazin, E. 2004. From wild wolf to domestic dog: Gene expression changes in the brain. *Brain Res. Mol. Brain Res.* **126:** 198–206.

Sargan, D. 2004. IDID: Inherited diseases in dogs: Web-based information for canine inherited disease genetics. *Mamm. Genome* **15:** 503–506.

Savolainen, P., Zhang, Y.P., Luo, J., Lundeberg, J., and Leitner, T. 2002. Genetic evidence for an East Asian origin of domestic dogs. *Science* **298:** 1610–1613.

Savolainen, P., Leitner, T., Wilton, A.N., Matisoo-Smith, E., and Lundeberg, J. 2004. A detailed picture of the origin of the Australian dingo, obtained from the study of mitochondrial DNA. *Proc. Natl. Acad. Sci.* **101:** 12387–12390.

Schmutz, S.M., Berryere, T.G., and Goldfinch, A.D. 2002. TYRP1 and MC1R genotypes and their effects on coat color in dogs. *Mamm. Genome* **13:** 380–387.

Stuehler, B., Reichert, J., Stremmel, W., and Schaefer, M. 2004. Analysis of the human homologue of the canine copper toxicosis gene MURR1 in Wilson disease patients. *J. Mol. Med.* **82:** 629–634.

Sundin, O.H., Yang, J.M., Li, Y., Zhu, D., Hurd, J.N., Mitchell, T.N., Silva, E.D., and Maumenee, I.H. 2000. Genetic basis of total colourblindness among the Pingelapese islanders. *Nat. Genet.* **25:** 289–293.

Sutter, N.B. and Ostrander, E.A. 2004. Dog star rising: The canine genetic system. *Nat. Rev. Genet.* **5:** 900–910.

Sutter, N.B., Eberle, M.A., Parker, H.G., Pullar, B.J., Kirkness, E.F., Kruglyak, L., and Ostrander, E.A. 2004. Extensive and breed-specific linkage disequilibrium in Canis familiaris. *Genome Res.* **14:** 2388–2396.

Switonski, M., Szczerbal, I., and Nowacka, J. 2004. The dog genome map and its use in mammalian comparative genomics. *J. Appl. Genet.* **45:** 195–214.

Thannickal, T.C., Moore, R.Y., Nienhuis, R., Ramanathan, L., Gulyani, S., Aldrich, M., Cornford, M., and Siegel, J. 2000. Reduced number of hypocretin neurons in human narcolepsy. *Neuron* **27:** 469–474.

Thomas, R., Bridge, W., Benke, K., and Breen, M. 2003a. Isolation and chromosomal assignment of canine genomic BAC clones representing 25 cancer-related genes. *Cytogenet. Genome Res.* **102:** 249–253.

Thomas, R., Fiegler, H., Ostrander, E.A., Galibert, F., Carter, N.P., and Breen, M. 2003b. A canine cancer-gene microarray for CGH analysis of tumors. *Cytogenet. Genome Res.* **102:** 254–260.

van de Sluis, B.J., Breen, M., Nanji, M., van Wolferen, M., de Jong, P., Binns, M.M., Pearson, P.L., Kuipers, J., Rothuizen, J., Cox, D.W., et al. 1999. Genetic mapping of the copper toxicosis locus in Bedlington terriers to dog chromosome 10, in a region syntenic to human chromosome region 2p13-p16. *Hum. Mol. Genet.* **8:** 501–507.

van De Sluis, B., Rothuizen, J., Pearson, P.L., van Oost, B.A., and Wijmenga, C. 2002. Identification of a new copper metabolism gene by positional cloning in a purebred dog population. *Hum. Mol. Genet.* **11:** 165–173.

Vignaux, F., Hitte, C., Priat, C., Chuat, J.C., Andre, C., and Galibert, F. 1999. Construction and optimization of a dog whole-genome radiation hybrid panel. *Mamm. Genome* **10:** 888–894.

Vila, C., Savolainen, P., Maldonado, J.E., Amorim, I.R., Rice, J.E., Honeycutt, R.L., Crandall, K.A., Lundeberg, J., and Wayne, R.K. 1997. Multiple and ancient origins of the domestic dog. *Science* **276:** 1687–1689.

Vila, C., Seddon, J., and Ellegren, H. 2005. Genes of domestic mammals augmented by backcrossing with wild ancestors. *Trends Genet.* **21:** 214–218.

Wade, C.M., Karlsson, E.K., Mikkelsen, T.S., Zody, M.C., and Lindblad-Toh, K. 2005. The dog genome: Sequence, evolution, and haplotype structure. In *The dog and its genome* (eds. E.A. Ostrander, U. Giger, and K. Lindblad-Toh). Cold Spring Harbor Press, Cold Spring Harbor, NY.

Wayne, R.K. 1986a. Cranial morphology of domestic and wild canids: The influence of development on morphological change. *Evolution* **40:** 243–261.

———. 1986b. Developmental constraints on limb growth in domestic and some wild canids. *J. Zool.* **210:** 381–399.

———. 1986c. Limb morphology of domestic and wild canids: The influence of development on morphologic change. *J. Morphol.* **187:** 301–319.

Wayne, R.K. and Jenks, S.M. 1991. Mitochondrial DNA analysis implying extensive hybridization of the endangered red wolf *Canis rufus. Nature* **351:** 565–568.

Wayne, R.K., Nash, W.G., and O'Brien, S.J. 1987a. Chromosomal evolution of the Canidae. I. Species with high diploid numbers. *Cytogenet. Cell Genet.* **44:** 123–133.

———. 1987b. Chromosomal evolution of the Canidae. II. Divergence from the primitive carnivore karyotype. *Cytogenet. Cell Genet.* **44:** 134–141.

Wayne, R.K., Van Valkenburgh, B., Kat, P.W., Fuller, T.K., Johnson, W.E., and O'Brien, S.J. 1989. Genetic and morphological divergence among sympatric canids. *J. Hered.* **80:** 447–454.

Wayne, R.K., Geffen, E., Girman, D.J., Koepfli, K.P., Lau, L.M., and Marshall, C.R. 1997. Molecular systematics of the Canidae. *Syst. Biol.* **46:** 622–653.

Wayne, R.K., Leonard, J.A., and Vila, C. 2006. Genetic analysis of dog domestication. In *Documenting domestication: New genetic and archeological paradigms* (ed. M.E. Zeder). Smithsonian Institution Press, Washington, DC. (in press).

Weiss, K.M. and Clark, A.G. 2002. Linkage disequilibrium and the mapping of complex human traits. *Trends Genet.* **18:** 19–24.

Wilcox, B. and Walkowicz, C. 1995. *Atlas of dog breeds of the world.* T.F.H. Publications, Neptune City, NJ.

Wilson, P.J., Grewal, S., Lawford, I.D., Heal, J.N.M., Granacki, A.G., Pennock, D., Theberge, J.B., Theberge, M.T., Voigt, D.R., Waddell, W., et al. 2000. DNA profiles of the eastern Canadian wolf and the red wolf provide evidence for a common evolutionary history independent of the gray wolf. *Canadian J. Zool.* **78:** 2156–2166.

Yuzbasiyan-Gurkan, V., Blanton, S.H., Cao, V., Ferguson, P., Li, J., Venta, P.J., and Brewer, G.J. 1997. Linkage of a microsatellite marker to the canine copper toxicosis locus in Bedlington terriers. *Am. J. Vet. Res.* **58:** 23–27.

Zajc, I., Mellersh, C.S., and Sampson, J. 1997. Variability of canine microsatellites within and between different dog breeds. *Mamm. Genome* **8:** 182–185.

14

Impact of Genomics on Research in the Rat

Jozef Lazar, Carol Moreno,
Howard J. Jacob, and Anne E. Kwitek
Medical College of Wisconsin
Milwaukee, Wisconsin 53226

IMPORTANCE OF THE RAT IN BIOMEDICAL RESEARCH

The dominant power of the laboratory rat is the biological characterization of the >500 strains (http://rgd.mcw.edu/strains/), most of which were developed as models for complex, common diseases. However, while the rat is primarily known as a "physiological" model, there has been a steady increase in the use of the rat for genomic and genetic studies over the last 14 years. Given the need to annotate the human genome with function, linking the rat into this process via its own genome project is a logical and necessary requirement for accelerating improvements in health care, as virtually every drug is tested in the rat before humans.

Since 1966, there have been more than 1 million publications using the rat, with nearly 37,000 published annually for the last eight years. While the vast majority of rat papers remain mechanistic in nature, there are increasing numbers of genetic studies. From 1966, when PubMed started its coverage of the literature, there were 9657 genetic papers that include rat. Since 1991, when the first quantitative trait locus (QTL) was mapped in the rat (Hilbert et al. 1991; Jacob et al. 1991), there have been 26,064 papers published with rat genetics as a component. A CRISP search of the National Institute's funded grants (http://crisp.cit.nih.gov/) identified 41 currently funded grants on topics such as alcohol, hypertension, cancer, and autoimmune disease. There have been 406 grants funded relating to rat genetics since 1991. Given that the publication and funding rates in the rat remain strong in mechanistic studies, concurrent deployment of the genetic infrastructure to place these critical biological parameters on the rat genome will enable us to translate our understanding and treatment of disease from rat to human, through comparative genomics.

The Rat Genome Project

In 1987, Robinson reported the current status of genetic linkage in the rat, the first major report (Robinson 1987). At that time, there were 10 identified linkage groups and four named chromosomes, constructed with 39 phenotypes (coat color, eye color, growth, tumors, teeth, etc.) and 33 electrophoretic and coat color markers. A major contributor to the genetic mapping revolution in the human genome project began with Weber and May's 1989 publication for the use of simple sequence length polymorphisms (SSLPs), also known as "CA-repeats" and "microsatellites" (Weber and May 1989). This new class of genetic markers was also deployed in the rat. Since then, the rat genome project has yielded a tremendous wealth of genomic resources including genetic maps; radiation hybrid (RH) cell lines and the associated RH maps (>5000 genetic markers and 19,500 genes and ESTs mapped); cDNA libraries generating >683,500 ESTs (with more being generated) clustered into >40,000 UniGenes; >10,033 genetic markers; and a published draft (~6.8×) sequence of the genome based on the inbred BN/NHsdMcwi (Brown Norway) strain (Gibbs et al. 2004). The novel sequencing strategy combined whole-genome shotgun (WGS) with bacterial artificial chromosome (BAC) sequencing and covered 90% of the rat genome. Moreover, Celera recently released another 1.5× of draft sequence from the Sprague-Dawley rat (Kaiser 2005). The sequenced rat genome is estimated to be 2.75 Gb, distributed across 21 of the 22 chromosomes (the Y-chromosome is not yet complete), and is predicted to encode ~20,973 genes, with 28,516 transcripts and 205,623 exons (Gibbs et al. 2004). The exact number of genes and transcripts will take several more years to resolve, but the bulk of the data is available for investigators to use now. Because of the success of the rat sequencing project and the value of the rat for functional genomics, the Mammalian Gene Collection (Gerhard et al. 2004) (full-length cDNA project) decided to sequence 6000 full-length genes from the same BN strain that was sequenced, with >4500 nonredundant genes completed. Most of these resources are publicly available through NCBI, the Rat Genome Database (RGD), RatMap, UCSC, Ensembl, and other genome databases (Table 1).

Rat Genomics 283

Table 1. List of major rat resources

Database	Data type	References	URL
Rat Genome Database (RGD)	Several	Twigger et al. 2005	http://rgd.mcw.edu
RatMap	Several	Petersen et al. 2005	http://ratmap.gen.gu.se
NCBI Rat Genome Resources	Several		http://www.ncbi.nlm.nih.gov/genome/guide/rat/index.html
UCSC Rat Browser	Several	Karolchik et al. 2003	http://genome.brc.mcw.edu/
Ensembl Rat Browser	Several	Hubbard et al. 2005	http://www.ensembl.org/Rattus_norvegicus/index.html
Baylor College of Medicine Rat Resources	Sequence, contigs, assembly	Gibbs et al. 2004	http://www.hgsc.bcm.tmc.edu/projects/rat/
Rat EST Project at the University of Iowa	Rat ESTs	Scheetz et al. 2004	http://ratest.uiowa.edu/
PhysGen Program for Genomics Applications (PGA)	Strains, phenotypes, genotypes	Jacob and Kwitek 2002	http://pga.mcw.edu
National Bio Resource Project Japan	Strains, phenotypes, genotypes	Mashimo et al. 2005	http://www.anim.med.kyoto-u.ac.jp/nbr/
TIGR Program for Genomics Applications (TREX)	Microarray		http://pga.tigr.org/
NIAMS: ARB Rat Genetic Database	Strains, maps, markers	Dracheva et al. 2000	http://www.niams.nih.gov/rtbc/ratgbase/
Wellcome Trust Centre: Rat Mapping Resources	Maps, markers	Wilder et al. 2004	http://www.well.ox.ac.uk/rat_mapping_resources/
RRRC: Rat Resource and Research Center	Strains		http://www.nrrrc.missouri.edu/
TIGR Gene Index	Genes	Lee et al. 2005	http://www.tigr.org/tigr-scripts/tgi/T_index.cgi?species=rat
ECR Comparative Genome Browser	Comparative genomics	Ovcharenko et al. 2004	http://ecrbrowser.dcode.org/index.php?db=rn3
VISTA: Comparative Sequence Alignment Browser	Comparative genomics	Frazer et al. 2004	http://pipeline.lbl.gov/cgi-bin/gateway2?bg=rn3&selector=vista
LONI Rat Atlas Image Database	Anatomy	Toga et al. 1995	http://www.loni.ucla.edu/Research/Atlases/RatAtlas.html

The data from the rat genome sequence provide researchers with a precise knowledge of the rat gene content, essential for the advance of biomedical research. It also improves physical and genetic map resolution, since chromosomal position no longer depends on recombination rates and statistical analysis. However, it must be noted that other lines of evidence may be required, that is, genetic linkage analysis or other forms of mapping, to ensure that local regions of the genome under investigation have been assembled correctly. The genomic tool box is now nearly complete and, as outlined throughout this review, is having an important impact in ongoing research using the rat.

Strain Characterization

Many rat strains have been selectively bred for multifactorial disease (polygenic with environmental influence) and then bred to isogeneity. Currently 1015 rat strains are found in the Rat Genome Database, >50% of which are inbred strains for complex traits (538 strains). The diseases studied in these strains range from arthritis to cancer, to hypertension, to multiple sclerosis, to seizures, including 168 different diseases and 393 phenotypes, as defined by RGD's strain disease and phenotype ontologies. In some cases, there are multiple inbred strains for a single multifactorial disease. For example, five different rat strains (BUF, DA, F344, LEW, and PVG) have an increased risk of multiple sclerosis (MS). Crosses between two of these disease strains (DA and LEW) and resistant control strains have resulted in the identification of 18 QTLs involved in experimental allergic encephalomyelitis, an animal model of MS (Dahlman et al. 1999a,b; Roth et al. 1999; Bergsteinsdottir et al. 2000). Overlapping QTL confidence intervals, for the same trait in multiple strains (e.g., Eae2 and Eae11), may then allow for identification of shared haplotypes between the disease strains, which can facilitate positional cloning of the disease allele.

As inbred strains are developed, multiple genes conferring disease may be concurrently fixed, resulting in multiple disease models within a single inbred strain, although some of these traits may remain unidentified. Because of this, there is a need to better characterize strains, both at the phenotypic and at the genomic levels, that is, generate a rat phenome resource. Major efforts are focusing on generating a rat phenome; Mashimo et al. (2005), from the National Bio Resource Project for the Rat (NBRP), have characterized 109 traits in 54 inbred rats, while PhysGen (http://pga.mcw.edu) has characterized 11 different strains (nine inbred and two outbred) for >280 different traits, and have generated and characterized two chromosome substitution panels (44 strains derived from the sequenced BN and the FHH and SS hypertensive strains) (Jacob and

Kwitek 2002). An important aspect of these types of studies is that all experiments are performed using the same methodology.

To complement the detailed phenotypes generated by these efforts, alleles of 48 common inbred strains have been determined for 4328 SSLPs spanning the rat genome, as part of the U.S. Rat Genome Project (Steen et al. 1999). Furthermore, the NBRP has determined 357 SSLP genotypes in 98 strains, including the 54 strains from their Rat Phenome Project (Mashimo et al. 2005). These data allow the construction of haplotypes across all major rat strains using publicly available tools such as the ACP Haplotyper (http://rgd.mcw.edu/ACPHAPLOTYPER/) in order to identify common haplotypes within models with similar diseases. From these data, one can determine the "evolutionary" relatedness of the various inbred strains of rats (Thomas et al. 2003). These allele data are now being greatly supplemented by the addition of >45,000 single nucleotide polymorphisms (SNPs) identified across multiple rat strains (http://www.ncbi.nlm.nih.gov/SNP/snp_summary.cgi; Zimdahl et al. 2004; Guryev et al. 2005). These numbers will grow rapidly with two major SNP discovery projects underway in Europe (the Functional Genomics Group in the Netherlands and the Max-Delbruck-Center for Molecular Medicine in Germany) and a recent NHGRI White Paper to identify SNPs in an additional eight strains. Integration of the detailed phenotype information with haplotypes will provide an incredibly powerful tool for complex disease gene discovery.

GENETIC MAPPING AND POSITIONAL CLONING

QTL Mapping

Quantitative trait locus (QTL) mapping is a proven useful resource to assign the biology of the rat onto the genomic sequence by identifying chromosomal regions that contain genes affecting complex phenotypes. While a QTL is a rather large genetic locus, the genes within this interval are responsible for a component of the trait variation, enabling the genome to be annotated with physiology. Importantly, most rat models reflect a clinical phenotype, and several comparative mapping studies have determined that common phenotypes often map to conserved genomic regions between rat and human (outlined in detail below). The ultimate goal of QTL mapping is to identify the genes, by positional cloning, that underlie complex phenotypes and diseases and to gain a better understanding of their physiology and pathophysiology.

To date, there have been 536 QTL papers published with >1000 QTLs reported for different physiological and pathophysiological traits. These papers include investigations of the genetic basis of blood pressure

(Rapp 2000), diabetes (Jacob et al. 1992; Galli et al. 1996; Pravenec et al. 1996), cardiovascular disease (Stoll et al. 2001; Moreno et al. 2003), stroke (Rubattu et al. 1996), ethanol preference (Murphy et al. 2002), behavioral conditioning and anxiety (Fernandez-Teruel et al. 2002; Flint 2003), fat accumulation (Tanomura et al. 2002), arthritis (Olofsson et al. 2003a), copper metabolism (de Wolf et al. 2002), pituitary tumor growth (Wendell and Gorski 1997), aerobic capacity (Ways et al. 2002), and chemical carcinogenesis (De Miglio et al. 2002). Most of the QTLs have been mapped in the last three years, mainly because of advances in technologies that allow for high-throughput genotyping and an accelerated development of genetically modified strains. Table 2 shows a classification by phenotype category of the QTLs found in rat to date.

Table 2. Quantitative trait loci (QTLs) mapped in rat

Major phenotype category	Phenotype	No. of QTLs
Cardiovascular system phenotypes	Blood pressure	263
	Heart rate	17
	Heart size, weight, morphology	39
	Renin concentration/activity	5
	Vasculature phenotypes	2
Immune system phenotypes	Joint swelling and inflammation	70
	IgE levels	8
	Type II hypersensitivity reaction	1
	White blood cell count	3
	Immune system organ morphology	2
	Natural killer cell alloreactivity	2
	Acute phase protein physiology	3
	Interleukin physiology	2
	Cytokine physiology	3
Renal/urinary system phenotypes	Kidney morphology	38
	Kidney physiology and urine chemistry	95
Lipid homeostasis	Phospholipids	7
	Triglycerides	22
	Cholesterol	35
Glucose homeostasis	Glucose levels	33
	Glucose tolerance	34
	Insulin levels	19
Digestive system physiology and morphology	Pancreas inflammation	4

Table 2. (*Continued*)

Major phenotype category	Phenotype	No. of QTLs
	Pancreas morphology	2
	Pancreas regeneration	1
Tumorigenesis/cancer-related phenotypes	Liver	27
	Tongue	7
	T-cell-derived lymphoma	3
	Pituitary gland	8
	Thymus	2
	Mammary	13
	Gastric	4
	Adrenal gland	1
Nervous system phenotype	Experimental autoimmune encephalomyelitis	20
	CNS ischemia	3
	Spike wave discharge	2
	Brain development	5
Consumption phenotypes	Alcohol preference and consumption	18
	Saline consumption	2
	Food consumption	1
	Saccharin preference and consumption	6
Body growth, size, and fat phenotypes	Body weight	59
	Body fat	13
Musculoskeletal phenotypes	Limb and digit morphology	6
Vision/eye phenotypes		5
Blood chemistry phenotypes	Thyroid hormones	2
	Adrenal gland hormones	5
	Leptin levels	4
	Atrial natriuretic factor levels	1
Mental-health-related phenotypes	Stress response	5
	Anxiety	16
	Depression	15
Locomotor/activity phenotypes		6
Other	X-ray sensitivity	3
	Liver copper content	4
	Male infertility	1

Source: Rat Genome Database, April 2005. The phenotypes and diseases with mapped QTLs cover the majority of the Institutes and Centers at the National Institutes of Health, demonstrating that rats are widely used for research.

QTL mapping is often followed by confirmation of the loci by the development of congenic lines, in order to evaluate the QTL in the absence of other mapped QTLs, and as a step in positional cloning (Flint et al. 2005). To date, 118 QTLs mapped in rat have been confirmed by congenic lines, many of which have narrowed the critical genomic interval to a handful of candidate genes. More than 50% of these congenics (59 strains) were developed for studying blood pressure control, followed by congenics for non-insulin-dependent diabetes mellitus (29 strains). From the 118 congenic lines developed following QTL mapping, 61 congenic lines have been published only since 2002. Following this trend, it is expected that we will see acceleration in the rate of gene discovery in the rat, reflecting the availability of the rat sequence, the accessibility to high-throughput sequencing for the search of sequence variants, as well as microarray technologies for gene identification, pathway analysis, and mapping of *cis* and *trans* regulatory elements. These resources will greatly facilitate the identification of genes underlying the hundreds of QTLs mapped for complex diseases and phenotypes.

Genes Positionally Cloned

Traditional positional cloning efforts in the rat have been coming to fruition in identifying disease genes—many over the past two years. Numerous genes have now been identified in the rat by positional cloning, concurrent with the great increase in rat genomic resources. These include genes for cancer (*Flcn*, *Tsc2*) (Yeung et al. 1994; Okimoto et al. 2004), type 1 diabetes (*Gimap5*, *Cblb*) (MacMurray et al. 2002; Yokoi et al. 2002), type 2 diabetes (*Cd36*) (Aitman et al. 1999), neurological disorders (*Cct4*, *Reln*, *Unc5h3*) (Lee et al. 2003; Yokoi et al. 2003; Kuramoto et al. 2004), arthritis (*Ncf1*) (Olofsson et al. 2003b), renal disease (*Pkhd1*, *Rab38*) (Ward et al. 2002), bleeding disorders (*Rab38*, *Vkorc1*) (Oiso et al. 2004; Rost et al. 2004), retinal degeneration (*Mertk*) (Gal et al. 2000), and hypotrichosis (*Dsg4*, *Foxn1*) (Segre et al. 1995; Jahoda et al. 2004). Many of these genes were cloned from spontaneous mutants with Mendelian inheritance of disease, for example, the *Pkdh1* mutation in the PCK rat causes autosomal recessive polycystic kidney disease (ARPKD). However, the number of identified genes involved in complex traits is on the rise.

One of the most challenging tasks in genomics is the prediction of gene function, and the study of the interactions between genes in the genome—what is called functional genomics. DNA microarray studies provide the potential to greatly enhance our knowledge of the genes and pathways involved in the physiological responses to physiological stressors, drugs, and environmental stimuli, and in pathogenesis of diseases. Microarray studies in rat have been used in conjunction with other genetic strategies, like QTL analysis, congenic mapping, or transgenic techniques to accelerate the search for genes underlying various phenotypes (Aitman et al. 1999; Monti et al. 2001; Liang et al. 2003; Vitt et al. 2004). The cloning of the *Cd36* gene is one of the first examples of cloning a complex trait gene in the rat using a combined approach of introgressing a QTL to generate congenic strains and profiling their expression patterns compared to the parental strain. A more recent study looked at gene expression in a panel of BXH/HXB recombinant inbred (RI) rat strains (Hubner et al. 2005) to identify eQTLs (expression QTLs) in the rat genome. eQTLs that overlap with previously identified QTLs for metabolic syndrome have provided nearly 76 candidate genes to be evaluated.

One application of using animal models for the purpose of gene identification of disease is the translation of that gene to the human clinical setting, although some are skeptical that disease genes translate well between humans and animals. For instance, mutations in the leptin/leptin receptor gene cause dramatic obesity in rodents but do not contribute to the common form of human obesity but, rather, to rare Mendelian forms (Montague et al. 1997; Clement et al. 1998; Strobel et al. 1998). However, the entire field of obesity research was opened up with the cloning of the *Lep* and *Lepr* genes. This scenario is not unlike the cloning of rare monogenic forms of hypertension and hypotension in humans (Lifton et al. 2001). Although the specific genes may not significantly contribute to essential hypertension, they nevertheless generate understanding of the pathways related to sodium handling in the kidney, which certainly play a role in the mechanisms leading to the more common form of the disease. The strength that the rat provides is a platform on which to study these mechanisms and pathways in much more detail.

TRANSGENETICS AND TRANSLATING

Transgenic Rats

For more than 15 years, genetics studies have followed two tracks: positional cloning from genetic mapping to gene identification, and transgenesis (random insertion of gene, knockouts, knock-ins, and conditional knockouts) to unravel gene function. Traditional rat transgenesis by pronuclear injection has been established since 1990 (Hammer et al. 1990; Mullins et al. 1990). However, because the rat lacks viable embryonic stem (ES) cell lines, knockout and knock-in technology is unavailable, somewhat limiting the adoption of the rat for gene-manipulation studies. Nonetheless, there have been >200 transgenic rats generated.

As in the mouse, the major purpose of generating rats via transgenesis was to study a particular gene of interest. It is now relatively straightforward to alter the expression of specific rat genes as well as to use rats as surrogate hosts for expression of genes from other species. Many transgenic rat strains have been "humanized," by using a gene from human, providing a bridge between genetic linkage studies in humans, and functional association of a (mutant) gene with particular pathological features. For example, humanized rats were used to dissect complex diseases such as heart hypertrophy (Tian et al. 2004), end-organ damage (Hocher et al. 1996), and hypertension (for review, see Pinto-Sietsma and Paul 1997; Liefeldt et al. 1999; Bohlender et al. 2000). These examples provide a proof of principle that human disease modeling in rats is valuable, and one can expect that the etiology of other diseases will be similarly illuminated through sequence knowledge and transgenesis. Furthermore, the availability of transgenic rats expressing human genes makes it possible to follow up disease progression in longitudinal in vivo studies, to monitor the effects of long-term treatments, cell implantation, or antisense approaches on the course of disease. Finally, the need to validate a gene cloned by position via transgenic rescue will further increase the use of transgenic technologies in two ways, functional cloning and target validation.

Functional Cloning

The availability of large insert clones enables their use for transgenesis in rats, with the advantage of screening multiple genes with their *cis*-regulatory elements intact. The rat is fortunate to have several different types of large insert clones from a variety of strains, including one PAC and two BAC libraries from BN (two substrains), and BAC libraries from SS, FHH, and F344 strains. Functional cloning was initially coined by Eddy Rubin, who used a human YAC to determine that there was a functional element within the human YAC that influenced asthma (Symula et al. 1999). The ability to modify the phenotype of an animal with a large insert clone, in advance of knowing the causal gene, demonstrates that a gene or element within that clone is capable of modifying function. Furthermore, it is becoming increasingly evident that not only are multiple QTLs responsible for complex disease, but also that what was thought to be a single QTL may actually contain multiple linked QTLs playing a role in a mapped phenotype. In this situation, a single gene transgenic or a knockout would be insufficient to affect a phenotype. If this is the case, then functional cloning may provide an alternative strategy to help identify the causal genes.

Target Validation

Once a gene has been positionally cloned or implicated to be causal, there is a need to prove the causality, a phase of gene discovery termed target validation. The gold standard for proving that a gene is causal is to knock in the particular allele or mutation responsible for the trait, or replace the defect with a "normal" allele to demonstrate that this specific substitution changes the phenotype. While there can be no doubt that conducting this type of experiment can provide conclusive proof, it is an onerous and expensive process that is not likely feasible for all 30,000+ genes. The release of the rat genome sequence has facilitated an alternative approach to target validation—transgenic rescue—whereby the phenotype is normalized via a transgene, particularly when the trait shows a recessive mode of inheritance (Pravenec et al. 2001; Jacob and Kwitek 2002). Recently, cloning of fertile adult rats has been achieved by nuclear transfer (Zhou et al. 2003), which opens the door for targeted gene manipulations techniques such as knockout technology. However, it will be some time before this can be done routinely as the efficiency is too low to be used as a general methodology.

Comparative Mapping

The primary motivation for the rat genome project was to leverage the deep biological history to annotate the human sequence with common complex diseases (Jacob and Kwitek 2002), with the goal of understanding the pathogenesis of human disease. Consequently, most rat research is ultimately translational, aimed at improving human health through the understanding of key genetic and physiological factors in common disease pathways. Using the homologous regions between genomes (Brudno et al. 2004; Gibbs et al. 2004; Wilder et al. 2004) to map disease-causing genes or regions from one organism to another has begun to bear fruit in humans, a result of the investments into the rat genome project.

The availability of the rat genomic sequence along with the genomic sequence of many other species introduces the possibility of comparative genome analysis at a nucleotide level rather than the ordering of large blocks of genes used previously. The utility of comparative analysis is based on the hypothesis that functionally important sequences will be conserved across species. In 2000, Stoll et al. (2000) reported that QTLs in the rat could be used to predict the locations at which human QTLs were likely to exist. Since then, numerous other studies have demonstrated that there are evolutionarily conserved regions in human, mouse, and rat that are linked to the same phenotype in all three species (Stoll et al. 2000; Sugiyama et al. 2001; Jacob and Kwitek 2002; Korstanje and DiPetrillo 2004). Over the next five years, we can expect a large increase in the number of studies using cross-species comparisons to find causes of common complex diseases (Glazier et al. 2002; Korstanje and Paigen 2002). As an example, Peter Harris's group at the Mayo Clinic had sought the gene responsible for autosomal recessive polycystic kidney disease (ARPKD) in humans. Comparative genomics showed ARPKD mapped to the conserved region of the PKC rat and human, and subsequent studies showed the same gene to be the cause of PKD in both species (Ward et al. 2002). Another example of a successful translation of a gene between rat and human includes *CD36* (Aitman et al. 1999; Febbraio et al. 2002) in non-insulin-dependent diabetes mellitus. Approximately 100 papers report that a particular disease trait maps to the same conserved region in rat and human, illustrating the near-term benefits of the rat genome project. However, one must keep in mind that QTLs are large and numerous for multifactorial traits, such as behavior and metabolic syndrome. Therefore, overlapping QTLs may sometimes be merely a chance event. To address this issue, the rat can be extensively evaluated for disease subphenotypes to better match QTLs by intermediate phenotypes. Furthermore, with the promise of a rat SNP map, finer-resolution mapping may reduce the size of a QTL. Finally, comparative mapping data from additional species such as dog, cow, or other models might be used to confirm conserved QTLs. Integration of the genome sequence with existing mapping data and the biological data attached to those maps, plus the creation and annotation of a comprehensive catalog of gene products, will increase the use of such comparative studies and the impact the rat has on translational research.

Gene Therapy

The rat has already served as a useful model for gene transfer experiments of optic nerve disease (for review, see Martin and Quigley 2004), Parkinson's disease (Klein et al. 2002; Lo Bianco et al. 2002; Yamada et al. 2005; Zheng et al. 2005), treatment of cerebral ischemia (Tsai et al. 2002), gene transfer into rat heart (Most et al. 2004; Schroder et al. 2004; O'Donnell and Lewandowski 2005; Ross et al. 2005; Schmidt et al. 2005), erectile dysfunction in diabetic rats (Bennett et al. 2005), and testing naked DNA gene transfer and therapy (Herweijer and Wolff 2003). Moreover, the application of RNA interference to efficiently and specifically knock down expression of mammalian gene products has opened a novel avenue for experimental and therapeutic applications, designed to reduce the levels of an undesirable protein (Shi 2003). A second RNA technology, mRNA reprogramming by *trans*-splicing, offers the ability to repair mRNAs, and thus proteins (Mansfield et al. 2000). In the future, we expect the rat to play a more active role in improving the relatively poor results in human clinical trials to date.

THE "NEW" GENETIC MODELS

Various approaches to generating novel animal models for complex disease provide an additional way to map phenotypes to the genome, and facilitate the process of gene identification, by narrowing the region where linkage to a phenotype resides and by fixing the effect of a gene in a homogeneous genetic background. In the sections below, we describe several means to generate new or "designer" rat models to follow up genetic linkage or QTL studies.

Congenic Strains

In order to validate the functional importance of a genomic region, initially identified by genetic linkage analysis, congenic techniques were originally developed to study the MHC in the mouse by Nobel Prize winner G. Snell (Snell 1948) at the Jackson Laboratory. This strategy remains a common way to study genes nearly 60 years later. Since congenic strains differ only in a short chromosomal segment from their background strain, it is possible to investigate the phenotypic effect of the locus, isolated from other effects caused by other loci on the original genetic background (Fig. 1). The development of congenic strains can be accelerated by genotyping the whole genome and selecting the breeders that, besides containing the target region of the donor strain, have a greater proportion of alleles from the recipient strain throughout the genome. This process of whole-genome marker-assisted selection, also called "speed congenics," can reduce the breeding time by half (Visscher 1999). With this method, generation of a congenic strain is reduced from 4–5 yr to 2–3 yr. In complex diseases, with multiple QTLs determining a trait, generation of double or triple congenic strains is sometimes necessary in

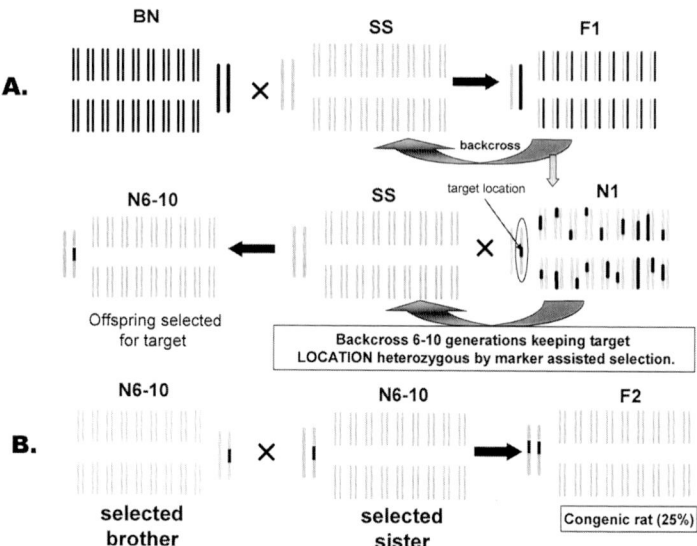

Figure 1. Schematic representation of the generation of a congenic strain from two genetically different rat strains. (A) Parental strains Brown Norway (BN) and Dahl salt-sensitive (SS) are intercrossed for the generation of a heterozygous F_1 population. The F_1 is then crossed with the parental background of interest (in this example, the SS) to generate an N_2 population. The N_2 rats are then backcrossed six to ten generations using marker-assisted selection of the offspring, in order to substitute a selected genomic region from the BN rat. (B) A male and female rat, selected by genotyping for this specific target region containing the phenotype of interest, are then mated. Twenty-five percent of the offspring from this cross will be homozygous for this region. These rats are then inbred to produce a stable inbred congenic strain. Reprinted with permission from Blackwell Publishing © 2004, from Cowley Jr. et al. (2004).

order to confirm a causative locus. These multiple congenics are typically constructed one at a time and then assembled onto multicongenics.

Consomic Strains

Consomic strains are rat strains in which a whole chromosome from one strain is transferred to the genomic background of another strain, by a methodology similar to that of congenic generation. The Medical College of Wisconsin has assembled two complete panels of consomic strains, using the BN strain that was sequenced as the donor strain. In these consomic strains, a chromosome from the BN rat was substituted, one at a time, into the genetic background of the SS/JrHsdMcwi (Dahl salt-sensitive; SS) or FHH/EurMcwi (Fawn Hooded Hypertensive; FHH) rats.

The SS rat is a model for salt-sensitive hypertension (Rapp 1982), insulin resistance (Kotchen et al. 1991), hyperlipidemia (Reaven et al. 1991), endothelial dysfunction (Luscher et al. 1987), cardiac hypertrophy (Ganguli et al. 1979), and glomerulosclerosis (Roman and Kaldunski 1991). The FHH rat is a model for systolic hypertension, renal disease, pulmonary hypertension, a bleeding disorder, alcoholism, and depression (Provoost 1994). The two consomic panels capture nearly 50% of the genetic variation in the rat (Steen et al. 1999), and provide a foundation of genetic resources for the study of disorders of heart, kidney, lung, and vasculature. Given the 50% genetic variability, it is reasonable to assume that a similar level of biological variability can be expected, making the consomic strains a powerful tool for mapping additional complex traits.

One major advantage of using consomic strains is to rapidly generate congenic lines. Generation of congenic rats from consomic rat strains takes at most three generations of breeding, following an intercross with the consomic and parental strain (Fig. 2). There are additional applications for consomic rat strains; for example, they can be used to assess the role of a genomic region in different backgrounds, assessing whether

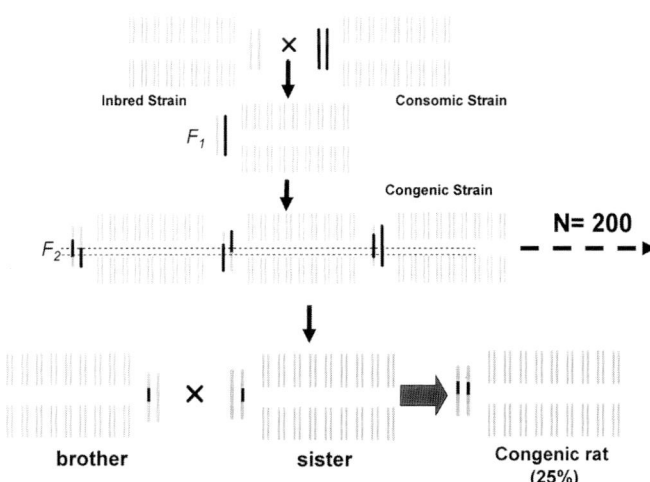

Figure 2. Generation of congenic rats from consomic strains. The parental strain is crossed with the consomic strain, to generate an F_1 population with identical genetic background and a heterozygous target chromosome. These F_1 rats are intercrossed to generate an F_2 population of rats whose target chromosome will be congenic, because of recombination events. Two similar F_2 rats are selected (by genotype) and mated to fix the region of interest. Reprinted with permission from Blackwell Publishing © 2004, from Cowley Jr. et al. (2004).

background effects modify gene function. Consomic rat strains can also be used to develop polygenic models to study gene–gene interactions. The contribution of genes on each chromosome to the observed traits cannot only be assessed both by phenotyping but also by expression profiling. Comparisons between the consomic and parental strains provide valuable insights into the genomic pathways (clustered gene expression patterns) that differ between strains and how these differences might be connected to a particular pathogenic phenotype of the animal.

Recombinant Inbred Strains

Recombinant inbred (RI) strains provide an additional tool for mapping phenotypes to the genome. This strategy is based on the generation of a panel of inbred strains derived from an F_2 population (Pravenec et al. 1996). The RI strategy generates lines that contain more than one QTL, permitting the analysis of gene interaction and detection of weak loci. However, the presence of a unique genome background in each strain prevents their use in rapid generation of congenic animals.

One of the largest rodent recombinant inbred panels is the reciprocal HXB/BXH recombinant inbred strains, derived form the Spontaneously Hypertensive (SHR) and the BN rat strains (Pravenec et al. 1989, 1999). These strains are a great resource for genetic analysis of cardiovascular and metabolic phenotypes. The use of this RI strain panel has facilitated the mapping of several traits, including blood pressure (Pravenec et al. 1995), reproductive traits (Zidek et al. 1999), metabolic traits (Pravenec et al. 2002), behavior (Conti et al. 2004), and susceptibility to cancer (Bila and Kren 1996). Other RIs have also been developed, such as the LExF (Shisa et al. 1997) and the SWXJ (Svenson et al. 1995) panels, although phenotypic characterization of this panel not been as extensive as that for the HXB/BXH.

ENU Mutagenesis

For many years, the genetics community relied on spontaneous mutations as a source of rat models, which is a major limiting factor in the use of rat for structure–function studies, particularly for disease gene validation. An alternative approach includes inducing mutation through the use of chemical mutagens such as N-ethyl-N-nitrosourea (ENU). While ENU mutagenesis has been used for many years, the last 10 years has seen an increase in its use to generate gene mutations (Guenet 2004). The typical ENU strategy is to treat males with ENU, inducing mutations in the spermatogonial stem cells (predominantly loss-of-function mutations), breed them to untreated females, and screen the offspring for phenotypic effects. As the ENU approach has advanced, phenotyping is now emerging as the rate-limiting step in such large-scale studies. These studies produce large numbers of animals having mutations with phenotypic effects. However, the vast majority of these mutants are never identified because sufficient phenotypic screening is not applied. There has been an increasing emphasis on large-scale ENU screens for systematic and comprehensive gene function analyses of the mouse genome (Hrabe de Angelis et al. 2000; Nadeau 2000; Nolan et al. 2000; Brown and Balling 2001). The mutagenesis programs are phenotype-driven, using appropriate screens on mutagenized animals to identify novel mutant phenotypes, particularly those that model human disease, followed by mapping and isolation of the underlying causal genes. However, large-scale phenotypic screens are very expensive and consume a tremendous amount of animal per diem charges. As such, there has not been a similar attempt at this type of screen using rats.

In 2003, two groups published on the use of ENU in combination with a gene screen to generate rat gene knockouts. The major difference in this strategy is that the genes of interest are screened to determine which animals have a mutation in a gene of interest (Zan et al. 2003; Smits et al. 2004). Only the animals that have a mutation are kept. In this way, per diem costs are minimized, although the gene screening also involves a cost. This strategy has been used to knock out several genes in the rat (Zan et al. 2003; Homberg et al. 2005). The strategy has one other major advantage; the ENU mutagenesis can be done in the strain of choice (following the determination of the appropriate ENU dosing), preventing problems associated with the genome background effects that are relatively common when knockouts are generated on a limited number of ES cell lines. We expect this methodology to gain in popularity until such time as there are ES cells for rats, and remain an alternative approach when genome backgrounds affect a phenotype.

Heterogeneous Stocks

Heterogeneous stocks (HS) are derived from crossing eight inbred strains followed by continuous outbreeding for several generations (Hansen and Spuhler 1984). While this strategy was developed before the human and rat genome projects, the resources of the rat genome project make this collection of rat strains extremely powerful and another alternative for investigators using rats in their research programs.

The chromosomes of the HS progeny represent a random mosaic of the founding animals with an average distance between recombination events close to a single centimorgan. This high degree of recombination enables the fine-mapping of QTLs into sub-centimorgan intervals and the identification of multiple QTLs within what was previously identified as a single QTL (e.g., Ariyarajah et al. 2004; Stylianou et al. 2004). The HS rat colony was derived by the NIH in 1984 for alcohol studies (Pandey et al. 2002). Fine-mapping of QTL to sub-centimorgan intervals has been successful for traits of anxiety (Mott et al. 2000), ethanol-induced locomotor activity (Demarest et al. 2001), and conditioned fear (Mott et al. 2000) in HS mice. Studies are currently underway to detect QTLs for multiple traits including behavior and diabetic traits using HS rats.

Embryonic Stem Cells in Rats

The development of ES cells is the final tool needed for the rat genomic and genetic toolbox. To some investigators this remains the holy grail of genetics and functional genomics. While having ES cells in the rat would facilitate some specific research questions, it is not an essential requirement, as evidenced by the continued use of the rat in the absence of this technology. Nonetheless, having the technology would remove the temptation to switch model systems based on the technology rather than the biology. At this point it is difficult to know when there will be ES cell lines for the rat. Over the last decade, numerous groups have attempted to develop ES cells, but thus far to no avail. However, given the current level of interest, it is likely that there will be ES cells developed over the next decade. Whether they are true ES cells or some other extrapolation from advances in stem cell research remains to be seen. Another potential solution to generating knockouts may come from conditionally expressed siRNA probes or similar inhibitors. In the interim, the rat will continue to be effectively deployed using available technologies and alternative study designs to the need for ES cells.

GENOME DATABASES

The rapid increase in rat genetic and genomic data facilitated the need for a centralized database to efficiently and effectively collect, manage, and distribute a rat-centric view of this data to researchers around the world (Table 1). The Rat Genome Database (http://rgd.mcw.edu, RGD) works in collaboration with the Mouse Genome Database (MGI), NCBI, UCSC, EBI, RGSC, Baylor College of Medicine, SWISS-PROT, BIND, RatMap, and the Phys-Gen PGA (de la Cruz et al. 2005). RGD coordinates all nomenclature for genes, strains and QTLs in collaboration with Ratmap (http://ratmap.gen.gu.se), a European database aimed to manage rat related data, and the International Rat Genome and Nomenclature Committee (http://rgnc.gen.gu.se). RGD curates and integrates rat genetic and genomic data and provides its free access to support research using the rat as a genetic model for the study of human disease. It was created to serve as a repository of rat genetic and genomic data, as well as mapping, strain, and physiological information. It also facilitates investigators' research efforts by providing tools to search, mine, and analyze these data sets. Curated data in RGD include rat genes, QTLs, microsatellite markers and expressed sequences (ESTs), rat strains, genetic and radiation hybrid maps, sequence information, and ontologies. Ontologies provide standardized vocabularies for annotating molecular function, biological processes, cellular components, phenotypes, and disease associations. These ontologies allow for searching across genes, QTLs, and strains, and provide a basis for cross-species comparisons. RGD also contains a variety of tools to search and analyze both the data incorporated into RGD and data generated in an investigator's own laboratory. The list below identifies the major tools provided at RGD.

Genome Browser

The Genome Browser (http://rgd.mcw.edu/sequenceresources/gbrowse.shtml) allows rapid visualization of different types of information beneath genome coordinate positions. Information includes genes, genetic markers, ESTs, and mapped QTLs for different phenotypes. Other genome browsers, with numerous tracks and links, are listed in Table 1.

Genome Viewer (GViewer) Tool

GViewer Tool (http://www.rgd.mcw.edu/gviewer/Gviewer.jsp) provides users with a complete genome view of genes and QTLs annotated with function, biological process, cellular component, phenotype, disease, or pathway information. GViewer will search for terms from the Gene Ontology, Mammalian Phenotype Ontology, Disease Ontology, or Pathway Ontology.

Virtual Comparative Map (VCMap)

VCMap (http://rgd.mcw.edu/VCMAP/) is a dynamic sequence-based homology tool that allows researchers of rat, mouse, and human to view mapped genes and sequences and their chromosomal locations in the other two organisms. The comparative maps are based either on radiation hybrid maps or genome sequence-based conservation between species, and allow sequence-related information derived from one species to be mapped to the conserved region in another. The maps are clickable, allowing the view of information on markers, chromosomal regions, and mapped QTLs.

ACP Haplotyper

ACP Haplotyper (http://rgd.mcw.edu/ACPHAPLOTYPER/) allows visualization of chromosomal segments that are conserved between 48 different inbred strains and provides a measurement of relatedness between these inbred strains. This tool creates a visual haplotype that can be used to identify conserved and nonconserved chromosomal regions between any of the 48 rat strains characterized as part of the Allele Characterization Project. For the selected chromosome, the tool compares the allele size data between selected strains, for microsatellite markers on genetic, RH, or sequence maps.

Genome Scanner

Genome Scanner (http://rgd.mcw.edu/GENOMESCANNER/) combines the strain allele data with the position of the genetic marker on the genetic and radiation hybrid maps. By selecting the strain of interests, Genome Scanner will select polymorphic markers between the selected rat strains, which can be further used to perform genetic linkage studies or to increase map density in a specific chromosomal region.

Gene Annotation Tool

The purpose of the Gene Annotation Tool (http://rgd.mcw.edu/gatool/) is to gather information about genes by parsing several databases available online (Entrez, SWISS-PROT, KEGG, and RGD) to provide the user with a comprehensive file of gene descriptions and annotations.

In addition to these tools, there is a wealth of genomic information and bioinformatics tools for the rat, freely available on the Web (Table 1).

CONCLUSIONS

Throughout this chapter, we have discussed the deployment of genetics and genomics in the rat. We have demonstrated the increase in use of these resources within the various sections. However, we believe the real numbers are much larger, as there is always a lag in science between adoption of new strategies and the publication of the results. We believe there will be expansive growth in the use of the rat over the next five years for the following reasons. First, the ability to clone genes by position is poised to explode, based on the availability of the genomic sequence and the large number of genetic mapping studies and related congenic strains already completed. The identity of these genes will drive interest in finding other genes for other diseases, and the wealth of rat models will continue to attract other investigators. Second, comparative genomics continues to show that QTLs in rats match the evolutionarily conserved regions where the QTLs map in human. This means that the genes found in the rat have an increased likelihood to contribute to the disease process in humans. With the rat's genomic sequencing being ~90% identical to that of the human at the gene level, we can expect the rat to continue to add value to the annotation of function onto the human genome. There will, of course, be instances where the results in the rat are unique to the rat. In these instances, the results are likely to be important to the evolutionary biologists. Third, where the disease gene is shown to play a role in both rat and human, we can predict that mechanism-based studies and the development of new therapeutics will quickly follow suit. Fourth, with respect to the mechanistic-based studies, the genetic models, notably the congenics, consomics, and ENU mutagenesis strains, offer the first real controls for mechanistic-based studies of common, complex diseases. The reason these models are advantageous is quite simple—the genome background of the disease strain and control strain are the same; the only issue for these models is how different are they, phenotypically, from the control. For example, in the case of a consomic, the maximum difference between the two strains studied would be the size of the largest chromosome, the minimum the size of the smallest chromosome. Therefore, the genetic relatedness of the consomic models to their respective control strain would range from ~90% to ~98% identical. For congenics and ENU mutagenesis on an inbred genome background, the identity will be even higher between the "case" and control strains. The ability to provide genome background control strains, rather then simply comparing two or more different strains or the use of outbred strains, will be a revolution. Fifth, once found, genes involved in complex traits will then be studied in a variety of model systems, for example, (1) transgenic rescue using genes (Pravenec et al. 2003) and large insert clones (Michalkiewicz et al. 2004); (2) mutagenesis using a combination of ENU and a gene-specific screening assay now provides the ability to create rat gene knockouts (KOs) for use in functional studies (Zan et al. 2003; Smits et al. 2004). H.J. Jacob and A.E. Kwitek and colleagues at the Medical College of Wisconsin (MCW) have been funded to generate KO rats to complement the consomic panels; M.N. Gould was funded to make an additional 20 KO rats; Ingenium Pharmaceuticals (http://www.ingenium-ag.com) has an ENU program for rats. (3) An alternative strategy involves knocking the gene out in cell culture and then cloning the rat (Zhou et al. 2003). While not yet efficient enough for mainstream research it offers promise. (4) Continued efforts are under way to develop ES cells for rats within both the academic and industry settings. (5) The availability of RNAi probes for the rat offers another means to alter phenotype using transgenics. Sixth, the sequenced strain (BN) has been extensively studied, physiologically, and compared to 10 commonly used strains of rats; furthermore, it has been used as the donor strain to produce 44 chromosome substitution strains (consomics). These consomics capture an estimated 50% of known genetic variance in the rat and offer the ability to genetically map new traits and to generate congenics at an unparalleled pace (http://pga.mcw.edu). Seventh, there is a single-nucleotide polymorphism (SNP) project underway in the European Union, and another similar project in the United States. This SNP project will generate 100,000s of SNPS and facilitate development of haplotype maps. For the U.S. project, a White Paper has been accepted by the NHGRI to generate SNPs from eight strains of rats. The strains were selected to capture the maximal degree of known genetic variation (Thomas et al. 2003) and significant numbers of QTLs. Table 3 shows how many QTLs were identified in crosses involving the most commonly used strains (and their derivatives). Clearly, the five most commonly used strains account for the majority of the QTLs. Finally, in the last five years, the rat community has grown from a handful of investigators with a biannual meeting for ~50 people to being between one-third and one-half the size of the mouse community in terms of data generation and grants, and holding a meeting annually with ~150 participants. The science and interest have grown to such a level that Cold Spring Harbor Laboratory hosts the rat meeting every other year.

In conclusion, the sequencing of the rat has resulted in an explosion of genetic research using the rat. Given that the rat is a model known and studied for its biology, it is likely that the biology of the rat will now

Table 3. Quantitative trait loci (QTLs) per strain for strains included in single-nucleotide polymorphism identification

Strain	No. of QTLs
BN	216
SHR/SHRSP	210
SS	155
SD	136
F344	112
WKY	94
BB	87
DA	74
LEW	76
OETF	58
ACI	20
GK	15
BUF	4
FHH	4
PVG	1

The strains considered were based on their diverse genetic phylogeny and the number of QTLs identified in genetic crosses involving them. Note that QTLs are mapped using two strains so that the numbers do not add up to 708 QTLs.

greatly facilitate the knowledge potential of the human genome project. Indeed, the rapid response of the community of investigators to the rat genomic tools and the important results generated to date using these resources are impressive. Perhaps it is time to consider a sequencing project of up to 10 more strains at much lower density coverage and the construction of a high-density SNP haplotype chip. It would markedly accelerate the discovery of genes in these critical rat models, thereby accelerating our understanding of disease in humans.

REFERENCES

Aitman, T.J., Glazier, A.M., Wallace, C.A., Cooper, L.D., Norsworthy, P.J., Wahid, F.N., Al-Majali, K.M., Trembling, P.M., Mann, C.J., Shoulders, C.C., et al. 1999. Identification of Cd36 (Fat) as an insulin-resistance gene causing defective fatty acid and glucose metabolism in hypertensive rats. *Nat. Genet.* 21: 76–83.

Ariyarajah, A., Palijan, A., Dutil, J., Prithiviraj, K., Deng, Y., and Deng, A.Y. 2004. Dissecting quantitative trait loci into opposite blood pressure effects on Dahl rat chromosome 8 by congenic strains. *J. Hypertens.* 22: 1495–1502.

Bennett, N.E., Kim, J.H., Wolfe, D.P., Sasaki, K., Yoshimura, N., Goins, W.F., Huang, S., Nelson, J.B., de Groat, W.C., Glorioso, J.C., et al. 2005. Improvement in erectile dysfunction after neurotrophic factor gene therapy in diabetic rats. *J. Urol.* 173: 1820–1824.

Bergsteinsdottir, K., Yang, H.T., Pettersson, U., and Holmdahl, R. 2000. Evidence for common autoimmune disease genes controlling onset, severity, and chronicity based on experimental models for multiple sclerosis and rheumatoid arthritis. *J. Immunol.* **164:** 1564–1568.

Bila, V. and Kren, V. 1996. The teratogenic action of retinoic acid in rat congenic and recombinant inbred strains. *Folia Biol (Praha)* **42:** 167–173.

Bohlender, J., Ganten, D., and Luft, F.C. 2000. Rats transgenic for human renin and human angiotensinogen as a model for gestational hypertension. *J. Am. Soc. Nephrol.* **11:** 2056–2061.

Brown, S.D. and Balling, R. 2001. Systematic approaches to mouse mutagenesis. *Curr. Opin. Genet. Dev.* **11:** 268–273.

Brudno, M., Poliakov, A., Salamov, A., Cooper, G.M., Sidow, A., Rubin, E.M., Solovyev, V., Batzoglou, S., and Dubchak, I. 2004. Automated whole-genome multiple alignment of rat, mouse, and human. *Genome Res.* **14:** 685–692.

Clement, K., Vaisse, C., Lahlou, N., Cabrol, S., Pelloux, V., Cassuto, D., Gourmelen, M., Dina, C., Chambaz, J., Lacorte, J.-M., et al. 1998. A mutation in the human leptin receptor gene causes obesity and pituitary dysfunction. *Nature* **392:** 398–401.

Conti, L.H., Jirout, M., Breen, L., Vanella, J.J., Schork, N.J., and Printz, M.P. 2004. Identification of quantitative trait Loci for anxiety and locomotion phenotypes in rat recombinant inbred strains. *Behav. Genet.* **34:** 93–103.

Cowley Jr., A.W., Roman, R.J., and Jacob, H.J. 2004. Application of chromosomal substitution techniques in gene-function discovery. *J. Physiol.* **554:** 46–55.

Dahlman, I., Jacobsson, L., Glaser, A., Lorentzen, J.C., Andersson, M., Luthman, H., and Olsson, T. 1999a. Genome-wide linkage analysis of chronic relapsing experimental autoimmune encephalomyelitis in the rat identifies a major susceptibility locus on chromosome 9. *J. Immunol.* **162:** 2581–2588.

Dahlman, I., Wallstrom, E., Weissert, R., Storch, M., Kornek, B., Jacobsson, L., Linington, C., Luthman, H., Lassmann, H., and Olsson, T. 1999b. Linkage analysis of myelin oligodendrocyte glycoprotein-induced experimental autoimmune encephalomyelitis in the rat identifies a locus controlling demyelination on chromosome 18. *Hum. Mol. Genet.* **8:** 2183–2190.

de la Cruz, N., Bromberg, S., Pasko, D., Shimoyama, M., Twigger, S., Chen, J., Chen, C.F., Fan, C., Foote, C., Gopinath, G.R., et al. 2005. The Rat Genome Database (RGD): Developments towards a phenome database. *Nucleic Acids Res.* **33:** D485–D491.

Demarest, K., Koyner, J., McCaughran Jr., J., Cipp, L., and Hitzemann, R. 2001. Further characterization and high-resolution mapping of quantitative trait loci for ethanol-induced locomotor activity. *Behav. Genet.* **31:** 79–91.

De Miglio, M.R., Pascale, R.M., Simile, M.M., Muroni, M.R., Calvisi, D.F., Virdis, P., Bosinco, G.M., Frau, M., Seddaiu, M.A., Ladu, S., et al. 2002. Chromosome mapping of multiple loci affecting the genetic predisposition to rat liver carcinogenesis. *Cancer Res.* **62:** 4459–4463.

de Wolf, I.D., Bonne, A.C., Fielmich-Bouman, X.M., van Oost, B.A., Beynen, A.C., van Zutphen, L.F., and van Lith, H.A. 2002. Quantitative trait loci influencing hepatic copper in rats. *Exp. Biol. Med. (Maywood)* **227:** 529–534.

Dracheva, S.V., Remmers, E.F., Chen, S., Chang, L., Gulko, P.S., Kawahito, Y., Longman, R.E., Wang, J., Du, Y., and Shepard, J. 2000. An integrated genetic linkage map with 1137 markers constructed from five F2 crosses of autoimmune disease-prone and -resistant inbred rat strains. *Genomics* **63:** 202–226.

Febbraio, M., Guy, E., Coburn, C., Knapp Jr., F.F., Beets, A.L., Abumrad, N.A., and Silverstein, R.L. 2002. The impact of overexpression and deficiency of fatty acid translocase (FAT)/CD36. *Mol. Cell. Biochem.* **239:** 193–197.
Fernandez-Teruel, A., Escorihuela, R.M., Gray, J.A., Aguilar, R., Gil, L., Gimenez-Llort, L., Tobena, A., Bhomra, A., Nicod, A., Mott, R., et al. 2002. A quantitative trait locus influencing anxiety in the laboratory rat. *Genome Res.* **12:** 618–626.
Flint, J. 2003. Analysis of quantitative trait loci that influence animal behavior. *J. Neurobiol.* **54:** 46–77.
Flint, J., Valdar, W., Shifman, S., and Mott, R. 2005. Strategies for mapping and cloning quantitative trait genes in rodents. *Nat. Rev. Genet.* **6:** 271–286.
Frazer, K.A., Pachter, L., Poliakov, A., Rubin, E.M., and Dubchak, I. 2004. VISTA: Computational tools for comparative genomics. *Nucleic Acids Res.* **32:** W273–W279.
Gal, A., Li, Y., Thompson, D.A., Weir, J., Orth, U., Jacobson, S.G., Apfelstedt-Sylla, E., and Vollrath, D. 2000. Mutations in MERTK, the human orthologue of the RCS rat retinal dystrophy gene, cause retinitis pigmentosa. *Nat. Genet.* **26:** 270–271.
Galli, J., Li, L.S., Glaser, A., Ostenson, C.G., Jiao, H., Fakhrai-Rad, H., Jacob, H.J., Lander, E.S., and Luthman, H. 1996. Genetic analysis of non-insulin dependent diabetes mellitus in the GK rat. *Nat. Genet.* **12:** 31–37.
Ganguli, M., Tobian, L., and Iwai, J. 1979. Cardiac output and peripheral resistance in strains of rats sensitive and resistant to NaCl hypertension. *Hypertension* **1:** 3–7.
Gerhard, D.S., Wagner, L., Feingold, E.A., Shenmen, C.M., Grouse, L.H., Schuler, G., Klein, S.L., Old, S., Rasooly, R., Good, P., et al. 2004. The status, quality, and expansion of the NIH full-length cDNA project: The Mammalian Gene Collection (MGC). *Genome Res.* **14:** 2121–2127.
Gibbs, R.A., Weinstock, G.M., Metzker, M.L., Muzny, D.M., Sodergren, E.J., Scherer, S., Scott, G., Steffen, D., Worley, K.C., Burch, P.E., et al. 2004. Genome sequence of the Brown Norway rat yields insights into mammalian evolution. *Nature* **428:** 493–521.
Glazier, A.M., Nadeau, J.H., and Aitman, T.J. 2002. Finding genes that underlie complex traits. *Science* **298:** 2345–2349.
Guenet, J.L. 2004. Chemical mutagenesis of the mouse genome: An overview. *Genetica* **122:** 9–24.
Guryev, V., Berezikov, E., and Cuppen, E. 2005. CASCAD: A database of annotated candidate single nucleotide polymorphisms associated with expressed sequences. *BMC Genomics* **6:** 10.
Hammer, R.E., Maika, S.D., Richardson, J.A., Tang, J.P., and Taurog, J.D. 1990. Spontaneous inflammatory disease in transgenic rats expressing HLA-B27 and human β 2m: An animal model of HLA-B27-associated human disorders. *Cell* **63:** 1099–1112.
Hansen, C. and Spuhler, K. 1984. Development of the National Institutes of Health genetically heterogeneous rat stock. *Alcohol Clin. Exp. Res.* **8:** 477–479.
Herweijer, H. and Wolff, J.A. 2003. Progress and prospects: Naked DNA gene transfer and therapy. *Gene Ther.* **10:** 453–458.
Hilbert, P., Lindpaintner, K., Beckmann, J.S., Serikawa, T., Soubrier, F., Dubay, C., Cartwright, P., De Gouyon, B., Julier, C., Takahasi, S., et al. 1991. Chromosomal mapping of two genetic loci associated with blood-pressure regulation in hereditary hypertensive rats. *Nature* **353:** 521–529.
Hocher, B., Liefeldt, L., Thone-Reineke, C., Orzechowski, H.D., Distler, A., Bauer, C., and Paul, M. 1996. Characterization of the renal phenotype of transgenic rats expressing the human endothelin-2 gene. *Hypertension* **28:** 196–201.

Homberg, J.R., Olivier, J.D., Smits, B., Mudde, J., Cools, A.R., Ellenbroek, B.A., and Cuppen, E. 2005. O9 phenotyping of the serotonin transporter knockout rat. *Behav. Pharmacol.* **16 Suppl 1:** S21.

Hrabe de Angelis, M.H., Flaswinkel, H., Fuchs, H., Rathkolb, B., Soewarto, D., Marschall, S., Heffner, S., Pargent, W., Wuensch, K., Jung, M., et al. 2000. Genome-wide, large-scale production of mutant mice by ENU mutagenesis. *Nat. Genet.* **25:** 444–447.

Hubbard, T., Andrews, D., Caccamo, M., Cameron, G., Chen, Y., Clamp, M., Clarke, L., Coates, G., Cox, T., Cunningham, F., et al. 2005. Ensembl 2005. *Nucleic Acids Res.* **33:** D447–D453.

Hubner, N., Wallace, C.A., Zimdahl, H., Petretto, E., Schulz, H., Maciver, F., Mueller, M., Hummel, O., Monti, J., Zidek, V., et al. 2005. Integrated transcriptional profiling and linkage analysis for identification of genes underlying disease. *Nat. Genet.* **37:** 243–253.

Jacob, H.J. and Kwitek, A.E. 2002. Rat genetics: Attaching physiology and pharmacology to the genome. *Nat. Rev. Genet.* **3:** 33–42.

Jacob, H.J., Lindpaintner, K., Lincoln, S.E., Kusumi, K., Bunker, R.K., Mao, V.P., Ganten, D., Dzau, V.J., and Lander, E.S. 1991. Genetic mapping of a gene causing hypertension in the stroke-prone spontaneously hypertensive rat. *Cell* **67:** 213–224.

Jacob, H.J., Pettersson, A., Wilson, D., Mao, Y., Lernmark, A., and Lander, E.S. 1992. Genetic dissection of autoimmune type I diabetes in the BB rat. *Nat. Genet.* **2:** 56–60.

Jahoda, C.A., Kljuic, A., O'Shaughnessy, R., Crossley, N., Whitehouse, C.J., Robinson, M., Reynolds, A.J., Demarchez, M., Porter, R.M., Shapiro, L., et al. 2004. The lanceolate hair rat phenotype results from a missense mutation in a calcium coordinating site of the desmoglein 4 gene. *Genomics* **83:** 747–756.

Kaiser, J. 2005. GENOMICS: Celera to end subscriptions and give data to public GenBank. *Science* **308:** 775.

Karolchik, D., Baertsch, R., Diekhans, M., Furey, T.S., Hinrichs, A., Lu, Y.T., Roskin, K.M., Schwartz, M., Sugnet, C.W., Thomas, D.J., et al. 2003. The UCSC Genome Browser Database. *Nucleic Acids Res.* **31:** 51–54.

Klein, R.L., King, M.A., Hamby, M.E., and Meyer, E.M. 2002. Dopaminergic cell loss induced by human A30P α-synuclein gene transfer to the rat substantia nigra. *Hum. Gene Ther.* **13:** 605–612.

Korstanje, R. and DiPetrillo, K. 2004. Unraveling the genetics of chronic kidney disease using animal models. *Am. J. Physiol. Renal Physiol.* **287:** F347–F352.

Korstanje, R. and Paigen, B. 2002. From QTL to gene: The harvest begins. *Nat. Genet.* **31:** 235–236.

Kotchen, T.A., Zhang, H.Y., Covelli, M., and Blehschmidt, N. 1991. Insulin resistance and blood pressure in Dahl rats and in one-kidney, one-clip hypertensive rats. *Am. J. Physiol.* **261:** E692–E697.

Kuramoto, T., Kuwamura, M., and Serikawa, T. 2004. Rat neurological mutations cerebellar vermis defect and hobble are caused by mutations in the netrin-1 receptor gene Unc5h3. *Brain Res. Mol. Brain Res.* **122:** 103–108.

Lee, M.J., Stephenson, D.A., Groves, M.J., Sweeney, M.G., Davis, M.B., An, S.F., Houlden, H., Salih, M.A., Timmerman, V., de Jonghe, P., et al. 2003. Hereditary sensory neuropathy is caused by a mutation in the δ subunit of the cytosolic chaperonin-containing t-complex peptide-1 (Cct4) gene. *Hum. Mol. Genet.* **12:** 1917–1925.

Lee, Y., Tsai, J., Sunkara, S., Karamycheva, S., Pertea, G., Sultana, R., Antonescu, V., Chan, A., Cheung, F., and Quackenbush, J. 2005. The TIGR Gene Indices: Clustering and assembling EST and known genes and integration with eukaryotic genomes. *Nucleic Acids Res.* **33:** D71–D74.

Liang, M., Yuan, B., Rute, E., Greene, A.S., Olivier, M., and Cowley Jr., A.W. 2003. Insights into Dahl salt-sensitive hypertension revealed by temporal patterns of renal medullary gene expression. *Physiol. Genomics* **12**: 229–237.
Liefeldt, L., Schonfelder, G., Bocker, W., Hocher, B., Talsness, C.E., Rettig, R., and Paul, M. 1999. Transgenic rats expressing the human ET-2 gene: A model for the study of endothelin actions in vivo. *J. Mol. Med.* **77**: 565–574.
Lifton, R.P., Gharavi, A.G., and Geller, D.S. 2001. Molecular mechanisms of human hypertension. *Cell* **104**: 545–556.
Lo Bianco, C., Ridet, J.L., Schneider, B.L., Deglon, N., and Aebischer, P. 2002. α-Synucleinopathy and selective dopaminergic neuron loss in a rat lentiviral-based model of Parkinson's disease. *Proc. Natl. Acad. Sci.* **99**: 10813–10818.
Luscher, T.F., Raij, L., and Vanhoutte, P.M. 1987. Endothelium-dependent vascular responses in normotensive and hypertensive Dahl rats. *Hypertension* **9**: 157–163.
MacMurray, A.J., Moralejo, D.H., Kwitek, A.E., Rutledge, E.A., Van Yserloo, B., Gohlke, P., Speros, S.J., Snyder, B., Schaefer, J., Bieg, S., et al. 2002. Lymphopenia in the BB rat model of type 1 diabetes is due to a mutation in a novel immune-associated nucleotide (Ian)-related gene. *Genome Res.* **12**: 1029–1039.
Mansfield, S.G., Kole, J., Puttaraju, M., Yang, C.C., Garcia-Blanco, M.A., Cohn, J.A., and Mitchell, L.G. 2000. Repair of CFTR mRNA by spliceosome-mediated RNA *trans*-splicing. *Gene Ther.* **7**: 1885–1895.
Martin, K.R. and Quigley, H.A. 2004. Gene therapy for optic nerve disease. *Eye* **18**: 1049–1055.
Mashimo, T., Birger, V., Kuramoto, T., and Serikawa, T. 2005. Rat Phenome Project: The untapped potential of existing rat strains. *J. Appl. Physiol.* **98**: 371–379.
Michalkiewicz, M., Michalkiewicz, T., Ettinger, R.A., Rutledge, E.A., Fuller, J.M., Moralejo, D.H., Van Yserloo, B., MacMurray, A.J., Kwitek, A.E., Jacob, H.J., et al. 2004. Transgenic rescue demonstrates involvement of the Ian5 gene in T cell development in the rat. *Physiol. Genomics* **19**: 228–232.
Montague, C.T., Farooqi, I.S., Whitehead, J.P., Soos, M.A., Rau, H., Wareham, N.J., Sewter, C.P., Digby, J.E., Mohammed, S.N., Hurst, J.A., et al. 1997. Congenital leptin deficiency is associated with severe early-onset obesity in humans. *Nature* **387**: 903–908.
Monti, J., Gross, V., Luft, F.C., Franca Milia, A., Schulz, H., Dietz, R., Sharma, A.M., and Hubner, N. 2001. Expression analysis using oligonucleotide microarrays in mice lacking bradykinin type 2 receptors. *Hypertension* **38**: E1–E3.
Moreno, C., Dumas, P., Kaldunski, M.L., Tonellato, P.J., Greene, A.S., Roman, R.J., Cheng, Q., Wang, Z., Jacob, H.J., and Cowley Jr., A.W. 2003. Genomic map of cardiovascular phenotypes of hypertension in female Dahl S rats. *Physiol. Genomics* **15**: 243–257.
Most, P., Pleger, S.T., Volkers, M., Heidt, B., Boerries, M., Weichenhan, D., Loffler, E., Janssen, P.M., Eckhart, A.D., Martini, J., et al. 2004. Cardiac adenoviral S100A1 gene delivery rescues failing myocardium. *J. Clin. Invest.* **114**: 1550–1563.
Mott, R., Talbot, C.J., Turri, M.G., Collins, A.C., and Flint, J. 2000. A method for fine mapping quantitative trait loci in outbred animal stocks. *Proc. Natl. Acad. Sci.* **97**: 12649–12654.
Mullins, J.J., Peters, J., and Ganten, D. 1990. Fulminant hypertension in transgenic rats harbouring the mouse Ren-2 gene. *Nature* **344**: 541–544.
Murphy, J.M., Stewart, R.B., Bell, R.L., Badia-Elder, N.E., Carr, L.G., McBride, W.J., Lumeng, L., and Li, T.K. 2002. Phenotypic and genotypic characterization of the Indiana University rat lines selectively bred for high and low alcohol preference. *Behav. Genet.* **32**: 363–388.

Nadeau, J.H. 2000. Muta-genetics or muta-genomics: The feasibility of large-scale mutagenesis and phenotyping programs. *Mamm. Genome* **11:** 603–607.

Nolan, P.M., Peters, J., Strivens, M., Rogers, D., Hagan, J., Spurr, N., Gray, I.C., Vizor, L., Brooker, D., Whitehill, E., et al. 2000. A systematic, genome-wide, phenotype-driven mutagenesis programme for gene function studies in the mouse. *Nat. Genet.* **25:** 440–443.

O'Donnell, J.M. and Lewandowski, E.D. 2005. Efficient, cardiac-specific adenoviral gene transfer in rat heart by isolated retrograde perfusion in vivo. *Gene Ther.* **12:** 958–964.

Oiso, N., Riddle, S.R., Serikawa, T., Kuramoto, T., and Spritz, R.A. 2004. The rat Ruby (R) locus is Rab38: Identical mutations in Fawn-hooded and Tester-Moriyama rats derived from an ancestral Long Evans rat sub-strain. *Mamm. Genome* **15:** 307–314.

Okimoto, K., Sakurai, J., Kobayashi, T., Mitani, H., Hirayama, Y., Nickerson, M.L., Warren, M.B., Zbar, B., Schmidt, L.S., and Hino, O. 2004. A germ-line insertion in the Birt-Hogg-Dube (BHD) gene gives rise to the Nihon rat model of inherited renal cancer. *Proc. Natl. Acad. Sci.* **101:** 2023–2027.

Olofsson, P., Holmberg, J., Pettersson, U., and Holmdahl, R. 2003a. Identification and isolation of dominant susceptibility loci for pristane-induced arthritis. *J. Immunol.* **171:** 407–416.

Olofsson, P., Holmberg, J., Tordsson, J., Lu, S., Akerstrom, B., and Holmdahl, R. 2003b. Positional identification of Ncf1 as a gene that regulates arthritis severity in rats. *Nat. Genet.* **33:** 25–32.

Ovcharenko, I., Nobrega, M.A., Loots, G.G., and Stubbs, L. 2004. ECR Browser: A tool for visualizing and accessing data from comparisons of multiple vertebrate genomes. *Nucleic Acids Res.* **32:** W280–W286.

Pandey, J., Cracchiolo, D., Hansen, F.M., and Wendell, D.L. 2002. Strain differences and inheritance of angiogenic versus angiostatic activity in oestrogen-induced rat pituitary tumours. *Angiogenesis* **5:** 53–66.

Petersen, G., Johnson, P., Andersson, L., Klinga-Levan, K., Gomez-Fabre, P.M., and Stahl, F. 2005. RatMap—Rat genome tools and data. *Nucleic Acids Res.* **33:** D492–D494.

Pinto-Sietsma, S.J. and Paul, M. 1997. Transgenic rats as models for hypertension. *J. Hum. Hypertens.* **11:** 577–581.

Pravenec, M., Klir, P., Kren, V., Zicha, J., and Kunes, J. 1989. An analysis of spontaneous hypertension in spontaneously hypertensive rats by means of new recombinant inbred strains. *J. Hypertension* **7:** 217–221.

Pravenec, M., Gauguier, D., Schott, J.-J., Buard, J., Kren, V., Bila, V., Szpirer, C., Szpirer, J., Wang, J.-M., Huang, H., et al. 1995. Mapping of quantitative trait loci for blood pressure and cardiac mass in the rat by genome scanning of recombinant inbred strains. *J. Clin. Invest.* **96:** 1973–1978.

———. 1996. A genetic linkage map of the rat derived from recombinant inbred strains. *Mamm. Genome* **7:** 117–127.

Pravenec, M., Kren, V., Krenova, D., Bila, V., Zidek, V., Simakova, M., Musilova, A., van Lith, H.A., and van Zutphen, L.F. 1999. HXB/Ipcv and BXH/Cub recombinant inbred strains of the rat: Strain distribution patterns of 632 alleles. *Folia Biol. (Praha)* **45:** 203–215.

Pravenec, M., Landa, V., Zidek, V., Musilova, A., Kren, V., Kazdova, L., Aitman, T.J., Glazier, A.M., Ibrahimi, A., Abumrad, N.A., et al. 2001. Transgenic rescue of defective Cd36 ameliorates insulin resistance in spontaneously hypertensive rats. *Nat. Genet.* **27:** 156–158.

Pravenec, M., Zidek, V., Musilova, A., Simakova, M., Kostka, V., Mlejnek, P., Kren, V., Krenova, D., Bila, V., Mikova, B., et al. 2002. Genetic analysis of metabolic defects in the spontaneously hypertensive rat. *Mamm. Genome* **13**: 253–258.
Pravenec, M., Landa, V., Zidek, V., Musilova, A., Kazdova, L., Qi, N., Wang, J., St Lezin, E., and Kurtz, T.W. 2003. Transgenic expression of CD36 in the spontaneously hypertensive rat is associated with amelioration of metabolic disturbances but has no effect on hypertension. *Physiol. Res.* **52**: 681–688.
Provoost, A.P. 1994. Spontaneous glomerulosclerosis: Insights from the fawn-hooded rat. *Kidney Int. Suppl.* **45**: S2–S5.
Rapp, J.P. 1982. Dahl salt-susceptible and salt-resistant rats. A review. *Hypertension* **4**: 753–763.
———. 2000. Genetic analysis of inherited hypertension in the rat. *Physiol. Rev.* **80**: 135–172.
Reaven, G.M., Twersky, J., and Chang, H. 1991. Abnormalities of carbohydrate and lipid metabolism in Dahl rats. *Hypertension* **18**: 630–635.
Robinson, R. 1987. Genetic linkage in the Norway rat. *Genetica* **74**: 137–142.
Roman, R.J. and Kaldunski, M. 1991. Pressure natriuresis and cortical and papillary blood flow in inbred Dahl rats. *Am. J. Physiol.* **261**: R595–R602.
Ross, M.T., Grafham, D.V., Coffey, A.J., Scherer, S., McLay, K., Muzny, D., Platzer, M., Howell, G.R., Burrows, C., Bird, C.P., et al. 2005. The DNA sequence of the human X chromosome. *Nature* **434**: 325–337.
Rost, S., Fregin, A., Ivaskevicius, V., Conzelmann, E., Hortnagel, K., Pelz, H.J., Lappegard, K., Seifried, E., Scharrer, I., Tuddenham, E.G., et al. 2004. Mutations in VKORC1 cause warfarin resistance and multiple coagulation factor deficiency type 2. *Nature* **427**: 537–541.
Roth, M.P., Viratelle, C., Dolbois, L., Delverdier, M., Borot, N., Pelletier, L., Druet, P., Clanet, M., and Coppin, H. 1999. A genome-wide search identifies two susceptibility loci for experimental autoimmune encephalomyelitis on rat chromosomes 4 and 10. *J. Immunol.* **162**: 1917–1922.
Rubattu, S., Volpe, M., Kreutz, R., Ganten, U., Ganten, D., and Lindpaintner, K. 1996. Chromosomal mapping of quantitative trait loci contributing to stroke in a rat model of complex human disease. *Nat. Genet.* **13**: 429–434.
Scheetz, T.E., Laffin, J.J., Berger, B., Holte, S., Baumes, S.A., Brown II, R., Chang, S., Coco, J., Conklin, J., Crouch, K., et al. 2004. High-throughput gene discovery in the rat. *Genome Res.* **14**: 733–741.
Schmidt, U., Zhu, X., Lebeche, D., Huq, F., Guerrero, J.L., and Hajjar, R.J. 2005. In vivo gene transfer of parvalbumin improves diastolic function in aged rat hearts. *Cardiovasc Res.* **66**: 318–323.
Schroder, J.N., Williams, M.L., and Koch, W.J. 2004. Gene delivery approaches to heart failure treatment. *Expert Opin. Biol. Ther.* **4**: 1413–1422.
Segre, J.A., Nemhauser, J.L., Taylor, B.A., Nadeau, J.H., and Lander, E.S. 1995. Positional cloning of the nude locus: Genetic, physical, and transcription maps of the region and mutations in the mouse and rat. *Genomics* **28**: 549–559.
Shi, Y. 2003. Mammalian RNAi for the masses. *Trends Genet.* **19**: 9–12.
Shisa, H., Lu, L., Katoh, H., Kawarai, A., Tanuma, J., Matsushima, Y., and Hiai, H. 1997. The LEXF: A new set of rat recombinant inbred strains between LE/Stm and F344. *Mamm. Genome* **8**: 324–327.
Smits, B.M., Mudde, J., Plasterk, R.H., and Cuppen, E. 2004. Target-selected mutagenesis of the rat. *Genomics* **83**: 332–334.

Snell, G. 1948. Methods for the study of histocompatibility genes. *J. Genet.* **49:** 87–108.
Steen, R.G., Kwitek-Black, A.E., Glenn, C., Gullings-Handley, J., Van Etten, W., Atkinson, O.S., Appel, D., Twigger, S., Muir, M., Mull, T., et al. 1999. A high-density integrated genetic linkage and radiation hybrid map of the laboratory rat. *Genome Res.* **9:** AP1–AP8, insert.
Stoll, M., Kwitek-Black, A.E., Cowley Jr., A.W., Harris, E.L., Harrap, S.B., Krieger, J.E., Printz, M.P., Provoost, A.P., Sassard, J., and Jacob, H.J. 2000. New target regions for human hypertension via comparative genomics. *Genome Res.* **10:** 473–482.
Stoll, M., Cowley Jr., A.W., Tonellato, P.J., Greene, A.S., Kaldunski, M.L., Roman, R.J., Dumas, P., Schork, N.J., Wang, Z., and Jacob, H.J. 2001. A genomic-systems biology map for cardiovascular function. *Science* **294:** 1723–1726.
Strobel, A., Issad, T., Camoin, L., Ozata, M., and Strosberg, A.D. 1998. A leptin missense mutation associated with hypogonadism and morbid obesity. *Nat. Genet.* **18:** 213–215.
Stylianou, I.M., Christians, J.K., Keightley, P.D., Bunger, L., Clinton, M., Bulfield, G., and Horvat, S. 2004. Genetic complexity of an obesity QTL (Fob3) revealed by detailed genetic mapping. *Mamm. Genome* **15:** 472–481.
Sugiyama, F., Churchill, G.A., Higgins, D.C., Johns, C., Makaritsis, K.P., Gavras, H., and Paigen, B. 2001. Concordance of murine quantitative trait loci for salt-induced hypertension with rat and human loci. *Genomics* **71:** 70–77.
Svenson, K.L., Cheah, Y.C., Shultz, K.L., Mu, J.L., Paigen, B., and Beamer, W.G. 1995. Strain distribution pattern for SSLP markers in the SWXJ recombinant inbred strain set: Chromosomes 1 to 6. *Mamm. Genome* **6:** 867–872.
Symula, D.J., Frazer, K.A., Ueda, Y., Denefle, P., Stevens, M.E., Wang, Z.E., Locksley, R., and Rubin, E.M. 1999. Functional screening of an asthma QTL in YAC transgenic mice. *Nat. Genet.* **23:** 241–244.
Tanomura, H., Miyake, T., Taniguchi, Y., Manabe, N., Kose, H., Matsumoto, K., Yamada, T., and Sasaki, Y. 2002. Detection of a quantitative trait locus for intramuscular fat accumulation using the OLETF rat. *J. Vet. Med. Sci.* **64:** 45–50.
Thomas, M.A., Chen, C.F., Jensen-Seaman, M.I., Tonellato, P.J., and Twigger, S.N. 2003. Phylogenetics of rat inbred strains. *Mamm. Genome* **14:** 61–64.
Tian, X.L., Pinto, Y.M., Costerousse, O., Franz, W.M., Lippoldt, A., Hoffmann, S., Unger, T., and Paul, M. 2004. Over-expression of angiotensin converting enzyme-1 augments cardiac hypertrophy in transgenic rats. *Hum. Mol. Genet.* **13:** 1441–1450.
Toga, A.W., Santori, E.M., Hazani, R., and Ambach, K. 1995. A 3D digital map of rat brain. *Brain Res. Bull.* **38:** 77–85.
Tsai, T.H., Chen, S.L., Xiao, X., Liu, D.W., and Tsao, Y.P. 2002. Gene therapy for treatment of cerebral ischemia using defective recombinant adeno-associated virus vectors. *Methods* **28:** 253–258.
Twigger, S.N., Pasko, D., Nie, J., Shimoyama, M., Bromberg, S., Campbell, D., Chen, J., dela Cruz, N., Fan, C., Foote, C., et al. 2005. Tools and strategies for physiological genomics—The Rat Genome Database. *Physiol. Genomics* **23:** 246–256.
Visscher, P.M. 1999. Speed congenics: Accelerated genome recovery using genetic markers. *Genet. Res.* **74:** 81–85.
Vitt, U., Gietzen, D., Stevens, K., Wingrove, J., Becha, S., Bulloch, S., Burrill, J., Chawla, N., Chien, J., Crawford, M., et al. 2004. Identification of candidate disease genes by EST alignments, synteny, and expression and verification of Ensembl genes on rat chromosome 1q43–54. *Genome Res.* **14:** 640–650.

Ward, C.J., Hogan, M.C., Rossetti, S., Walker, D., Sneddon, T., Wang, X., Kubly, V., Cunningham, J.M., Bacallao, R., Ishibashi, M., et al. 2002. The gene mutated in autosomal recessive polycystic kidney disease encodes a large, receptor-like protein. *Nat. Genet.* **30:** 259–269.

Ways, J.A., Cicila, G.T., Garrett, M.R., and Koch, L.G. 2002. A genome scan for Loci associated with aerobic running capacity in rats. *Genomics* **80:** 13–20.

Weber, J.L. and May, P.E. 1989. Abundant class of human DNA polymorphisms which can be typed using the polymerase chain reaction. *Am. J. Hum. Genet.* **44:** 388–396.

Wendell, D.L. and Gorski, J. 1997. Quantitative trait loci for estrogen-dependent pituitary tumor growth in the rat. *Mamm. Genome* **8:** 823–829.

Wilder, S.P., Bihoreau, M.-T., Argoud, K., Watanabe, T.K., Lathrop, M., and Gauguier, D. 2004. Integration of the rat recombination and EST maps in the rat genomic sequence and comparative mapping analysis with the mouse genome. *Genome Res.* **14:** 758–765.

Yamada, M., Mizuno, Y., and Mochizuki, H. 2005. Parkin gene therapy for α-synucleinopathy: A rat model of Parkinson's disease. *Hum. Gene Ther.* **16:** 262–270.

Yeung, R.S., Xiao, G.H., Jin, F., Lee, W.C., Testa, J.R., and Knudson, A.G. 1994. Predisposition to renal carcinoma in the Eker rat is determined by germ-line mutation of the tuberous sclerosis 2 (TSC2) gene. *Proc. Natl. Acad. Sci.* **91:** 11413–11416.

Yokoi, N., Komeda, K., Wang, H.Y., Yano, H., Kitada, K., Saitoh, Y., Seino, Y., Yasuda, K., Serikawa, T., and Seino, S. 2002. Cblb is a major susceptibility gene for rat type 1 diabetes mellitus. *Nat. Genet.* **31:** 391–394.

Yokoi, N., Namae, M., Wang, H.Y., Kojima, K., Fuse, M., Yasuda, K., Serikawa, T., Seino, S., and Komeda, K. 2003. Rat neurological disease creeping is caused by a mutation in the reelin gene. *Brain Res. Mol. Brain Res.* **112:** 1–7.

Zan, Y., Haag, J.D., Chen, K.S., Shepel, L.A., Wigington, D., Wang, Y.R., Hu, R., Lopez-Guajardo, C.C., Brose, H.L., Porter, K.I., et al. 2003. Production of knockout rats using ENU mutagenesis and a yeast-based screening assay. *Nat. Biotechnol.* **21:** 645–651.

Zheng, J.S., Tang, L.L., Zheng, S.S., Zhan, R.Y., Zhou, Y.Q., Goudreau, J., Kaufman, D., and Chen, A.F. 2005. Delayed gene therapy of glial cell line-derived neurotrophic factor is efficacious in a rat model of Parkinson's disease. *Brain Res. Mol. Brain Res.* **134:** 155–161.

Zhou, Q., Renard, J.P., Le Friec, G., Brochard, V., Beaujean, N., Cherifi, Y., Fraichard, A., and Cozzi, J. 2003. Generation of fertile cloned rats by regulating oocyte activation. *Science* **302:** 1179.

Zidek, V., Pintir, J., Musilova, A., Bila, V., Kren, V., and Pravenec, M. 1999. Mapping of quantitative trait loci for seminal vesicle mass and litter size to rat chromosome 8. *J. Reprod. Fertil.* **116:** 329–333.

Zimdahl, H., Nyakatura, G., Brandt, P., Schulz, H., Hummel, O., Fartmann, B., Brett, D., Droege, M., Monti, J., Lee, Y.A., et al. 2004. A SNP map of the rat genome generated from cDNA sequences. *Science* **303:** 807.

15

The Mouse Genome

Jean Louis Guénet
Département de Biologie du Développement
Institut Pasteur
75724 Paris Cedex 15, France

BECAUSE OF ITS MANY ADVANTAGES AS AN ANIMAL MODEL, geneticists have used the house mouse since the early days of genetics. Historical records indicate that Mendel himself bred and crossed mice, segregating for coat color mutations, until he was requested by the ecclesiastical hierarchy to stop experimenting with animals and to resume working with garden peas (Paigen 2003). In 1902, shortly after Mendel's laws were rediscovered, Lucien Cuénot used mice to demonstrate that the laws in question applied to mammals as they did to plants (Cuénot 1902). Since these initial observations, and if we exclude interruptions in progress due to the world wars, it is not an exaggeration to say that the advances in mouse genetics have been growing exponentially, even booming over the last 15 years. In this anniversary review, I will cover the most important achievements that occurred in the period spanning 1990 to 2005 and will discuss the anticipated developments in the years to come.

Before we start reviewing these achievements, it is important to note that the context in which mouse genetics has been evolving during the period covered by this review can in no way be compared with the period before it. For the majority of the 20th century, the community of mouse geneticists operated like a club of friends, with occasional meetings and the exchanging of ideas and animals (mostly mutant strains) in a very informal way. During this period, research projects were run on a small scale and were carried out independently. With the advent of molecular techniques for mouse genetics, and in particular after the development of in vitro transgenic techniques that made it possible to manipulate the genome almost at will, the situation changed dramatically. The number of scientists working with mice increased abruptly, probably because the community realized that engineering the mouse genome was perhaps the most efficient way to study gene function and to generate animal models for human pathologies. During the same period, several large integrated and concerted projects, sometimes on an international scale, were undertaken, and these resulted in an enormous increase in our knowledge of the species. These projects included the sequencing of the mouse genome, the production of thousands of new mutations with chemicals or by gene trapping, the accurate and systematic phenotyping of many inbred strains, the development of tools aimed at a better analysis of complex traits, as well as a few other projects, all of which have opened a new era in mouse genetics. There is no reason to believe that, after such a boom, the situation will enter into a recession. Another radical change was that an enormous amount of new information was made available to the community immediately after being gathered through a network of databases that were easily accessible on the Web and free of charge (see Table 1 for a listing of some of these databases). Indeed, the last 15 years have been crucial for the development of mouse genetics, and they have certainly paved the way for a few decades to come.

SEQUENCING OF THE MOUSE GENOME AND ITS CONSEQUENCES

The recent publication of the nearly complete mouse genome sequence (Waterston et al. 2002) can be regarded as a major event for two main reasons. First, the availability of this sequence has provided direct access to the blueprint of a living creature that is relatively close to our own species. This has allowed the identification of similarities and differences between humans and mice, from which it is possible to gather information about genome evolution and gene function at the molecular level. Second, easy access to the sequence of the mouse genome assists scientists in designing more efficient genetic alterations in embryonic stem (ES) cells. In addition to these two points, which were expected and even listed among the arguments in favor of sequencing, the accessibility of a nearly complete and reliable sequence of the mouse genome has had several other important consequences for geneticists by providing them with an enormous number of new polymorphisms.

Table 1. Some useful Web sites

Web site description	Web address
Mouse genome database MGD[c]	http://www.informatics.jax.org/
Mouse genome sequence browsers	
Ensembl mouse genome server	http://www.ensembl.org/Mus_musculus/
NCBI mouse genome resource	http://www.ncbi.nih.gov/genome/guide/mouse/
UCSC mouse genome browser gateway	http://genome.ucsc.edu/cgi-bin/hgGateway?org=mouse
Gene expression databases	
Mouse Genome Informatics—gene expression	http://www.informatics.jax.org/menus/expression_menu.shtml
Edinburgh Mouse Atlas Project (emap)	http://genex.hgu.mrc.ac.uk
The RIKEN FANTOM2 cDNA resource	http://fantom2.gsc.riken.go.jp/
ENU mutagenesis programs	
Baylor College	http://www.mouse-genome.bcm.tmc.edu/ENU/ENUHome.asp
German ENU Mouse Mutagenesis Screen Project	http://www.gsf.de/isg/groups/enu-mouse.html
MRC Harwell	http://www.mut.har.mrc.ac.uk/
The Jackson Laboratory (neurological mutations)	http://www.jax.org/nmf/
The Jackson Laboratory (heart, lung, blood, and sleep disorders)	http://pga.jax.org/
Mouse Mutagenesis Program Core Facility	http://mmp.sinica.edu.tw/mmp/english/
McLaughlin Research Institute	http://www.montana.edu/wwwmri/index.html
Northwestern University	http://www.genome.northwestern.edu/
The RIKEN Institute	http://www.gsc.riken.go.jp/Mouse/main.htm
Tennessee Mouse Genome Consortium	http://www.tnmouse.org/
The Toronto Centre for Modeling Human Disease	http://www.cmhd.ca/

(*Continued*)

Table 1. (Continued)

Web site description	Web address
Gene-trap databases	
Bay Genomics gene trap resource	http://baygenomics.ucsf.edu/
Centre for Modeling Human Diseases gene trap core	http://www.cmhd.ca/genetrap/index.html
Frontier in BioSciences	http://www.bioscience.org/knockout/knochome.htm
German GeneTrap Consortium	http://www.genetrap.de
The Jackson Laboratory (Induced Mutant Resource)	http://www.jax.org/imr/index.html
Lexicon Genetics	http://omnibank.lexgen.com/blast_frame.htm
Mammalian Functional Genomics Centre	http://www.escells.ca
SNP resources	
SNPview	http://snp.gnf.org/
Whitehead Institute mouse SNP data	http://www.genome.wi.mit.edu/snp/mouse
Roche mouse SNP database	http://mouseSNP.roche.com
Phenotyping resources	
Cornell University Phenotyping Core Facility	http://www.med.cornell.edu/research/cores/gem/
EUMORPHIA	http://www.eumorphia.org/
Mouse Metabolic Phenotyping Center	http://www.mmpc.org/index.htm
The Jackson Laboratory phenotyping service	http://jaxmice.jax.org/services/phenotyping.html
Monterotondo Phenotyping Service	http://www.db.embl.de/jss/EmblGroupsMR/g_271.html
Toronto Centre for Phenogenomics	http://www.phenogenomics.ca/services/phenotyping.html
The Mouse Phenome Database	http://aretha.jax.org/pub-cgi/phenome/mpdcgi?rtn=docs/home

[a]Bult et al. (2004), Hill et al. (2004), and Eppig et al. (2005) are helpful for browsing the Mouse Genome Databases efficiently.

The first drafts of the mouse sequence, released shortly after the turn of this century, were of excellent quality with error rates lower than 10^{-4}. However, they contained several gaps that were due to inherent limitations in the sequencing protocol (whole-genome shotgun) (Waterston et al. 2002) and to technical difficulties encountered for certain genomic segments, such as duplications or highly repeated regions (Bailey et al. 2004). Many of these imperfections are now being corrected, and the latest assembly released by the Mouse Genome Sequencing Consortium (NCBI build 34) has a length of 2.6 Gb, of which about 1.9 Gb (73%) is finished with less than one sequencing error per 10^{-5} base pairs. A few chromosomes (Chr 2, 4, 11, and X) are entirely sequenced, allowing comparisons with homologous regions of the human genome to be performed at a very high resolution. Such comparisons, revealing similarities and differences, potentially are a rich source of information. Similarities, for example, allow us to detect regions that are under selective pressure (genomicists sometimes say "purifying selection") and have remained unchanged or nearly so for several millions of years because they are genetically important and, accordingly, have resisted random drift. Differences at the sequence level may be even more interesting a priori, because they may contain keys explaining how speciation proceeds.

Mouse and Human Genomes Are Very Similar

When comparing the mouse genomic sequence with that of human, the overall impression is one of similarity (Pennacchio 2003). NCBI build 34 indicates that the mouse sequence is about 14% shorter than the human sequence, but a number of segmental duplications, repeats, and nonalignments are still being analyzed, which, when completed, may make the difference less noticeable (Cheung et al. 2003; Fitzgerald and Bateman 2004). In the same way, the genome that has been chosen for sequencing is the one of the C57BL/6 inbred strain, which is the most widely used strain, but it would not be surprising to discover in the end that different inbred strains in fact have a slightly different number of genes. Sequence comparisons at high resolution, made by matching orthologous segments spanning a few tens of base pairs from each genome and scoring the number of nucleotide mismatches at variable stringency, reveal that, on average, 40% of the mouse sequence can be aligned to the human sequence (Waterston et al. 2002; Schwartz et al. 2003). The stringency of sequence conservation is, however, unevenly distributed (Yap and Patcher 2004). Coding sequences appear to be highly conserved, but the degree of conservation varies depending on the function of the protein. This peculiarity, which is associated with our knowledge of the genetic code, has been exploited by informatics experts to validate their estimations of the coding fractions of the mouse and human genomes. They concluded that the number of exons in both species is almost equal (245,200). They also postulated the existence of a very similar number of protein-coding genes (~25,000 in the mouse and ~24,200 in human). This is much less than expected a few years ago, but one can trust that these numbers are close to reality since, after major improvements in the computerized analysis of genome sequences (e.g., after detection of the exonic sequences and validation of the actual coding potential of a sample, by RT-PCR amplification in a panel of RNA libraries), these estimations seem to have stabilized (Guigó et al. 2003; Parra et al. 2003).

Another interesting observation is that about 90% of the mouse and human genomes can be partitioned into regions of conserved synteny, reflecting the structural organization of the chromosome in the common ancestor. In fact, the two genomes share about 350 segments of conserved synteny, whose sizes range from 300 kb to 65 Mb, with a mean of 7 Mb. About 99% of mouse genes have a homolog in the human genome, and for 80% of these genes, the best match in the human genome has, in turn, its best match against the orthologous mouse gene in the conserved syntenic interval. This one-to-one ratio allows us to define a set of genes that are mammal-specific: a basic kit of genes for creating a basic mammal! Of course, the precise delineation of this set of genes requires that comparisons are made with several other mammalian genome sequences as they become available (rat, dog, pig, cattle, macaque, chimpanzee, etc.).

At high stringency, the percentage of conservation between the human and mouse sequences is close to 5%, which indicates that other sequences, in addition to those encoding proteins and representing roughly 1.5% of the genome, are under selective pressure. The function(s) of these conserved noncoding sequences (CNSs) is the subject of intense research at the moment, and there is little doubt that CNSs will keep geneticists busy for another few years. It makes sense to speculate that many of these sequences are essential for controlling the correct spatial and temporal expression of genes (regulatory elements). Other CNSs are known to encode a variety of essential nontranslated RNAs. Some of these RNAs, such as tRNAs (about 350 units), ribosomal RNAs, and microRNAs, have relatively well-known functions. The functions of many others are totally unknown. Finally, some CNSs probably play an important role in the organization of chromosome structure, which is inherited, and also in the determination of imprinting. The regions with the highest levels of sequence conservation are found in certain domains of genes that encode proteins playing an important role in the patterning of development (e.g., the *Hox* and *Pax* series). They are also found in some CNSs whose function is yet unknown. Obviously, a precise inventory of these sequences, as well as knowledge regarding their structure and function, will be critical for making comparisons with other species and for engineering specific alterations in vitro in embryonic stem (ES) cells.

The discovery of all these similarities is not so surprising if we consider the relatively short evolutionary distance between the mouse and human species (75 ± 15 Myr). This is important to know because the similarities can be used for making predictions in one species in the regions where the sequence is incomplete or less reliable in the other. They can also be used for making comparisons with a third species (e.g., rat or chimp) when the sequence is complete and reliable but different in human and mouse. This would be especially helpful for scientists whose aim is to decide whether a gene present in one species and absent in the other results from the addition to or deletion from the ancestral chromosome. Finally, similarities at the sequence level will be important to consider for the analysis of other mammalian genomes whose sequences will remain a draft.

Human and Mouse Genomes Exhibit Some Interesting Differences

Comparisons concerning the number of genes in the two genomes need to be interpreted with care because they do not take into account the existence of, for example, nonprocessed pseudogenes (D'Errico et al. 2004). These genes, which originate from a common ancestor either by gene duplication or, less frequently, by unequal crossing over, are more frequent in the mouse genome than in the human genome, and some are functional while others are not (Cheung et al. 2003; D'Errico et al. 2004). A well-known example involves the gene(s) coding for the hormone renin (*Ren1* and *Ren2*). Some mouse strains (and humans) have only one copy of the gene (*Ren1*, coding for renin, primarily in the kidney and, at a very low level, in the submaxillary glands), while other mouse strains have an extra gene (*Ren2*, which is closely linked to *Ren1* and encodes an isoform of renin in submaxillary glands only). These two genes originated by tandem duplication of an ancestral gene (Panthier et al. 1984). *Ren1* is essential, since its invalidation by knockout has deleterious effects, but *Ren2* is not. The two genes, however, have different promoters.

A better example to illustrate the complexity in this matter is the case of the *OAS* (*oligoadenylate synthetase*) gene cluster, whose function is important in the innate mechanisms of defense against viral infections

(Mashimo et al. 2002; Perelygin et al. 2002). In the human, it is a cluster of three genes on chromosome 12, designated respectively as *OAS2*, *OAS3*, and *OAS1*. Human *OAS2* and *OAS3* correspond to mouse orthologs *Oas2* and *Oas3* (Chr 5), and the transcription products of these genes are very similar, with two alternatively spliced isoforms encoded in both mouse *Oas2* and human *OAS2* and only one transcript from human *OAS3* and mouse *Oas3*. However, the structural organization of *OAS1* is very different in the two species, with only one *OAS1* gene encoding four different OAS proteins (p42, p44, p46, and p48) in human and no less than eight transcription units in the mouse, which are all orthologous of the *OAS1* gene and arranged in tandem with the following order: *Oas1e*, *Oas1c*, *Oas1b*, *Oas1f*, *Oas1h*, *Oas1g*, *Oas1a*, and *Oas1d*. For all of these eight genes, a specific interferon-inducible promoter regulates the transcription of a single product, except for *Oas1a*, where there are two alternatively spliced transcripts (Mashimo et al. 2003). It is likely that a few of these eight genes are nonfunctional pseudogenes that have been generated by duplication of the ancestral *OAS1* gene; others, however, look like bona fide genes. We know that a mutation in *Oas1b* that leads to gene invalidation is correlated with an extreme susceptibility of the affected mice to flavivirus infections (Mashimo et al. 2002; Perelygin et al. 2002). We also know that mice homozygous for a knockout allele of *Oas1d* ($Oas1d^{-/-}$) display reduced fertility because of defects in ovarian follicle development (Yan et al. 2005). These two phenotypes are apparently totally unrelated, and the situation is far from being clear.

Situations where the mouse genome harbors genes that are orthologous to human genes but variable in terms of copy number are common. Genes encoding olfactory or taste receptors are such an example; they are arranged in clusters and are at least three times more numerous in mouse than in human. Similar observations have also been made for genes encoding proteins with an immunological function and genes encoding proteins involved in the metabolism of drugs. It has been suggested that these variations resulted from different selective environmental pressures experienced by the ancestors of modern rodents and primates that contributed to "genome shaping" (Godfrey et al. 2004).

If comparisons of the mouse and human genomes allow for detecting such examples of "gene birth," they also allow for detecting cases of "gene death." On human chromosome 4, for example, the gene encoding Interleukin 8 (*IL8*) has no ortholog in the homologous segment of mouse Chr 5 (Fig. 1). IL8 is a chemokine that is secreted by several cell types and is one of the major mediators of the inflammatory response. Surprisingly, although a mouse ortholog of human *IL8* is not detectable, *Il8rb*, which

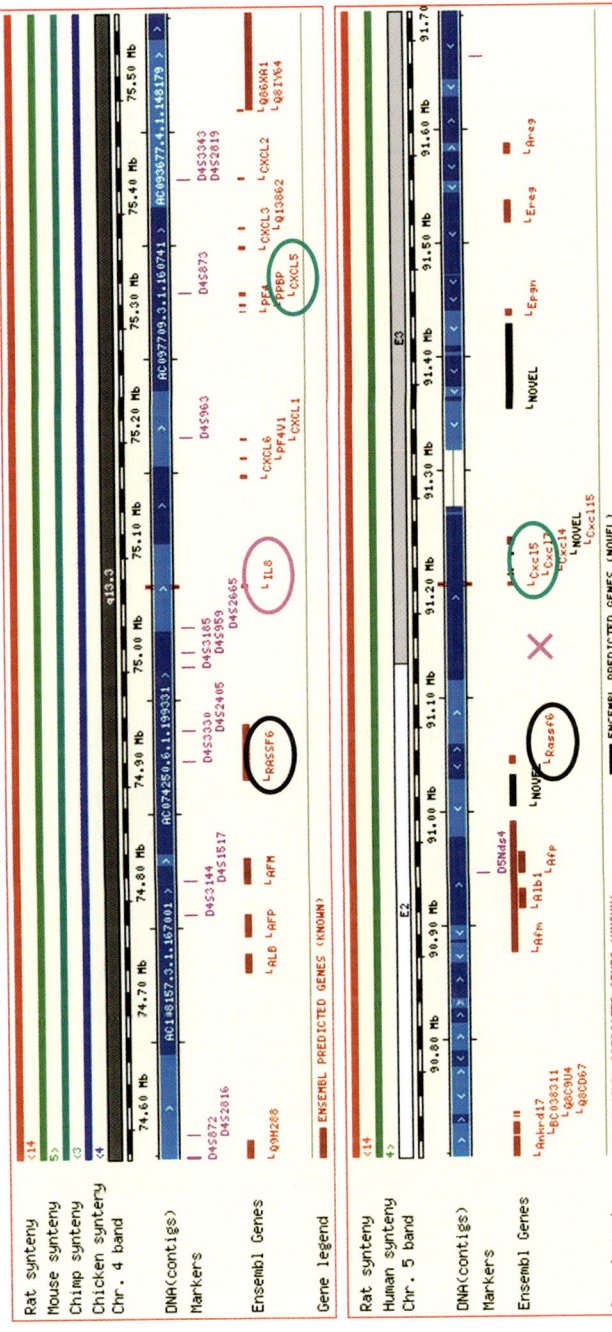

Figure 1. About 90% of the mouse and human genomes can be partitioned into regions of conserved synteny with similar linear arrangements of the genes. Sometimes, however, some segments are either deleted or duplicated. In this case, the gene coding for Interleukin 8 (*IL8*) is deleted from the mouse genome while the human ortholog (*IL8*) is still present.

is an ortholog of the gene coding for human *IL8* receptor, has been cloned and mapped on mouse Chr 1, and the protein encoded by this *Il8rb* gene has a very strong similarity to human IL8. For the time being, immunologists have no answer to this puzzling situation. Maybe the mouse uses another gene to achieve the same function or, maybe, the ortholog of human *IL8* is transposed elsewhere in the mouse genome. The second explanation seems unlikely because there is no ortholog of human *IL8* in the rat genome either. Other examples of gene death have also been reported (Fitzgerald and Bateman 2004).

Comparisons concerning the absolute number of genes in the mouse and human genomes must also be interpreted with care because of important differences in the RNA splicing mechanism. The physical entities that computer scientists detect at the sequence level and label "gene" may encode a different number of proteins in the two species by arranging their set of exons in the transcription products differently. However, what is more important for the biologist is the set of cDNAs (encoding proteins or not) rather than a mere inventory of the genes. To document this point, an international consortium organized and led by the RIKEN Institute in Yokohama, Japan, has undertaken the task of establishing a comprehensive inventory of the cDNA collection encoded in the mouse genome, including functional annotation for each component (Hayashizaki 2003a,b; Carninci et al. 2005). "Deciphering the logic of the transcriptome" is the ultimate goal of the consortium, and the database is known as FANTOM (an acronym for functional annotation of the mouse genome). FANTOM has had three successive releases: FANTOM1, FANTOM2, and FANTOM3. *Genome Research* published a special issue about FANTOM2 (Vol. 13(6b), June 2003), in which a library gathering a total of 60,770 full-length cDNA clones prepared from 263 mouse tissues that were collected at different stages of development was thoroughly analyzed by the consortium (Okazaki et al. 2002; Kasukawa et al. 2003). This considerable amount of work led to the identification of regions that are potential promoters for 8637 known genes, with an estimation of 63,000 putative transcriptional starting points (Dike et al. 2004). FANTOM3, an even more advanced generation of the database, is now in progress; it aims to identify full-length cDNAs in a total of 103,000 clones (Carninci et al. 2005). All of these databases are regularly updated, are publicly available, and are freely accessible.

Other differences between the mouse and human genomes are worth considering in the frame of this review. Among these, an ensemble of repeated elements known as long interspersed elements (LINES) and short interspersed elements (SINES), as well as the long terminal repeats (LTRs) of retroviruses, are probably the most important or are, at least, the best known (McCarthy and McDonald 2004). These sequences represent about 38% of the total amount of mouse genomic DNA and are active retrotransposons. Some elements move in the genome; they increase in number as they transpose and are occasionally lost in blocks. Genomicists believe that the difference in size between the mouse and human genomes (~0.3 Gb) might be due, at least in part, to a more efficient mechanism of "transposon cleaning" in the mouse. In fact, transposition of these elements is much more active in the mouse than in the human genome, where transposons have been found to be responsible for about 10% of the spontaneous mutations when they insert inside genes and interfere with the splicing mechanics. The role of these elements as mutagenic agents is quite clear. However, in addition to this role, some scientists in the community believe that transposons may also assume regulatory functions by altering the expression of neighboring genes after transposition (Allen et al. 2004; Medstrand et al. 2005). This point, however, is not yet universally accepted and should be the matter of future investigations. Another amazing peculiarity of these elements is their very heterogeneous distribution in the genomes of both human and mouse, with some homologous regions being totally preserved and others "overcrowded." This is very puzzling for elements that are, by definition, mobile.

Single-Nucleotide Polymorphisms: An Unmatched Wealth of Polymorphisms

While the genome of the mouse was being sequenced, genomicists discovered that, when homologous regions originating from different laboratory inbred strains were aligned and compared, base-pair mismatches were rather common yet unevenly distributed. Comparing segments of the C57BL/6 inbred strain (the reference strain) with a panel of other inbred strains yielded either high (≥ 40 mismatches per 10 kb) or low (~0.5 mismatch per 10 kb) rates of polymorphism with an abrupt delineation between the segments (Wade et al. 2002; Wiltshire et al. 2003). In all strain-to-strain comparisons examined, about one third of the genome appeared to be composed of long regions (≥ 1 Mb) containing high single-nucleotide polymorphism (SNP) rates, while the rest of it exhibited low SNP rates (Fig. 2).

Another important observation was that the distribution of high/low SNP rates is unique to a particular pair of strains, when considered on a genome-wide scale. This unexpected heterogeneity found a logical explanation when matching it to historical records about the origin of laboratory-inbred strains and the way these strains were developed during the 20th century. As hypothesized by geneticists, including ourselves in the late

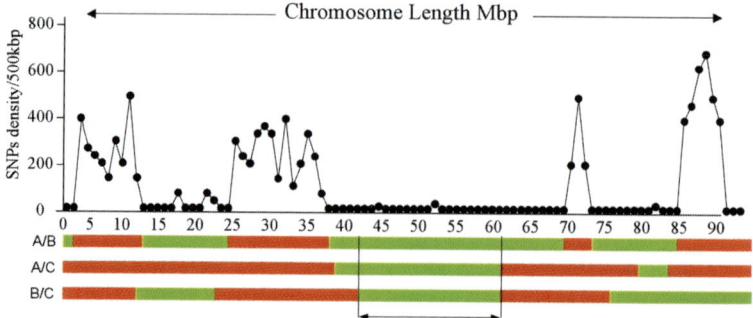

Figure 2. The SNP density between any two inbred strains of mice varies according to the chromosomal region concerned and changes abruptly when passing from one region to the next. This observation is in agreement with historical data on the origin of laboratory inbred strains indicating that they are derived from a small pool of wild ancestors belonging to different subspecies of the genus *Mus*. Because of this polyphyletic origin, the mouse genome can be regarded as a mosaic of chromosomal segments of various sizes. When the SNP density is low, the segments in question share the same ancestral origin and the few observed SNPs are those resulting from recent mutations. When the SNP density is high, on the contrary, the chromosomal segments have a different origin. When three strains are compared, as on the diagram represented here, one can perfectly observe that three homologous segments have a high SNP density on pairwise comparisons, if all three of them have an independent origin stemming, for example, in three different subspecies of the genus *Mus*. When a particular region with low SNP density cosegregates with a particular phenotype, the region in question may harbor the genetic determinants for the phenotype in question. Redrawn with permission from *PNAS* © 2003, Wiltshire et al. (2003).

1980s, inbred strains have a polyphyletic origin stemming from three subspecies of the genus *Mus*: *Mus m. domesticus* (the occidental wild mice), *Mus m. musculus* (the oriental wild mice), and *Mus m. castaneus* (the Asiatic wild mice) (Bonhomme et al. 1987). The genome of a given inbred strain is a unique mosaic, with variable proportions of these three components but with the vast majority of segments deriving from either *M. m. domesticus* or *M. m. musculus* (Fig. 3). This is reflected in the genome-wide distribution of SNPs, even if a small proportion of these polymorphisms are of recent origin (mutations). The discovery of these SNPs, which are extremely abundant and very easy to characterize (dense genotyping can be processed automatically), has important implications for the development of several aspects of mouse genetics, including the establishment of pedigrees, the cloning of genes or quantitative trait loci (QTLs), and the comparisons of aligned sequences (Lindblad-Toh et al. 2000; Ideraabdullah et al. 2004).

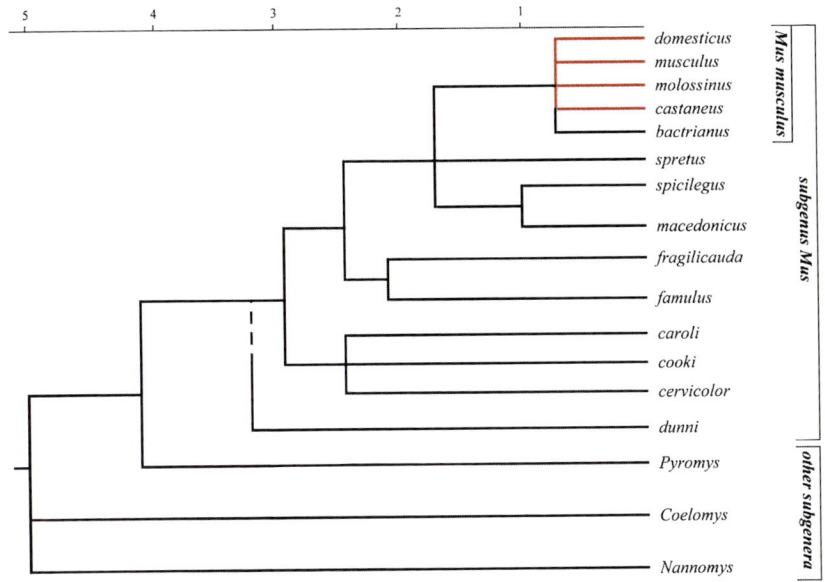

Figure 3. Evolutionary tree of the genus *Mus* (the time scale is in millions of years, Myrs). The branch leading to the subgenus *Mus* encompasses all of the species with a basic karyotype of 2N = 40 acrocentric chromosomes. Today's classical laboratory strains are recombinant strains derived (in unequal percentages) from three parental components: *Mus musculus domesticus*, *Mus musculus musculus*, and *Mus musculus castaneus*. Hybrids between mice of the *Mus musculus* complex are viable and most are also fertile in both sexes. Hybrids of the *Mus musculus* complex with *Mus spretus* or *Mus spicilegus* are viable but male sterile. Redrawn with permission from Elsevier © 2003, Guénet and Bonhomme (2003).

Where Do We Go Now?

The decision to systematically and comprehensively sequence the mouse genome was certainly a wise decision. There is no doubt that the genome of this important species would have been eventually sequenced one day, but this would probably have been achieved by small "bits," with many redundancies and, accordingly, at a greater cost for the community. This was also a very democratic decision because laboratories that did not have easy access to sequencing facilities, for whatever reason, can now use this public resource for designing their experiments.

The future of mouse genomics/genetics in the after-sequencing era is relatively clear. The sequence will progressively become complete and precise, and as I have already mentioned, this will aid geneticists who are making interspecific comparisons at high resolution. We will then have another way to understand the forces that shape the genomes. Comparing specific CNSs in a wider panel of species (rat, dog, pig, cattle, *Maccaca* monkeys, chimpanzee), and matching the results with tissue-specific transcriptomes, will be especially helpful for understanding gene regulation. The design of genetic alterations to be performed in ES cells will also be greatly facilitated by a precise knowledge of the genome sequence. Finally, it seems clear that the genome of several other mouse inbred strains will be sequenced, if not entirely, at least by 0.2- to 1-Mbp nonoverlapping and evenly distributed segments, to enrich the SNPs collection. Comparisons of the SNPs distribution across the genome may be relevant to understanding common polymorphisms, especially those causing strain phenotypic variations and diseases.

GENERATING MANY, MANY NEW MUTATIONS IN THE MOUSE GENOME

Mutations have been collected by mouse fanciers since well before the study of genetics began, and these mutations have been instrumental for the expansion of the discipline. Mouse mutations have also provided, and continue to provide, helpful models to human geneticists. In the postsequencing era, mutations will still be very useful, contributing efficiently to gene annotation by allowing matching a specific alteration in the DNA sequence to a particular mutant phenotype. Indeed, when a gene becomes nonfunctional after a mutation has occurred, the careful comparison of the mutant phenotype with the normal one, considered together with the molecular defect generated by the mutational event and the spatiotemporal expression of the gene, is an excellent way for assessing the function(s) of the gene in question.

Spontaneous mutations unfortunately have a few major drawbacks. First, they occur at very low frequency (on the average 5×10^{-6} per locus and per gamete) and the number of genotype stored in the repositories of the community, although quite large (around 1200 mutant alleles), was considered by researchers as largely insufficient compared with the expected number of genes in the mouse genome. For example, among the many mutations reported in human that lead to a pathological condition, few have an orthologous counterpart in the mouse and vice versa. This is either because the mutant allele in question never occurred or because it occurred but was not recognized as such and accordingly got lost. To give another example of the shortage of mutant alleles in the mouse, we found that most (over 60%) of the mutations that were discovered spontaneously in our laboratory in the decades 1970 to 2000 turned out to be mutations in a gene that previously had no other alleles rather than new alleles at a locus where mutations already occurred. This clearly indicates that the mouse genome is far from being saturated by mutations, a situation that British colleagues defined as a "phenotype gap." Another major drawback is that the collection of mutations available corresponds mainly to alleles that are viable ab utero, even if only for a few hours, and have a clear and easily identifiable phenotype. Lethal alleles and those with a weak phenotype or a phenotype with a very late onset (e.g., a hypothetical and eagerly awaited model of Alzheimer's disease) were in general not detected, although they probably represent quite a substantial proportion of all mutations.

To improve the situation, several projects were launched between 1998 and 2005, in different countries, with the aim to generate large amounts of new genetic alterations leading ideally to the production of at least one mutation per gene. Most of these ambitious projects have been extensively described (Durick et al. 1999; Hrabe de Angelis et al. 2000; Nolan et al. 2000, 2002; Nadeau et al. 2001; Hansen et al. 2003; Austin et al. 2004; Auwerx et al. 2004; Clark et al. 2004; Masuya et al. 2004; Schnutgen et al. 2005); we will only briefly outline them here.

Inducing Mutations with Chemicals

The first initiatives were making use of the powerful chemical mutagen ethyl-nitroso-urea (ENU). ENU induces point mutations (mostly base pair substitutions) and increases the spontaneous mutation rate by a factor of 130 to 160 depending on the experimental conditions (Guénet 2004). Now, at least twelve ENU projects are in progress over the five continents and thousands of mice, offspring of mutagenized progenitors and potentially affected by mutations (dominant or recessive), are carefully analyzed each year for a large number of phenotypic criteria. If we consider the progress reports released by some of the consortia involved in ENU mutagenesis, the strategy appears to be extremely rewarding and many new mutations have been generated this way (Hrabe de Angelis et al. 2000; Nolan et al. 2000; Rathkolb et al. 2000; Clark et al. 2004; Masuya et al. 2004; Hoebe and Beutler 2005; Wilson et al. 2005). These mutations, once identified, are, in general, roughly described phenotypically and then their description is posted on specific websites announcing to the community the availability of new mutant genotypes. Finally, deep-frozen sperm cells are stored to ensure long-term preservation of the genotypes.

The advantages of the strategy making use of ENU are a relatively high efficiency and, mostly, the fact that ENU generates a great variety of mutations (i.e., hypomorphs, hypermorphs, neomorphs, etc.) and not only null alleles. A major drawback of the strategy is that one cannot target the occurrence of mutation hits at a specific site. In fact, mutations occur randomly and, accordingly, characterization of the molecular defect requires the use of the tedious (and expensive) process of positional cloning (forward genetics). Another drawback is that each and every cell of a mouse embryo originating from a mutant gamete (a spermatozoon, in most instances) is affected by the mutation. This may have negative consequences on the survival of the embryo when, for example, the mutant allele impedes normal development for one reason or another. For example, a mutation producing a cleft palate, among other phenotypes, would remain undetected because a baby mouse with such a minor malformation cannot suckle and would die, even if the other tissues of the organism are only weakly affected by the mutation.

The first and most important of these drawbacks has been circumvented by using a clever strategy that consists of the induction of mutations by chemical mutagenesis as usual, and then storing independently DNA samples and spermatozoa from a large number of male G1 offspring of a mutagenized G0 male mouse. In a second step, mutations in a given gene of interest are sought among the collection of G1 DNA samples, using powerful PCR-based high-throughput methods to detect DNA strands mismatches in the targeted region (Coghill et al. 2002; Zan et al. 2003; Quwailid et al. 2004). This region can be a coding sequence or any other sort of sequence of interest, like those coding for RNAs or even the CNSs discussed earlier. INGENIUM, for example, claims that by scoring an archive of some 16,000 G1 male mice (DNAs) they are able to identify over 10 potentially interesting mutations per average-sized gene, and this within only one week. The alleles available in the archive can be evaluated and scored by in silico or in vitro assays for structural or biochemical effects. In a final step, the selected alleles are revitalized from thawed spermatozoa by in vitro fertilization. A resource encompassing over 300,000 unique gene-specific alterations is now available on the C3HeB/FeJ background strain (Augustin et al. 2005). Compared with others, the strategy is fast, dependable, and relatively cheap. In addition, all sorts of mutations are produced, allowing the investigators to choose in an allelic series.

Chemical mutagenesis and radiations have also been used for the massive induction of mutations in vitro in ES cells (Goodwin et al. 2001; Munroe et al. 2004). Here again, the detection of mutant alleles can be performed at the DNA level and the relevant ES cells can then be selected and used for producing heterozygous animals or stored deep frozen for further use. Some of the chemicals and most of the radiations that were used in these experiments produced deletions of various sizes. These deletions, provided they are viable in the heterozygous state, are very precious tools for a detailed analysis of some specific parts of the genome (Rinchik et al. 2002).

Inducing Point Mutations by Genetic Engineering in ES Cells

As mentioned earlier, chemical mutagenesis is "blind." This is, at the same time, a blessing and a curse—a blessing because, with this technique of mutagenesis, all sorts of mutant alleles are produced, but a curse because the ambitious aim of producing one chemically induced mutation per gene would certainly be extremely costly if not unattainable. This intrinsic limitation is probably worsened if, as suspected, we consider that chemical mutagens are not equally active on all parts of the genome, leaving some regions less affected than others. Alternatives to chemical mutagenesis have been suggested to complete the "one-mutation-per-gene" challenge. Among these alternatives, the one that consists of making knockouts in vitro, in ES cells, was the first that came to mind.

Engineering knockouts in ES cells is a reverse genetics approach in the sense that one first selects a gene of interest, then makes a specific mutation in it, and finally observes the phenotype of the mutant mouse, if any. To date, it is estimated that knockout alleles potentially exist somewhere throughout the world for only about 10% of mouse genes, but unfortunately many of these knockouts are limited in utility because they have not been phenotyped in standardized ways or are not freely available. However, with the genomic sequence now available, and accordingly the vast majority of genes identified, one could, at least in theory, decide to systematically knock out all the genes in the mouse genome. In addition, considering the wide arsenal of techniques the geneticists now have at their disposition, they could also decide to make all these knockouts conditional and/or tissue specific. To this end, large-scale efforts have recently been launched in Europe (the European Conditional Mouse Mutagenesis Program or EUCOMM project, Auwerx et al. 2004) and in the USA (the knockout mouse project or KOMP project, Austin et al. 2004), and it is likely that other such projects will also be undertaken in other countries with the aim to produce a comprehensive collection of mouse ES cells heterozygous, in some cases homozygous, for a conditional null mutant allele in every gene in the genome. All the mouse strains generated in these projects would then be phenotyped using a battery of sophisticated and standardized tests, including transcriptome-based phenotyping (microarrays), and will finally be made freely available to the community (Schnutgen et al. 2005).

Producing knockout mutations in ES cells with the classical methods is very reliable but rather cumbersome and not easily amenable to high throughput. Gene trapping is an alternative to gene targeting with some interesting advantages. Like homologous recombination, it is performed in ES cells, but, unlike homologous recombination, it is a random approach that produces, with high throughput, a large number of insertional mutations across the mouse genome (Durick et al. 1999; Stanford et al. 2001; Forrai and Robb 2005). Gene trapping disrupts genes by inserting an engineered DNA element (a promoterless reporter/selectable cassette flanked by an upstream 3' splice acceptor and a downstream adenylation sequence). When the DNA vector inserts in an intron of an endogenous gene and the gene is expressed, a fusion mRNA is transcribed to produce a nonfunctional version of the cellular protein fused to the reporter/ selectable marker. The gene trap strategy has then three advantages over other techniques: (1) it inactivates the genes randomly; (2) it reports the expression of the trapped gene when the latter is expressed; and (3) it provides a DNA tag for PCR identification of the disrupted gene. This approach has been used by a few laboratories, members of the former *Gene Trap Consortium*, to generate a public resource of roughly 36,000 characterized ES cells, harboring null mutations in approximately one third of the genes (10,000) (Hansen et al. 2003; Stryke et al. 2003; Skarnes et al. 2004). The resources generated by this consortium, although very abundant and very useful, suffer from a drawback that I have already mentioned: They do not produce conditional alleles and, accordingly, the analysis of mutants generated from these resources is limited to the earliest developmental functions of the trapped gene. The aim of the EUCOMM consortium is to generate, in the five years to come, a library of up to 20,000 different engineered ES cell strains, with roughly 12,000 of these cells being randomly gene trapped (Adams et al. 2004; Austin et al. 2004; Schnutgen et al. 2005). This initiative is extremely interesting because, unlike the former program and thanks to a clever design of the trapping vector, the new mutations will be conditional. The mutations in the new library will enable analysis of gene function(s) in a temporally and spatially restricted manner and will be either in the germ line or in somatic cells/ tissues, depending on the deletion system used. In addition to the massive random production of inactivated genes by gene trapping, some other genes (8000), selected a priori for their relevance to human pathology and that may have been missed by the gene trapping project, will be inactivated by gene targeting using a new kind of insertional targeting vector with high targeting efficiency (Adams et al. 2004). Here again, the mutant mice generated will be thoroughly phenotyped, archived, and made available to the public. Using the two strategies is interesting because they are really complementary. Gene trapping is fast and very efficient but, as mentioned above, those genes that are not expressed in vitro, in ES cells, would not be "trappable," but targeting could inactivate all these genes.

Based on what we described in the previous lines, the collection of mouse mutations may turn plethoric in the forthcoming years. If we add the number of chemically induced mutations, in vivo or in vitro, to those resulting from the complementary efforts of the "gene trappers," we can seriously expect a vast majority of the mouse genes to have at least one mutant allele (in most instances a knockout) in the next five years. Mouse geneticists may then feel like they are in Wonderland!

Just like for the sequencing of the genome, the decisions to provide funding for the projects aimed at systematic mutagenesis of the mouse genome were also wise decisions. These projects will undoubtedly contribute to a greater efficiency in modern genome research and, in the end, they will also save a lot of public funding that could be put on other important themes. Many laboratories with very little or no experience in ES cells technology will certainly appreciate being able to order a conditional knockout allele of their favorite gene.

Unfortunately Wonderland does not exist, and many genes will escape chemical mutagenesis or gene trapping. The efficiency of chemical mutagenesis depends, in part, on the size of the gene and, for this reason, the very small-sized genes will be difficult to hit. In the same way, those genes with only one exon, which are so far mostly unknown, won't be "trappable." Mutant alleles with a cellular dominant lethal effect will be extremely difficult, not to say impossible, to analyze. Another important point is that, while only one null allele can be induced per gene, an infinite number of alleles resulting from missense mutations are possible. Replacing any base pair with one of the other three may, in some cases, lead to amino acid substitution with totally unpredictable consequences in the protein. We found, for example, that the mouse mutation *pmn*, which is a Trp524Gly substitution in the sequence of the chaperone molecule TBCE (Tubulin folding co-factor E), resulted in a severe neurological syndrome with relatively late onset, whereas a 2-bp deletion, which probably inactivates the orthologous human gene, leads to a totally different pathology (Kenny Caffey syndrome OMIM 244460 or Sanjad-Sakati syndrome; OMIM 241410) (Martin et al. 2002). It is presumably because the mutant protein is unstable in the mouse and absent in humans that the phenotype is so different. In this case at least, analysis of a null allele of the *Tbce* gene would have been insufficient for the complete annotation of the functions of this gene. This clearly indicates that spontaneous mutations with interesting and unique phenotypes will always be interesting to consider for positional cloning. Finally, the mutations that can be induced in the CNSs are innumerable, and so far, we have no way of predicting their effects and very few strategies for generating them with high throughput.

Engineering Chromosomal Rearrangements in Mice

As discussed above, the modern techniques of ES cell technology have revolutionized our ability to produce mouse genomic alterations and point mutations in particular. They also have permitted us to produce a virtually unlimited number of chromosomal rearrangements that will be very useful either as tools for mouse geneticists or, maybe more importantly, for modeling human diseases. Using the Cre/*loxP* site-specific recombination system and the knowledge we now have of the mouse genome sequence, one can very precisely delete, duplicate, or invert chromosomal segments of various size, almost at will, just by inserting the appropriate number of *loxP* sequences at specific sites, in *cis* or in *trans*, and in the appropriate orientation (Yu and Bradley 2001). For the production of deletions there is a limit (in particular in size), which is related to the impossibility of putting some chromosomal segments in the haploid state. Deletions and nested deletions can be produced that are useful for generating resources for regional recessive genetic screens and for facilitating the functional analysis of the mouse genome (Nobrega et al. 2004). Deletions and duplications are useful for studying the effect of gene dosage on cell physiology because any gene in the mouse genome can be put in different copy numbers (Yan et al. 2004). Chromosomal engineering can also be used to generate duplications in mouse chromosomal regions conserved with the trisomic human chromosome 21 regions to identify the genetic domain(s) and causative gene(s) that are responsible for the clinical characteristics of the Down syndrome (Reeves et al. 2001). This strategy is an alternative to the one of targeted meiotic recombination (TAMERE) but seems to be more efficient (Herault et al. 1998; Olson et al. 2005). Inversions are useful because they are cross-over suppressors and allow the creation of balancer chromosomes (Kile et al. 2003).

Reciprocal translocations can also be engineered in ES cells and can be produced to develop mouse models for certain forms of human cancer resulting from the creation of novel fusion genes with cellular oncogenes.

THE PHENOTYPING PROGRAMS AND THE GENETIC ANALYSIS OF COMPLEX TRAITS: A CHALLENGE FOR THE FUTURE

The sequencing of the mouse genome and the massive production of new mutations have been, undoubtedly, the two most important projects of the last decade in mouse genetics, and I already have emphasized that these two projects were indeed complementary in the sense that, while sequencing identifies the genes through their DNA structure, mutations identify the genes through their function and contribute to gene annotation. The development of these two projects has triggered other initiatives and I would like to mention two other projects, which are also complementary and will probably mark an important change in the analysis of gene functions. The aim of the first project is to perform accurate and comprehensive phenotypings of the different inbred and mutant strains. The aim of the second is to develop new tools and strategies for the genetic analysis of heritable traits with a complex or multigenic determinism.

The Phenotyping Programs

The Mouse Phenome Project was created in May 1999, with the aim to establish a collection of baseline phenotypic data of the mouse inbred strains (Paigen and Eppig 2000). In a first step, the most commonly used inbred strains have been classified in four groups by a panel of specialists, and for each group, reliable phenotypic data are now being progressively collected and stored in a central, web-accessible database (the Mouse Phenome Database or MPD), housed at The Jackson Laboratory and cross-linked to the Mouse Genome Database. This large-scale collaborative project, mainly based on the free, willing participation of expert scientists in diverse fields of biomedical science, was another wise decision because any laboratory embarking on a new research project can now select, in only a couple of "clicks," the best strains to work with based on the existing collection of phenotyping data. As I will discuss further, the creation of this database will progressively become an essential resource for realizing the full utility of information that emerged from the sequencing of the mouse genome.

Since the foundation of the Phenome database, other initiatives with the same aim of accurate phenotyping have been or will be developed in several countries, in particular, in Europe and North America. EUMORPHIA, for example, is a consortium of 18 research institutes from across Europe that is developing a comprehensive, robust, and validated phenotyping platform in which mutant strains will be thoroughly scrutinized, using a broad range of state-of-the-art technologies for detecting even the most subtle phenotypic changes. The new screen, known as EMPReSS (European Mouse Phenotyping Resource for Standardized Screens), will incorporate more than 150 standard operating procedures, covering all of the main body systems (Auwerx et al. 2004; Brown et al. 2005). These mouse "phenotyping workshops" are, of course, logical corollaries of the projects aimed at the induction of new mutations and represent enormous progress for a better characterization of gene functions. After accurate and comprehensive phenotyping, it is likely that many knockouts that were generated in the last 20 years and so far have been classified as "normal" (i.e., with no detectable phenotype and accordingly useless for genome annotation) will be found to be "abnormal" on a second examination with more refined protocols. It would be even more interesting if, in the future, all these workshops or "clinics" could develop a network of exceptional expertise and work on an integrated basis to resolve the difficult cases. This would save a lot of money but would require that interesting genotypes be freely and easily exchanged worldwide, with these exchanges not being impeded by sanitary or legal issues. This point should be seriously considered by the community.

The Genetic Analysis of Complex Traits

Until now, and in the vast majority of cases, correlations between a particular phenotype (generally one with deleterious aspects) and its determinism at the genome level have been established only for monogenic traits. Positional cloning of a mouse mutation is a typical example in which a single Mendelian character determines a pathology and a causal relationship can be established by finding a correlated alteration in the sequence. Many such genes have been cloned and, presumably, many more will be cloned in the years to come, given that, after the sequencing of the genome, the strategy is greatly facilitated and, after the massive production of new alleles, many interesting phenotypes that were missing will be available. The problem however is that most of the pathologies are not "monogenic" but, on the contrary, are influenced by multiple genes with additive or synergistic effects. In the same way, most alterations that have been found to account for a deleterious phenotype in the mouse have been found to affect the coding regions of the mutated gene (base pair substitutions, deletions, insertions, splicing abnormalities, etc.), but mutations with an effect on the quantitative or spatiotemporal expression of a gene are not well known although they are probably quite common. Finally, many genes with a modifier effect, increasing, for example, the severity of a phenotype or making a certain inbred strain of mouse more or less susceptible to an infectious disease or a certain type of cancer, have been identified only exceptionally. In fact, in our analysis of the genotype/phenotype relationships so far, we have probably considered only the "tip of the iceberg," because we have had no tools suitable for assessing the genetic analysis of complex traits. As a consequence of the mouse genome sequencing effort, and in particular after the discovery of so many SNPs in the various inbred strains, and as a direct consequence of the development of better phenotyping strategies, one can expect the situation to change dramatically in the forthcoming years. An extremely dynamic consortium dealing with all these problems, the *Complex Trait Consortium* or CTC, was created in the spring of 2002, by a group of expert scientists who decided to identify research priorities and tools to tackle the problems related to quantitative inheritance in the mouse model. The goals of the consortium have been published in several scientific journals (Glazier et al. 2002; Abiola et al. 2003; Nishimura et al. 2003; Churchill et al. 2004; Pletcher and Wiltshire 2004; Singer et al. 2004). In short, they are preparing and collecting the tools that would allow relating specific haplotypes (or segments) of the mouse genome, identified by a particular set of SNPs, with a particular phenotype identified by a QTL (Petkov et al. 2004; Pletcher et al. 2004). Among the strategies suggested by the CTC, the most impressive and certainly the most innovative is the implementation of a resource known as *Collaborative Cross* (Churchill et al. 2004). The Collaborative Cross will consist of a total of 1000 recombinant inbred strains (RIS), each derived from an initial eight-way cross involving very different and unrelated inbred strains (Fig. 4). Theoretical computations indicate that the genome of each RIS in such a cross will capture ~135 unique recombination events (135,000 for the whole set of RIS) and each of these RIS will then have a unique genomic constitution representing a patchwork of 135 elements with, roughly, an equal proportion of the eight founder genotypes. Each strain of the Collaborative Cross will capture an abundance of polymorphisms every 100–200 bp that will be sufficient to drive phenotypic diversity in almost any trait of interest, provided it segregates among the eight parental strains. Finally, the very large number of RIS will guarantee high mapping resolution (achieved by SNPping) of any QTL segregating in at least two of the eight parental strains.

The two projects discussed above—the phenotyping programs and the analysis of complex traits by the partners of the CTC—are really innovative. Unlike the projects aimed at sequencing the mouse genome and those aimed at the production of new mutations, they have not yet

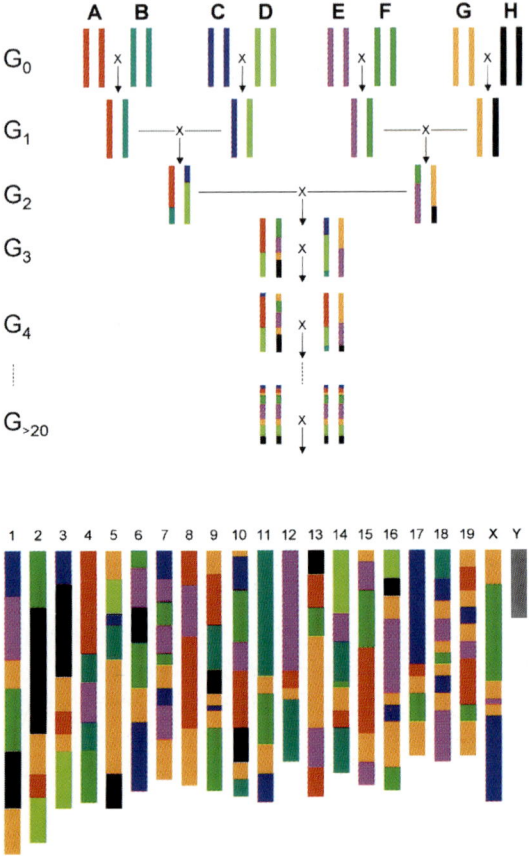

Figure 4. The collaborative cross stems from an eight-way interstrain cross. Eight carefully selected inbred strains are crossed to produce four F1 hybrids. These hybrids are then crossed together; one thousand independent inbreds are derived after 20 additional generations of brother × sister mating. In the end, this cross will be materialized by a resource of about one thousand different inbred strains (actually recombinant inbred strains, RIS) whose genome will be a "patchwork" with a roughly equivalent contribution of the original inbred strains. For each strain the "patchwork" will be unique. Altogether, the resource of 1000 strains will represent 135,000 recombination events in the mouse genome and will segregate for a large quantity of polymorphisms. Redrawn with permission from *Nature* © 2004, The Complex Trait Consortium (2004).

reached their "full speed," since the essential tools are not yet available, but there is no doubt that with these research programs we will certainly have a better idea of quantitative inheritance. Even if the eight strains that have been selected as founder strains of the eight-way cross represent only a sample of the polymorphisms that may segregate in the mouse species, this will probably be more than enough to allow unraveling of at least some elementary mechanisms of quantitative inheritance in mammals. Another conclusion we must draw from these last two initiatives is that they both were the consequence of a very high level of interactive and constructive consultation inside the community. This is a big change in the mentalities to the benefit of the progress in mouse genomics.

CONCLUSIONS

The aim of this review was to summarize the most important advances in mouse genetics that occurred over the last 15 years. Among these advances, the release on public databases, in 2002, of a first draft of the mouse genome sequence is certainly the one that had the greatest impact on the community. The other projects described that are aimed at understanding the function(s) of genes or the complex nature of quantitative inheritance will certainly also influence dramatically the future of mouse genetics and change the way we have been approaching human health and disease.

This review, however, is not comprehensive, and many other subjects would also have deserved to be discussed. For example, the extraordinary progress made by molecular geneticists for engineering alterations in ES cells should be mentioned. While homologous recombination was only in its infancy 15 years ago, it has now reached a very high level of sophistication with the possibility of making, almost at will, a conditional or inducible and tissue specific mutation in each and every gene (Kos 2004). The advent of siRNA-directed technology for gene silencing, which is only in an early step of development at the moment but will certainly play an important role in the arsenal of mouse geneticists in the future, is also worthy of mention (Hasuwa et al. 2002). Finally, I should mention the development of new inbred strains of mice, derived from recently trapped wild specimens of the same genus *Mus* but from different species or subspecies. These strains offer geneticists a virtually unlimited amount of polymorphisms of all kinds and an endless variety of new alleles that have been selected by chance (or by necessity?) under the only pressure of natural selection (Guénet and Bonhomme 2003; Ideraabdullah et al. 2004). It would be interesting, for example, to set interspecific crosses and study the consequences of bringing together, in the same cell, the products of genes separated by divergent evolution. This could help to identify genetic functions that are subject to rapid divergence and may help to pinpoint the functions that eventually promote speciation. Questions concerning epistatic interactions may also find an answer by analyzing the phenotype of offspring of interspecific crosses at the genomic level and assessing the consequences of "packing" into the same genome alleles stemming from distantly related species. So far, we have no clear answers to these questions, but data exist indicating that some combinations of alleles are strongly counterselected in the offspring of some interspecific crosses (Montagutelli et al. 1996) and deleterious phenotypes, such as diabetes, autoimmune diseases, or male sterility, are common. Our attempts to develop interspecific consomic (or chromosome substitution) strains with complete chromosomes of *Mus spretus* introgressed into a C57BL/6 background substituting for the original chromosome have failed in most instances, while consomic strains have been easily created between strains of the same species (Singer et al. 2004). Again, we have no explanation for this, but deleterious epistatic interactions due to genetic divergence from the ancestral alleles are highly suspected.

As discussed, the last 15 years have obviously been extremely important, providing the community with an enormous amount of new information, but what is also remarkable is that all of this information and these tools, thanks to the Internet, have been made available to the public very rapidly and at no cost. Then, if the microcosm of mouse geneticists has changed so dramatically (and for the better!) over the last 15 years, maybe it is time to thank the legion of anonymous curators taking care of all these databases. They have made the world a bit better.

ACKNOWLEDGMENTS

I thank Professor Steve D.M. Brown, for reading this manuscript and making interesting suggestions. This review is dedicated to Professor François Jacob who, 35 years ago, convinced me as a young vet that the mouse was really an interesting animal model for geneticists. It is also dedicated to the scientists at Harwell and Bar Harbor who gave generously of their time to teach me the basics of Mouse Genetics.

REFERENCES

Abiola, O., Angel, J.M., Avner, P., Bachmanov, A.A., Belknap, J.K., Bennett, B., Blankenhorn, E.P., Blizard, D.A., Bolivar, V., Brockmann, G.A., et al. 2003. The nature and identification of quantitative trait loci: A community's view. *Nat. Rev. Genet.* **4:** 911–916.

Adams, D.J., Biggs, P.J., Cox, T., Davies, R., van der Weyden, L., Jonkers, J., Smith, J., Plumb, B., Taylor, R., Nishijima, I., et al. 2004. Mutagenic insertion and chromosome engineering resource (MICER). *Nat. Genet.* **36:** 867–871.

Allen, T.A., Von Kaenel, S., Goodrich, J.A., and Kugel, J.F. 2004. The SINE-encoded mouse B2 RNA represses mRNA transcription in response to heat shock. *Nat. Struct. Mol. Biol.* **11**: 816–821.
Augustin, M., Sedlmeier, R., Peters, T., Huffstadt, U., Kochmann, E., Simon, D., Schoniger, M., Garke-Mayerthaler, S., Laufs, J., Mayhaus, M., et al. 2005. Efficient and fast targeted production of murine models based on ENU mutagenesis. *Mamm. Genome* **16**: 405–413.
Austin, C.P., Battey, J.F., Bradley, A., Bucan, M., Capecchi, M., Collins, F.S., Dove, W.F., Duyk, G., Dymecki, S., Eppig, J.T., et al. 2004. The knockout mouse project. *Nat. Genet.* **36**: 921–924.
Auwerx, J., Avner, P., Baldock, R., Ballabio, A., Balling, R., Barbacid, M., Berns, A., Bradley, A., Brown, S., Carmeliet, P., et al. 2004. The European dimension for the mouse genome mutagenesis program. *Nat. Genet.* **36**: 925–927.
Bailey, J.A., Church, D.M., Ventura, M., Rocchi, M., and Eichler, E.E. 2004. Analysis of segmental duplications and genome assembly in the mouse. *Genome Res.* **14**: 789–801.
Bonhomme, F., Guénet, J.L., Dod, B., Moriwaki, K., and Bulfield, G. 1987. The polyphyletic origin of laboratory inbred mice and their rate of evolution. *Biol. J. Linnean Soc.* **30**: 51–58.
Brown, S.D.M., Chambon, P. and de Angelis, M.M. 2005. EMPReSS: Standardized phenotype screens for functional annotation of the mouse genome. *Nat. Genet.* **37**: 1155.
Bult, C.J., Blake, J.A., Richardson, J.E., Kadin, J.A., Eppig, J.T., Baldarelli, R.M., Barsanti, K., Baya, M., Beal, J.S., Boddy, W.J., et al. 2004. The Mouse Genome Database (MGD): Integrating biology with the genome. *Nucleic Acids Res.* **32**: D476–D481.
Carninci, P., Kasukawa, T., Katayama, S., Gough, J., Frith, M.C., Maeda, N., Oyama, R., Ravasi, T., Lenhard, B., Wells, C., et al. 2005. The transcriptional landscape of the mammalian genome. *Science* **309**: 1559–1563.
Cheung, J., Wilson, M.D., Zhang, J., Khaja, R., MacDonald, J.R., Heng, H.H., Koop, B.F., and Scherer, S.W. 2003. Recent segmental and gene duplications in the mouse genome. *Genome Biol.* **4**: R47.
Churchill, G.A., Airey, D.C., Allayee, H., Angel, J.M., Attie, A.D., Beatty, J., Beavis, W.D., Belknap, J.K., Bennett, B., Berrettini, W., et al. 2004. The Collaborative Cross, a community resource for the genetic analysis of complex traits. *Nat. Genet.* **36**: 1133–1137.
Clark, A.T., Goldowitz, D., Takahashi, J.S., Vitaterna, M.H., Siepka, S.M., Peters, L.L., Frankel, W.N., Carlson, G.A., Rossant, J., Nadeau, J.H., et al. 2004. Implementing large-scale ENU mutagenesis screens in North America. *Genetica* **122**: 51–64.
Coghill, E.L., Hugill, A., Parkinson, N., Davison, C., Glenister, P., Clements, S., Hunter, J., Cox, R.D., and Brown, S.D. 2002. A gene-driven approach to the identification of ENU mutants in the mouse. *Nat. Genet.* **30**: 255–256.
The Complex Trait Consortium. 2004. The Collaborative Cross, a community resource for the genetic analysis of complex traits. *Nature Gen.* **36**: 1133–1137.
Cuénot, L. 1902. La loi de Mendel et l'hérédité de la pigmentation chez la souris. *Arch. Zool. Exp. Gén. 3ème série* **3**: 27/30.
D'Errico, I., Gadaleta, G., and Saccone, C. 2004. Pseudogenes in metazoa: Origin and features. *Brief. Funct. Genomic Proteomic* **3**: 157–167.
Dike, S., Balija, V.S., Nascimento, L.U., Xuan, Z., Ou, J., Zutavern, T., Palmer, L.E., Hannon, G., Zhang, M.Q., and McCombie, W.R. 2004. The mouse genome: Experimental examination of gene predictions and transcriptional start sites. *Genome Res.* **14**: 2424–2429.

Durick, K., Mendlein, J., and Xanthopoulos, K.G. 1999. Hunting with traps: Genome-wide strategies for gene discovery and functional analysis. *Genome Res.* **9:** 1019–1025.

Eppig, J.T., Bult, C.J., Kadin, J.A., Richardson, J.E., Blake, J.A., Anagnostopoulos, A., Baldarelli, R.M., Baya, M., Beal, J.S., Bello, S.M., et al. 2005. The Mouse Genome Database (MGD): From genes to mice—A community resource for mouse biology. *Nucleic Acids Res.* **33:** D471–D475.

Fitzgerald, J. and Bateman, J.F. 2004. Why mice have lost genes for COL21A1, STK17A, GPR145 and AHRI: Evidence for gene deletion at evolutionary breakpoints in the rodent lineage. *Trends Genet.* **20:** 408–412.

Forrai, A. and Robb, L. 2005. The gene trap resource: A treasure trove for hemopoiesis research. *Exp. Hematol.* **33:** 845–856.

Glazier, A.M., Nadeau, J.H., and Aitman, T.J. 2002. Finding genes that underlie complex traits. *Science* **298:** 2345–2349.

Godfrey, P.A., Malnic, B., and Buck, L.B. 2004. The mouse olfactory receptor gene family. *Proc. Natl. Acad. Sci.* **101:** 2156–2161.

Goodwin, N.C., Ishida, Y., Hartford, S., Wnek, C., Bergstrom, R.A., Leder, P., and Schimenti, J.C. 2001. DelBank: A mouse ES-cell resource for generating deletions. *Nat. Genet.* **28:** 310–311.

Guénet, J.L. 2004. Chemical mutagenesis of the mouse genome: An overview. *Genetica* **122:** 9–24.

Guénet, J.L. and Bonhomme, F. 2003. Wild mice: An ever-increasing contribution to a popular mammalian model. *Trends Genet.* **19:** 24–31.

Guigó, R., Dermitzakis, E.T., Agarwal, P., Ponting, C.P., Parra, G., Reymond, A., Abril, J.F., Keibler, E., Lyle, R., Ucla, C., et al. 2003. Comparison of mouse and human genomes followed by experimental verification yields an estimated 1,019 additional genes. *Proc. Natl. Acad. Sci.* **100:** 1140–1145.

Hansen, J., Floss, T., Van Sloun, P., Fuchtbauer, E.M., Vauti, F., Arnold, H.H., Schnutgen, F., Wurst, W., von Melchner, H., and Ruiz, P. 2003. A large-scale, gene-driven mutagenesis approach for the functional analysis of the mouse genome. *Proc. Natl. Acad. Sci.* **100:** 9918–9922.

Hasuwa, H., Kaseda, K., Einarsdottir, T., and Okabe, M. 2002. Small interfering RNA and gene silencing in transgenic mice and rats. *FEBS Lett.* **532:** 227–230.

Hayashizaki, Y. 2003a. Mouse Genome Encyclopedia Project. *Cold Spring Harb. Symp. Quant. Biol.* **68:** 195–204.

———. 2003b. The Riken mouse genome encyclopedia project. *C. R. Biol.* **326:** 923–929.

Herault, Y., Rassoulzadegan, M., Cuzin, F., and Duboule, D. 1998. Engineering chromosomes in mice through targeted meiotic recombination (TAMERE) *Nat. Genet.* **20:** 381–384.

Hill, D.P., Begley, D.A., Finger, J.H., Hayamizu, T.F., McCright, I.J., Smith, C.M., Beal, J.S., Corbani, L.E., Blake, J.A., Eppig, J.T., et al. 2004. The mouse Gene Expression Database (GXD): Updates and enhancements. *Nucleic Acids Res.* **32:** D568–D571.

Hoebe, K. and Beutler, B. 2005. Unraveling innate immunity using large scale N-ethyl-N-nitrosourea mutagenesis. *Tissue Antigens* **65:** 395–401.

Hrabe de Angelis, M.H., Flaswinkel, H., Fuchs, H., Rathkolb, B., Soewarto, D., Marschall, S., Heffner, S., Pargent, W., Wuensch, K., Jung, M., et al. 2000. Genome-wide, large-scale production of mutant mice by ENU mutagenesis. *Nat. Genet.* **25:** 444–447.

Ideraabdullah, F.Y., de la Casa-Esperon, E., Bell, T.A., Detwiler, D.A., Magnuson, T., Sapienza, C., and de Villena, F.P. 2004. Genetic and haplotype diversity among wild-derived mouse inbred strains. *Genome Res.* **14:** 1880–1887.

Kasukawa, T., Furuno, M., Nikaido, I., Bono, H., Hume, D.A., Bult, C., Hill, D.P., Baldarelli, R., Gough, J., Kanapin, A., et al. 2003. Development and evaluation of an automated annotation pipeline and cDNA annotation system. *Genome Res.* **13:** 1542–1551.

Kile, B.T., Hentges, K.E., Clark, A.T., Nakamura, H., Salinger, A.P., Liu, B., Box, N., Stockton, D.W., Johnson, R.L., Behringer, R.R., et al. 2003. Functional genetic analysis of mouse chromosome 11. *Nature* **425:** 81–86.

Kos, C.H. 2004. Cre/*loxP* system for generating tissue-specific knockout mouse models. *Nutr. Rev.* **62:** 243–246.

Lindblad-Toh, K., Winchester, E., Daly, M.J., Wang, D.G., Hirschhorn, J.N., Laviolette, J.P., Ardlie, K., Reich, D.E., Robinson, E., Sklar, P., et al. 2000. Large-scale discovery and genotyping of single-nucleotide polymorphisms in the mouse. *Nat. Genet.* **24:** 381–386.

Martin, N., Jaubert, J., Gounon, P., Salido, E., Haase, G., Szatanik, M., and Guénet, J.L. 2002. A missense mutation in Tbce causes progressive motor neuronopathy in mice. *Nat. Genet.* **32:** 443–447.

Mashimo, T., Lucas, M., Simon-Chazottes, D., Frenkiel, M.P., Montagutelli, X., Ceccaldi, P.E., Deubel, V., Guénet, J.L., and Despres, P. 2002. A nonsense mutation in the gene encoding 2′-5′-oligoadenylate synthetase/L1 isoform is associated with West Nile virus susceptibility in laboratory mice. *Proc. Natl. Acad. Sci.* **99:** 11311–11316.

Mashimo, T., Glaser, P., Lucas, M., Simon-Chazottes, D., Ceccaldi, P.E., Montagutelli, X., Despres, P., and Guénet, J.L. 2003. Structural and functional genomics and evolutionary relationships in the cluster of genes encoding murine 2′,5′-oligoadenylate synthetases. *Genomics* **82:** 537–552.

Masuya, H., Nakai, Y., Motegi, H., Niinaya, N., Kida, Y., Kaneko, Y., Aritake, H., Suzuki, N., Ishii, J., Koorikawa, K., et al. 2004. Development and implementation of a database system to manage a large-scale mouse ENU-mutagenesis program. *Mamm. Genome* **15:** 404–411.

McCarthy, E.M. and McDonald, J.F. 2004. Long terminal repeat retrotransposons of *Mus musculus*. *Genome Biol.* **5:** R14.

Medstrand, P., van de Lagemaat, L.N., Dunn, C.A., Landry, J.R., Svenback, D., and Mager, D.L. 2005. Impact of transposable elements on the evolution of mammalian gene regulation. *Cytogenet. Genome Res.* **110:** 342–352.

Montagutelli, X., Turner, R., and Nadeau, J.H. 1996. Epistatic control of non-Mendelian inheritance in mouse interspecific crosses. *Genetics* **143:** 1739–1752.

Munroe, R.J., Ackerman, S.L., and Schimenti, J.C. 2004. Genomewide two-generation screens for recessive mutations by ES cell mutagenesis. *Mamm. Genome* **15:** 960–965.

Nadeau, J.H., Balling, R., Barsh, G., Beier, D., Brown, S.D., Bucan, M., Camper, S., Carlson, G., Copeland, N., Eppig, J., et al. 2001. Sequence interpretation. Functional annotation of mouse genome sequences. *Science* **291:** 1251–1255.

Nishimura, I., Drake, T.A., Lusis, A.J., Lyons, K.M., Nadeau, J.H., and Zernik, J. 2003. ENU large-scale mutagenesis and quantitative trait linkage (QTL) analysis in mice: Novel technologies for searching polygenetic determinants of craniofacial abnormalities. *Crit. Rev. Oral Biol. Med.* **14:** 320–330.

Nobrega, M.A., Zhu, Y., Plajzer-Frick, I., Afzal, V., and Rubin, E.M. 2004. Megabase deletions of gene deserts result in viable mice. *Nature* **431:** 988–993.

Nolan, P.M., Peters, J., Strivens, M., Rogers, D., Hagan, J., Spurr, N., Gray, I.C., Vizor, L., Brooker, D., Whitehill, E., et al. 2000. A systematic, genome-wide, phenotype-driven mutagenesis programme for gene function studies in the mouse. *Nat. Genet.* **25:** 440–443.

Nolan, P.M., Hugill, A., and Cox, R.D. 2002. ENU mutagenesis in the mouse: Application to human genetic disease. *Brief. Funct. Genomic Proteomic* **1:** 278–289.

Okazaki, Y., Furuno, M., Kasukawa, T., Adachi, J., Bono, H., Kondo, S., Nikaido, I., Osato, N., Saito, R., Suzuki, H., et al. 2002. Analysis of the mouse transcriptome based on functional annotation of 60,770 full-length cDNAs. *Nature* **420:** 563–573.

Olson, L.E., Tien, J., South, S., and Reeves, R.H. 2005. Long-range chromosomal engineering is more efficient in vitro than in vivo. *Transgenic Res.* **14:** 325–332.

Paigen, K. 2003. One hundred years of mouse genetics: An intellectual history. I. The classical period (1902–1980). *Genetics* **163:** 1–7.

Paigen, K. and Eppig, J.T. 2000. A mouse phenome project. *Mamm. Genome* **11:** 715–717.

Panthier, J.J., Dreyfus, M., Roux, T.L., and Rougeon, F. 1984. Mouse kidney and submaxillary gland renin genes differ in their 5' putative regulatory sequences. *Proc. Natl. Acad. Sci.* **81:** 5489–5493.

Parra, G., Agarwal, P., Abril, J.F., Wiehe, T., Fickett, J.W., and Guigó, R. 2003. Comparative gene prediction in human and mouse. *Genome Res.* **13:** 108–117.

Pennacchio, L.A. 2003. Insights from human/mouse genome comparisons. *Mamm. Genome* **14:** 429–436.

Perelygin, A.A., Scherbik, S.V., Zhulin, I.B., Stockman, B.M., Li, Y., and Brinton, M.A. 2002. Positional cloning of the murine flavivirus resistance gene. *Proc. Natl. Acad. Sci.* **99:** 9322–9327.

Petkov, P.M., Ding, Y., Cassell, M.A., Zhang, W., Wagner, G., Sargent, E.E., Asquith, S., Crew, V., Johnson, K.A., Robinson, P., et al. 2004. An efficient SNP system for mouse genome scanning and elucidating strain relationships. *Genome Res.* **14:** 1806–1811.

Pletcher, M. and Wiltshire, T. 2004. Can we find the genes involved in complex traits? *Genome Biol.* **5:** 347.

Pletcher, M.T., McClurg, P., Batalov, S., Su, A.I., Barnes, S.W., Lagler, E., Korstanje, R., Wang, X., Nusskern, D., Bogue, M.A., et al. 2004. Use of a dense single nucleotide polymorphism map for in silico mapping in the mouse. *PLoS Biol.* **2:** e393.

Quwailid, M.M., Hugill, A., Dear, N., Vizor, L., Wells, S., Horner, E., Fuller, S., Weedon, J., McMath, H., Woodman, P., et al. 2004. A gene-driven ENU-based approach to generating an allelic series in any gene. *Mamm. Genome* **15:** 585–591.

Rathkolb, B., Fuchs, E., Kolb, H.J., Renner-Muller, I., Krebs, O., Balling, R., Hrabe de Angelis, M., and Wolf, E. 2000. Large-scale N-ethyl-N-nitrosourea mutagenesis of mice—from phenotypes to genes. *Exp. Physiol.* **85:** 635–644.

Reeves, R.H., Baxter, L.L., and Richtsmeier, J.T. 2001. Too much of a good thing: Mechanisms of gene action in Down syndrome. *Trends Genet.* **17:** 83–88.

Rinchik, E.M., Carpenter, D.A., and Johnson, D.K. 2002. Functional annotation of mammalian genomic DNA sequence by chemical mutagenesis: A fine-structure genetic mutation map of a 1- to 2-cM segment of mouse chromosome 7 corresponding to human chromosome 11p14-p15. *Proc. Natl. Acad. Sci.* **99:** 844–849.

Schnutgen, F., De-Zolt, S., Van Sloun, P., Hollatz, M., Floss, T., Hansen, J., Altschmied, J., Seisenberger, C., Ghyselinck, N.B., Ruiz, P., et al. 2005. Genomewide production of multipurpose alleles for the functional analysis of the mouse genome. *Proc. Natl. Acad. Sci.* **102:** 7221–7226.

Schwartz, S., Kent, W.J., Smit, A., Zhang, Z., Baertsch, R., Hardison, R.C., Haussler, D., and Miller, W. 2003. Human–mouse alignments with BLASTZ. *Genome Res.* **13:** 103–107.

Singer, J.B., Hill, A.E., Burrage, L.C., Olszens, K.R., Song, J., Justice, M., O'Brien, W.E., Conti, D.V., Witte, J.S., Lander, E.S., et al. 2004. Genetic dissection of complex traits with chromosome substitution strains of mice. *Science* **304:** 445–448.

Skarnes, W.C., von Melchner, H., Wurst, W., Hicks, G., Nord, A.S., Cox, T., Young, S.G., Ruiz, P., Soriano, P., Tessier-Lavigne, M., et al. 2004. A public gene trap resource for mouse functional genomics. *Nat. Genet.* **36:** 543–544.

Stanford, W.L., Cohn, J.B., and Cordes, S.P. 2001. Gene-trap mutagenesis: Past, present and beyond. *Nat. Rev. Genet.* **2:** 756–768.

Stryke, D., Kawamoto, M., Huang, C.C., Johns, S.J., King, L.A., Harper, C.A., Meng, E.C., Lee, R.E., Yee, A., L'Italien, L., et al. 2003. BayGenomics: A resource of insertional mutations in mouse embryonic stem cells. *Nucleic Acids Res.* **31:** 278–281.

Wade, C.M., Kulbokas III, E.J., Kirby, A.W., Zody, M.C., Mullikin, J.C., Lander, E.S., Lindblad-Toh, K., and Daly, M.J. 2002. The mosaic structure of variation in the laboratory mouse genome. *Nature* **420:** 574–578.

Waterston, R.H., Lindblad-Toh, K., Birney, E., Rogers, J., Abril, J.F., Agarwal, P., Agarwala, R., Ainscough, R., Alexandersson, M., An, P., et al. 2002. Initial sequencing and comparative analysis of the mouse genome. *Nature* **420:** 520–562.

Wilson, L., Ching, Y.H., Farias, M., Hartford, S.A., Howell, G., Shao, H., Bucan, M., and Schimenti, J.C. 2005. Random mutagenesis of proximal mouse chromosome 5 uncovers predominantly embryonic lethal mutations. *Genome Res.* **15:** 1095–1105.

Wiltshire, T., Pletcher, M.T., Batalov, S., Barnes, S.W., Tarantino, L.M., Cooke, M.P., Wu, H., Smylie, K., Santrosyan, A., Copeland, N.G., et al. 2003. Genome-wide single-nucleotide polymorphism analysis defines haplotype patterns in mouse. *Proc. Natl. Acad. Sci.* **100:** 3380–3385.

Yan, J., Keener, V.W., Bi, W., Walz, K., Bradley, A., Justice, M.J., and Lupski, J.R. 2004. Reduced penetrance of craniofacial anomalies as a function of deletion size and genetic background in a chromosome engineered partial mouse model for Smith-Magenis syndrome. *Hum. Mol. Genet.* **13:** 2613–2624.

Yan, W., Ma, L., Stein, P., Pangas, S.A., Burns, K.H., Bai, Y., Schultz, R.M., and Matzuk, M.M. 2005. Mice deficient in oocyte-specific oligoadenylate synthetase-like protein OAS1D display reduced fertility. *Mol. Cell. Biol.* **25:** 4615–4624.

Yap, V.B. and Pachter, L. 2004. Identification of evolutionary hotspots in the rodent genomes. *Genome Res.* **14:** 574–579.

Yu, Y. and Bradley, A. 2001. Engineering chromosomal rearrangements in mice. *Nat. Rev. Genet.* **2:** 780–790.

Zan, Y., Haag, J.D., Chen, K.S., Shepel, L.A., Wigington, D., Wang, Y.R., Hu, R., Lopez-Guajardo, C.C., Brosc, H.L., Porter, K.I., et al. 2003. Production of knockout rats using ENU mutagenesis and a yeast-based screening assay. *Nat. Biotechnol.* **21:** 645–651.

16

Genomics of the Future: Identification of Quantitative Trait Loci in the Mouse

Lorraine Flaherty, Bruce Herron, and Derek Symula
Genomics Institute
Wadsworth Center
Troy, New York 12180

IDENTIFYING GENETIC LOCI CONTROLLING COMPLEX TRAITS (quantitative trait loci or QTLs) is one of the biggest challenges confronting genetics. These genes influence such traits as growth, morphology, and behavior and determine susceptibility and severity for nearly every disease. In particular, QTLs represent a gateway to the genetic factors controlling common, non-Mendelian diseases, such as heart disease and cancer, and affect many more people than the classic single gene diseases studied in the early days of positional cloning. These diseases have clear genetic components, yet the underlying genes have proven difficult to identify. Unfortunately, these genes are usually subtle in their expression and effect on the general phenotype of the organism; they interact with other genes and environmental effects making them difficult to isolate and they often have a low penetrance. Positional cloning of these QTLs in rodents has proved to be one of the most powerful tools for the functional identification of these genes (Georges 1997). The first step in this procedure is to map these loci to particular chromosomal regions, usually spanning anywhere from 10–40 cm. Over two thousand QTLs have been mapped (http://www.pubmed.com); however, the future narrowing of these regions and the subsequent identification of these QTLs are not easy (Flint et al. 2005). Sometimes, these QTLs contain polymorphisms resulting in obvious and deleterious effects on expression and function. More often, sequence differences are difficult to reconcile with the predicted phenotypes or cause expression and/or phenotypic variations that are subtler. Often there is no "smoking gun." Given the importance of QTLs in common

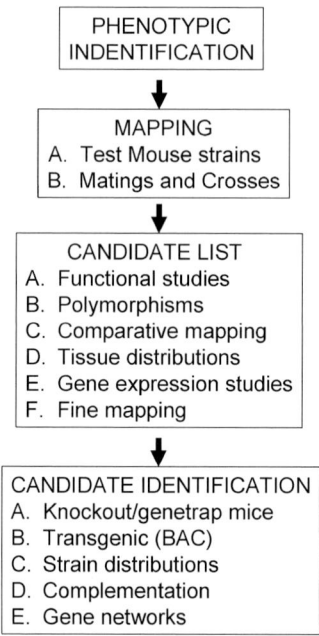

Figure 1. Scheme for the identification of quantitative trait loci.

diseases and in pharmacogenomics, recent genetic studies have focused on the identification of these loci.

Genomic sequence and related tools promised to help us find these genes by improving existing approaches and making possible new approaches. Here, we discuss some of the present and future techniques for making identification of these QTLs an easier task. Now that we are several years into the post-genome, we can begin to evaluate some of these strategies. These fall into three categories: (1) making candidate intervals as small as possible, (2) efficiently evaluating large numbers of genes in candidate intervals, and (3) testing candidate genes in a more powerful and efficient manner. These strategies are schemed in Figure 1. It is important to note that, with the possible exception of a targeted knock-in mutation of a QTL, none of the methods yet described can provide explicit proof of a gene's involvement in a trait. Rather, each approach tests a gene or (ideally) multiple genes in a different way until the weight of evidence is considered sufficient to prove causation (Abiola et al. 2003).

DETECTION AND LOCALIZATION

In 1992, Dietrich and colleagues (Dietrich et al. 1992) forged a major tool for mapping QTLs—a dense, highly polymorphic linkage map based on molecular markers. Not only did this map become a wonderful tool to map QTLs in standard crosses, but it also made it practical to construct more specialized resources for use in linkage studies. These consisted of new mouse constructs, including large sets of recombinant inbred lines, heterogeneous stocks, consomics, advanced inbred lines, and congenic strains (Kruglyak and Lander 1995a,b; Lander and Kruglyak 1995; Darvasi 1998; Nadeau et al. 2000). For mapping studies, recombinant inbred lines and consomics have been the most useful and have led to the rough localization of a large number of QTLs (see http://www.informatics.jax.org/searches/allele_form.shtml). Perhaps, even more important, the complete C57BL/6 genome and low coverage sequence for 129/S1, 129X1, A/J, and DBA/2J (Celera/Applera) became available on the Web, and 15 strains of mice are now being resequenced by the Center for Rodent Genetics (http://www.niehs.nih.gov/crg/). The single nucleotide polymorphisms (SNPs) revealed by these sequence databases provide an extremely dense map for linkage mapping.

What excited the QTL community more than improved mapping resolution, however, was the potential to use these SNPs for in silico localization using linkage disequilibrium (LD) analysis. Because most inbred mouse strains have a common ancestral heritage and ancestral genotypes account for most of the genetic variation among inbred strains (Wade et al. 2002; Frazer et al. 2004; Yalcin et al. 2004b), QTL genes are likely (though not necessarily) to be found in regions of different ancestry in strains with different phenotypes. Whole-genome association studies might be used to identify ancestral haplotype blocks, i.e., blocks of DNA inherited from a common ancestor, which correlate with different values of a trait among a large collection of inbred strains. This approach is attractive because it can use previously generated phenotypes, such as from the JAX Phenome Database, and requires no further crosses. Thus, in the hypothetical array of strains in Figure 2, sex-linked traits that follow strains A, C, F, and G would be due to the region between markers 5 and 6. For example, Pletcher et al. (2004) used nearly 11,000 SNPs to derive inferred haplotypes for 48 strains and applied these haplotypes to existing phenotype data. Previously identified loci for several monogenic and several quantitative traits showed highly significant associations with the appropriate phenotypes (Pletcher et al. 2004). A similar approach would use haplotypes within a previously mapped QTL to potentially refine the QTL to a smaller haplotype block that associates with the trait (Manenti et al. 2004).

Despite the potential power of LD analysis, the emerging body of studies suggests major challenges to its application in mice (Frazer et al.

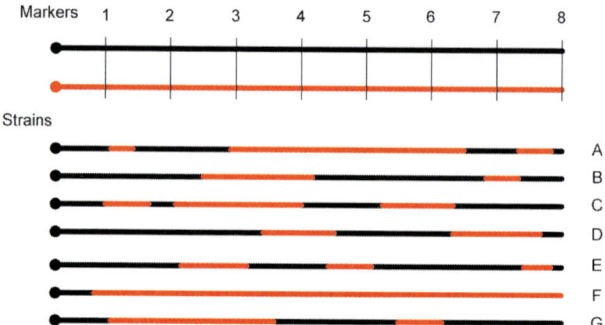

Figure 2. Haplotype block mapping of inbred strains to determine phenotype associated with red but not black chromosome.

2004; Yalcin et al. 2004a). Complete resequencing has revealed greater detail than the whole genome shotgun reads used initially. Haplotype blocks in mice are not consistent among inbred strains; a given region may have different boundaries in different strains. There also are concerns that in silico analysis in mice may be impractical for precise QTL gene localization because of the large number of strains needed for good statistical power even under favorable conditions. It has been estimated that detecting a QTL would require between 40 and 150 strains, with 40 strains needed to determine a major gene controlling the majority of the phenotypic variance (Darvasi 2001). There is still hope for in silico strategies, though. While the structure of the mouse genome is more complex than originally thought, there are still blocks of sequence with clear phylogenetic relationships among the inbred strains (Frazer et al. 2004; Yalcin et al. 2004a). This information still allows for association tests that use a small set of SNPs to represent a given region. Alternatively, we can compare the strain distribution pattern (SDP) of each SNP with the SDP of the trait without regard for the structure of the genome. While this strategy requires complete sequence information for each strain, the limited number of SNP SDPs suggested by results of Yalcin and colleagues may have better statistical power than might be expected (Yalcin et al. 2004b). It seems likely that, given complete sequence coverage for many strains and analysis methods that balance power and resolution, some form of association will be important in QTL localization.

CANDIDATE GENE IDENTIFICATION

Genomic tools, techniques, and databases have transformed the hunt for candidate genes. Perhaps the greatest effect concerns the initial characterization of the candidate interval. Database searches focusing on the candidate interval have replaced arduous physical mapping and sequencing that, very recently, were state-of-the-art. In addition, multi-species sequence comparisons have identified conserved noncoding regions that are important in gene regulation (Boffelli et al. 2004; Loots et al. 2005). It is now possible, through database searches and multi-species sequence comparisons, to produce a nearly comprehensive list of candidate functional units from Web-based databases (Ahituv et al. 2005). Thus, genomic tools focus the search for causative genes and polymorphisms to those regions most likely to impact phenotype. Evaluating these genes and regulatory regions for appropriate function and changes in sequence and expression remains a challenge but the efficiency of the techniques used has been increased vastly.

Expression profiling would seem to be a perfect tool for evaluating candidates. As for Mendelian traits, an investigator can assay the entire interval to identify genes whose expression correlates with phenotype. The simplest application of expression profiling is to compare two groups of mice with different phenotypes and ask whether any genes in the candidate region are differentially expressed. The clearest reported success using this approach is susceptibility of mice to allergen-induced-airway hyper-responsiveness. Wills-Karp and colleagues (Ewart et al. 2000; Karp et al. 2000) mapped two QTLs for this trait with one differentially expressed gene, hemolytic component *Hc* (previously known as C5) as a prime candidate. Subsequently, differences in the *Hc* sequence were shown to be the cause of this hyper-responsiveness. As simple and attractive as this strategy is, it does not work very often. The investigator must assay the appropriate tissue, cell type, environmental conditions, and developmental stage to evaluate the correct and "bottleneck" phenotype, i.e., the place and time where the QTL makes a difference. Moreover, few cloned QTLs determine expression differences of a magnitude that would be detected by this approach (Flint et al. 2005). Therefore, recent efforts have focused on detecting subtle expression differences and building "gene networks" to take advantage of the tremendous throughput of expression profiling.

One such approach, genetical genomics, treats gene expression levels as quantitative traits and uses a segregating population to map loci that regulate expression of a given gene (Schadt et al. 2003; Bystrykh et al. 2005; Chesler et al. 2005; Hubner et al. 2005; Li et al. 2005). Statistical analyses can then determine if a biological QTL co-segregates with an expression QTL (eQTL), which would suggest that the same gene is responsible for both the biological and expression phenotypes. It is possible to detect subtler expression differences by linkage mapping rather than through a simple two-group comparison (Schadt et al. 2003), in part because recombinant inbred lines, for example, will show trait segregation correlating with a particular genotype, thus increasing the statistical power through replication (Bystrykh et al. 2005; Chesler et al. 2005; Hubner et al. 2005).

Cis-acting eQTLs are defined as those that are closely linked to the target gene, while *trans*-acting eQTLs map elsewhere in the genome. A *cis*-acting eQTL is of particular interest for candidate gene identification since it is likely to regulate the closely linked structural gene and implicate it as a strong candidate. *Trans*-acting eQTLs are more difficult to understand since they suggest candidate genes more indirectly. The genes regulated by a given *trans*-acting eQTL probably represent downstream effects of the regulator and, as might be expected, can have similar functions or act in the same pathway (Chesler et al. 2005). Thus, gene networks defined by *trans*-acting eQTLs will broaden our interpretations of quantitative traits and provide a wealth of information for understanding gene regulation and coordinately regulated genes. Several reports recently demonstrated the power of this approach in linking expression of specific genes to behavioral (Chesler et al. 2005) and diabetes-related phenotypes in mice (Doss et al. 2005), hypertension in rats (Hubner et al. 2005), and turnover of mouse hematopoetic stem cells (Bystrykh et al. 2005). Importantly, these candidate genes frequently were polymorphic or mapped to regions of differing haplotypes between the parental strains, and thus were strong candidates for regulatory QTLs.

Like the LD studies described above, genetical genomics is a highly in silico method that has increased power to compare phenotypes from existing databases with emerging expression data and sequencing variations so that probable candidates can be identified. Some of this analysis software is available on the WebQTL Web site (http://www.genenetwork.org/home). While the authors of these initial studies generated their own expression data, data for other experiments are becoming increasingly available in expression databases such as NCBI GEO (http://www.ncbi.nlm.nih.gov/geo/). This

Mouse Quantitative Trait Loci 351

approach is a powerful one and is likely to become a common one to use for QTL studies.

CAUSATIVE GENE IDENTIFICATION

Once strong candidates are identified, it is crucial to test them. The main problems include the difficulty in predicting the effects of existing polymorphisms on gene expression and function of the candidate gene, in extrapolating these effects to biological impact, and in ascribing phenotypic differences between two strains to one of many genes polymorphic between those strains. In some cases, the primary candidate gene, e.g., *Rgs2* (Yalcin et al. 2004b) in the case of behavior, has not been shown to have a structural defect. In others, the only observed polymorphism may have no clear effect on candidate gene function. For example, the *Apoa2* gene was a strong functional and positional candidate for a QTL controlling plasma HDL level but only one *Apoa2* polymorphism was significantly associated with plasma HDL level when the coding region was examined, a conserved alanine that is changed to valine in high HDL strains (Wang et al. 2004). Furthermore, the common origin of the inbred strains has contributed to several polymorphisms in a candidate interval and may have nothing causative to do with the expected complex trait phenotype. For most QTLs, investigators must explicitly test new alleles of the candidate gene. They may generate null alleles, complement an existing allele, or, in the most recent twist, find new alleles of the candidate's ortholog in a different species (Lyons et al. 2005; Wang et al. 2005).

Targeted knock-in mutations of a candidate gene remain the gold standard for QTL gene identification (Abiola et al. 2003). A central reason for the utilization of mice in complex trait analysis has been the ability to isolate and genetically manipulate mouse embryonic stem (ES) cells that can be used to create mice with specific gene defects. Historically, generating a targeted mutation in a specific candidate gene was a major undertaking and the resources required to do so on a large scale were unlikely to be found in academic laboratories. Publicly available targeted mutations covered only a fraction of genome until recently. The availability of the mouse genome sequence and the clones used in the public sequencing effort make generating a targeting construct easier (Cotta-de-Almeida et al. 2003). One can identify the target site in a gene using a sequence database and generate homology arms by PCR of a purified BAC template. More sophisticated targeting constructs for conditional knockouts can be made by recombineering (Copeland et al. 2001). In addition, it soon may become unnecessary to build a single targeting construct. Instead, one can use MICER, a library of pre-made targeting vectors for insertional mutations (Adams et al. 2004). In addition, there is an initiative to make and screen a massive library of ES cells that contain a reporter gene in place of each putative mouse open reading frame (Austin et al. 2004). Thus, heterozygous mice created from each of these lines would provide spatial and temporal information about where the gene is expressed, while homozygous null mice could provide insight to gene function in vivo. A tiered strategy can also be used for mutant line characterization. Here multiple levels of phenotype ascertainment will be used to funnel specific loci into more focused studies. For example, if heterozygous mice show strong reporter expression in brain, a battery behavioral testing might be warranted for that mutant line (Austin et al. 2004).

Moreover, there is a wealth of genetrap mutations available to the rodent research community. These are mutations that result from the insertion of an expression vector into the target gene. This expression vector contains a reporter gene, a selectable marker, poly(A) site, and translation stop site, but no promoter sequences. The vector is electroporated into mouse ES cells and, if it inserts in a transcribed sequence and produces a fusion transcript composed of part of the endogenous transcribed sequence (i.e., the "trapped" gene), the reporter, and the selectable marker, it will allow these ES cell clones to be selected. The trapped gene then is identified by 5' gene sequencing. The insertion often serves as a hypomorphic or null mutation, since it prematurely terminates transcription of the endogenous gene, and contains a reporter for the expression patterning of the endogenous gene. More recent genetraps feature site-specific recombinase sequences to create conditional mutations (Schnütgen et al. 2005). Several genetrap ES cell libraries have been compiled and over 45,000 are available publicly (see http://www.sanger.ac.uk/PostGenomics/genetrap/) covering more than one-third of the genes in the mouse genome (Skarnes et al. 2004). These genetrap libraries are particularly useful to investigators who are not mouse geneticists but wish to test a QTL candidate of interest.

Classical complementation is another powerful method for gene testing that has been made more practical by genomics. Correlation of the QTL with the presence of a single BAC would greatly refine the search for the causal mutation. BACs have become broadly utilized in functional mouse genomics and are the vector of choice for large scale sequencing because of its superior stability and DNA yield compared with other large insert vectors (Haldi et al. 1994). BACs also broaden the spectrum of candidate gene alleles available for testing and through the availability of libraries from different strains and through recent advances in BAC engineering that may facilitate the use of BAC transgenic mice to test for strain-specific dominant effects. The limitation of this approach is the availability of the proper BAC library and considerably large interval that most QTLs currently cover. BAC libraries from many strains are now publicly available with other strain libraries nearing completion (http://bacpac.chori.org/). Mapping of these BACs on the mouse genome will greatly facilitate their utilization.

SNP data from other species make possible a new, powerful candidate gene test, the cross-species comparison. These SNPs make it possible to ask the following question: Do different alleles of the human ortholog (or rat, platypus, etc.) of a candidate mouse gene correlate with phenotypic differences caused by the QTL? This procedure has been successfully used by Wang and colleagues (Wang et al. 2005), who identified *Tnfsf4*, as a candidate for diet-induced atherosclerosis based on position, expression, and sequence polymorphism. In addition, the authors found a SNP in the human *TNFSF4* gene that was associated with elevated risk of coronary artery disease and myocardial infarction. This approach provides new and independent populations in which to test the original hypothesis. This can be important given the limited polymorphism present in the inbred mouse strains. It further points to the conservation of sequence (though probably not individual polymorphisms) across millions of years of evolutionary distance as a very persuasive demonstration of normal and variant function. However, the failure to recapitulate the effects of a rodent gene in humans (or any other species) is more difficult to interpret, as the results could be attributed to any number of complicating genetic or environmental factors. There is also concern about the low reproducibility of human association studies (Ioannidis 2005). Still, such cross-species comparisons provide powerful evidence of causation when successful and may yet prove to be important.

Mapping the location of the QTL is an extremely important first step in QTL identification. While it is clear that location is not *sufficient* to identify a QTL gene, it is an integral part of the investigation and substantially limits the candidate gene pool. Given the rapid increases in high-throughput evaluation of genes and functional databases and the difficulties in studying genes of small effect in heterogeneous segregating populations, is it possible to robustly identify candidate genes without positional information? Our hope is that approaches such as genetic transcription networks will soon be refined in a manner that is sufficient to "guess" what genes may be involved in a physiological function. The insertion and/or the deletion of these same candidates could then lead to their identification as QTLs that influence function. For example,

Soriano and colleagues demonstrated a position-independent approach to identify genetrap clones mutant in genes important in PDGF signaling (Chen et al. 2004). Such position-independent strategies will certainly be used in the future especially when expression profiling becomes reproducible and available for a large number of tissues and strains.

CONCLUSIONS

SNP mappings of 15 strains of mice (sponsored by the Center for Rodent Genetics) will only be the beginning. The mouse investigator will be able to compare and contrast different chromosomal regions to identify new genes influencing the fate of QTLs affecting minor functions. It is likely that the contributions from each strain will focus on narrow regions for identification of these QTLs, leading to ~100–500 genes with a known tissue-determining effect. Microarray analyses of gene activity should then narrow these genes down to a short list, functionally distinguishing them from their neighbors. Thus, QTL identifications should become easier tasks and ones that are easily accomplished by the amplification of our current mouse and molecular genetic techniques.

REFERENCES

Abiola, O., Angel, J.M., Avner, P., Bachmanov, A.A., Belknap, J.K., Bennett, B., Blankenhorn, E.P., Blizard, D.A., Bolivar, V., Brockmann, G.A., et al. 2003. The nature and identification of quantitative trait loci: A community's view. *Nat. Rev. Genet.* **4:** 911–916.

Adams, D.J., Biggs, P.J., Cox, T., Davies, R., van der Weyden, L., Jonkers, J., Smith, J., Plumb, B., Taylor, R., Nishijima, I., et al. 2004. Mutagenic insertion and chromosome engineering resource (MICER). *Nat. Genet.* **36:** 867–871.

Ahituv, N., Prabhakar, S., Poulin, F., Rubin, E.M., and Couronne, O. 2005. Mapping cis-regulatory domains in the human genome using multi-species conservation of synteny. *Hum. Mol. Genet.* **14:** 3057–3063.

Austin, C.P., Battey, J.F., Bradley, A., Bucan, M., Capecchi, M., Collins, F.S., Dove, W.F., Duyk, G., Dymecki, S., Eppig, J.T., et al. 2004. The knockout mouse project. *Nat. Genet.* **36:** 921–924.

Boffelli, D., Weer, C.V., Weng, L., Lewis, K.D., Shoukry, M.I., Pachter, L., Keys, D.N., and Rubin, E.M. 2004. Intraspecies sequence comparisons for annotating genomes. *Genome Res.* **14:** 2406–2411.

Bystrykh, L., Weersing, E., Dontje, B., Sutton, S., Pletcher, M.T., Wiltshire, T., Su, A.I., Vellenga, E., Wang, J., Manly, K.F., et al. 2005. Uncovering regulatory pathways that affect hematopoietic stem cell function using 'genetical genomics'. *Nat. Genet.* **37:** 225–232.

Chen, W.V., Delrow, J., Corrin, P.D., Frazier, J.P., and Soriano, P. 2004. Identification and validation of PDGF transcriptional targets by microarray-coupled gene-trap mutagenesis. *Nat. Genet.* **36:** 304–312.

Chesler, E.J., Lu, L., Shou, S., Qu, Y., Gu, J., Wang, J., Hsu, H.C., Mountz, J.D., Baldwin, N.E., Langston, M.A., et al. 2005. Complex trait analysis of gene expression uncovers polygenic and pleiotropic networks that modulate nervous system function. *Nat. Genet.* **37:** 233–242.

Copeland, N.G., Jenkins, N.A., and Court, D.L. 2001. Recombineering: A powerful new tool for mouse functional genomics. *Nat. Rev. Genet.* **2:** 769–779.

Cotta-de-Almeida, V., Schonhoff, S., Shibata, T., Leiter, A., and Snapper, S.B. 2003. A new method for rapidly generating gene-targeting vectors by engineering BACs through homologous recombination in bacteria. *Genome Res.* **13:** 2190–2194.

Darvasi, A. 1998. Experimental strategies for the genetic dissection of complex traits in animal models. *Nat. Genet.* **18:** 19–24.

———. 2001. In silico mapping of mouse quantitative trait loci. *Science* **294:** 2423.

Dietrich, W., Katz, H., Lincoln, S.E., Shin, H.S., Friedman, J., Dracopoli, N.C., and Lander, E.S. 1992. A genetic map of the mouse suitable for typing intraspecific crosses. *Genetics* **131:** 423–447.

Doss, S., Schadt, E.E., Drake, T.A., and Lusis, A.J. 2005. Cis-acting expression quantitative trait loci in mice. *Genome Res.* **15:** 681–691.

Ewart, S.L., Kuperman, D., Schadt, E., Tankersley, C., Grupe, A., Shubitowski, D.M., Peltz, G., and Wills-Karp, M. 2000. Quantitative trait loci controlling allergen-induced airway hyperresponsiveness in inbred mice. *Am. J. Respir. Cell Mol. Biol.* **23:** 537–545.

Flint, J., Valdar, W., Shifman, S., and Mott, R. 2005. Strategies for mapping and cloning quantitative trait genes in rodents. *Nat. Rev. Genet.* **6:** 271–286.

Frazer, K.A., Tao, H., Osoegawa, K., de Jong, P.J., Chen, X., Doherty, M.F., and Cox, D.R. 2004. Noncoding sequences conserved in a limited number of mammals in the SIM2 interval are frequently functional. *Genome Res.* **14:** 367–372.

Georges, M. 1997. QTL mapping to QTL cloning: Mice to the rescue. *Genome Res.* **7:** 663–665.

Haldi, M., Perrot, V., Saumier, M., Desai, T., Cohen, D., Cherif, D., Ward, D., and Lander, E.S. 1994. Large human YACs constructed in a rad52 strain show a reduced rate of chimerism. *Genomics* **24:** 478–484.

Hubner, N., Wallace, C.A., Zimdahl, H., Petretto, E., Schulz, H., Maciver, F., Mueller, M., Hummel, O., Monti, J., Zidek, V., et al. 2005. Integrated transcriptional profiling and linkage analysis for identification of genes underlying disease. *Nat. Genet.* **37:** 243–253.

Ioannidis, J.P. 2005. Why most published research findings are false. *PLoS Med.* **2:** e124.

Karp, C.L., Grupe, A., Schadt, E., Ewart, S.L., Keane-Moore, M., Cuomo, P.J., Kohl, J., Wahl, L., Kuperman, D., Germer, S., et al. 2000. Identification of complement factor 5 as a susceptibility locus for experimental allergic asthma. *Nat. Immunol.* **1:** 221–226.

Kruglyak, L. and Lander, E.S. 1995a. A nonparametric approach for mapping quantitative trait loci. *Genetics* **139:** 1421–1428.

———. 1995b. High-resolution genetic mapping of complex traits. *Am. J. Hum. Genet.* **56:** 1212–1223.

Lander, E. and Kruglyak, L. 1995. Genetic dissection of complex traits: Guidelines for interpreting and reporting linkage results. *Nat. Genet.* **11:** 241–247.

Li, H., Lu, L., Manly, K.F., Chesler, E.J., Bao, L., Wang, J., Zhou, M., Williams, R.W., and Cui, Y. 2005. Inferring gene transcriptional modulatory relations: A genetical genomics approach. *Hum. Mol. Genet.* **14:** 1119–1125.

Loots, G.G., Kneissel, M., Keller, H., Baptist, M., Chang, J., Collette, N.M., Ovcharenko, D., Plajzer-Frick, I., and Rubin, E.M. 2005. Genomic deletion of a long-range bone enhancer misregulates *sclerostin* in Van Buchem disease. *Genome Res.* **15:** 928–935.

Lyons, M.A., Korstanje, R., Li, R., Sheehan, S.M., Walsh, K.A., Rollins, J.A., Carey, M.C., Paigen, B., and Churchill, G.A. 2005. Single and interacting QTLs for cholesterol gallstones revealed in an intercross between mouse strains NZB and SM. *Mamm. Genome* **16:** 152–163.

Manenti, G., Galbiati, F., Gianni-Barrera, R., Pettinicchio, A., Acevedo, A., and Dragani, T.A. 2004. Haplotype sharing suggests that a genomic segment containing six genes accounts for the pulmonary adenoma susceptibility 1 (Pas1) locus activity in mice. *Oncogene* **23:** 4495–4504.

Nadeau, J.H., Singer, J.B., Matin, A., and Lander, E.S. 2000. Analysing complex genetic traits with chromosome substitution strains. *Nat. Genet.* **24:** 221–225.

Pletcher, M.T., McClurg, P., Batalov, S., Su, A.I., Barnes, S.W., Lagler, E., Korstanje, R., Wang, X., Nusskern, D., Bogue, M.A., et al. 2004. Use of a dense single nucleotide polymorphism map for in silico mapping in the mouse. *PLoS Biol.* **2:** e393.

Schadt, E.E., Monks, S.A., Drake, T.A., Lusis, A.J., Che, N., Colinayo, V., Ruff, T.G., Milligan, S.B., Lamb, J.R., Cavet, G., et al. 2003. Genetics of gene expression surveyed in maize, mouse and man. *Nature* **422:** 297–302.

Schnütgen, F., De-Zolt, S., Van Sloun, P., Hollatz, M., Floss, T., Hansen, J., Altschmied, J., Seisenberger, C., Ghyselinck, N.B., Ruiz, P., et al. 2005. Genomewide production of multipurpose alleles for the functional analysis of the mouse genome. *Proc. Natl. Acad. Sci.* **102:** 7221–7226.

Skarnes, W.C., von Melchner, H., Wurst, W., Hicks, G., Nord, A.S., Cox, T., Young, S.G., Ruiz, P., Soriano, P., Tessier-Lavigne, M., et al. 2004. A public gene trap resource for mouse functional genomics. *Nat. Genet.* **36:** 543–544.

Wade, C.M., Kulbokas III, E.J., Kirby, A.W., Zody, M.C., Mullikin, J.C., Lander, E.S., Lindblad-Toh, K., and Daly, M.J. 2002. The mosaic structure of variation in the laboratory mouse genome. *Nature* **420:** 574–578.

Wang, X., Korstanje, R., Higgins, D., and Paigen, B. 2004. Haplotype analysis in multiple crosses to identify a QTL gene. *Genome Res.* **14:** 1767–1772.

Wang, X., Ishimori, N., Korstanje, R., Rollins, J., and Paigen, B. 2005. Identifying novel genes for atherosclerosis through mouse–human comparative genetics. *Am. J. Hum. Genet.* **77:** 1–15.

Yalcin, B., Fullerton, J., Miller, S., Keays, D.A., Brady, S., Bhomra, A., Jefferson, A., Volpi, E., Copley, R.R., Flint, J., et al. 2004a. Unexpected complexity in the haplotypes of commonly used inbred strains of laboratory mice. *Proc. Natl. Acad. Sci.* **101:** 9734–9739.

Yalcin, B., Willis-Owen, S.A., Fullerton, J., Meesaq, A., Deacon, R.M., Rawlins, J.N., Copley, R.R., Morris, A.P., Flint, J., and Mott, R. 2004b. Genetic dissection of a behavioral quantitative trait locus shows that Rgs2 modulates anxiety in mice. *Nat. Genet.* **36:** 1197–1202.

17

Comparing the Human and Chimpanzee Genomes: Searching for Needles in a Haystack

Ajit Varki and Tasha K. Altheide
Glycobiology Research and Training Center
Departments of Medicine and Cellular & Molecular Medicine
University of California at San Diego
La Jolla, California 92093

HUMANS (*HOMO SAPIENS*) AND CHIMPANZEES (*Pan troglodytes*) last shared a common ancestor ~5–7 million years ago (Mya) (Chen and Li 2001; Brunet et al. 2002). What makes humans different from their closest evolutionary relatives, and how, why, and when did these changes occur? These are fascinating questions, and a major challenge is to explain how genomic differences contributed to this process (Goodman 1999; Gagneux and Varki 2001; Klein and Takahata 2002; Carroll 2003; Olson and Varki 2003; Enard and Pääbo 2004; Gagneux 2004; Ruvolo 2004; Goodman et al. 2005; Li and Saunders 2005; McConkey and Varki 2005). Most genome projects focus on elucidating the sequence and structure of a species' genome and then identifying conserved functionally important genes and genomic elements. The finished human genome (International Human Genome Sequencing Consortium 2004) provides such a catalog of genomic features that ultimately interact with the environment to determine our biology, physiology, and disease susceptibility. Completion of the draft chimpanzee genome sequence (The Chimpanzee Sequencing and Analysis Consortium 2005) provides a genome-wide comparative catalog that can be used to identify genes or genomic regions underlying the many features that distinguish humans and chimpanzees.

As humans, we have an inherent interest in understanding and improving the human condition. We also believe that we have many characteristics that are uniquely human. Table 1 lists some of the definite and possible

Table 1. Some phenotypic traits of humans for comparison with those of great apes

LIFE HISTORY	ANATOMY
Secondary Altriciality	Sagittal Crest of Skull
Helplessness of the Newborn	Brow Ridge
Prolonged Helplessness of Young	Protuberantia Menti (Chin)
Extended Care of Young	Length of Sphenoid Sinus
Childhood	Choroid Plexus Biondi Bodies
Adolescence	Inner Ear Canal Orientation
Age at First Reproduction	Apical Phalangeal Tufts
Longevity	Age of Pelvic Bone Fusion
	Bone Cortex Thickness
REPRODUCTIVITY BIOLOGY	Laryngeal Position
Concealed Ovulation	Pharyngeal Air Sacs
Virgin Breast Development	Earlobes
Female Pituitary Menopause	Sexual Body Size Dimorphism
Placentophagy	Lacrimal Gland Structure
Female Labia Majora	Visible Whites of the Eyes
Vaginal Hymen	Small/Large Intestine Length Ratio
Baculum (Penis Bone)	Meningeal Artery Source
Sperm Count	
Copulatory Plug	**BIOMECHANICS**
	Bipedal Gait
EMBRYOLOGY	Adductive Thumb
Early Fetal Wastage/Aneuploidy	Skeletal Muscle Strength
Hydatiform Molar Pregnancy	Hand–Eye Coordination
Umbilical Cord Length	Fine Motor Coordination
PREGNANCY/PARTURITION	**ORGAN PHYSIOLOGY**
Cephalo-pelvic Disproportion	Aldosterone Response to Posture
Duration of Labor	Salt-Wasting Kidneys
Maternal Mortality in Childbirth	Ability For Sustained Running
Pain During Childbirth	Voluntary Control of Breathing
Need for Assistance with Childbirth	Ability to Dive Underwater
Neonatal Cephalhematoma	Diving Reflex
	Ability to Float/Swim
POSTNATAL DEVELOPMENT	Emotion Lacrimation
Late Closure of Cranial Sutures	Salt Content of Tears
Duration of Infant Arousal	Olfactory Sense
Inconsolable Infant Crying	
Infant–Caregiver Attunement	**CELL BIOLOGY**
Maternal–Infant Eye-To-Eye Gaze	No Differences Are Known?

Table 1. (*Continued*)

BIOCHEMISTRY
Placental Alkaline Phosphatase
N-Glycolylneuraminic Acid
  Expression
Alpha 2-6–Linked Sialic Acid Expression

ENDOCRINOLOGY
Thyroid Hormone Metabolism

PHARMACOLOGY
Methylation of Inorganic Arsenic

ANATOMIC PATHOLOGY
Cortical Neurofibrillary Tangles

CLINICAL PATHOLOGY
Erythrocyte Sedimentation Rate
Serum Alkaline Phosphatase Level
RBC and Serum Folate
Serum Vitamin B12/B12 Binding
Total Leukocyte Count
Absolute Neutrophil Count
Absolute Lymphocyte Count

DENTAL BIOLOGY/DISEASE
Canine Tooth Diastema
Canine Tooth Dysmorphism
Tooth Enamel Thickness
Retromolar Gap
Third Molar Impaction
Dental Eruption Sequence/Timing

MEDICAL/SURGICAL DISEASES
HIV Progression to AIDS
P. falciparum Malaria
Viral Hepatitis B/C Complications
Influenza A Infection Severity
Incidence of Carcinomas
Hemorrhoids
Varicose Veins
Pelvic Phleboliths

Foamy Virus (Spumavirus) Infections
Sexually Transmitted Diseases

IMMUNOLOGY
Sialoadhesin on Macrophages

SKIN BIOLOGY AND DISEASE
Eyebrows
Eccrine Sweat Glands
Acne Vulgaris
Subcutaneous Fat
Body Lice

NUTRITION
Frugivory
Carnivory
Aquatic Foods
Underground Foods
Cooking

NEUROANATOMY
Relative Brain Size
Direct Cortical Projections
Relative Volume of Frontal Cortex
Relative Volume of Corpus Callosum
Relative Volume of Cerebellum
% of Brain Growth Complete at Birth
Rate of Postnatal Brain Growth

NEUROBIOLOGY
Population Distribution of
  Handedness
Postnatal Dendritic Growth
Postnatal Synapse Formation
Cortical Synapse Density
Cortical Neuron Density
Dendrites per Neuron
Synapses per Neuron
Adult Neurogenesis
Cingulate Cortical Spindle Neurons
Fingertip Sensory Nerve Endings

(*Continued*)

Table 1. (*Continued*)

NEUROCHEMISTRY
Brain Aromatization of Testosterone
Tyrosine Hydroxylase Heterogeneity

MENTAL DISEASE
Schizophrenia
Bipolar Psychosis
Autism
Suicide

BEHAVIOR
Control of Facial Expressions
Planning Ahead
Intentional Deception
Deliberately Delaying Gratification
Long-Range Transport of Materials
Secondary Tool-Making
Mechanical Multi-Tasking
Physical Abuse of the Young
Torture
Organized Warfare
Adult Play
Symbolic Play
Abuse of Other Animals
Inter-Group Coalition Formation
Use of Containers
Care of Infirm and Elderly
Grandparenting
Home Base
Control of Fire
Food Preparation
Organized Gathering of Food
Domestication of Animals
Domestication of Plants
Altruistic Punishment
Peace-Making
Somnambulism
Mind-Altering Drug Use

COGNITIVE CAPACITY
Declarative Memory
Imitative Learning
Teaching

Symbolic Representation
Awareness of Death
Awareness of the Past
Awareness of the Future
Theory of Mind
Theory of Other Minds
Empathy
Numeracy

COMMUNICATION
"Parentese" Sounds
Infant "Protoconversations"
Gestural Communication
Symbolic Communication
Semantics
Grammar and Syntax
Recursion
Writing

SOCIAL ORGANIZATION
Institutions
Social Conventions
Governments
Enforcement through Sanctions

CULTURE
Composition of Art
Composition of Music
Composition of Rhythms
Death Rituals
Clothing (Covering of Body Parts)
Rites of Passage
Genocide
Competitive Sports
Practicing of Skills
Physical Modifications of the Body
Inheritance of Resources and Status
Rhythmic Dance
Sculpture
Belief in Supernatural/Religion
Body Adornment
Childbirth Customs
Sexual Intercourse in Private

Table 1. (*Continued*)

Gift-Giving	Taboos
Hospitality	Taxonomy of Species
Intertwining (e.g., Weaving)	Trade
Meal Times	Measurement of Time
Poetry	Weapons
Property	Toys
Construction of Shelters	

A major limitation in translating genomic comparative information into an understanding of "humanness" is that we know relatively little about the basic phenotypic features of the great apes, relative to humans. This table lists topic areas in which there are real or claimed "differences" between humans and the great apes (as a group). A given "difference" listed here could be a suggested gain or loss in humans, with respect to the great apes. This is a partial listing of topics that will appear later at a Web-based "Museum of Comparative Anthropogeny" (http://origins.ucsd.edu/gapp1.html).

phenotypic traits that appear to differentiate us from chimpanzees and other "great apes."[1] For the most part, we do not know which genetic features interact with the environment to generate these differences between the "phenomes"[2] of our two species. The chimpanzee has also long been seen as a model for human diseases because of its close evolutionary relationship. This is indeed the case for a few disorders. Nevertheless, it is a striking paradox that chimpanzees are in fact *not* good models for many major human diseases/conditions (see Table 2) (Varki 2000; Olson and Varki 2003). In retrospect, this should not be too surprising. After all, at least some major diseases of a species are likely related to (mal)adaptations during the recent evolutionary past of that species (Nesse and Williams 1995). Thus, comparisons with the chimpanzee genome could shed important light on the uniquely human pathogenic mechanisms of serious diseases. This, in turn, could point to novel approaches toward prevention or treatment. Opportunities to address broader questions in evolutionary biology also arise, i.e., to examine the evolutionary forces that underlie recent speciation and phenotypic divergence between two closely related mammalian taxa, as well as the mechanisms by which evolutionary novelties are generated. It is this intertwining of anthropogeny (the study of human origins), biomedical

---

[1] The term "great apes" is used here in the now colloquial sense, as genomic information no longer supports this species grouping (Goodman 1999). Under the currently more common classification, these species are now grouped together with humans in the family Hominidae.

[2] The term "phenome" has been used in multiple publications (e.g., Mahner and Kary 1997; Varki et al. 1998; Paigen and Eppig 2000; Nevo 2001; Walhout et al. 2002; Freimer and Sabatti 2003), but still lacks an accepted definition. Discussions with researchers who have used the term suggest the following definition: "The body of information describing an organism's phenotypes, under the influences of genetic and environmental factors."

Table 2. Differences between humans and apes in incidence or severity of medical conditions

Medical condition	Humans	Great apes
Definite		
HIV progression to AIDS	Common	Very rare
Hepatitis B/C late complications	Moderate to severe	Mild
P. falciparum malaria	Susceptible	Resistant
Myocardial infarction	Common	Very rare
Endemic infectious retroviruses	Rare	Common
Influenza A symptomatology	Moderate to severe	Mild
Probable		
Menopause	Universal	Rare?
Alzheimer's disease pathology	Complete	No neurofibrillary tangles
Epithelial cancers	Common	Rare?
Atherosclerotic strokes	Common	Rare?
Hydatiform molar pregnancy	Common	Rare?
Possible		
Rheumatoid arthritis	Common	Rare?
Endometriosis	Common	Rare?
Toxemia of pregnancy	Common	Rare?
Early fetal wastage (aneuploidy)	Common	Rare?
Bronchial asthma	Common	Rare?
Autoimmune diseases	Common	Rare?
Major psychoses	Common	Rare?

See Varki (2000); Olson and Varki (2003); and references therein. This list excludes disease states explained by anatomical differences (e.g., difficult labor, varicose veins, spine disorders, hemorrhoids, hernias).

interests, and general evolutionary principles that make the chimpanzee genome such an invaluable resource.

Here, we briefly mention some of the initial findings from sequencing the chimpanzee genome, and describe how these data might be used to address some of the above questions. This pursuit requires the involvement and cooperation of scientists from a wide variety of fields, far beyond the scope of genomics. Thus, our comments are focused not so much on genomics per se, but are rather addressed to the broader scientific community of "genome users."

SEQUENCING OF THE CHIMPANZEE GENOME

Less than a decade ago, sequencing the chimpanzee genome was not even on the "radar screen" of the major sequencing centers. Repeated public statements of interest from many other sectors of the scientific community (McConkey and Goodman 1997; McConkey and Varki 2000; Varki 2000) and increasing interest within the genome community eventually led to the writing of "white papers"[3] and the assignment of high priority to this effort (http://www.genome.gov/10002851). The recent analysis of the draft chimpanzee genome sequence (The Chimpanzee Sequencing and Analysis Consortium 2005), and the many "companion" papers (Cheng et al. 2005; Hughes et al. 2005; Linardopoulou et al. 2005) now provide researchers with a wealth of comparative genetic information.

The published sequence was from a single captive-born male of the *Pan troglodytes verus* subspecies. Sequence data obtained via a whole-genome shotgun approach to a BAC library generated an ~3.6× coverage, i.e, ~3.6-fold redundancy in sequencing reads from the autosomes (sex chromosomes have half that redundancy in a male). The assembled sequence covers ~94% of the genome, with 98% of the sequence having an estimated error rate of $\leq 10^{-4}$. Several additional chimpanzees (including other subspecies) were sequenced at lower coverage. In addition to identifying polymorphisms within chimpanzees, these data confirmed the high quality and completeness of the human genome sequence and established ancestral states of human single-nucleotide polymorphisms (SNPs).

DEFINING THE IMPORTANT DIFFERENCES: SEARCHING FOR NEEDLES IN A HAYSTACK

With the two genome sequences in hand, one can begin a systematic identification of genes, regulatory elements, and other functionally relevant genomic regions that differentiate humans and chimpanzees. Of course, what we are really exploring is a complex interplay between multiple genetic differences, interacting with diverse physiological, environmental, and cultural factors, eventually resulting in the observed phenotypic differences. Single-nucleotide divergence was estimated at ~1.23%, with ~1% corresponding to fixed species divergence and the remainder representing species-specific polymorphisms (The Chimpanzee

---

[3]Olson, M.V., Eichler, E.E., Varki, A., Myers, R.M., Erwin, J.M., and McConkey, E.H.A. 2002. White paper advocating complete sequencing of the genome of the common chimpanzee, Pan troglodytes (white paper submitted to NHGRI, February 2002).

Reich, D.E., Lander, E.S., Waterston, R., Pääbo, S., Ruvolo, M., and Varki, A. 2002. Sequencing the chimpanzee genome (white paper submitted to NHGRI, February 2002).

Sequencing and Analysis Consortium 2005). While insertion-deletion (indel) events were fewer, they represented 40–45 Mb in each species, i.e., ~90 Mb difference between the two, giving an ~3% divergence in this category. Thus, the overall divergence between the genomes is closer to 4%, in keeping with two recent studies (Britten 2002; Watanabe et al. 2004), but far greater than most previous estimates, which were made using shorter alignable sequence fragments. Fortunately, orthologous proteins are still extremely similar, with almost a third being identical, and the typical protein differing only by two amino acids between human and chimpanzees. Thus, the oft-repeated "<1% difference" still applies to amino acid sequences (The Chimpanzee Sequencing and Analysis Consortium 2005). However, a substantial proportion of the differences will likely be neutral with respect to understanding the human condition. The search for functionally important differences is further complicated because many of the important ones may not be within known coding sequences. In addition to protein evolution (Li and Saunders 2005), two other major hypotheses have been put forth to explain human-specific changes, i.e., changes in gene regulation (King and Wilson 1975) and loss-of-function changes (Olson and Varki 2003). We suggest simultaneous genomic and candidate gene approaches toward narrowing the field to the functionally relevant changes captured under these hypotheses.

GENOMIC APPROACH 1: NARROWING THE SEARCH TO THE IMPORTANT DIFFERENCES

Using Outgroups to Define Human-specific Changes

Noting a difference between the two genome sequences does not indicate which lineage experienced the change. We can also assume that roughly half of all the differences occurred on the lineage leading to chimpanzees. One or more "outgroups" are needed to determine the ancestral state (nucleotide or otherwise) at any given locus, and thus establish the subset of human-specific changes. The large divergence time between primates and rodents (>60 Myr) means that some mutational events and/or orthology between loci will be obscured when using rodents as outgroups. Sequencing additional primate genomes is thus important, as is the careful choice of an appropriate outgroup species. A primate close enough to humans and chimpanzees is needed to reliably determine substitutional polarity; however, some evolutionary distance is also useful to define regions of functional sequence conservation. The orangutan (*Pongo*) provides the appropriate level of sequence divergence—unlike *Gorilla*, which may be too closely related to humans and chimpanzees to provide a consistent signal of polarity (Satta et al. 2000; Klein and Takahata 2002). Sequencing of *Pongo* is already underway at the Genome Sequencing Centers (GSC) at Washington University, St. Louis, and at the Baylor College of Medicine. Rhesus macaque (*Macaca mulatta*) genome sequencing is also underway, led by the Baylor GSC, in collaboration with the J. Craig Venter Institute Joint Technology Center, and Washington University GSC. The latter genome, representing an Old World monkey, is useful not only for its greater phylogenetic distance from humans (~25 Myr divergence) (Goodman et al. 2005), but also because the long tradition of rhesus macaque use in biomedical research provides comparative biological information and easier access to tissues and biological materials. Eventually, we will need multiple, additional primate genomes to fully understand the changes that have contributed to traits distinguishing humans (Goodman et al. 2005).

More speculatively, studying the gorilla genome could help define some genetic features related to cognitive function. Until recent human encroachment, the gorilla probably faced less social and cognitive challenges, with a single male overseeing a harem of females, in a relatively predator-free and food-rich environment. This contrasts to the more complex nature of chimpanzee and orangutan environments and behaviors. Also, unlike chimpanzees and orangutans, gorillas fail to pass the "mirror self-recognition test" (Shillito et al. 1999) and have only rarely been observed to use tools (Breuer et al. 2005). It is possible that the gorilla lost some genetic endowments related to cognitive abilities that the great ape common ancestor already had, and comparison to the human, chimpanzee, and orangutan genomes may help identify such changes.

A confounding factor with respect to the ~3.6× chimpanzee genome sequence is the variable quality of the genome sequence and assembly process. Individual low-quality nucleotide positions can result in sites that appear falsely divergent between humans and chimpanzees. Sequence data from multiple individuals allows some such sites to be identified and eliminated. In addition, some higher-order discrepancies may be introduced by difficulties in assembling certain regions, resulting in false differences between chimpanzees and humans. All of these issues should be resolved as the "polishing" phase of the chimpanzee genome sequencing proceeds (E. Mardis, pers. comm.).

Excluding Intra-Species Polymorphisms

One must also ensure that an apparent genomic difference between humans and chimpanzees is not simply due to a polymorphism in one of the species (Ruvolo 2004). While such polymorphisms are interesting in their own right (e.g., informing us about regions that have undergone recent selective sweeps), they are by definition not contributors to species-specific differences. Sequences from multiple individuals of both species are needed to set aside intra-specific polymorphism and focus only on fixed differences. Some genome-wide polymorphism data for chimpanzees already exists, and human polymorphism can be initially assessed using the numerous SNPs defined in current databases (http://www.ncbi.nlm.nih.gov/projects/SNP/). Surveying sequence variation in a minimum of 10 globally distributed humans has also been suggested to ensure a high probability that a given sequence is fixed (Enard and Pääbo 2004). This minimum number will be higher for chimpanzees because of their greater intra-specific diversity (Gagneux et al. 1999; Kaessmann et al. 1999; Stone et al. 2002; Yu et al. 2003).

GENOMIC APPROACH 2: FOCUSING THE SEARCH

Examine Sites of Human-specific Chromosomal Changes

In addition to the few previously known chromosomal inversions and rearrangements between humans and chimpanzees (Yunis et al. 1980; Yunis and Prakash 1982; Nickerson and Nelson 1998; Fan et al. 2002a,b; Dennehey et al. 2004), several new smaller chromosomal regions containing likely inversions and rearrangements were detected (The Chimpanzee Sequencing and Analysis Consortium 2005; Newman et al. 2005). Targeted comparisons with other great ape genomes will help to define which of these events are human specific. One can then examine each breakpoint region in detail, searching for potential changes in local genes or regulatory elements, as has already been done in a few instances (Nickerson and Nelson 1998; Fan et al. 2002a,b; Dennehey et al. 2004).

Examine Sites of Human-specific Insertions and Deletions

Insertions or deletions (indels) can range from a few nucleotides to tens of kilobases, and have major impacts on gene structure, expression, and/or function. While the number of indel differences between human and chimpanzee genomes is lower than the number of single-nucleotide divergences (SNDs), fixation of such events could suggest that these losses/gains have been adaptive. There are already a few known examples of indels with potential functional consequences differentiating humans and chimpanzees (Table 3), including the human loss of *CMAH* gene function in humans due to a 92-bp exon deletion (Chou et al. 1998); the loss of two coding exons in the human *ELN* gene, which contributes to

Table 3. Some candidate genes and gene families that may contribute to phenotypic differences between humans and apes

Gene(s)	Gene product(s)	Unusual hominid or human-specific features	Potential relevance to the human condition
Individual genes			
FOXP2	Putative transcription factor with polyglutamine tract and forkhead DNA binding domain	Two human-specific amino acid changes	Mutant humans have motoric speech disorder (developmental verbal dyspraxia). Region positively selected and fixed in humans <200,000 years ago. (Enard et al. 2002b; Zhang et al. 2002)
MYH16	Myosin heavy chain 16	Human-specific 2-bp deletion causing frameshift—predicted 76-kD unstable head domain	Claimed to be cause of reduction in the type II fibers of human jaw muscle (Stedman et al. 2004; Perry et al. 2005)
CMAH	CMP-Neu5Ac hydroxylase	92-bp deletion of exon 6 causing frameshift and inactive enzyme. Fixed in modern humans	Absence of sialic acid Neu5Gc. Change in resistance or susceptibility to pathogens. Loss of ligand for some Siglecs. Dated ~2.5–3 Mya. Dietary Neu5Gc in meat became foreign antigen. (Chou et al. 1998, 2002; Irie et al. 1998; Hayakawa et al. 2001)

(*Continued*)

Table 3. (*Continued*)

Gene(s)	Gene product(s)	Unusual hominid or human-specific features	Potential relevance to the human condition
MAOA	Monoamine oxidase A	Human-specific nonconservative change Glu151Lys in active site	Substitution affects protein dimerization according to a 3D structural model and predicts functional change. (Andres et al. 2004)
ASPM	Modulator of mitotic spindle in neural progenitors?	Accelerated evolution in ape and human lineages	Deletions in ASPM lead to microcephaly. Presumed to be related to increased brain size and/or other features of human brain. (Zhang 2003; Dorus et al. 2004; Evans et al. 2004; Kouprina et al. 2004b; Mekel-Bobrov et al. 2005)
MCPH1	Microcephalin	As above	As above (Dorus et al. 2004; Evans et al. 2004, 2005)
TTR	Transthyretin	Decreased expression in humans in blood and brain	May be related to altered thyroid hormone metabolism in humans versus chimpanzees (Gagneux et al. 2001)
ST6GAL1	Alpha 2–6 sialyltransferase	Apparent human-specific up-regulation on epithelia	Can explain relative resistance of chimpanzees to human influenza A virus. Other consequences unknown. (Gagneux et al. 2003)

EMR4	EGF-TM7 receptor family	Human-specific deletion in exon 8. Frameshift	Predominantly expressed by immune system cells. Functional significance unknown. (Hamann et al. 2003)
PCDH11Y	Protocadherin XY	Duplicated onto Y in Yp11.2/Xq21.3 pseudoautosomal region only in humans	Expressed from Y and escapes X-inactivation? Y copy has undergone structural changes. Selectively expressed in brain. Probable adhesion molecule. Significance unknown, hypothesized to be involved in brain development, lateralization, and schizophrenia risk. (Ross et al. 2003; Blanco-Arias et al. 2004)
IL9R (Y)	Interleukin-9 receptor	As above	Expressed from Y and escapes X-inactivation? Growth factor for T cells, mast cells, and macrophages. Significance unknown. Related to asthma? (Vermeesch et al. 1997)
SPRY3	Sprouty 3	As above	Expressed from Y and escapes X-inactivation? Cysteine-rich protein—homolog of *Drosophila* antagonist of FGF

(*Continued*)

Table 3. (Continued)

Gene(s)	Gene product(s)	Unusual hominid or human-specific features	Potential relevance to the human condition
			signaling that patterns apical airways branching. Significance unknown. (Vermeesch et al. 1997)
SYBL1	Synaptobrevin-like	As above	Inactive on Y chromosome? Significance unknown. (Vermeesch et al. 1997)
KRTHAP1	Type 1 acidic hair keratin	Human-specific single bp substitution and termination codon	Different hair keratin expression pattern noted in the hair follicle. Inactivated 0.25 Mya? Possibly related to human:ape differences in hair. (Winter et al. 2001)
RLN	Relaxin hormone	Human-specific expression in placenta and corpus luteum	Possibly related to differences in reproductive biology (Evans et al. 1994)
ELN	Tropoelastin	2 exons deleted. Open reading frame maintained	Extracellular matrix component, including vascular wall. Alteration in vascular wall structure? (Szabo et al. 1999)
SIGLEC11	Siglec-11	Human-specific gene conversion by adjacent pseudogene, maintaining ORF	Change in binding specificity for sialic acids. Human-specific expression in brain microglia. Biological

CASP12P1	Caspase-12—cysteine protease related to ICE subfamily	Human-specific disruption of SHG box required for activity. Premature stop codon also in most humans	consequences unknown. (Hayakawa et al. 2005) In rodents, *Casp12* mediates apoptosis in response to ER stress. Human SNP can restore full-length caspase proenzyme which confers hypo-responsiveness to LPS-stimulated cytokine production but has no significant effect on apoptotic sensitivity. (Fischer et al. 2002)
Gene families			
OR (17p13, etc.)	Olfactory receptors	Many more human pseudogenes and fewer active genes in this large family	Related to diminished human olfactory capabilities? However, some intact genes show evidence of positive selection. (Gilad et al. 2003a,b 2004)
TAS2R (12p13, 7q31, 7q34, etc.)	Bitter taste receptors	Fixation of loss-of-function mutations	Proposed relaxation of selective constraint and loss of function (Wang et al. 2004; Fischer et al. 2005)
SIGLEC (19q13)	CD33-related innate immune system regulating genes	Mutations, deletions, gene conversions, expression changes	Sialic acid recognizing signaling receptors. Changes in binding, expression patterns, etc. Could

(*Continued*)

Table 3. (Continued)

Gene(s)	Gene product(s)	Unusual hominid or human-specific features	Potential relevance to the human condition
			be partly a secondary consequence of human loss of Neu5Gc. (Angata et al. 2001; Sonnenburg et al. 2004; The Chimpanzee Sequencing and Analysis Consortium 2005)
COX (multiple locations)	Mitochondria cytochrome oxidase subunits	Multiple genes show rapid evolution in hominids. COX5A specifically in humans	Altered electron transport chain. Enhanced oxidative phosphorylation postulated to support increased brain energy consumption? (Grossman et al. 2004; Goodman et al. 2005)
SPANX (Xq27.1)	Sperm proteins associated with nucleus—genes on X chromosome	SPANX-C is specific to humans. SPANX-B has duplicated in humans	Rapidly evolving in all hominids. Expressed in normal testis, and in some cancers. (Kouprina et al. 2004a)
Morpheus (multiple locations)	Proteins not characterized yet	Contained within large duplicated regions in humans and apes	Evidence of rapid evolution. Most extreme case of positive selection among hominids. Some human-specific sequences. Functional significance uncertain. (Johnson et al. 2001)

Gene (location)	Name	Observation	Comments
LILR (19q13.4)	Leukocyte Ig-like receptors	Rapid evolution, only few clear orthologs between chimpanzee and human	Part of a larger family of genes. Involved in recognizing "self" via molecules like MHC. (Canavez et al. 2001)
KIR (19q13.4)	Killer inhibitory receptors	Rapid evolution, only few orthologs clear between chimpanzee and human	Expressed in NK cells. Recognize "self" molecules like MHC. (Hao and Nei 2005; Sambrook et al. 2005)
TRG (7p14)	T cell receptors	4 TCRs are pseudogenes in humans	Part of a larger family of genes. Functional significance uncertain. (Meyer-Oslon et al. 2003)
FCGR1 (1p and 1q)	High affinity IgG-Fc receptors	Pericentric inversion, distinguishing human from chimpanzee chromosome 1	Functional significance of inversion uncertain. (Maresco et al. 1998)
IGKV (2p11.2)	κ light chains of immunoglobulins	Possible human specific duplication	Part of a larger family of genes. Functional significance uncertain. (Ermert et al. 1995)
GYP (4q28-q31, 2q14-q21)	Glycophorins	Accelerated evolution in humans	Red blood cell proteins. Rapid evolution of extracellular domain, likely due to selection pressure by merozoite stage of *Plasmodium falciparum*. (Rearden et al. 1990; Baum et al. 2002; Wang et al. 2003)

(*Continued*)

Table 3. (*Continued*)

Gene(s)	Gene product(s)	Unusual hominid or human-specific features	Potential relevance to the human condition
LCE (1q21)	Epidermal differentiation complex	High density of rapidly evolving genes	Proteins that help form the cornified layer of the skin barrier (Marshall et al. 2001; The Chimpanzee Sequencing and Analysis Consortium 2005)
CST (20p11)	Cystatins	As above	Physiological cysteine proteinase inhibitors (The Chimpanzee Sequencing and Analysis Consortium 2005)
PSG (19q13)	Pregnancy-specific β-1-glycoproteins	As above	High quantities secreted by placental trophoblasts. Exact physiologic role during pregnancy unknown. (The Chimpanzee Sequencing and Analysis Consortium 2005)

KRT (17q21)	Hair keratins and keratin-associated proteins	As above. Major components of the cytoskeleton in hair and skin epithelial cells (The Chimpanzee Sequencing and Analysis Consortium 2005)
WFDC (20q13)	Protein domains with homology to whey acidic protein (WAP)	As above. Postulated protease inhibitors. Possible host defense against invading micro-organisms or regulation of endogenous proteolytic enzymes. (The Chimpanzee Sequencing and Analysis Consortium 2005)

This list is not meant to be exhaustive. It also does not include genes that have specifically changed in chimpanzees, but not in humans (e.g., *MICA/B*, *HLA*); genes that are polymorphic within humans (e.g., *APOE*, *COMT*); or instances in which a human disease-causing amino acid mutation appears to be the wild-type state in the chimpanzee (e.g., *AIRE*, *MKKS*, *MLH1*, *MYOC*, *OTC*, and *PRSS1*).

extracellular matrix structure (Szabo et al. 1999), and the complete deletion of *SIGLEC13* in humans (Angata et al. 2004).

The idea that gene loss was a major contributor to human evolution remains an intriguing one (Olson 1999; Olson and Varki 2003). Interestingly, ~50 known or predicted human genes were found to be missing partially or entirely in the chimpanzee genome, and some of these differences were confirmed by PCR or Southern blotting (The Chimpanzee Sequencing and Analysis Consortium 2005). Confirmation of the ancestral state of these loci and reciprocal analysis of genes disrupted exclusively in humans requires additional primate outgroup data and further "polishing" of the chimpanzee genome sequence.

Examine Gene Duplications and Retroposed Genes

Gene duplication via segmental duplication or retrotransposition of mRNA sequences is an evolutionary mechanism for creating new genes with new biological functions. Duplicated genes can become nonfunctional (pseudogenes), neofunctional (acquire a new function), or subfunctional (adopt a portion of the previous function) (Ohno 1999; Hurles 2004). Such species-specific changes in copy number of gene families may allow for the evolution of new functions unique to the species—and are thus pertinent loci for investigation. A recent study reported that 33% of human duplications are human specific (Cheng et al. 2005); and with an estimated 200–300 species-specific retroposed gene copies in humans and chimpanzees (The Chimpanzee Sequencing and Analysis Consortium 2005), there is an ample landscape to explore. Of note, previous work suggests that humans have experienced more copy-number changes than the great apes (Fortna et al. 2004), such as appears to be the case for the *PRAME* cluster (Birtle et al. 2005) and the *SPANX-B* genes (Kouprina et al. 2004a). Also, some neofunctional retroposed loci such as *GLUD2* are thought to be involved in hominid brain function (Burki and Kaessmann 2004).

Identify Genes and Gene Families Showing Evidence of Human-specific Rapid Evolution

Genes that have the signature of accelerated evolution (Clark et al. 2003; Nielsen et al. 2005), i.e., a high ratio of nonsynonymous to synonymous substitutions (K_a/K_s ratios), are good candidates for further study. In particular, genes that show $K_a/K_s > 1$ are possible targets of positive selection (Messier and Stewart 1997; Yang and Bielawski 2000). Several loci with relatively high K_a/K_s ratios between the human and chimpanzee genomes were reported (The Chimpanzee Sequencing and Analysis Consortium 2005). Since a majority of nonsynonymous substitutions are considered deleterious (Enard and Pääbo 2004), a high rate of nonsynonymous substitution between taxa can suggest either adaptive evolution or relaxation of functional constraint. However, this approach is generally conservative (Yang and Bielawski 2000). For example, a K_a/K_s value of <1 does not rule out that a gene has undergone positive selection (Dorus et al. 2004). Also, a protein could have only one or a few important amino acid changes, perhaps confined to a critical domain, motif, or site (Andres et al. 2004; Sonnenburg et al. 2004), and thus not have an elevated K_a relative to K_s. Careful examination of the specific types or positions of amino acid changes such as radical amino acid substitutions (hydrophobic vs. hydrophilic, acidic vs. basic, etc.) in conserved regions is another potential way to identify important changes in protein sequence.

For genes that have zero synonymous changes between humans and chimpanzees, one has to use the adjacent genome sequence to estimate a local intergenic/intronic substitution rate, K_i. Of ~13,000 human–chimpanzee orthologs studied, ~4% had an observed $K_a/K_i > 1$ (The Chimpanzee Sequencing and Analysis Consortium 2005). However, given the low divergence between humans and chimpanzees, about half of these are predicted to occur simply by chance if purifying selection is allowed to act nonuniformly across genes.

Examine Sites of Human-specific Repetitive Element Insertion

Repetitive elements such as LINEs (long interspersed elements) and SINEs (short interspersed elements) can duplicate and spread throughout the genome by reverse transcription, causing potentially important functional changes in coding and flanking sequences (Smit 1999; Carroll et al. 2001). *Alu* elements are the most abundant class of SINEs in humans, making up ~10% of the genome (Lander et al. 2001), where they apparently expanded up to three times more than in the chimpanzee genome (The Chimpanzee Sequencing and Analysis Consortium 2005). In addition, most human-specific *Alu* elements belong to two subfamilies (Ya5 and Yb8) not found in great apes (Carroll et al. 2001). Identification of these human-specific loci makes them candidates for further inquiry. It is possible that some of these elements inserted into functional genes or flanking regions became alternatively spliced introns or promotor regulators, or either deleted or shuffled genomic regions via *Alu–Alu* recombination.

Look for Human-specific Gene Conversions

Another potential source of differences arises from species-specific gene conversion events that become fixed. Gene conversion homogenizes coding or noncoding sequences between adjacent paralogous gene copies within a species. Conversion may also introduce harmful mutations from a pseudogenized gene copy into a functional copy, or conversely, restore function to a former pseudogene. For example, the 5' end of human *Siglec-11* was converted by an adjacent pseudogene after the common ancestor with chimpanzees (Hayakawa et al. 2005). This resulted in a change in sialic acid-binding properties, as well as new expression in human brain microglia. The gene-converted *Siglec-11* can thus be considered the first example of a human-specific protein. More such examples might be found by systematically screening genomic regions, wherein genes and paralogous pseudogenes are nearby one another.

Look for Changes in Noncoding Regions

A majority of comparative genomic studies have focused on coding regions at the expense of examining regulatory sequences (Carroll 2005). However, given the relatively few protein-sequence differences between human and chimpanzees, differential regulation of gene and protein expression is a likely mechanism for explaining human:chimpanzee differences (King and Wilson 1975; Enard et al. 2002a; Caceres et al. 2003; Carroll 2003; Preuss et al. 2004; Uddin et al. 2004). Functional noncoding regions such as promoters, enhancers, flanking sequences, and introns can regulate the expression of genes (Wray et al. 2003), and thus play a role in human evolution. The wealth of new information being generated about noncoding RNA sequences also makes them an intriguing candidates for potential differences (Eddy 2001; Dykxhoorn et al. 2003; Mello and Conte 2004; Kim 2005; Tang 2005).

GENOMIC APPROACH 3: LOOKING FOR HUMAN-SPECIFIC GENE EXPRESSION DIFFERENCES

As mentioned above, species-specific changes in genomic sequence can be manifested in regulatory processes such as timing and location of expression of genes or of functional noncoding sequences, such as siRNAs. However, it is difficult to predict changes in expression simply by comparing genomic sequences (Carroll 2005). Differences in expression pattern between humans and chimpanzees are being investigated using microarray analyses, which allow for a rapid screen of multiple loci expressed in a single tissue at a given time point. Several such analyses and reanalyses have been carried out (Enard et al. 2002a, 2004; Caceres et al. 2003; Gu and Gu 2003; Hsieh et al. 2003; Khaitovich et al. 2004a; Preuss et al. 2004; Uddin et al. 2004). While the rate of brain gene expression changes appears increased in the human lineage, gene expression in the brain is overall more conserved than in other tissues, perhaps because of functional constraints in this complex organ (Enard et al. 2002a; Caceres et al. 2003; Gu and Gu 2003; Preuss et al. 2004). However, there are several caveats. For example, microarrays based on human oligonucleotide sequences may not accurately detect levels of expression in nonhuman primates nor detect significant alternative splicing of mRNAs (Modrek and Lee 2002; Hsieh et al. 2003; Preuss et al. 2004; Steinmetz and Davis 2004). Additionally, mRNA levels are not always good predictors of the actual levels of the gene product found in a cell (Gygi et al. 1999). Moreover, a recent study suggests that most expression differences have little or no significance, and are likely due to neutral evolution (Khaitovich et al. 2004b). Finally, many of the ultimate "gene products" are not the proteins themselves, but result from their enzymatic activity (e.g., lipids, glycans, and bioactive small molecules). Thus, gene expression studies must be complemented by a variety of other "omic" approaches, e.g., proteomics, lipomics, glycomics, etc. Any differences found need to be confirmed by focused biochemical studies on the molecules in question.

CANDIDATE GENE APPROACHES

In parallel with the above genomic studies, it is important to continue the more traditional candidate gene approach—as the genomic approach can miss many biologically significant differences. The candidate approach focuses on specific genes, based on some a priori knowledge about which loci or system(s) might be expected to show functionally significant differences between humans and chimpanzees.

Candidate Gene Approach 1: Making Choices on the Basis of Comparative Phenomics

Humans and chimpanzees differ in many morphological, cognitive, and physiological arenas. When attempting to identify the genetic mechanisms responsible, it is logical to focus on genes known or predicted to contribute in some way to the phenotypic differences, i.e., differences in the "phenome."

There are many morphological and physiological traits for which we have some knowledge of the responsible genetic pathways. This can, in turn, allow us to identify appropriate candidate loci underlying the traits. We can then test them for their contribution to uniquely human traits affecting organs such as the skin, brain, and female reproductive system (Table 1). Additionally, many diseases and pathological conditions appear to be unique to humans, and genes involved in some of these disease pathways are known or can be predicted (Table 2). It makes sense to focus first on phenotypes or diseases that appear most directly relevant to explaining the human condition. For example, recent work has suggested that two genes involved in the regulation of brain size appear to have undergone human-specific adaptive evolution (Evans et al. 2005; Mekel-Bobrov et al. 2005). However, we would recommend against a purely "brain-centric" approach that assumes that the only major differences of interest are in the nervous system. A single genetic change may have had an impact on multiple organs, and such a change may be easier to study in organs other than the brain. For example, there are organs such as the skin and its derivatives (e.g., the female breast and the sweat glands) that show at least as many morphological and functional differences as the brain and are easier to study. Genetic differences found in such systems may then help predict which molecules, pathways, or mechanisms have also undergone the most drastic changes during the evolution of the human brain.

Candidate Gene Approach 2: Choices Based on Naturally Occurring Human Mutations

A population size of 6 billion humans suggests that many postnatally viable genetic diseases affecting "uniquely human" traits are likely to exist somewhere on the planet. Identifying such defects in the human population, particularly in families, provides an approach for directly linking genotype to phenotype and for choosing genes for human and chimpanzee comparisons. The medical community in particular should be educated and vigilant about such opportunities. A striking outcome of this type of approach is *FOXP2*, a transcription factor shown to be associated with an inherited human disorder of speech production (Enard et al. 2002b; Zhang et al. 2002). Intriguingly, this putative transcription factor was found to have two human-specific amino acid changes, and the genomic region in question appears to have been positively selected and fixed in humans <200,000 years ago (Enard et al. 2002b). The next step is to look at the consequences of such abnormal genotypes in vitro and by developing transgenic mice that manifest symptoms of the condition.

Indeed, mice with a disruption in a single copy of the murine *Foxp2* gene manifest a modest developmental delay and a significant alteration in ultrasonic vocalizations that are normally elicited when pups are removed from their mothers (Shu et al. 2005).

Another intriguing finding is that some amino acid sequence variants that cause disease in humans turn out to be a reversion to the conserved ancestral state, still present in the normal chimpanzee (The Chimpanzee Sequencing and Analysis Consortium 2005). This phenomenon has been explained as being due to a high rate of compensatory mutations at other sites in the same protein. Assuming that such mutations are more likely to be fixed by positive selection than by neutral drift, these genes are candidates for adaptive differences between humans and chimpanzees.

Candidate Gene Approach 3: Making Choices Based on Sequence Data

Both "top-down" and "bottom-up" approaches have been successfully used to identify genes potentially involved in uniquely human phenotypes. As discussed above, the "phenome-down" candidate approach involves selecting a candidate gene based on phenotypic information and doing a first-pass genomic workup before proceeding to functional analyses of the gene product. Conversely, a "genome-up" approach (see genomic approach 2 above) can identify genes involved in a particular pathway or system that may be diverged enough (i.e., high K_a/K_s values) or expressed in the organs of interest, or harbor amino acid mutations (i.e., in a conserved domain) to be of interest in a functional screen. Following a search for such candidate genes, a narrowed-down list of loci can then be prioritized according to putative function and position in the pathway of interest (as opposed to a complete list of loci generated from a purely genome-based search, with little functional knowledge linked to them). For example, the human sequence could be "chimpanized" and the gene product compared with that of the native human and chimpanzee gene product via in vitro or transgenic mouse studies in order to investigate the effect of particular sequence changes on a given phenotype.

One example of a functional genetic difference discovered through candidate genomic sequence analysis is *MYH16*, initially identified as a putative member of the myosin heavy-chain family (Stedman et al. 2004). The *MHY16* gene product is most prominently expressed in the jaw muscles of vertebrates. Humans are homozygous for a defective frame-shifted allele. Stedman et al. (2004) dated the mutation to ~2.4 Mya, approximately the time of origin of the genus *Homo*, and hypothesized that the human-specific loss of *MYH16* function may have affected craniofacial morphology and/or selected for the evolution of larger brain size (Currie 2004; Stedman et al. 2004). Alternatively, since human ancestors switched to a less herbivorous diet at about that time, the loss of *MHY16* and jaw muscle strength might simply have been inconsequential and thus drifted to fixation (Currie 2004).

A subsequent analysis of a much larger region of exonic and intronic sequence data flanking the deletion (30,000 vs. 1000 bp) estimated the age of the mutation as ~5 Mya (Perry et al. 2005), consistent with the timing of human–chimpanzee divergence rather than with the origin of *Homo*. While this may cast doubt on *MYH16*'s role in the evolution of *Homo*, the fact remains that the frameshift is human specific and belongs in the repertoire of human–chimpanzee genetic differences. Whether it contributed in some meaningful way to species-specific character differences is a question for continued investigation. Regardless, the availability of the human and chimpanzee genome sequences facilitated the expanded analysis, and underscores the important role that genome sequences can play in our understanding about evolutionary history.

Candidate Gene Approach 4: A "Systems Approach" to Promising Groups of Genes

The traditional and powerful approach to genome-wide analysis has been to either consider homologous gene families or genes that are grouped together by similar functions, as in the Gene Ontology (GO) System (Ashburner et al. 2000; The Chimpanzee Sequencing and Analysis Consortium 2005). Of course, since most genes are highly interrelated and function in multiple pathways and systems, no single classification system can do justice to all of the possibilities. One complementary approach is to select a biological process or system not defined under a traditional GO category and focus attention on groups of genes that are thought to be involved. Taking such an approach, Dorus et al. (2004) recently found that a group of genes involved in nervous-system development and function showed evidence of accelerated evolution, when compared with other "housekeeping" genes. A related approach is to assume that a major change in a single gene is likely to affect the evolution of other genes that are functionally connected. For example, following up on the discovery of the *CMAH* mutation affecting synthesis of one type of sialic acid (Chou et al. 1998), multiple functional genetic differences in the biology of sialic acids have been identified between humans and great apes (Angata et al. 2001; Gagneux et al. 2003; Sonnenburg et al. 2004). Since

<60 genes are directly involved in all of the major processes of sialic acid biology, it is reasonable to suggest that this system underwent multiple related changes at some point(s) in human evolution. A systematic comparative analysis of all of these genes between humans and chimpanzees is underway. However, the genes in question are not identified in the GO system as belonging to a single category. By exercising both caution and creativity in how they identify loci united in a biological process, researchers will likely come up with new and novel insights into human and chimpanzee evolution.

CONCLUSIONS

Sequencing of the chimpanzee genome signals not an end, but rather a beginning for researchers across diverse fields. The impressive array of data and analyses that have come from this sequencing has provided researchers with new and novel insights into rates and results of molecular processes such as nucleotide substitutions, gene duplications, insertions and deletions, retrotranspositions, and potential karyotypic changes. These data will provide the springboard for understanding the potential consequences of changes in these attributes between humans and chimpanzees. Over the years, scientists have proposed many theories about what makes humans different from the great apes, ranging from subtle changes in regulatory regions (King and Wilson 1975) all the way to the differential loss of gene activity in humans (Olson 1999; Olson and Varki 2003). In fact, given the rather complex series of events evident in the hominid fossil record (Wood and Collard 1999; Cela-Conde and Ayala 2003), every one of these hypothesized genetic mechanisms likely contributed to some degree to human–chimpanzee differences. Understanding what makes us evolutionarily, biomedically, and cognitively different from chimpanzees will require extensive comparative phenomics to complement the comparative genomics now possible using the chimpanzee genome. However, despite decades of research on wild and captive chimpanzees, our overall knowledge about the chimpanzee phenome is very incomplete (Gagneux 2004; Olson and Varki 2004; McConkey and Varki 2005). Studies of intra-specific variation among great apes are in their infancy, and biomedical and physiological data are few. This lack of comparative phenotypic data represents a serious knowledge imbalance. Better phenomic data would enhance our ability to make additional, focused choices for candidate gene studies, and also increase our understanding of the biochemical consequences of any genomic changes we do find. One step to extend the utility of the genome project is to have the phenome much better defined, not only through morphological and anatomical studies, but also via systematic collection of existing data in all fields relevant to understanding the human condition (practically speaking, most of the biological and social sciences). A recently initiated "Great Ape Phenome Project" will begin this process (Varki et al. 1998; Gagneux 2004; Olson and Varki 2004). Of course, the critically endangered status of great apes in the wild, and the fiscal, logistical, and ethical issues of studying great apes in captivity (Gagneux et al. 2005; McConkey and Varki 2005) create a situation wherein new data and resources will not be easy to come by. Regardless, for the purposes of comparison, there is no point in doing any study on a captive great ape that one would not also do on a human subject (Gagneux et al. 2005). Also, all studies on captive apes should try to financially contribute toward their conservation in the wild, e.g., via a proposed Great Apes Conservation Trust, which would receive a 10% overage on all grant funds awarded by various agencies for research projects on ape genomes, phenomes, or behavior (McConkey and Varki 2005).

In the absence of adequate comparative phenomic data between humans and chimpanzees, genomic data provides only part of the blueprint for the phenotype. It is crucial, after identifying differences in the genomic data, to ascertain which ones are important by studying their biological consequences in the laboratory. For example, the relatively limited genomic differences between humans and chimpanzees mean that identifying statistically meaningful differences in rates of evolution are difficult. This limitation will hamper our ability to identify genes or regions of biological interest and importance. Additional primate outgroups will be important for detecting selection over longer time periods and for eliminating false positives. Also, genomic data alone cannot predict epistatic interactions between various loci, nor can it reveal the pleiotropic effects of changes that have occurred in a single gene. Comparative functional studies are necessary to reap the full potential of the genomic data, to translate the observed genetic changes into tangible quantitative differences. However, even such systematic functional studies may not capture the full magnitude of a difference's importance by examining only a single player in a multiplayer interaction. It is likely that, while there may be single-gene changes of large consequence, there will also be synergistic effects of many minor changes at multiple loci. That is, the human condition is likely to be the result of many small effect changes, not just a few large effect mutations. These smaller, subtle changes will be difficult to detect by genomic methods. On the other hand, even clearly identifiable genomic and phenomic differences between humans and chimpanzees may not be directly related to speciation nor to the question of "what makes us human." Such differences may be a simple byproduct of neutral divergence or genetic drift.

Also, what might seem an important phenotypic difference between humans and great apes might not actually be the most critical factor in determining unique features of the human condition. For example, despite the frequent attention given to big brain size (Wood and Collard 1999; Preuss 2005), there is little evidence for causative connections between brain size and human cognitive abilities (Preuss 2005). Additionally, maximum brain size was achieved long before the emergence of modern human behaviors (Klein 1999; Wood and Collard 1999). Thus, while increased brain size is an impressively human-specific phenotypic difference from great apes, it may well have been just one step (like bipedalism) that occurred earlier, along the way to the emergence of uniquely human cognitive features. Conversely, apparently small phenotypic differences could turn out to play major roles. For example, a small (approximately twofold) difference in the level of a thyroid hormone-binding protein and associated differences in thyroid hormone metabolism between humans and apes (Gagneux et al. 2001) could turn out to be as important as brain-expressed genes in altering the trajectory and mechanisms of human brain development.

Explaining "humanness" is a vague and broadly philosophical question, not easily approached using the genome alone. We prefer to use the term "the human condition" to refer to the entire suite of characters that makes humans different from the great apes. What it means to be human involves quantitative aspects of biochemistry, physiology, and morphology, as well as more qualitative arenas such as cognition, behavior, symbolic communication, and culture. However, unlike typical biological questions, the great majority of experiments one might propose for studying the consequences of species-specific genetic changes are unethical and/or impractical to do, either in humans or in great apes (Gagneux et al. 2005; McConkey and Varki 2005). Meanwhile, studies in mice may not provide sufficient answers. Thus, we suggest that many answers must come from a logical inductive approach that synthesizes many various "clues" to arrive at the best possible "diagnosis". Also, apparently minor differences between humans and great apes could turn out to be critical. For all of these reasons, we must keep an open mind, and leave no clue unattended to, even if it may appear trivial at first glance. It may well be that findings made from systems that are more ethically accessible and practical to study (such as the blood and the skin) will reveal clues that will eventually allow generation of testable hypotheses about organs like the brain. The other reason to take this type of broad approach to the "human condition" is that there are major biomedical lessons to be learned, which will benefit both humans and great apes, even though they may not be useful in explaining "humanness" in its philosophical sense.

Because of the many limitations mentioned above, we will have to arrive at many of our conclusions by considering all of the facts in aggregate, including some circumstantial evidence. In the final analysis, the best long-term approach to understanding human–chimpanzee differences is to ensure that the next generation of biologists interested in the evolution of the human phenotype is a cross-trained and collaborative one, with an interdisciplinary focus. Interactions among a great many disciplines, such as genomics, biochemistry, physiology, neurobiology, cognitive science, medicine, pathology, anthropology, ecology, primatology, and evolutionary biology, will be essential in dissecting out the key genetic features that contribute to making us human.

ACKNOWLEDGMENTS

We thank three anonymous reviewers, Anders Aannestad, Sandra Diaz, Pascal Gagneux, Hopi Hoekstra, Elaine Mardis, Tarjei Mikkelsen, Jennifer Stevenson, and Nissi Varki for valuable comments and suggestions. We also thank Jim Else, Liz Strobert, and Dan Anderson at the Yerkes Primate Center, Atlanta, GA, for helpful discussions about great ape diseases. A.V. was supported by grants from the NIGMS, NHLBI, and NCI and by the Harold G. and Leila Y. Mathers Charitable Foundation, and T.K.A. was supported by a postdoctoral fellowship from the American Cancer Society.

REFERENCES

Andres, A.M., Soldevila, M., Navarro, A., Kidd, K.K., Oliva, B., and Bertranpetit, J. 2004. Positive selection in MAOA gene is human exclusive: Determination of the putative amino acid change selected in the human lineage. *Hum. Genet.* **115:** 377–386.

Angata, T., Varki, N.M., and Varki, A. 2001. A second uniquely human mutation affecting sialic acid biology. *J. Biol. Chem.* **276:** 40282–40287.

Angata, T., Margulies, E.H., Green, E.D., and Varki, A. 2004. Large-scale sequencing of the CD33-related Siglec gene cluster in five mammalian species reveals rapid evolution by multiple mechanisms. *Proc. Natl. Acad. Sci.* **101:** 13251–13256.

Ashburner, M., Ball, C.A., Blake, J.A., Botstein, D., Butler, H., Cherry, J.M., Davis, A.P., Dolinski, K., Dwight, S.S., Eppig, J.T., et al. 2000. Gene ontology: Tool for the unification of biology. The Gene Ontology Consortium. *Nat. Genet.* **25:** 25–29.

Baum, J., Ward, R.H., and Conway, D.J. 2002. Natural selection on the erythrocyte surface. *Mol. Biol. Evol.* **19:** 223–229.

Birtle, Z., Goodstadt, L., and Ponting, C.P. 2005. Duplication and positive selection among hominin-specific *PRAME* genes. *BMC Genomics* **6**: 120.

Blanco-Arias, P., Sargent, C.A., and Affara, N.A. 2004. A comparative analysis of the pig, mouse, and human *PCDHX* genes. *Mamm. Genome* **15**: 296–306.

Breuer, T., Ndoundou-Hockemba, M., and Fishlock, V. 2005. First observation of tool use in wild gorillas. *PLoS Biol.* **3**: e380.

Britten, R.J. 2002. Divergence between samples of chimpanzee and human DNA sequences is 5%, counting indels. *Proc. Natl. Acad. Sci.* **99**: 13633–13635.

Brunet, M., Guy, F., Pilbeam, D., Mackaye, H.T., Likius, A., Ahounta, D., Beauvilain, A., Blondel, C., Bocherens, H., Boisserie, J.R., et al. 2002. A new hominid from the Upper Miocene of Chad, Central Africa. *Nature* **418**: 145–151.

Burki, F. and Kaessmann, H. 2004. Birth and adaptive evolution of a hominoid gene that supports high neurotransmitter flux. *Nat. Genet.* **36**: 1061–1063.

Caceres, M., Lachuer, J., Zapala, M.A., Redmond, J.C., Kudo, L., Geschwind, D.H., Lockhart, D.J., Preuss, T.M., and Barlow, C. 2003. Elevated gene expression levels distinguish human from non-human primate brains. *Proc. Natl. Acad. Sci.* **100**: 13030–13035.

Canavez, F., Young, N.T., Guethlein, L.A., Rajalingam, R., Khakoo, S.I., Shum, B.P., and Parham, P. 2001. Comparison of chimpanzee and human leukocyte Ig-like receptor genes reveals framework and rapidly evolving genes. *J. Immunol.* **167**: 5786–5794.

Carroll, S.B. 2003. Genetics and the making of *Homo sapiens*. *Nature* **422**: 849–857.

———. 2005. Evolution at two levels: On genes and form. *PLoS Biol.* **3**: e245.

Carroll, M.L., Roy-Engel, A.M., Nguyen, S.V., Salem, A.H., Vogel, E., Vincent, B., Myers, J., Ahmad, Z., Nguyen, L., Sammarco, M., et al. 2001. Large-scale analysis of the *Alu* Ya5 and Yb8 subfamilies and their contribution to human genomic diversity. *J. Mol. Biol.* **311**: 17–40.

Cela-Conde, C.J. and Ayala, F.J. 2003. Genera of the human lineage. *Proc. Natl. Acad. Sci.* **100**: 7684–7689.

Chen, F.C. and Li, W.H. 2001. Genomic divergences between humans and other hominoids and the effective population size of the common ancestor of humans and chimpanzees. *Am. J. Hum. Genet.* **68**: 444–456.

Cheng, Z., Ventura, M., She, X., Khaitovich, P., Graves, T., Osoegawa, K., Church, D., DeJong, P., Wilson, R.K., Pääbo, S., et al. 2005. A genome-wide comparison of recent chimpanzee and human segmental duplications. *Nature* **437**: 88–93.

The Chimpanzee Sequencing and Analysis Consortium. 2005. Initial sequence of the chimpanzee genome and comparison with the human genome. *Nature* **437**: 69–87.

Chou, H.H., Takematsu, H., Diaz, S., Iber, J., Nickerson, E., Wright, K.L., Muchmore, E.A., Nelson, D.L., Warren, S.T., and Varki, A. 1998. A mutation in human CMP-sialic acid hydroxylase occurred after the *Homo-Pan* divergence. *Proc. Natl. Acad. Sci.* **95**: 11751–11756.

Chou, H.H., Hayakawa, T., Diaz, S., Krings, M., Indriati, E., Leakey, M., Pääbo, S., Satta, Y., Takahata, N., and Varki, A. 2002. Inactivation of CMP-N-acetylneuraminic acid hydroxylase occurred prior to brain expansion during human evolution. *Proc. Natl. Acad. Sci.* **99**: 11736–11741.

Clark, A.G., Glanowski, S., Nielsen, R., Thomas, P.D., Kejariwal, A., Todd, M.A., Tanenbaum, D.M., Civello, D., Lu, F., Murphy, B., et al. 2003. Inferring nonneutral evolution from human–chimp–mouse orthologous gene trios. *Science* **302**: 1960–1963.

Currie, P. 2004. Human genetics: Muscling in on hominid evolution. *Nature* **428**: 373–374.
Dennehey, B.K., Gutches, D.G., McConkey, E.H., and Krauter, K.S. 2004. Inversion, duplication, and changes in gene context are associated with human chromosome 18 evolution. *Genomics* **83**: 493–501.
Dorus, S., Vallender, E.J., Evans, P.D., Anderson, J.R., Gilbert, S.L., Mahowald, M., Wyckoff, G.J., Malcom, C.M., and Lahn, B.T. 2004. Accelerated evolution of nervous system genes in the origin of *Homo sapiens*. *Cell* **119**: 1027–1040.
Dykxhoorn, D.M., Novina, C.D., and Sharp, P.A. 2003. Killing the messenger: Short RNAs that silence gene expression. *Nat. Rev. Mol. Cell. Biol.* **4**: 457–467.
Eddy, S.R. 2001. Non-coding RNA genes and the modern RNA world. *Nat. Rev. Genet.* **2**: 919–929.
Enard, W. and Pääbo, S. 2004. Comparative primate genomics. *Annu. Rev. Genomics Hum. Genet.* **5**: 351–378.
Enard, W., Khaitovich, P., Klose, J., Zollner, S., Heissig, F., Giavalisco, P., Nieselt-Struwe, K., Muchmore, E., Varki, A., Ravid, R., et al. 2002a. Intra- and interspecific variation in primate gene expression patterns. *Science* **296**: 340–343.
Enard, W., Przeworski, M., Fisher, S.E., Lai, C.S., Wiebe, V., Kitano, T., Monaco, A.P., and Pääbo, S. 2002b. Molecular evolution of *FOXP2*, a gene involved in speech and language. *Nature* **418**: 869–872.
Enard, W., Fassbender, A., Model, F., Adorjan, P., Pääbo, S., and Olek, A. 2004. Differences in DNA methylation patterns between humans and chimpanzees. *Curr. Biol.* **14**: R148–R149.
Ermert, K., Mitlohner, H., Schempp, W., and Zachau, H.G. 1995. The immunoglobulin k locus of primates. *Genomics* **25**: 623–629.
Evans, B.A., Fu, P., and Tregear, G.W. 1994. Characterization of two relaxin genes in the chimpanzee. *J. Endocrinol.* **140**: 385–392.
Evans, P.D., Anderson, J.R., Vallender, E.J., Choi, S.S., and Lahn, B.T. 2004. Reconstructing the evolutionary history of microcephalin, a gene controlling human brain size. *Hum. Mol. Genet.* **13**: 1139–1145.
Evans, P.D., Gilbert, S.L., Mekel-Bobrov, N., Vallender, E.J., Anderson, J.R., Vaez-Azizi, L.M., Tishkoff, S.A., Hudson, R.R., and Lahn, B.T. 2005. Microcephalin, a gene regulating brain size, continues to evolve adaptively in humans. *Science* **309**: 1717–1720.
Fan, Y., Linardopoulou, E., Friedman, C., Williams, E., and Trask, B.J. 2002a. Genomic structure and evolution of the ancestral chromosome fusion site in 2q13-2q14.1 and paralogous regions on other human chromosomes. *Genome Res.* **12**: 1651–1662.
Fan, Y., Newman, T., Linardopoulou, E., and Trask, B.J. 2002b. Gene content and function of the ancestral chromosome fusion site in human chromosome 2q13-2q14.1 and paralogous regions. *Genome Res.* **12**: 1663–1672.
Fischer, H., Koenig, U., Eckhart, L., and Tschachler, E. 2002. Human caspase 12 has acquired deleterious mutations. *Biochem. Biophys. Res. Commun.* **293**: 722–726.
Fischer, A., Gilad, Y., Man, O., and Pääbo, S. 2005. Evolution of bitter taste receptors in humans and apes. *Mol. Biol. Evol.* **22**: 432–436.
Fortna, A., Kim, Y., MacLaren, E., Marshall, K., Hahn, G., Meltesen, L., Brenton, M., Hink, R., Burgers, S., Hernandez-Boussard, T., et al. 2004. Lineage-specific gene duplication and loss in human and great ape evolution. *PLoS Biol.* **2**: E207.
Freimer, N. and Sabatti, C. 2003. The human phenome project. *Nat. Genet.* **34**: 15–21.
Gagneux, P. 2004. A *Pan*-Oramic view: Insights into hominid evolution through the chimpanzee genome. *Trends Ecol. Evol.* **19**: 571–576.

Gagneux, P. and Varki, A. 2001. Genetic differences between humans and great apes. *Mol. Phylogenet. Evol.* **18:** 2–13.

Gagneux, P., Wills, C., Gerloff, U., Tautz, D., Morin, P.A., Boesch, C., Fruth, B., Hohmann, G., Ryder, O.A., and Woodruff, D.S. 1999. Mitochondrial sequences show diverse evolutionary histories of African hominoids. *Proc. Natl. Acad. Sci.* **96:** 5077–5082.

Gagneux, P., Amess, B., Diaz, S., Moore, S., Patel, T., Dillmann, W., Parekh, R., and Varki, A. 2001. Proteomic comparison of human and great ape blood plasma reveals conserved glycosylation and differences in thyroid hormone metabolism. *Am. J. Phys. Anthropol.* **115:** 99–109.

Gagneux, P., Cheriyan, M., Hurtado-Ziola, N., Brinkman van der Linden, E.C., Anderson, D., McClure, H., Varki, A., Varki, N.M. 2003. Human-specific regulation of Alpha2-6 linked sialic acids. *J. Biol. Chem.* **278:** 48245–48250.

Gagneux, P., Moore, J.J., and Varki, A. 2005. The ethics of research on great apes. *Nature* **437:** 27–29.

Gilad, Y., Bustamante, C.D., Lancet, D., and Pääbo, S. 2003a. Natural selection on the olfactory receptor gene family in humans and chimpanzees. *Am. J. Hum. Genet.* **73:** 489–501.

Gilad, Y., Man, O., Pääbo, S., and Lancet, D. 2003b. Human specific loss of olfactory receptor genes. *Proc. Natl. Acad. Sci.* **100:** 3324–3327.

Gilad, Y., Wiebe, V., Przeworski, M., Lancet, D., and Pääbo, S. 2004. Loss of olfactory receptor genes coincides with the acquisition of full trichromatic vision in primates. *PLoS Biol.* **2:** E5.

Goodman, M. 1999. The genomic record of Humankind's evolutionary roots. *Am. J. Hum. Genet.* **64:** 31–39.

Goodman, M., Grossman, L.I., and Wildma, D.E. 2005. Moving primate genomics beyond the chimpanzee genome. *Trends Genet.* **9:** 511–517.

Grossman, L.I., Wildman, D.E., Schmidt, T.R., and Goodman, M. 2004. Accelerated evolution of the electron transport chain in anthropoid primates. *Trends Genet.* **20:** 578–585.

Gu, J. and Gu, X. 2003. Induced gene expression in human brain after the split from chimpanzee. *Trends Genet.* **19:** 63–65.

Gygi, S.P., Rochon, Y., Franza, B.R., and Aebersold, R. 1999. Correlation between protein and mRNA abundance in yeast. *Mol. Cell. Biol.* **19:** 1720–1730.

Hamann, J., Kwakkenbos, M.J., de Jong, E.C., Heus, H., Olsen, A.S., and van Lier, R.A. 2003. Inactivation of the EGF-TM7 receptor EMR4 after the *Pan-Homo* divergence. *Eur. J. Immunol.* **33:** 1365–1371.

Hao, L. and Nei, M. 2005. Rapid expansion of killer cell immunoglobulin-like receptor genes in primates and their coevolution with MHC Class I genes. *Gene* **347:** 149–159.

Hayakawa, T., Satta, Y., Gagneux, P., Varki, A., and Takahata, N. 2001. *Alu*-mediated inactivation of the human CMP-N-acetylneuraminic acid hydroxylase gene. *Proc. Natl. Acad. Sci.* **98:** 11399–11404.

Hayakawa, T., Angata, T., Lewis, A.L., Mikkelsen, T.S., Varki, N.M., and Varki, A. 2005. A human-specific gene in microglia. *Science* **309:** 1693.

Hsieh, W.P., Chu, T.M., Wolfinger, R.D., and Gibson, G. 2003. Mixed-model reanalysis of primate data suggests tissue and species biases in oligonucleotide-based gene expression profiles. *Genetics* **165:** 747–757.

Hughes, J.F., Skaletsky, H., Pyntikova, T., Minx, P.J., Graves, T., Rozen, S., Wilson, R.K., and Page, D.C. 2005. Conservation of Y-linked genes during human evolution revealed by comparative sequencing in chimpanzee. *Nature* **437:** 100–103.

Hurles, M. 2004. Gene duplication: The genomic trade in spare parts. *PLoS Biol.* **2:** E206.
International Human Genome Sequencing Consortium. 2004. Finishing the euchromatic sequence of the human genome. *Nature* **431:** 931–945.
Irie, A., Koyama, S., Kozutsumi, Y., Kawasaki, T., and Suzuki, A. 1998. The molecular basis for the absence of N-glycolylneuraminic acid in humans. *J. Biol. Chem.* **273:** 15866–15871.
Johnson, M.E., Viggiano, L., Bailey, J.A., Abdul-Rauf, M., Goodwin, G., Rocchi, M., and Eichler, E.E. 2001. Positive selection of a gene family during the emergence of humans and African apes. *Nature* **413:** 514–519.
Kaessmann, H., Wiebe, V., and Pääbo, S. 1999. Extensive nuclear DNA sequence diversity among chimpanzees. *Science* **286:** 1159–1162.
Khaitovich, P., Muetzel, B., She, X., Lachmann, M., Hellmann, I., Dietzsch, J., Steigele, S., Do, H.H., Weiss, G., Enard, W., et al. 2004a. Regional patterns of gene expression in human and chimpanzee brains. *Genome Res.* **14:** 1462–1473.
Khaitovich, P., Weiss, G., Lachmann, M., Hellmann, I., Enard, W., Muetzel, B., Wirkner, U., Ansorge, W., and Pääbo, S. 2004b. A neutral model of transcriptome evolution. *PLoS Biol.* **2:** E132.
Kim, V.N. 2005. MicroRNA biogenesis: Coordinated cropping and dicing. *Nat. Rev. Mol. Cell. Biol.* **6:** 376–385.
King, M.C. and Wilson, A.C. 1975. Evolution at two levels in humans and chimpanzees. *Science* **188:** 107–116.
Klein, R.G. 1999. *The human career: Human biological and cultural origins.* University of Chicago Press, Chicago, IL.
Klein, J. and Takahata, N. 2002. *Where do we come from? The molecular evidence for human descent.* Springer, New York.
Kouprina, N., Mullokandov, M., Rogozin, I.B., Collins, N.K., Solomon, G., Otstot, J., Risinger, J.I., Koonin, E.V., Barrett, J.C., and Larionov, V. 2004a. The SPANX gene family of cancer/testis-specific antigens: Rapid evolution and amplification in African great apes and hominids. *Proc. Natl. Acad. Sci.* **101:** 3077–3082.
Kouprina, N., Pavlicek, A., Mochida, G.H., Solomon, G., Gersch, W., Yoon, Y.H., Collura, R., Ruvolo, M., Barrett, J.C., Woods, C.G., et al. 2004b. Accelerated evolution of the ASPM gene controlling brain size begins prior to human brain expansion. *PLoS Biol.* **2:** 653–663.
Lander, E.S., Linton, L.M., Birren, B., Nusbaum, C., Zody, M.C., Baldwin, J., Devon, K., Dewar, K., Doyle, M., Fitzhugh, W., et al., 2001. Initial sequencing and analysis of the human genome. *Nature* **409:** 860–921.
Li, W.H. and Saunders, M.A. 2005. News and views: The chimpanzee and us. *Nature* **437:** 50–51.
Linardopoulou, E.V., Williams, E.M., Fan, Y., Friedman, C., Young, J.M., and Trask, B.J. 2005. Human subtelomeres are hot spots of interchromosomal recombination and segmental duplication. *Nature* **437:** 94–100.
Mahner, M. and Kary, M. 1997. What exactly are genomes, genotypes and phenotypes? And what about phenomes? *J. Theor. Biol.* **186:** 55–63.
Maresco, D.L., Blue, L.E., Culley, L.L., Kimberly, R.P., Anderson, C.L., and Theil, K.S. 1998. Localization of FCGR1 encoding Fcgamma receptor class I in primates: Molecular evidence for two pericentric inversions during the evolution of human chromosome 1. *Cytogenet. Cell Genet.* **82:** 71–74.
Marshall, D., Hardman, M.J., Nield, K.M., and Byrne, C. 2001. Differentially expressed late constituents of the epidermal cornified envelope. *Proc. Natl. Acad. Sci.* **98:** 13031–13036.

McConkey, E.H. and Goodman, M. 1997. A human genome evolution project is needed. *Trends Genet.* **13:** 350–351.

McConkey, E.H. and Varki, A. 2000. A primate genome project deserves high priority. *Science* **289:** 1295–1296.

———. 2005. Thoughts on the future of great ape research. *Science* **309:** 1499–1501.

Mekel-Bobrov, N., Gilbert, S.L., Evans, P.D., Vallender, E.J., Anderson, J.R., Hudson, R.R., Tishkoff, S.A., and Lahn, B.T. 2005. Ongoing adaptive evolution of *ASPM*, a brain size determinant in *Homo sapiens*. *Science* **309:** 1720–1722.

Mello, C.C. and Conte, D.J. 2004. Revealing the world of RNA interference. *Nature* **431:** 338–342.

Messier, W. and Stewart, C.B. 1997. Episodic adaptive evolution of primate lysozymes. *Nature* **385:** 151–154.

Meyer-Olson, D., Brady, K.W., Blackard, J.T., Allen, T.M., Islam, S., Shoukry, N.H., Hartman, K., Walker, C.M., and Kalams, S.A. 2003. Analysis of the TCR b variable gene repertoire in chimpanzees: Identification of functional homologs to human pseudogenes. *J. Immunol.* **170:** 4161–4169.

Modrek, B. and Lee, C. 2002. A genomic view of alternative splicing. *Nat. Genet.* **30:** 13–19.

Nesse, R.M. and Williams, G.C. 1995. *Why we get sick: The new science of Darwinian medicine*. Times Books, New York.

Nevo, E. 2001. Evolution of genome-phenome diversity under environmental stress. *Proc. Natl. Acad. Sci.* **98:** 6233–6240.

Newman, T.L., Tuzun, E., Morrison, V.A., Hayden, K.E., Ventura, M., McGrath, S.D., Rocchi, M., and Eichler, E.E. 2005. A genome-wide survey of structural variation between human and chimpanzee. *Genome Res.* **15:** 1344–1356.

Nickerson, E. and Nelson, D.L. 1998. Molecular definition of pericentric inversion breakpoints occurring during the evolution of humans and chimpanzees. *Genomics* **50:** 368–372.

Nielsen, R., Bustamante, C., Clark, A.G., Glanowski, S., Sackton, T.B., Hubisz, M.J., Fledel-Alon, A., Tanenbaum, D.M., Civello, D., White, T.J., et al. 2005. A scan for positively selected genes in the genomes of humans and chimpanzees. *PLoS Biol.* **3:** E170.

Ohno, S. 1999. Gene duplication and the uniqueness of vertebrate genomes circa 1970–1999. *Semin. Cell. Dev. Biol.* **10:** 517–522.

Olson, M.V. 1999. When less is more: Gene loss as an engine of evolutionary change. *Am. J. Hum. Genet.* **64:** 18–23.

Olson, M.V. and Varki, A. 2003. Sequencing the chimpanzee genome: Insights into human evolution and disease. *Nat. Rev. Genet.* **4:** 20–28.

———. 2004. Genomics. The chimpanzee genome—a bittersweet celebration. *Science* **305:** 191–192.

Paigen, K. and Eppig, J.T. 2000. A mouse phenome project. *Mamm. Genome* **11:** 715–717.

Perry, G.H., Verrelli, B.C., and Stone, A.C. 2005. Comparative analyses reveal a complex history of molecular evolution for human *MYH16*. *Mol. Biol. Evol.* **22:** 379–382.

Preuss, T.M. 2005. What is it like to be a human? In *The cognitive neurosciences*, 3rd ed. (ed. M.S. Gazzaniga), pp. 5–22. MIT Press, Cambridge, MA.

Preuss, T.M., Caceres, M., Oldham, M.C., and Geschwind, D.H. 2004. Human brain evolution: Insights from microarrays. *Nat. Rev. Genet.* **5:** 850–860.

Rearden, A., Phan, H., Kudo, S., and Fukuda, M. 1990. Evolution of the glycophorin gene family in the hominoid primates. *Biochem. Genet.* **28:** 209–222.

Ross, N.L., Mavrogiannis, L.A., Sargent, C.A., Knight, S.J., Wadekar, R., DeLisi, L.E., and Crow, T.J. 2003. Quantitation of X-Y homologous genes in patients with schizophrenia by multiplex polymerase chain reaction. *Psychiatr. Genet.* **13:** 115–119.

Ruvolo, M. 2004. Comparative primate genomics: The year of the chimpanzee. *Curr. Opin. Genet. Dev.* **14:** 650–656.

Sambrook, J.G., Bashirova, A., Palmer, S., Sims, S., Trowsdale, J., Abi-Rached, L., Parham, P., Carrington, M., and Beck, S. 2005. Single haplotype analysis demonstrates rapid evolution of the killer immunoglobulin-like receptor (*KIR*) loci in primates. *Genome Res.* **15:** 25–35.

Satta, Y., Klein, J., and Takahata, N. 2000. DNA Archives and our nearest relative: The trichotomy problem revisited. *Mol. Phylogenet. Evol.* **14:** 259–275.

Shillito, D.J., Gallup, G.G.J., and Beck, B.B. 1999. Factors affecting mirror behaviour in western lowland gorillas, *Gorilla gorilla*. *Anim. Behav.* **57:** 999–1004.

Shu, W., Cho, J.Y., Jiang, Y., Zhang, M., Weisz, D., Elder, G.A., Schmeidler, J., De Gasperi, R., Sosa, M.A., Rabidou, D., et al. 2005. Altered ultrasonic vocalization in mice with a disruption in the *Foxp2* gene. *Proc. Natl. Acad. Sci.* **102:** 9643–9648.

Smit, A.F. 1999. Interspersed repeats and other mementos of transposable elements in mammalian genomes. *Curr. Opin. Genet. Dev.* **9:** 657–663.

Sonnenburg, J.L., Altheide, T.K., and Varki, A. 2004. A uniquely human consequence of domain-specific functional adaptation in a sialic acid-binding receptor. *Glycobiology* **14:** 339–346.

Stedman, H.H., Kozyak, B.W., Nelson, A., Thesier, D.M., Su, L.T., Low, D.W., Bridges, C.R., Shrager, J.B., Minugh-Purvis, N., and Mitchell, M.A. 2004. Myosin gene mutation correlates with anatomical changes in the human lineage. *Nature* **428:** 415–418.

Steinmetz, L.M. and Davis, R.W. 2004. Maximizing the potential of functional genomics. *Nat. Rev. Genet.* **5:** 190–201.

Stone, A.C., Griffiths, R.C., Zegura, S.L., and Hammer, M.F. 2002. High levels of Y-chromosome nucleotide diversity in the genus *Pan*. *Proc. Natl. Acad. Sci.* **99:** 43–48.

Szabo, Z., Levi-Minzi, S.A., Christiano, A.M., Struminger, C., Stoneking, M., Batzer, M.A., and Boyd, C.D. 1999. Sequential loss of two neighboring exons of the tropoelastin gene during primate evolution. *J. Mol. Evol.* **49:** 664–671.

Tang, G. 2005. siRNA and miRNA: An insight into RISCs. *Trends Biochem. Sci.* **30:** 106–114.

Uddin, M., Wildman, D.E., Liu, G., Xu, W., Johnson, R.M., Hof, P.R., Kapatos, G., Grossman, L.I., and Goodman, M. 2004. Sister grouping of chimpanzees and humans as revealed by genome-wide phylogenetic analysis of brain gene expression profiles. *Proc. Natl. Acad. Sci.* **101:** 2957–2962.

Varki, A. 2000. A chimpanzee genome project is a biomedical imperative. *Genome Res.* **10:** 1065–1070.

Varki, A., Wills, C., Perlmutter, D., Woodruff, D., Gage, F., Moore, J., Semendeferi, K., Benirschke, K., Katzman, R., Doolittle, R., et al. 1998. Great Ape Phenome Project? *Science* **282:** 239–240.

Vermeesch, J.R., Petit, P., Kermouni, A., Renauld, J.C., Van Den Berghe, H., and Marynen, P. 1997. The IL-9 receptor gene, located in the Xq/Yq pseudoautosomal region, has an autosomal origin, escapes X inactivation and is expressed from the Y. *Hum. Mol. Genet.* **6:** 1–8.

Walhout, A.J., Reboul, J., Shtanko, O., Bertin, N., Vaglio, P., Ge, H., Lee, H., Doucette-Stamm, L., Gunsalus, K.C., Schetter, A.J., et al. 2002. Integrating interactome, phenome, and transcriptome mapping data for the *C. elegans* germline. *Curr. Biol.* **12**: 1952–1958.

Wang, H.Y., Tang, H., Shen, C.K., and Wu, C.I. 2003. Rapidly evolving genes in human. I. The glycophorins and their possible role in evading malaria parasites. *Mol. Biol. Evol.* **20**: 1795–1804.

Wang, X., Thomas, S.D., and Zhang, J. 2004. Relaxation of selective constraint and loss of function in the evolution of human bitter taste receptor genes. *Hum. Mol. Genet.* **13**: 2671–2678.

Watanabe, H., Fujiyama, A., Hattori, M., Taylor, T.D., Toyoda, A., Kuroki, Y., Noguchi, H., BenKahla, A., Lehrach, H., Sudbrak, R., et al. 2004. DNA sequence and comparative analysis of chimpanzee chromosome 22. *Nature* **429**: 382–388.

Winter, H., Langbein, L., Krawczak, M., Cooper, D.N., Jave-Suarez, L.F., Rogers, M.A., Praetzel, S., Heidt, P.J., and Schweizer, J. 2001. Human type I hair keratin pseudogene *phihHaA* has functional orthologs in the chimpanzee and gorilla: Evidence for recent inactivation of the human gene after the *Pan-Homo* divergence. *Hum. Genet.* **108**: 37–42.

Wood, B. and Collard, M. 1999. Anthropology—the human genus. *Science* **284**: 65–66.

Wray, G.A., Hahn, M.W., Abouheif, E., Balhoff, J.P., Pizer, M., Rockman, M.V., and Romano, L.A. 2003. The evolution of transcriptional regulation in eukaryotes. *Mol. Biol. Evol.* **20**: 1377–1419.

Yang, Z. and Bielawski, J.P. 2000. Statistical methods for detecting molecular adaptation. *Trends Ecol. Evol.* **15**: 496–503.

Yu, N., Jensen-Seaman, M.I., Chemnick, L., Kidd, J.R., Deinard, A.S., Ryder, O., Kidd, K.K., and Li, W.H. 2003. Low nucleotide diversity in chimpanzees and bonobos. *Genetics* **164**: 1511–1518.

Yunis, J.J. and Prakash, O. 1982. The origin of man: A chromosomal pictorial legacy. *Science* **215**: 1525–1530.

Yunis, J.J., Sawyer, J.R., and Dunham, K. 1980. The striking resemblance of high-resolution G-banded chromosomes of man and chimpanzee. *Science* **208**: 1145–1148.

Zhang, J. 2003. Evolution of the human *ASPM* gene, a major determinant of brain size. *Genetics* **165**: 2063–2070.

Zhang, J., Webb, D.M., and Podlaha, O. 2002. Accelerated protein evolution and origins of human-specific features. *FOXP2* as an example. *Genetics* **162**: 1825–1835.

18

Structure and Function of the Human Genome

Peter F.R. Little
School of Biotechnology and Biomolecular Sciences
University of New South Wales
Sydney 2074, New South Wales, Australia

THE PAST DECADE IN BIOLOGICAL RESEARCH HAS SURELY been the decade of genome research—from the scientific perspective, in the public imagination, and even in the minds of international politicians. It is therefore timely to use this 10th anniversary of *Genome Research* to take stock of where we are and where we might be in another decade in our understanding of the human genome.

The scale of the human DNA sequence must mean that no reviewer can capture all of the information it contains and therefore I concentrate on what novel information emerges from the completed sequence rather than on the detail of what we learned from each gene or each base.

The Human Genome Project (HGP) has had scientific and political impacts on biological research; scientifically, it has provided a novel conceptual dimension to human biology, that of "completeness." This word captures the idea that we now have finite bounds to research because the genome sequence contains all of the information that is used in making human cells and organisms. We can soon legitimately claim to study the behavior of all of our genes in a way that was quite inconceivable prior to the availability of the sequence. Politically, the HGP is changing our perspectives on how biological research can be organized in our institutions. This review inevitably focuses on the scientific outcomes, but toward the end of the review, I discuss the idea that perhaps the HGP's significant long-term impact will be on the organization of scientific research.

The original inception of the HGP included optimistic views of the impact of knowledge of our genome on biomedical research (see, e.g., Collins et al. 1998), and the first biomedical impacts of the HGP are fundamental insights rather than pharmaceutical outcomes. For example, sequence analysis has led to the identification of new oncogenes (for review, see Strausberg et al. 2004), and microRNA composition is being used as a novel classifier of human tumors (He et al. 2005; Lu et al. 2005), but such information is presently distant from therapeutic outcome. The lack of immediate application of HGP data is unsurprising given the >10-yr drug development pipeline (Dickson and Gagnon 2004). Over the next decade we will see an accumulation of basic knowledge derived from the genome sequence, and this will then inform therapeutics, suggesting that benefit must necessarily be deferred.

If the biomedical goals of the HGP are in the future, the immediate outcomes expected by the scientific community were perhaps more pragmatic; a description of the gross structure of the human DNA sequence, the number of genes, and the proteins these might encode. Along with these reasonable expectations was the hope that the primary DNA sequence would reveal clues as to the control of gene expression. Secondary outcomes included describing the sequence variation between humans and, closely related to this, insights into the evolutionary and population history of our genome.

The present (assembly number 35, May 2004) human DNA sequence contains ~3,100,000,000 bp (depending on the actual source of the assembled DNA sequence) that covers most of the nonheterochromatic portions of the genome and contains some 250 gaps (see Fig. 1). Its analysis has produced both predictable and novel insights. In the predictable category are the complete description of base compositional bias, the variation of rates of recombination in relation to the physical DNA length, the high proportion of the genome comprising repetitive DNA sequences, and, more ambivalently, the identification of many genes of known and unknown function (Venter et al. 2001; International Human Genome Sequencing Consortium 2004). Essentially, in these areas, the HGP has simply extended what we already knew without adding wholly novel insight. In contrast, and the primary focus of this review, unexpected insights are being gained from the identification and analysis of genes and their distribution, the amount of transcription of non-protein-coding regions, and the large-scale duplication structure of the genome.

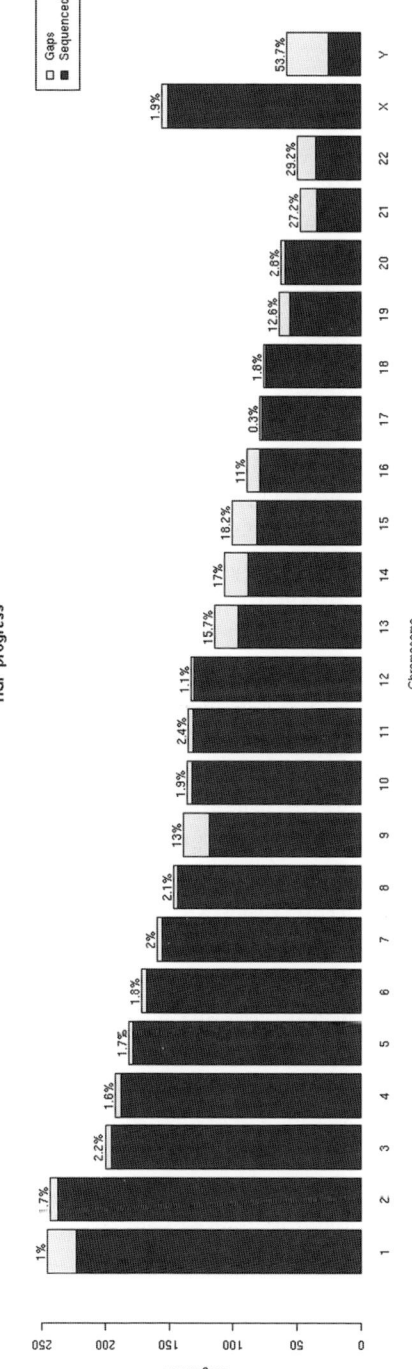

Figure 1. The sequenced (gray filled) and unsequenced (white) portions of the human genome, listed by chromosome; numbers in % are the proportion of chromosomes that are heterochromatic and unsequenced for this reason. Statistics are from the NCBI Build 35, UCSC assembly of May 2004, Assembly 17; data from http://www.genome.ucsc.edu/goldenPath/stats.html#hg17.

GENES IN THE HUMAN GENOME

Perhaps the most publicly discussed result of the HGP was the realization that we have ~20,000–25,000 genes (International Human Genome Sequencing Consortium 2004), somewhat fewer than estimates based on the preliminary reports of the human sequence (International Human Genome Sequencing Consortium 2001; Venter et al. 2001). Identifying genes—the process known as "annotation"—has predominantly been achieved through bioinformatics, most particularly by homology analyses and some de novo gene predictions. These data are readily accessible through several large genome "browsers" (for review, see Karolchik et al. 2003; Birney et al. 2004). The recent detailed analysis of 1% of the human genome under the ENCODE (ENCyclopedia Of DNA Elements) project (ENCODE Project Consortium 2004; http://www.genome.gov/10005107) indicates that these approaches have a relatively high success rate at identifying the presence of a gene within a region but a much lower success in predicting the gene's structure correctly (see, e.g., Brent and Guigo 2004); this suggests that annotation may underestimate gene number but not substantially. The low gene number prompted press comment on the difficulty of equating human complexity with apparent genetic simplicity; such comment seems to ignore the extraordinary combinatorial possibilities that can be generated from the interaction of even small numbers of gene products, a fact noted well before the final figure had been released (Ewing and Green 2000).

The extensive annotation process also confirmed the importance of alternative splicing in creating proteome diversity. Presently, estimates for the per gene frequency of alternative splicing range from 35% to ~60% (Johnson et al. 2003), but there remains substantial uncertainty in determining the extent to which these estimates reflect functionally significant splices or splice errors (for review, see Sorek et al. 2004). The influence of alternative splicing on proteome complexity (for review, see Southan 2004) is a matter of substantial biological importance, and lack of precision in predicting genes, gene structures, and alternative splices necessarily limits the present utility of genomic information; these are areas that must see substantial direct experimentation before a nearly complete data set can emerge.

Non-Protein-coding RNA Transcripts: The Relationship of Genes and Transcribed Regions

In parallel with the low gene number, there is accumulating evidence that there are many transcripts that appear to be non-protein-coding and of no known function (Cheng et al. 2005; Kapranov et al. 2005; for review, see Johnson et al. 2005), an observation that is mirrored in the mouse (for review, see Suzuki and Hayashizaki 2004). In humans, the original observations were controversial both because the level of RNA produced from these so-called transfrags (transcribed fragments) can be low and also because transfrags are often not annotated as genes; both concerns prompted doubts about the biological importance of such transcription. There are several reasons that these concerns may be unnecessary. J. Manak and T. Gingeras (pers. comm.) have shown that in early development in *Drosophila*, many of these transfrags are, in fact, alternative unannotated 5' start sites of otherwise annotated genes. If this finding is true for humans, it is tempting to believe that the transcription may be involved, for example, in reorganizing a chromatin domain so that it can subsequently be transcribed in a controlled fashion later in development. Secondly, these transfrags necessarily sequester RNA polymerase and relevant accessory proteins, and it is possible that the biological relevance of transcription might simply be in relation to the control of availability of the basal and cell-specific transcription factors. These speculations are as yet untested.

There has also been considerable speculation that noncoding RNAs might have a regulatory function, and in part these proposals have been influenced by the increasing evidence that the DNA of many genes is transcribed off both coding and noncoding strands (see, e.g., Kapranov et al. 2005). An essential role for some noncoding RNA transcripts in early embryonic development had been demonstrated by transgenesis long before the more general analysis of the genome (Brunkow and Tilghman 1991), and the role of antisense transcripts in regulating human genes is well documented (for recent review, see O'Neill 2005). The challenge of studying the function of the many new examples of antisense and noncoding transcripts is considerable, since it will require sophisticated manipulation of relevant regions to establish likely function; some of these analyses may emerge from the ENCODE project discussed below (ENCODE Project Consortium 2004).

MicroRNAs are a class of noncoding RNA that are the focus of increasing attention since their initial description in animals (see, e.g., Lagos-Quintana et al. 2001). The number of human microRNA genes in the genome may be >800 (Bentwich et al. 2005), and a significant majority of these are of unknown function. The increasing data that support a fundamental role for this class of noncoding RNAs (see, e.g., He et al. 2005; Lu et al. 2005) are driving research in this area, and the next few years will see progressively clearer descriptions of the number and biological role of these RNAs.

Emerging from these data is the realization that our concept of a gene is becoming somewhat unclear at a molecular level. In particular, the relationship of transcription to gene expression, to control of gene expression, and even to control of translation has become more complex, and by this measure a greater proportion of the genome is functional than we previously understood. It is important to recognize that function in these cases is being used in two different senses; in one extreme use, function resides entirely in the specific DNA sequence of a region (e.g., a transcription factor binding site), but at the other extreme, "structural" function can be quite independent of sequence (e.g., spacer DNAs). This view has critical implications for interpreting patterns of sequence conservation that show that overall only ~5% of our sequence is subject to selective evolutionary pressure and therefore "functional" (for review, see Miller et al. 2004).

THE DISTRIBUTION OF GENES WITHIN DNA

The gene distribution in the full sequence provided two surprises: firstly, striking gene-poor "deserts"; regions of up to 3Mb (Venter et al. 2001) that are devoid of genes, with a statistically high probability that these are not the tails of a random distribution of genes. In the mouse, deletion of two deserts had no immediate phenotypic consequences (Nobrega et al. 2004). Presently, there is no satisfactory explanation for the existence of gene deserts, but the varying pattern of conservation within deserts suggest some function; Nobrega et al. 2003, Ovcharenko et al. (2005), and de la Calle-Mustienes et al. (2005) show that some deserts contain enhancers distant to flanking genes.

Secondly, prior to the results of the HGP, the location of genes along the DNA was known to be functionally important; clusters of coordinately expressed genes such as the *HOX* or globin clusters were well studied, but it was clear that these clusters were products of gene duplication events in deep evolutionary time. However, Yamashita et al. (2004) identified large-scale functional clustering of genes that were coexpressed in specific human tissues. Boon et al. (2004) and Petkov et al. (2005) reported similar results in the mouse, and Caron et al. (2001) reported clustering of genes expressed at high levels into specific chromosomal regions. Importantly, the clusters do not appear to be the products of evolutionary duplications of an ancestral gene(s), and the implication is that clustering reflects some level of coordinate control, speculatively, such as enhancer sharing or open chromatin conformation.

ELEMENTS THAT CONTROL GENE EXPRESSION

The identification of *cis*-acting promoter sequences that control gene expression has inevitably become the focus both of intensive bioinformatics analysis (see, e.g., Liu and States 2002 or Zhang 2003) and experimental research (Kim et al. 2005). Perhaps the most difficult aspect of bioinformatics predictions is testing the results in practical experimentation, and here the ENCODE project is a key development. Presently, the ENCODE project has the goal "to identify all functional elements in the human genome sequence" (ENCODE Project Consortium 2004; http://www.genome.gov/10005107) by using a mix of different direct experimental and computational approaches. The challenge of these studies is considerable; many promoters function bidirectionally (Trinklein et al. 2004), and the relationship of transcription to "gene" expression is, as noted above, becoming more complex.

It is here that we can perhaps predict the next significant development of the HGP as a collaborative project because we face a severe technical and biological challenge—technically because evidence to date suggests that no one approach to elucidating gene control is satisfactory, and biologically because the tissue specificity of gene expression requires us to study its control, ultimately, in all human tissues. To meet these challenges is a task that will require coordination; perhaps systematic genome research (as opposed to research using genome information) should initially be concentrated on multiple technical approaches, targeted at a collaboratively agreed small number of well-studied cell types. Ideally, these should include the genetically well-characterized CEPH lymphoblastoid cell lines that have been extensively characterized for genetic variation in the Human Haplotype Map (the "HapMap") project (International HapMap Consortium 2003; http://www.hapmap.org). Such a project would certainly synergize cellular biological, genetic, and clinical studies to an unprecedented extent.

Arguably, one of the most surprising results of the HGP was the identification (Bejerano et al. 2004; Siepel et al. 2005) of regions of the genome, called "ultra-conserved elements" (UCEs), that were extraordinarily highly conserved between evolutionarily distant species. The human genome contains 481 such regions that are >200 bp in length (see Fig. 2) and are 100% invariant between the human, rat, and mouse sequences. This conservation is far greater than can be accounted for by protein-coding constraints of an absolutely conserved protein or by requirements of RNA secondary structure. Recently S. Salama and D. Haussler (pers. comm.) have shown that some UCEs are enhancer elements of nearby genes, and this suggests a potential solution to the puzzle of their ultra-conserved nature. Enhancers contain multiple transcription-factor-binding sites, and any given factor can bind to a family of short DNA sequences consisting of a mix of highly invariant or relatively

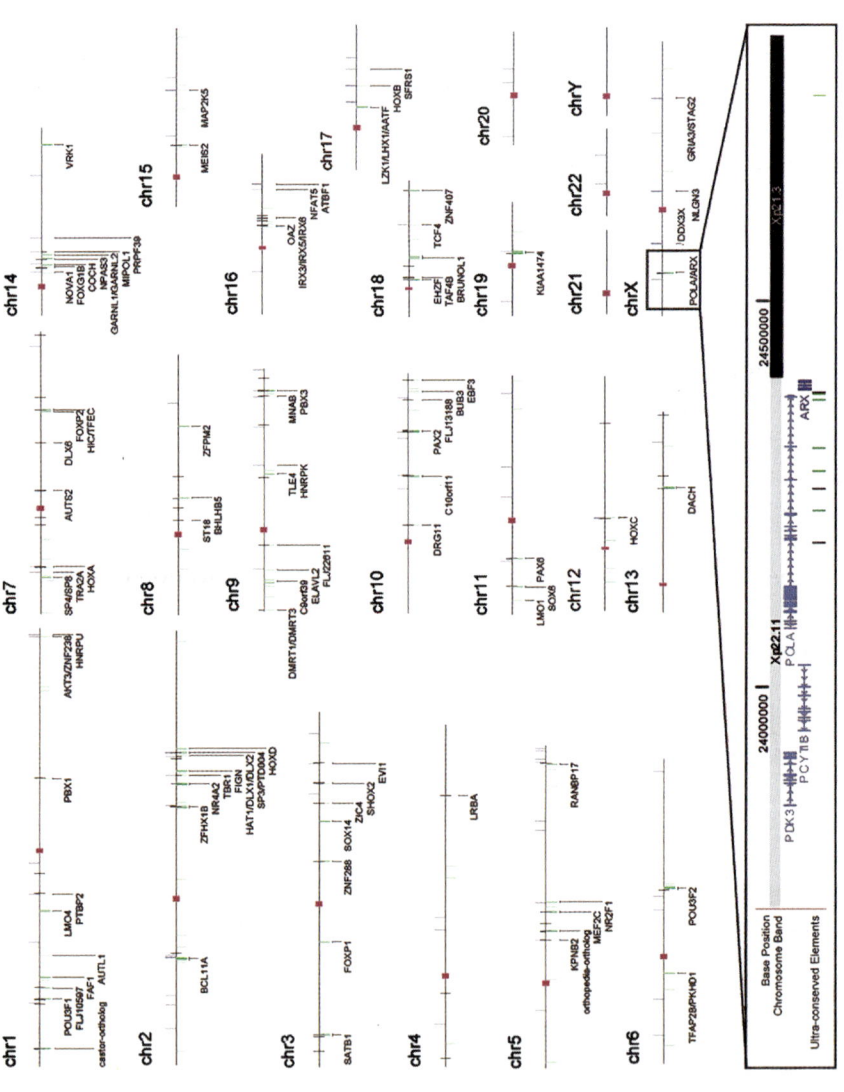

Figure 2. The location of 481 ultra-conserved elements (UCEs) in the human genome with a detailed display of the UCEs within the *POLA* gene. Reproduced with permission from *Science* © 2004, Bejarano et al. (2004).

unconstrained bases (Transfac database at http://www.gene-regulation.com/pub/databases.html). A testable hypothesis to explain the extraordinary conservation of some UCEs is to suggest that they consist of clusters of transcription-factor-binding sites organized as partially overlapping sets, such that the invariance of a base in one binding site defines the identity of an otherwise variable base in a second partially overlapping factor-binding site. The overall result of the overlap of factor-binding sites would be a DNA sequence that could not be altered, since variation of a base would disrupt the function of one or more transcription factors; such a sequence would therefore be highly resistant to evolutionary change.

LARGE-SCALE STRUCTURES IN DNA

The sequence revealed the full extent to which human DNA is comprised of abundant interspersed repeats, extending and completing what was already known; fully 45% of our DNA consists of repetitive elements interspersed within nonrepetitive sequences. Interestingly, the extent and diversity of gene repetitions contained in low copy number repeats were greater than expected; very extensive duplications of regions of DNA both within and between chromosomes were identified by the International Human Genome Sequencing Consortium (2001) and Venter et al. (2001).

Some years prior to the HGP and based on the identification of genes in multiples of four in our DNA, the suggestion was made that the human genome was a quadrupalized derivative of a smaller ancestral genome (see, e.g., Spring 1997). Analysis of the complete sequence fails to support this hypothesis, because there is no significant increase in fourfold repeated regions in the genome.

More recently She et al. (2004) have extended the initial analyses to define the full duplication landscape of the genome, and Tuzun et al. (2005) have shown that there are significant copy number polymorphisms between individuals, the phenotypic consequences of which in many cases are unknown. The location of deletions, insertions, and inversions are shown in Figure 3.

HUMAN GENETIC VARIATION

The Human Haplotype Map (the "HapMap") project is a key component of realizing the genetic potential of the HGP (The International HapMap Consortium 2003; http://www.hapmap.org). This project is based on identifying DNA sequence variations, predominantly single-nucleotide polymorphisms (SNPs), in a target of 270, ethnically diverse human

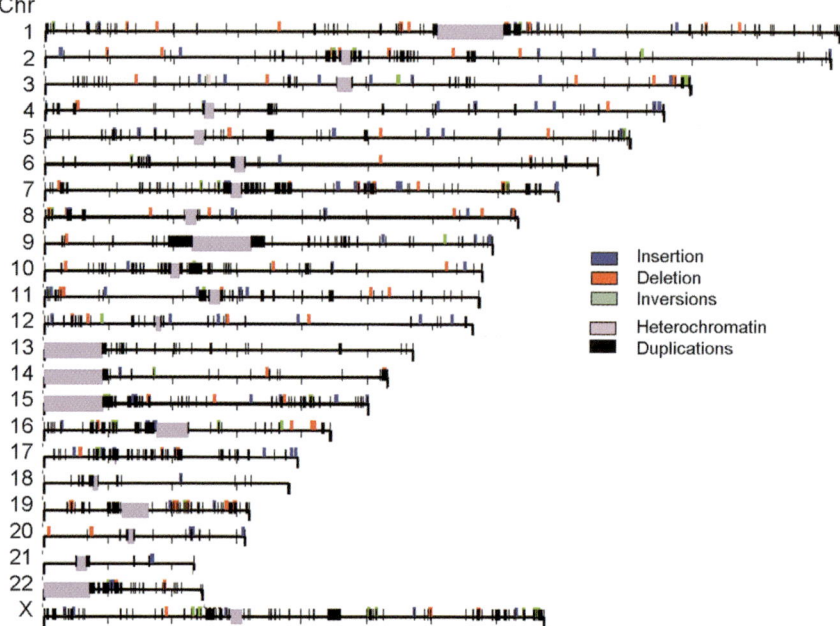

Figure 3. Location of 139 insertions, 102 deletions, and 56 inversions on each human chromosome, showing the location positioned against the DNA sequence. Reproduced and modified with permission from *Nature Genetics* © 2005, Tuzun et al. (2005).

beings. The SNPs are grouped into haplotypes to provide a descriptive framework on which human phenotypic (and further genotypic) variation can be mapped. The extent to which haplotypes capture human variation is still unclear (see, e.g., Evans and Cardon 2005; Sawyer et al. 2005), and we are still far from having reasonable estimates of how explicitly haplotypic variation influences, or is correlated with, phenotypic variation. The potential for the HapMap to inform analysis of human complex genetic disorders was one of the founding principles of the project, and the next five years will see the full application of this research. More conjecturally, the reduced complexity of haplotype sharing between two individuals, when compared to the full sequence difference, may allow us to introduce wholly novel genotypic classifications of human diversity. Such a development would have an important impact on population- and cohort-based research such as clinical trials and on the genetic basis of personalized medicine.

RECREATING HUMAN ANCESTRY

It was always understood that the HGP would provide the framework for the study of human diversity from a biomedical perspective but that these data could equally be applied to the study of human history through tracing historical patterns of migration and population structure. We can certainly anticipate that the HapMap will provide an enormous intellectual platform to power these analyses; the present expense of genotyping by resequencing will necessarily limit the extent to which human populations may be studied. There is every reason to suppose that the driver of biomedical research will force sequencing costs down to levels where large-scale study of populations by resequencing will become cost tractable; the outcome will be the most detailed description of human origins that these technologies and human history can allow.

Human ancestry can also be studied by comparison with the great ape DNA sequences, and this is an emerging area of research that has captured imaginations widely, particularly in respect to the evolution of human higher cognitive functions such as language (Enard et al. 2002). The difficulty of these studies is that the divergence of human and chimpanzee DNA at ~1.23% is so small (The International Chimpanzee Chromosome 22 Consortium 2004; Chimpanzee Sequencing and Analysis Consortium 2005) that it does not easily allow statistically robust identification of selection, and our understanding of the genetic basis for higher cognitive function remains sketchy.

The role of comparative sequence analysis in annotation has been amply demonstrated in the HGP and in many other genome projects. A novel use of this information allowed Blanchette et al. (2004) to attempt to recreate the molecular genetic ancestors of humans and other vertebrates (see Fig. 4). Conventional DNA sequence phylogenetic analyses use statistical approaches to establish a likely order of changes to DNA in the course of evolution and thus recreate an evolutionary history. Blanchette et al. (2004) used the ~1.8-Mb *CFTR* gene region DNA sequence from 18 mammalian species, but instead of focusing on the order of changes, they attempted to recreate the ancestral DNA sequence by statistical modeling to "reverse" base changes to the evolutionary basal state creating the eutherian ancestral genome (Fig. 4). That there are statistical limitations to this approach is certainly recognized, but perhaps in the not-too-distant future, we may be able to combine our theoretical knowledge of ancestral DNA sequences with our knowledge of evolutionary developmental biology to put flesh on the "bones" of the DNA sequence of an unknowable distant common ancestor!

BROADENING THE IMPACT OF THE HGP

So far this review has focused on the achievements of the HGP in terms of novel information and concepts generated from within the project

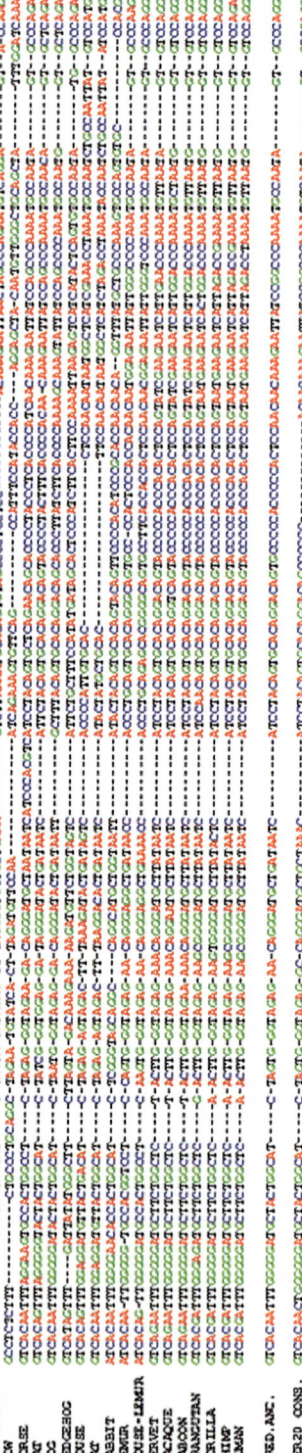

Figure 4. The recreation of an ancestral DNA sequence. This is an example based upon a MER20 retrotransposon. The sequence labeled "PRED.ANC." at the bottom is the prediction of the ancestral MER20 sequence. Reproduced and modified with permission from Cold Spring Harbor Laboratory Press © 2004, Blanchette et al. (2004).

itself, but, of course, one of the founding principles underlying the HGP was that the DNA sequence would inform a very wide range of research. Has this goal been achieved? It is clear that even the incomplete knowledge of the genic and therefore protein composition of humans has, indeed, supported much research but has not, perhaps, produced the flood of new therapies and concepts that more enthusiastic supporters had proclaimed.

Can the impact be broadened? Completing the annotation of genes would certainly contribute to increased impact by facilitating the technical exploitation of genome information, for example, enhancing our ability to define canonical DNA probes on microarrays, and contributing to biological study of human genes. Of course, the ethical limitations of research on humans has restricted the scope of experimental descriptions of tissue specificity of gene expression and of alternative splice forms. Human embryogenesis is a particularly difficult area of study, and it is likely that human embryonic (and other) stem cells will become very important surrogate targets for experimental analysis of the human gene complement and its control.

THE HUMAN SEQUENCE DISPLAY—GENOME BROWSERS

The HGP data is exceptionally rich in information but most critically, this richness is very much in the eye of the beholder; the genome sequence holds very different information for different biologists. For this reason, the second area where the HGP impact may be widened is in the difficult task of presenting the DNA sequence for use by the scientific community. In the introduction to "A User's Guide to the Human Genome" (Wolfsberg et al. 2003), it was noted that "many investigators whose research programs stand to benefit in a tangible way from the availability of this information have not been able to capitalize on its potential." In part, as the guide tried to argue, this was because of user unfamiliarity with the complex data available through the normal browsers that display the sequence, those of the UCSC (http://genome.ucsc.edu/; Kent et al. 2005), NCBI (http://www.ncbi.nlm.nih.gov/genome/guide/human/; Wheeler et al. 2005), and Ensembl (http://www.ensembl.org; Hubbard et al. 2005). However a potentially more significant factor to be considered is that these interfaces are all designed around the view of the DNA sequence as the ultimate genetic map, which is a geneticist/genomicist view of the genome. Most experimental biologists are not interested in genetic organization but rather are interested in biological organization—for example, proteins expressed or functioning in the same space or same time in an organism. The descriptive language of gene function is becoming formalized around the terms defined in the Genome Ontology project (The Gene Ontology Consortium 2000; http://www.geneontology.org), and this provides an important framework for description. Displays based on this more complex knowledge are presently intended for more specialist users (e.g., the proteomics community), and until they become more generic, it is likely biologists will remain somewhat distant from genome information. The next decade will most assuredly see enormous strides in this area, particularly under the integrative drive of systems-biology-based approaches.

THE ORGANIZATIONAL LEAD OF THE HGP

The HGP was one of the few examples in biology of an attempt to coordinate and focus research to specific goals by a strongly directed program of investigation. The HapMap and the ENCODE projects are good examples of the second generation of directed projects, and while both may have their scientific critics, both are clearly generating novel information that will support much hypothesis-driven research in the future. A continuing role for directed research is ultimately a political decision of some complexity if only because of the mix of individual creativity and more collective endeavor that is characteristic of the best of genome research. In any event, the HGP has in many respects led the way in modern biological research; it has altered the politics of science funding very substantially by virtue of its scale, of the scale of the information that biologists can now access and of the complexity of that information. Genomic-scale analyses have started to revolutionize biological research, drawing computer scientist, mathematician, biologist, clinician, chemist, and physicist into complex collaborative projects. This is surely one of the realized beneficial outcomes of the HGP, realized far in advance of the impact of increased biological knowledge itself. I would argue that one of the achievements of the HGP has been to alter the way in which we study ourselves, and this, it seems to me, is as profound an impact as one can hope for in a field as complex as human biology.

CONCLUSIONS

I have focused on novelty in this review, but the undeniable reality is that the human sequence presently, and for decades to come, underpins an extraordinary range of research that ultimately is only limited by the interests of those who use its information. This is the real success of the

HGP, but it is a success that does not readily lend itself to headlines. Genome information does not allow us to escape from the extraordinary complexity of our biology, and thus it is not a golden source of drugs, drug targets, cures, and insights. Its information cannot be read like a book because that is not the logic of living cells. Post-genome sequence, science is quite unlike anything we have previously encountered, but the brutal reality is that our own biology remains as difficult to study as it has ever been; perhaps, therefore, the greatest contribution the HGP has made is to show us just how complex we really are.

ACKNOWLEDGMENTS

Space limitations do not allow me to cite many of the relevant references in this review; I hope the individuals concerned can forgive the necessary omissions! I am indebted to the past and present members of my lab and colleagues for their help in forming my understanding of genome biology: this work was supported by a grant from the Australian ARC to P.F.R.L. and through the expertise of The Clive and Vera Ramaciotti Centre for Gene Function Analysis at the University of New South Wales.

REFERENCES

Bejerano, G., Pheasant, M., Makunin, I., Stephen, S., Kent, W.J., Mattick, J.S., and Haussler, D. 2004. Ultraconserved elements in the human genome. *Science* **304:** 1321–1325.

Bentwich, I., Avniel, A., Karov, Y., Aharonov, R., Gilad, S., Barad, O., Barzilai, A., Einat, P., Einav, U., Meiri, E., et al. 2005. Identification of hundreds of conserved and nonconserved human microRNAs. *Nat. Genet.* **37:** 766–770.

Birney, E., Andrews, T.D., Bevan, P., Caccamo, M., Chen, Y., Clarke, L., Coates, G., Cuff, J., Curwen, V., Cutts, T., et al. 2004. An overview of Ensembl. *Genome Res.* **14:** 925–928.

Blanchette, M., Green, E.D., Miller, W., and Haussler, D. 2004. Reconstructing large regions of an ancestral mammalian genome in silico. *Genome Res.* **14:** 2412–2423.

Boon, W.M., Beissbarth, T., Hyde, L., Smyth, G., Gunnersen, J., Denton, D.A., Scott, H., and Tan, S.S. 2004. A comparative analysis of transcribed genes in the mouse hypothalamus and neocortex reveals chromosomal clustering. *Proc. Natl. Acad. Sci.* **101:** 14972–14977.

Brent, M.R. and Guigo, R. 2004. Recent advances in gene structure prediction. *Curr. Opin. Struct. Biol.* **14:** 264–272.

Brunkow, M.E. and Tilghman, S.M. 1991. Ectopic expression of the H19 gene in mice causes prenatal lethality. *Genes & Dev.* **5:** 1092–1101.

Caron, H., van Schaik, B., van der Mee, M., Baas, F., Riggins, G., van Sluis, P., Hermus, M.C., van Asperen, R., Boon, K., Voute, P.A., et al. 2001. The human transcriptome map: Clustering of highly expressed genes in chromosomal domains. *Science* **291:** 1289–1292.

Cheng, J., Kapranov, P., Drenkow, J., Dike, S., Brubaker, S., Patel, S., Long, J., Stern, D., Tammana, H., Helt, G., et al. 2005. Transcriptional maps of 10 human chromosomes at 5-nucleotide resolution. *Science* **308:** 1149–1154.

Chimpanzee Sequencing and Analysis Consortium. 2005. Initial sequence of the chimpanzee genome and comparison with the human genome. *Nature* **437:** 69–87.

Collins, F.S., Patrinos, A., Jordan, E., Chakravarti, A., Gesteland, R., and Walters, L. 1998. New goals for the U.S. Human Genome Project: 1998–2003. *Science* **282:** 682–689.

de la Calle-Mustienes, E., Feijoo, C.G., Manzanares, M., Tena, J.J., Rodriguez-Seguel, E., Leticia, A., Allende, M.L., and Gomez-Skarmeta, J.L. 2005. A functional survey of the enhancer activity of conserved non-coding sequences from vertebrate Iroquois cluster gene deserts. *Genome Res.* **15:** 1061–1072.

Dickson, M. and Gagnon, J.P. 2004. Key factors in the rising cost of new drug discovery and development. *Nat. Rev. Drug Discov.* **3:** 417–429.

Enard, W., Przeworski, M., Fisher, S.E., Lai, C.S., Wiebe, V., Kitano, T., Monaco, A.P., and Pääbo, S. 2002. Molecular evolution of FOXP2, a gene involved in speech and language. *Nature* **418:** 869–872.

ENCODE Project Consortium. 2004. The ENCODE (ENCyclopedia Of DNA Elements) Project. *Science* **306:** 636–640.

Evans, D.M. and Cardon, L.R. 2005. A comparison of linkage disequilibrium patterns and estimated population recombination rates across multiple populations. *Am. J. Hum. Genet.* **76:** 681–687.

Ewing, B. and Green, P. 2000. Analysis of expressed sequence tags indicates 35,000 human genes. *Nat. Genet.* **25:** 232–234.

The Gene Ontology Consortium. 2000. Gene Ontology: Tool for the unification of biology. *Nat. Genet.* **25:** 25–29.

He, L., Thomson, J.M., Hemann, M.T., Hernando-Monge, E., Mu, D., Goodson, S., Powers, S., Cordon-Cardo, C., Lowe, S.W., Hannon, G.J., et al. 2005. A microRNA polycistron as a potential human oncogene. *Nature* **435:** 828–833.

Hubbard, T., Andrews, D., Caccamo, M., Cameron, G., Chen, Y., Clamp, M., Clarke, L., Coates, G., Cox, T., Cunningham, F., et al. 2005. Ensembl. *Nucleic Acids Res.* **33:** D447–D453.

The International Chimpanzee Chromosome 22 Consortium. 2004. DNA sequence and comparative analysis of chimpanzee chromosome 22. *Nature* **429:** 382–388.

The International HapMap Consortium. 2003. The International HapMap Project. *Nature* **426:** 789–796.

International Human Genome Sequencing Consortium. 2001. Initial sequencing and analysis of the human genome. *Nature* **409:** 860–921.

———. 2004. Finishing the euchromatic sequence of the human genome. *Nature* **431:** 931–945.

Johnson, J.M., Castle, J., Garrett-Engele, P., Kan, Z., Loerch, P.M., Armour, C.D., Santos, R., Schadt, E.E., Stoughton, R., and Shoemaker, D.D. 2003. Genome-wide survey of human alternative pre-mRNA splicing with exon junction microarrays. *Science* **302:** 2141–2144.

Johnson, J.M., Edwards, S., Shoemaker, D., and Schadt, E.E. 2005. Dark matter in the genome: Evidence of widespread transcription detected by microarray tiling experiments. *Trends Genet.* **21:** 93–102.

Kapranov, P., Drenkow, J., Cheng, J., Long, J., Helt, G., Dike, S., and Gingeras, T.R. 2005. Examples of the complex architecture of the human transcriptome revealed by RACE and high-density tiling arrays. *Genome Res.* **15:** 987–997.

Karolchik, D., Baertsch, R., Diekhans, M., Furey, T.S., Hinrichs, A., Lu, Y.T., Roskin, K.M., Schwartz, M., Sugnet, C.W., Thomas, D.J., et al. 2003. The UCSC Genome Browser database. *Nucleic Acids Res.* **31:** 51–54.

Kent, W.J., Hsu, F., Karolchik, D., Kuhn, R.M., Clawson, H., Trumbower, H., and Haussler, D. 2005. Exploring relationships and mining data with the UCSC Gene Sorter. *Genome Res.* **15:** 737–741.

Kim, T.H., Barrera, L.O., Zheng, M., Qu, C., Singer, M.A., Richmond, T.A., Wu, Y., Green, R.D., and Ren, B. 2005. A high-resolution map of active promoters in the human genome. *Nature* **436:** 876–880.

Lagos-Quintana, M., Rauhut, R., Lendeckel, W., and Tuschl, T. 2001. Identification of novel genes coding for small expressed RNAs. *Science* **294:** 853–858.

Liu, R. and States, D.J. 2002. Consensus promoter identification in the human genome utilizing expressed gene markers and gene modeling. *Genome Res.* **12:** 462–469.

Lu, J., Getz, G., Miska, E.A., Alvarez-Saavedra, E., Lamb, J., Peck, D., Sweet-Cordero, A., Ebert, B.L., Mak, R.H., Ferrando, A.A., et al. 2005. MicroRNA expression profiles classify human cancers. *Nature* **435:** 834–838.

Miller, W., Makova, K.D., Nekrutenko, A., and Hardison, R.C. 2004. Comparative genomics. *Annu. Rev. Genomics Hum. Genet.* **5:** 15–56.

Nobrega, M.A., Ovcharenko, I., Afzal, V., and Rubin, E.M. 2003. Scanning human gene deserts for long-range enhancers. *Science* **302:** 413.

Nobrega, M.A., Zhu, Y., Plajzer-Frick, I., Afzal, V., and Rubin, E.M. 2004. Megabase deletions of gene deserts result in viable mice. *Nature* **431:** 988–993.

O'Neill, M.J. 2005. The influence of non-coding RNAs on allele-specific gene expression in mammals. *Hum. Mol. Genet.* **14: 1:** R113–R120.

Ovcharenko, I., Loots, G.G., Nobrega, M.A., Hardison, R.C., Miller, W., and Stubbs, L. 2005. Evolution and functional classification of vertebrate gene deserts. *Genome Res.* **15:** 137–145.

Petkov, P.M., Graber, J.H., Churchill, G.A., Dipetrillo, K., King, B.L., and Paigen, K. 2005. Evidence of a large-scale functional organization of mammalian chromosomes. *PLoS Genet.* **1:** e33.

Sawyer, S.L., Mukherjee, N., Pakstis, A.J., Feuk, L., Kidd, J.R., Brookes, A.J., and Kidd, K.K. 2005. Linkage disequilibrium patterns vary substantially among populations. *Eur. J. Hum. Genet.* **13:** 677–686.

She, X., Jiang, Z., Clark, R.A., Liu, G., Cheng, Z., Tuzun, E., Church, D.M., Sutton, G., Halpern, A.L., and Eichler, E.E. 2004. Shotgun sequence assembly and recent segmental duplications within the human genome. *Nature* **431:** 927–930.

Siepel, A., Bejerano, G., Pedersen, J.S., Hinrichs, A.S., Hou, M., Rosenbloom, K., Clawson, H., Spieth, J., Hillier, L.W., Richards, S., et al. 2005. Evolutionarily conserved elements in vertebrate, insect, worm, and yeast genomes. *Genome Res.* **15:** 1034–1050.

Sorek, R., Shamir, R., and Ast, G. 2004. How prevalent is functional alternative splicing in the human genome? *Trends Genet.* **20:** 68–71.

Southan, C. 2004. Has the yo-yo stopped? An assessment of human protein-coding gene number. *Proteomics* **4:** 1712–1726.

Spring, J. 1997. Vertebrate evolution by interspecific hybridisation—Are we polyploid? *FEBS Lett.* **400:** 2–8.

Strausberg, R.L., Simpson, A.J., Old, L.J., and Riggins, G.J. 2004. Oncogenomics and the development of new cancer therapies. *Nature* **429:** 469–474.

Suzuki, M. and Hayashizaki, Y. 2004. Mouse-centric comparative transcriptomics of protein coding and non-coding RNAs. *Bioessays* **26:** 833–843.

Trinklein, N.D., Aldred, S.F., Hartman, S.J., Schroeder, D.I., Otillar, R.P., and Myers, R.M. 2004. An abundance of bidirectional promoters in the human genome. *Genome Res.* **14:** 62–66.

Tuzun, E., Sharp, A.J., Bailey, J.A., Kaul, R., Morrison, V.A., Pertz, L.M., Haugen, E., Hayden, H., Albertson, D., Pinkel, D., et al. 2005. Fine-scale structural variation of the human genome. *Nat. Genet.* **37:** 727–732.

Venter, J.C., Adams, M.D., Myers, E.W., Li, P.W., Mural, R.J., Sutton, G.G., Smith, H.O., Yandell, M., Evans, C.A., Holt, R.A., et al. 2001. The sequence of the human genome. *Science* **291:** 1304–1351.

Wheeler, D.L., Barrett, T., Benson, D.A., Bryant, S.H., Canese, K., Church, D.M., DiCuccio, M., Edgar, R., Federhen, S., Helmberg, W., et al. 2005. Database resources of the National Center for Biotechnology Information. *Nucleic Acids Res.* **33:** D39–D45.

Wolfsberg, T.G., Wetterstrand, K.A., Guyer, M.S., Collins, F.S., and Baxevanis, A.D. 2003. A user's guide to the human genome. *Nat. Genet.* **35:** 4.

Yamashita, T., Honda, M., Takatori, H., Nishino, R., Hoshino, N., and Kaneko, S. 2004. Genome-wide transcriptome mapping analysis identifies organ-specific gene expression patterns along human chromosomes. *Genomics* **84:** 867–875.

Zhang, M.Q. 2003. Prediction, annotation, and analysis of human promoters. *Cold Spring Harb. Symp. Quant. Biol.* **68:** 217–225.

19

Emerging Technologies in DNA Sequencing

Michael L. Metzker
Human Genome Sequencing Center and
 Department of Molecular and Human Genetics
Baylor College of Medicine
Houston, Texas 77030

MORE THAN JUST A MAPPING AND SEQUENCING ENDEAVOR, the Human Genome Project (HGP) has altered the mindset and approach to many basic and applied research efforts. Early skepticism and controversy (Koshland 1989; Luria et al. 1989; Roberts 1989b; Fox et al. 1990) were soon laid to rest by well-developed strategies (Roberts 1989a; Collins and Galas 1993; Collins et al. 1998) that led to the successful execution of mankind's largest biology project. At the core of the HGP was technology development that advanced the pace of sequencing a mammalian-size genome from years to months. Along the way, numerous strategies emerged that hold promise for rapid, efficient, and inexpensive delivery of DNA sequence information. For the HGP, a brute-force approach was adopted for completing the job by coupling the core technologies of Sanger sequencing and fluorescence detection. The completion of the sequencing phase could not have been accomplished without major innovations in recombinant protein engineering, fluorescent dye development, capillary electrophoresis, automation, robotics, informatics, and process management. The result was completion of a high-quality, reference sequence of the human genome in April, 2003 (Collins et al. 2003), marking the 50-year anniversary of the discovery of the double-helix structure. For many outside the genome community, that heroic milestone signaled the end of this international scientific project, but for the rest of us, it only marked the beginning of things to come.

The need for sequencing has never been greater than it is today, with applications spanning diverse research sectors including comparative genomics and evolution, forensics, epidemiology, and applied medicine for diagnostics and therapeutics. Arguably, the strongest rationale for ongoing sequencing is the quest for identification and interpretation of human sequence variation as it relates to health and disease. The most common form of variation is the single-nucleotide polymorphism (SNP). Although two unrelated people share, on average, 99.9% sequence identity (i.e., one difference in a thousand base pairs), the average occurrence of an SNP in the general population is once every few hundred base pairs. As such, more than nine million unique SNPs have been cataloged in the public database, dbSNP (Crawford and Nickerson 2005), with many more expected to be found in large-scale resequencing efforts.

A great deal of attention has been focused on common SNPs with a minor allele frequency >5% and their potential role in common disease (Lander 1996; Risch and Merikangas 1996; Collins et al. 1997). Recent, large-scale genotyping efforts of these common SNPs have shown that much of the human genome can be parsed into common haplotype blocks (Daly et al. 2001; Patil et al. 2001; Gabriel et al. 2002). The International HapMap Consortium (2003) was formed to characterize common patterns of sequence variation by determining allele frequencies and the degree of association between SNPs among geographically distinct groups, leading to the identification of "tagSNPs" for genomewide, disease-based association studies. With this method of characterization, however, rare SNPs/haplotypes may be overlooked, as highlighted by Liu et al. (2005), who described an association of rare variants/haplotypes with osteoporosis.

A shift in large-scale strategies from genotyping to resequencing is currently taking place to explore the significance of less common SNPs to human biology and disease. The "re" in this approach is the sequencing of additional genomes related to a reference genome for de novo SNP discovery and comparative genomics application. The ENCODE Project Consortium (2004) has described significant efforts toward resequencing megabase-sized blocks of the human genome. Consequently, genome centers are now diverting at least 10%–20% of their resources, which currently translates to ~5% capacity, to resequencing hundreds to thousands of gene regions. This increase in momentum for high-throughput resequencing will greatly facilitate studies to determine the genetic basis of susceptibility to common disease, cancer biology, and disease association in model and non-model organisms.

Current sequencing technologies are too expensive, labor intensive, and time consuming for broad application in human sequence variation studies.

Genome center cost is calculated on the basis of dollars per 1000 Q_{20} bases (defined below) and can be generally divided into the categories of instrumentation, personnel, reagents and materials, and overhead expenses. Currently, these centers are operating at less than one dollar per 1000 Q_{20} bases, with at least 50% of the cost resulting from DNA sequencing instrumentation alone. Developments in novel detection methods, miniaturization in instrumentation, microfluidic separation technologies, and an increase in the number of assays per run will most likely have the biggest impact on reducing cost. It should be emphasized, however, that new sequencing strategies will be needed to use these high-throughput platforms effectively. In September, 2004, the National Human Genome Research Institute (NHGRI) initiated two new programs aimed at bringing the cost of whole-genome sequencing down to $100,000 (http://grants.nih.gov/grants/guide/rfa-files/RFA-HG-04-002.html), with the eventual goal being $1000 (http://grants.nih.gov/grants/guide/rfa-files/RFA-HG-04-003.html).

Numerous strategies and platforms for ultrafast DNA sequencing currently under development include sequencing-by-hybridization (SBH), nanopore sequencing, and sequencing-by-synthesis (SBS), the latter of which encompasses many different DNA polymerase-dependent strategies. Use of the term SBS has become increasingly ambiguous in the literature; therefore, I propose a classification of DNA polymerase-dependent strategies into three major categories: *Sanger sequencing, single-nucleotide addition (SNA),* and *cyclic reversible termination (CRT)* (Text Box 1). In this review, I will focus only on DNA polymerase-dependent strategies, which represent the broadest area of research and development. For the SNA and CRT strategies, I will emphasize the chemistry in an effort to illustrate the advantages and challenges of these methods. Because of the competitive nature of technology development, the exchange of scientific ideas is often thwarted, as many companies do not readily publish results. Although this review will highlight recent advances reported in the literature, readers are directed to the Web sites of companies who are active in the sequencing field (Table 1). A recent review by Shendure et al. (2004) provides a comprehensive overview of SBH and nanopore sequencing technologies. Important issues surrounding whole-genome sequencing, such as ownership, consent, privacy, and legal, ethical, and social implications, will not be addressed here (Foster and Sharp 2002; Robertson 2003; Bonham et al. 2005).

SANGER SEQUENCING: STATE-OF-THE-ART TECHNOLOGY

The Sanger method is a mixed-mode process involving synthesis of a complementary DNA template using natural 2′-deoxynucleotides (dNTPs)

> **BOX 1 ■ DNA polymerase-dependent strategies**
>
> In the broadest sense, all methods involving a DNA polymerase could be considered a SBS approach, if synthesis alone was the defining process. The defining element of these DNA polymerase-dependent methods, however, is not really synthesis at all but rather the means by which DNA synthesis terminates. From this point of view, the DNA sequencing approaches highlighted here have been organized according to their termination strategies.
>
> *Sanger sequencing* and "dideoxy" sequencing are frequently used as synonymous terms. These unnatural ddNTP terminators replace the OH with an H at the 3′-position of the deoxyribose molecule and irreversibly terminate DNA polymerase activity, unless the nucleotide is removed by the process of phosphorolysis. This process is mediated by high concentrations of pyrophosphate or ATP and is a major cause of "drop-outs" in DNA sequence data.
>
> *Single-nucleotide addition* (SNA) methods such as pyrosequencing use limiting amounts of individual natural dNTPs to cause DNA synthesis to pause, which, unlike the Sanger method, can be resumed with the addition of natural nucleotides. Limiting the amount of a given dNTP is required to minimize misincorporation effects observed at higher concentrations. A major drawback with the SNA approach is the incomplete extension through homopolymer repeats.
>
> *Cyclic reversible termination* (CRT) uses reversible terminators containing a protecting group attached to the nucleotide that terminates DNA synthesis. For the reversible terminator, removal of the protecting group restores the natural nucleotide substrate, allowing subsequent addition of reversible terminating nucleotides. One example of a reversible terminator is a 3′-*O*-protected nucleotide (Fig. 4B), although protecting groups can be attached to other sites on the nucleotide as well. This stepwise base addition approach, which cycles between coupling and deprotection, mimics many of the steps of automated DNA synthesis of oligonucleotides.

and termination of synthesis using 2′,3′-dideoxynucleotides (ddNTPs) by DNA polymerase (Sanger et al. 1977). Balanced appropriately, competition between synthesis and termination processes results in the generation of a set of nested fragments, which differ in nucleoside monophosphate units. The ratio of dNTP/ddNTP in the sequencing reaction determines the frequency of chain termination, and hence the distribution of lengths of terminated chains. The nested fragments are then separated by their size using high-resolution gel electrophoresis and analyzed to reveal the DNA sequence. Advancements in fluorescence detection (Smith et al. 1986; Prober et al. 1987), enzymology (Tabor and Richardson 1989, 1995), fluorescent dyes (Ju et al. 1995; Metzker et al. 1996; Lee et al. 1997), dynamic-coating polymers and their derivatives (Ruiz-Martinez et al. 1993; Carrilho et al. 1996; Madabhushi

Table 1. Companies involved in DNA sequencing technology development

Company names	Web site addresses
454 Life Sciences Corp.	www.454.com
Agencourt Biosciences Corp.	www.agencourt.com
GE Healthcare, formerly Amersham Biosciences	www.amershambiosciences.com
Applied Biosystems, Inc.	www.appliedbiosystems.com
Genovoxx	www.genovoxx.de
Helicos Bioscience Corp.	www.helicosbio.com
LaserGen, Inc.	www.lasergen.com
Li-Cor, Inc.	www.licor.com
Microchip Biotechnologies, Inc.	www.mcbiotech.com
Nanofluidics	www.nanofluidics.com
SeqWright	www.seqwright.com
Solexa-Lynx	www.solexa.com
Visigen Biotechnologies, Inc.	www.visigenbio.com

et al. 1996, 1999; Madabhushi 1998; Salas-Solano et al. 1998; Guttman 2002a, 2002b), and capillary array electrophoresis (CAE) (Takahashi et al. 1994; Kheterpal et al. 1996) have helped to define current DNA sequencing platforms.

For automated Sanger sequencing, either the primer or the terminating ddNTP is tagged with a specific fluorescent dye (e.g., ddATP is labeled with the green dye). As these dye-labeled fragments pass through the detection region, fluorophores are excited by the laser in the DNA sequencer, producing fluorescence emissions of four different colors. The determination of the color is the underlying method for assigning a base call, and the order of the fluorescent fragments reveals the DNA sequence. The "raw" fluorescence signals, however, must be transformed. Removal of cross-talk, correction for dye mobility alterations, and normalization of emission intensities must be performed before readable DNA sequence information can be obtained (Smith et al. 1987). Base-calling and error probability assignment (Ewing and Green 1998; Ewing et al. 1998) applications are then used to call the DNA sequence and assess the accuracy of the call. A $Phred_{20}$ or Q_{20} score, equivalent to an error probability of 1% for a given base call, is considered a high-quality base and serves as the commodity standard throughout the sequencing community.

High-throughput DNA sequencing is conducted primarily at large genome centers that continue to refine the sequencing process and strive for Q_{20} bases at lower cost. For example, the Baylor College of Medicine

Human Genome Sequencing Center (BCM-HGSC) produces approximately four million sequencing reactions per month (R.A. Gibbs, pers. comm.). The current production efficiency or pass rate is approximately 89% (after removal of failed reactions, vector sequences, etc.), with sequencing reads averaging 805 Q_{20} bases in length. These metrics translate into the equivalent of sequencing one mammalian-size genome per month. Redundancy is required to improve the base-calling accuracy and contiguity of assembled genomes, resulting in the generation of six times the genome size in Q_{20} bases for production of a draft-quality sequence. Thus, delivery of a mammalian-size, draft-quality sequence requires approximately six months and $12 million. Ongoing advances in new technologies will be critical to meet the goal of rapid, genome-scale sequencing for the price of $100,000 and, ultimately, $1000 per genome.

SANGER SEQUENCING: RECENT ADVANCES

Microfluidic Separation Platforms

Technology development remains active for the fluorescence-based Sanger approach with emphasis on producing faster and cheaper sequencing reads. One key area of research is the application of microfluidic separation devices to DNA sequencing. These microfluidic devices can be fabricated using a variety of substrate materials, with several molecular biology processes integrated onto a single device (e.g., lab-on-a-chip). A number of reviews have been devoted to microfluidic devices (Becker and Gartner 2000; Carrilho 2000; McDonald et al. 2000; Quake and Scherer 2000; Boone et al. 2002; Paegel et al. 2003; Kan et al. 2004), recent advances of which I will highlight as they relate to DNA sequencing. These miniature devices have several advantages over CAE, including improved sample injection and faster separation times.

The separation principles of microfabricated devices are similar to those of conventional CAE, however, their injection methods are very different. With CAE, the sample is introduced by electrokinetic injection into the capillary. The injection time, which defines the length of the sample plug, is typically short and allows only a minute fraction of the sample to be analyzed. A further drawback is that data quality is compromised with increasing impurities in the sample and an intrinsic bias in favor of shorter DNA fragments over longer ones. Microfluidic devices, on the other hand, are less susceptible to these injection problems because the sample is introduced via a channel network by a variety of process strategies (Zhang and Manz 2001). Although early microfabricated chips employed

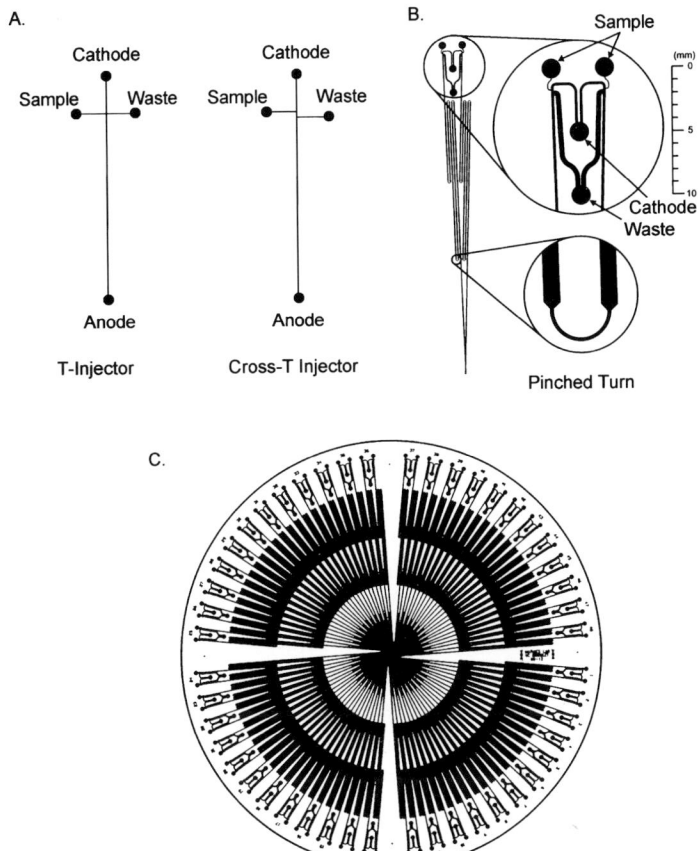

Figure 1. Microfabricated technologies. (A) Examples of a T-injector and cross-T injector layout. (B) Expanded view of the sample injector and pinched turn. (C) Schematic of the 96 channels in a radial chip design. (B,C) Reprinted with permission from National Academy of Sciences, U.S.A. © 2002, Paegel et al. (2002).

a "T"-injector design (Harrison et al. 1992), the cross-T design (Harrison et al. 1993) is widely used today because of its superior sample control (Fig. 1A). The narrow width of the injector affords greater control in selection of sample plug size, which contributes to higher resolution separations with shorter separation lengths compared with CAE.

Most microfabricated devices use borofloat glass or fused silica substrates, which have the advantages of (1) high-quality optical properties, (2) good thermal conductivity, (3) well-documented surface chemistry, and (4) effective translation of capillary innovations. Woolley and Mathies (1995) demonstrated the first application of DNA sequencing using a microfabricated glass device in 1995, reporting single-base resolution using their four-color scanner technology. Data quality and read-lengths have improved significantly since then, because of an increase in the effective separation lengths with run times of 30 minutes or less (Table 2) (Woolley and Mathies 1995; Liu et al. 1999; Schmalzing et al. 1999; Backhouse et al. 2000; Koutny et al. 2000; Liu et al. 2000; Salas-Solano et al. 2000; Simpson et al. 2000; Boone et al. 2002; Paegel et al. 2002; Shi and Anderson 2003). For example, Liu et al. (1999) reported 99.4% accuracy over 500 bases in 20 minutes, with an increase in separation length from 3.5 cm to 6.5 cm. More recent developments by Boone et al. (2002) and Shi and Anderson (2003) have shown the first DNA sequencing applications on plastic chips (Table 2). These chips can be fabricated with high geometric aspect ratios (i.e., deep and narrow channels) at significantly lower cost. Deep and narrow channel structures have the advantages of improved electrophoretic resolution (i.e., longer read-length) and better detection sensitivity.

While single-channel devices are useful for demonstrating feasibility, the construction of multiple channel arrays is essential for high-throughput DNA sequencing. A summary of DNA sequence metrics from several microfabricated multiple channel array devices is presented in Table 2. While Backhouse et al. (2000) and Koutny et al. (2000) reported improved read-lengths by increasing the effective separation lengths to 46.5 cm and 40 cm, respectively, these microfabricated channels were constructed on glass plates ≥50 cm in length, which is out of line with current efforts to miniaturize devices. One approach to circumvent this dilemma has been the introduction of turns along the length of the separation channel. Early studies, however, reported lower separation efficiency in channel turns due to band broadening (Jacobson et al. 1994) and differential field strength effects (Culbertson et al. 1998). Paegel et al. (2000) introduced a "pinched-turn" design (Fig. 1B) with an effective separation length of 15.9 cm on a 15-cm-diameter silica disc, which has been multiplexed into a 96-channel radial device (Fig. 1C) showing tremendous potential for increasing throughput in DNA sequencing applications (Paegel et al. 2002). Most of the data shown in Table 2, however, were derived using the standard M13mp18 vector as the sequencing template, and similar performance is not typically observed under the same conditions with "real-world" samples such as those from genome center production lines.

FLUORESCENCE DETECTION

The most widely used detection method for four-color DNA sequencing was initially described almost 20 years ago (Smith et al. 1986; Prober et al. 1987). This method is based on resolution of the emission signal from a

Emerging Technologies in DNA Sequencing   421

Table 2. Summary of microfabricated devices for DNA sequencing applications

Research group	Number of channels	Template source	Separation length (cm)	Accuracy (%)	Read-length (bp)	Read-out time (min)
Single channel						
Woolley & Mathies (1995)	1	M13mp18	3.5	97	147	9
Liu et al. (1999)	1	M13mp18	6.5	99.4	500	20
Schmalzing et al. (1999)	1	M13 clones[a]	11.5	99	505	27
Salas-Solano et al. (2000)	1	M13mp18	11.5	98.5	640	30
Boone et al. (2002)	1	M13mp18	18.0	98	640	30
Shi & Anderson (2003)	1	Unknown	4.5	99.1	320	13
Multiple channel arrays						
Liu et al. (2000)	16	M13mp18	7.5	99	457	16
Simpson et al. (2000)	48	M13mp18	10.0	97	400	25–45
Backhouse et al. (2000)	48	BigDye Std	46.5	98	640	150
Koutny et al. (2000)	32	M13 clones[a]	40.0	98	800	78
Paegel et al. (2002)	96	M13mp18	15.9	99	430	24

[a]Mixture of M13mp18 vector or twelve M13 clones from human chromosome 17 project.

dye-labeled nucleotide into color, with subsequent assignment in the DNA sequence. While successful for the sequencing of numerous higher and lower eukaryotic and prokaryotic genomes, these four-color systems have several disadvantages, including inefficient excitation of the fluorescent dyes, significant spectral overlap, and inefficient collection of the emission signals. The issue of inefficient excitation has been partially addressed by the use of fluorescence resonance energy-transfer (FRET) dyes (Ju et al. 1995; Metzker et al. 1996; Lee et al. 1997). At present, FRET dye-labeled ddNTP terminators are widely used throughout the sequencing community. The resulting improvements in acceptor dye signal intensities, however, are suboptimal compared with those of single dyes excited at their absorption maxima by the appropriate laser source.

To overcome these deficiencies, some investigators have proposed strategies using additional properties such as fluorescence life-time (Nunnally et al. 1997; Lieberwirth et al. 1998; Lassiter et al. 2000; Zhu et al. 2003, 2004) and radio frequency (RF) modulation (Alaverdian et al. 2002). For DNA sequencing applications, fluorescence life-time measurements have been described using pulsed lasers with high repetition rates (picosecond timescale) with detection in the photon-counting mode. Soper and colleagues have recently demonstrated a combined approach of emission wavelength and fluorescence lifetime measurements, with the potential to increase the number of fluorescent components in DNA sequencing assays (Zhu et al. 2003, 2004). Alaverdian et al. (2002) proposed using four continuous wave (CW) mode lasers, which are modulated at different RFs. To estimate the fluorescence signal for each dye, however, the resulting emission intensity pattern must be demodulated, which introduces a significant computational load for each capillary signal channel. Coupled with repetition rates on the order of ≥ 100 Hz, the RF method does not appear to be compatible with conventional CCD technology, limiting its scalability for detection of high-density capillary arrays.

Recently, Lewis et al. (2005) described a simple but effective method for multifluorescence discrimination called pulsed multiline excitation (PME). The underlying principle of this four-laser system is the correlation of sequential laser pulses with detector response (Fig. 2A). Advantages of PME are such that (1) absorption maxima for the four fluorescent dyes are matched to the excitation sources yielding maximum signal intensities, (2) temporal separation of the laser pulses and expansion of the dye set across the visible spectrum eliminate cross-talk between the dyes, and (3) collection of emission signals is improved by eliminating the requirement for dispersing elements (prisms or gratings) in color separation. In other words, PME measures multicomponent fluorescence assays in a color-blind

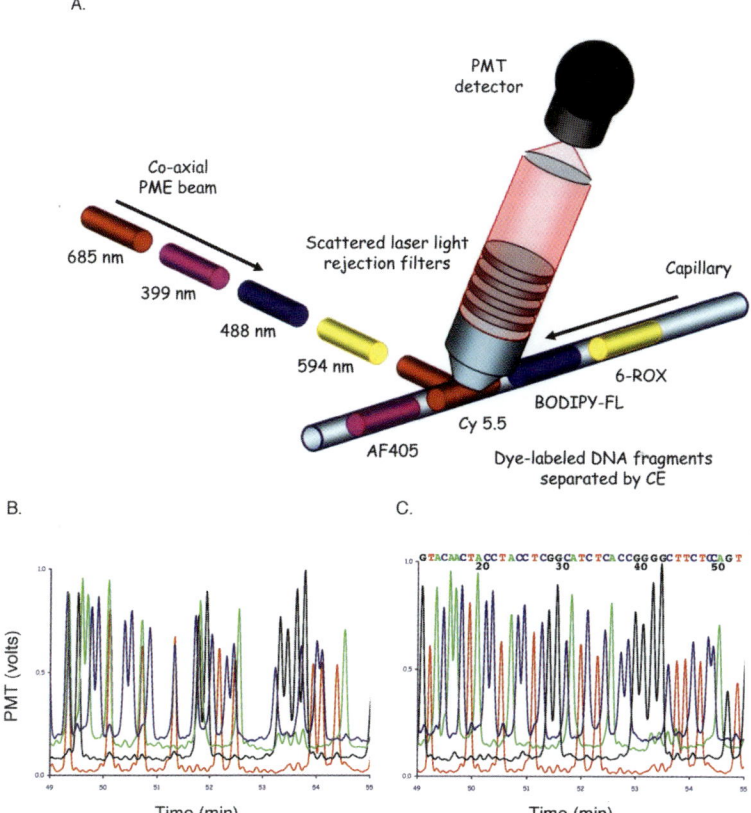

Figure 2. (A) Illustration of the PME technology. Here, each laser operates in a CW mode with mechanical shutters pulsing the different excitation beams in sequential order. The single coaxial PME beam interrogates the fluorescently labeled DNA fragments, which are separated by capillary gel electrophoresis. Scattered laser light is rejected via specific long-pass or wavelength notch filters, with pulsed emission signals from the dye-labeled DNA fragments being detected by the photomultiplier tube (PMT) without use of any dispersing elements. (B) Unprocessed fluorescence data are obtained during the electrophoretic run for the TCF1 exon 10 gene region using PME dye-primers. Blue, green, black, and red traces are AF-405, BODIPY-FL, 6-ROX, and Cy5.5 dye-primers terminated with ddCTP, ddATP, ddGTP, and ddTTP respectively. (C) Raw trace data derived from the experiment described in B is transformed into readable, DNA sequence data using mobility software correction. Reprinted with permission from National Academy of Sciences, U.S.A. © 2005, Lewis et al. (2005).

manner. To demonstrate these advantages, Lewis et al. (2005) applied the PME technology to capillary electrophoresis for DNA sequencing. Figure 2B shows the unprocessed signals from the four PME laser waveforms for a portion of the PCR amplicon for the *TCF1* (formerly known as *HNF1A*) exon 10. Transformation of the data into unambiguous sequence data (Fig. 2C) is accomplished by applying only dye mobility correction software, eliminating the need for cross-talk and signal normalization software transformation. The PME technology holds promise for real-time field applications for DNA sequencing.

SNA METHODOLOGIES

Pyrosequencing

Arguably the most successful non-Sanger method developed to date is pyrosequencing, first described in the literature by Hyman (1988). Pyrosequencing is a nonfluorescence technique that measures the release of inorganic pyrophosphate, which is proportionally converted into visible light by a series of enzymatic reactions (Ronaghi et al. 1996, 1998). Unlike other sequencing approaches that use 3′-modified dNTPs to terminate DNA synthesis, the pyrosequencing assay manipulates DNA polymerase by single addition of dNTPs in limiting amounts. Upon addition of the complementary dNTP, DNA polymerase extends the primer and pauses when it encounters a noncomplementary base. DNA synthesis is reinitiated following the addition of the next complementary dNTP in the dispensing cycle. The light generated by the enzymatic cascade is recorded as a series of peaks called a pyrogram, which corresponds to the order of complementary dNTPs incorporated and reveals the underlying DNA sequence. Applications for pyrosequencing have been reviewed by Ronaghi (2001) and Langaee and Ronaghi (2005).

Although elegant in design, the pyrosequencing approach has several limitations. For example, sequence reads are typically fewer than 100 bases in length, which has application in sequence tag identification such as serial analysis of gene expression (SAGE) (Velculescu et al. 1995), minisequencing for known SNPs, and mapping related genomes to a reference sequence, but limited application for whole-genome sequencing. Recent reports describe the use of single-stranded binding protein (Ronaghi 2000) and the isomeric Sp form of the dATPαS nucleotide (Gharizadeha et al. 2002), which may improve read-lengths up to 100 bases in routine settings. Secondly, homopolymer repeats greater than five nucleotides cannot be quantitatively measured. This is attributed to incomplete extension by DNA polymerase, which results from limiting the dNTP concentration to minimize nucleotide misincorporation effects. It has been suggested that re-addition of the same dNTP may be performed to ensure complete polymerization (Ronaghi 2001), although its practicality for high-throughput sequencing is unclear. Finally, the dispensing order of dNTPs determines the pyrogram profile, which must be carefully designed to avoid asynchronistic extensions of heterozygous sequences.

For a given dispensing order, approximately one half of all heterozygous sequences will result in asynchronistic extensions past the variable site. A survey of heterozygous variants detected by direct DNA sequencing of the *TCF1* gene revealed that 16 of 37 SNPs would result in nonsynchronistic extension after the heterozygous base (data not shown). If one allele extends past the heterozygous base position before the other and advances to the next nucleotide cycle, the nonsynchronicity becomes permanent. An illustration of the effect of dispensing order on asynchronistic extension is shown in Figure 3A. This observation is further highlighted by Entz et al. (2005) with the identification of more than 40 unique dispensing

A.

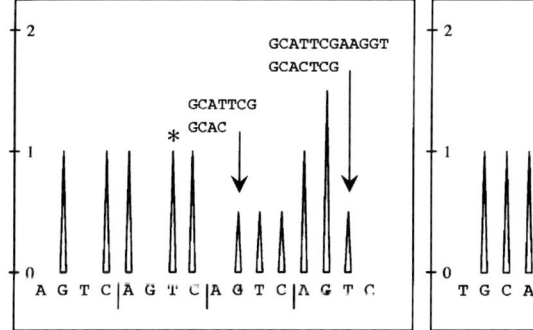
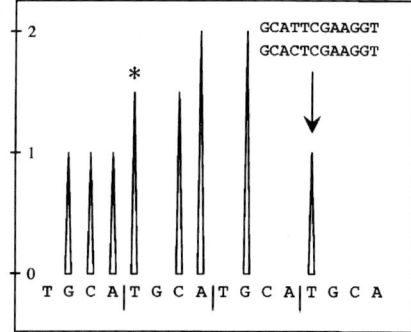

B.

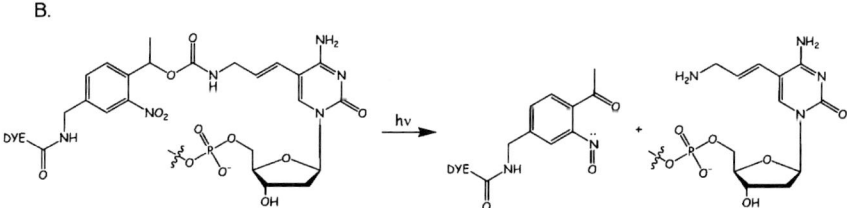

Figure 3. SNA technologies. (*A*) Simulated effects of two different dNTP dispensing orders on the outcome of the pyrogram profile. (*B*) The photocleavage reaction of a fluorescently labeled dNTP coupled with a photocleavable linker.

orders for the accurate typing of *HLA-DQB1* and *HLA-DRB1* alleles. Pyrosequencing may, therefore, be suited for pattern matching of known SNP profiles, while its application for de novo SNP discovery is less certain. Not surprisingly, base-calling for de novo SNPs is problematic and still performed manually (Langaee and Ronaghi 2005).

The 454 Corporation has recently introduced a whole-genome sequencing strategy by integrating pyrosequencing with their PicoTiterPlate (PTP) platform, which has been shown to amplify and image approximately 300,000 PCR templates captured on Sepharose beads (Leamon et al. 2003). The PTP is manufactured by anisotropic etching of a fiber optic faceplate with a well diameter of approximately 40 μm. The 454 group has developed a solution-based emulsion strategy to create microreactors for clonal amplification of single DNA molecules and attachment to these beads. One advantage of the clonal amplification strategy is that it addresses the dependence issue of dispensing order for sequencing of heterozygous bases discussed above. Following an enrichment step, DNA positive beads are loaded into individual PTP wells, which contain additional beads coupled with the necessary enzymes to perform the pyrosequencing chemistry (Margulies et al. 2005). Recently, the company announced its first complete genome sequencing of a recombinant adenoviral construct and the shotgun sequencing of the *Mycoplasma genitalium* genome.

The assembly of non-Sanger sequencing data will represent new challenges because the input read will differ in length, quantity, and quality. The complexity of the genome under analysis may also prove more difficult for assemblies compared with Sanger data, even when the offset is higher coverage of shorter reads. Chaisson et al. (2004) recently performed a simulated assembly study (short, error-free reads sampled at 30× coverage) using genome sequences from adenovirus, two mouse BACs, and two bacteria: *Campylobacter jejuni*, which contains very few repeat sequences (Parkhill et al. 2000b), and *Neisseria meningitidis*, which contains several hundred repetitive elements (Parkhill et al. 2000a). Compared with Sanger data, Chaisson et al. (2004) found that the read-length was inversely proportional to the number of contigs in the assembly (i.e., longer reads gave fewer contigs). Increasing genome complexity, on the other hand, directly increases the number of contigs. Here, they found that 95% of the genome was contained within 9–10 contigs for the BAC clones, and the number of contigs increased from 21 to 344 for *C. jejuni* and *N. meningitidis* genome sequences, respectively. Observed errors for real sequence data will undoubtedly decrease assembly performance for short reads. Thus, the success of the non-Sanger strategies for whole-genome sequencing applications will be highly dependent on the degree of its complexity, which appears to traverse all three phylogenetic domains.

OTHER SINGLE-ADDITION dNTP STRATEGIES

Methods other than pyrophosphate detection can be used to monitor single dNTP additions. For example, Braslavsky et al. (2003) used the technique of single-pair FRET (spFRET) to determine the order of nonconsecutive nucleotide additions. With this single-molecule approach, Cy3-labeled-UTP was initially incorporated into the primer strand, serving as the donor dye. Subsequent incorporation of a complementary Cy5-labeled-UTP or Cy5-labeled-dCTP substrate resulted in the spFRET signal. Following photobleaching of the Cy5 dye, the natural nucleotides dATP and dGTP were added to increase the nucleotide distance between subsequent Cy5-labeled dNTP additions, which would otherwise have resulted in a significant reduction in incorporation efficiencies due to steric hindrance effects. For the DNA template sequence, written 3'-**A**TC**G**TC**A**TC**G**-5' for convenience, the read-out would be the fingerprint sequence of 5'-UCUC. Levene et al. (2003) have recently described a zero-mode waveguide approach to single-molecule detection of R110-labeled-dCTP and coumarin-labeled-dCTP incorporation events by DNA polymerase.

Taking advantage of the steric effects observed in consecutively incorporated dye-labeled dNTPs, Mitra et al. (2003) introduced fluorescently labeled dNTPs, which contained cleavable linkers, to remove the bulky fluorescent group following incorporation by DNA polymerase. This method, called *fluorescent in situ sequencing* (FISSEQ), used linkers containing either a disulfide bridge, which is efficiently cleaved with a reducing agent, or a photocleavable group (Fig. 3B). Using the polony technology (Mitra and Church 1999), Church and colleagues elegantly demonstrated the addition of single Cy5-SSdNTPs followed by dye cleavage for accurate DNA sequencing of several templates. The presence of a fluorescence signal corresponding to the dispersing order of the Cy5-SS-dNTPs revealed the DNA sequence. Although read-lengths up to eight bases were demonstrated, several miscalls were reported. One such call resulted from nucleotide read-through. That is, consecutive incorporations of dye-labeled dNTPs can occur (e.g., the sequence 5'-CAGCC was read as 5'-CAGC), presumably with different efficiencies that are dependent on the local DNA sequence context. A second error occurred as a result of a single-nucleotide insertion (e.g., the sequence 5'-ATGT was read as 5'-AGTGT). Although more difficult to interpret, it is possible that the residual linker structure, remaining on the nucleobases following dye cleavage, could alter nucleotide specificity and incorporation efficiency of subsequent incoming dNTPs in a sequence-dependent manner. More recently, Seo et al. (2004, 2005) described a similar strategy using four different dye-labeled dNTPs with photocleavable linkers (Fig. 3B) and reported read-lengths of 12 bases. A key advantage of the four-color approach is that all four dNTPs can be assayed simultaneously, although both reports demonstrated use of the single-dNTP-addition method.

Kartalov and Quake (2004) proposed a different approach to overcome the steric effects of consecutive dye-labeled bases by use of single-addition, same-nucleobase mixtures (e.g., dCTP/TAMRA-labeled ddCTP) as a method for DNA sequencing. The nucleobase mixture strategy serves the dual purpose of dye-labeling for fluorescence detection (reporter phase) and ongoing DNA synthesis of the complementary nucleotide (extension phase). The dNTP and dye-labeled ddNTP concentrations are balanced appropriately so that only a fraction of the primer strands incorporate the dye-labeled ddNTP. The presence of a fluorescence signal reveals the complementary nucleotide in the DNA sequence, but reporters are eliminated from subsequent dNTP additions. With each nucleotide addition, signal loss is inversely proportional to the increased accumulation of termination products. The fluorescence is then quenched by photobleaching before the next nucleobase mixture is dispensed to repeat the process. Configured in a microfluidic device, the average read-length for the mixed nucleobase addition scheme was three bases, which can be partially attributed to signal loss with subsequent base additions. The accuracy of the method is highly dependent on the reporter phase mimicking the extension phase. For example, a simple homopolymer repeat of two bases will be undercalled in the DNA sequence, as the reporter phase will reflect a single-base addition while the extension phase will incorporate two bases.

CRT

While CRT technology represents tremendous potential for whole-genome sequencing, this strategy still faces significant challenges in its implementation. The CRT cycle is comprised of three steps: *incorporation, imaging,* and *deprotection,* as illustrated in Figure 4A. The advantages of CRT over Sanger are (1) elimination of gel electrophoresis and (2) formatting of the CRT assay in a highly parallel fashion. Its advantages over pyrosequencing are that (1) all four bases are present during the incorporation phase, (2) step-wise control allows for single-base additions through

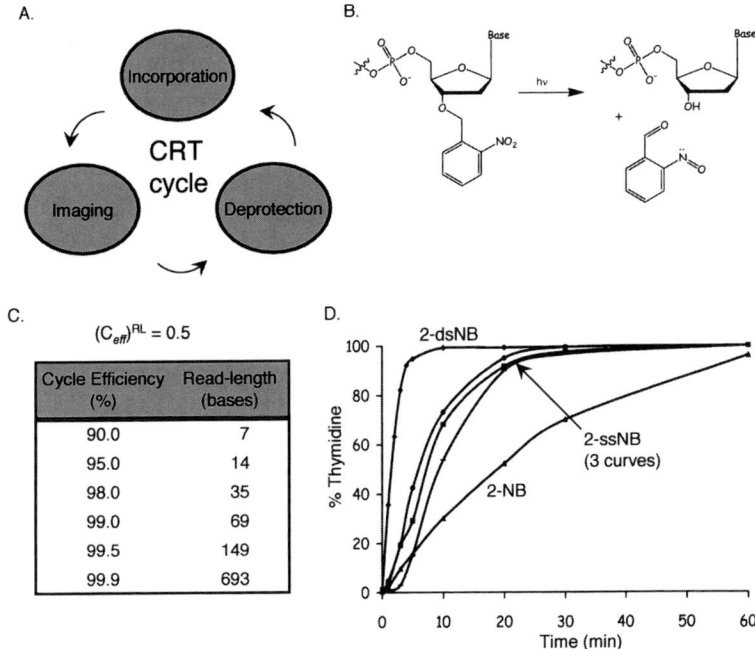

Figure 4. CRT technologies. (A) The CRT cycle. (B) The photocleavage reaction of a 3'-O-2-nitrobenzyl-nucleoside. (C) Effect of cycle efficiency on CRT read-length. (D) Kinetic study of protocleavage reaction for single substituted (2-SSNB) and double substituted (2-dsNB) 2-nitrobenzyl thymidine analogs. Percentage thymidine (%Thy) was calculated according to the equation: %Thy = $A_{Thy}/(A_{Thy} + A_{s2NB})$, where A_{Thy} and A_{s2NB} are the integrated peak areas from RP-HPLC analysis for thymidine and substituted 2-nitrobenzyl thymidine analogs, respectively.

homopolymer repeats, and (3) synchronistic extensions are maintained past heterozygous bases. An additional advantage is that unlike the pyrosequencing assay, which must be contained within a defined reaction well, the CRT assay can be performed on a number of highly parallel platforms, such as high-density oligonucleotide arrays (Pease et al. 1994; Albert et al. 2003), PTP arrays, (Leamon et al. 2003), polony arrays (Mitra and Church 1999), or random dispersion of single molecules. Albert et al. (2003) have demonstrated the 5'→3' synthesis of oligonucleotide on a high-density array and the application of incorporation of dye-labeled ddNTPs by DNA polymerase. These advantages of the CRT technology could represent significant improvements in speed, throughput, and accuracy over Sanger and SNA approaches.

At the center of the CRT chemistry is the reversible terminator. Ideally, these terminators should exhibit fast and efficient deprotection kinetics, efficient incorporation kinetics by DNA polymerase, and labels with desired characteristics, such as fluorophores with good fluorescence properties. Of the challenges associated with CRT for high-throughput genome sequencing, creating these reversible terminators with the desired properties and identifying DNA polymerases that recognize these substrates with high affinities are the most demanding aspects of the technology. The latter point is exemplified by the presence of competing natural nucleotides, which can readily cause asynchronistic base extensions (Metzker et al. 1998). The first examples of reversible terminators using commercially available DNA polymerases were reported by Canard and Sarfati (1994) and Metzker et al. (1994).

For CRT terminators to function properly, the protecting group must be efficiently cleaved under mild conditions while coupled to the primer. Removal of the protecting group generally involves either treatment with strong acid or base, catalytic or chemical reduction, or a combination of these methods. Unfortunately, these conditions may chemically perturb the DNA polymerase, nucleotides, oligonucleotide-primed template, or the solid support. Use of photocleavable protecting groups is an attractive alternative to rigorous chemical treatment and can be employed in a noninvasive manner. Of the various photocleavable protecting groups (Pillai 1980), the light-sensitive 2-nitrobenzyl group has been widely used. For example, it has been applied to natural nucleotides (Metzker et al. 1994, 1998), to the linker structure coupling a fluorescent dye to nucleobases (Li et al. 2003; Mitra et al. 2003), and to other nucleic acid structures as well (Ohtsuka et al. 1974; Pease et al. 1994; Chaulk and MacMillan 1998; Singh-Gasson et al. 1999). Under appropriate deprotection conditions (e.g., ultraviolet light >300 nm), the 2-nitrobenzyl group can be efficiently cleaved (Fig. 4B) without affecting either the pyrimidine or purine bases (Bartholomew and Broom 1975; Pease et al. 1994).

Other protecting groups have been described for reversible terminators as well. For example, Metzker et al. (1994) first described the synthesis and incorporation of a 3'-O-allyl-dATP by DNA polymerase, with the O-allyl group being removed using the well-known palladium (Pd) catalyst chemistry (Hayakawa et al. 1986, 1993; Honda et al. 1997). Recently, Ruparel et al. (2005) reported the synthesis of the first fluorescently labeled 3'-O-allyl-dNTPs. These unique reversible terminators require dual deprotection steps using UV light to cleave the fluorophore from the nucleotide (Fig. 3B), and the Pd catalyst reaction to restore the natural 3'-OH substrate. At this year's *Advances in Genome Biology and Technology/Automation in Mapping and Sequencing* meeting, Solexa reported on a similar CRT chemistry with a sequence read-length of approximately 20 bases (http://www.agbt.org) and recently reported the complete sequencing of the $\varphi\chi 174$ genome (http://www.solexa.com).

Earlier concerns regarding short read-lengths and assemblies for SNA strategies will prove relevant to CRT as well. To overcome this issue, research efforts in CRT technology development will continue to focus on the cycle efficiency. The CRT read-length is governed by the overall cycle efficiency, which is highly dependent on the product of deprotection and incorporation efficiencies. For example, if one considers the conservative loss of 50% signal as the assay's end-point, the read-length is a function of the cycle efficiency (C_{eff}) (Fig. 4C). Here, a read-length of only seven bases will be achieved with an overall cycle efficiency of 90% and can be increased beyond 100 bases in length by improving cycle efficiency to >99%. Figure 4D illustrates the effect that chemical modifications of the 2-nitrobenzyl ring system have on deprotection efficiency and thymidine production (V.A. Litosh, W. Wu, B. Stupi, and M. Metzker, unpubl.). Thus, recent improvements in chemical engineering of reversible terminators are important developments for CRT as an emerging technology for DNA sequencing applications.

CONCLUSIONS

Recent developments in DNA polymerase-dependent strategies highlight the central role these methods play in determination of the overall success of the sequencing assay. Although the standards for current Sanger technology have set the mark for emerging SNA and CRT technologies, these measures have evolved over several decades and from numerous research laboratories. The integration of additional technologies will be key for development of robust DNA sequencing platforms, including instrumentation, microfluidics, robotics, automation, software control, data acquisition, and informatics.

Beyond the integrated instrumentation built around the chemistry, the method by which genomes are sequenced will be important. Most strategies described in this review will employ the random approach of whole-genome shotgun sequencing and assembly (Weber and Myers 1997), including resequencing efforts for human sequence variation studies. While the random approach has the advantage of simplicity, it will require a tremendous number of sequence reads (i.e., a minimum of 900 million, 100-base reads will be needed to achieve a 30× assembly for a mammalian-size genome) to produce comprehensive sequence data for comparative studies between genomes. A directed approach, which targets specific regions across the genome, can effectively reduce genome size and complexity and, therefore, the number of sequencing reads needed to produce these comprehensive data sets. One example of a directed strategy for human resequencing could be the application of the CRT method to $5' \rightarrow 3'$ synthesized high-density oligonucleotide arrays (Albert et al. 2003) by relying on the reference sequence as anchor points along the genome. The careful selection of unique and functional priming sites would represent an oligonucleotide tiling path across the genome. Priming CRT reactions from these anchor points and sequencing to adjacent priming sites would provide contiguous coverage of the targeted regions of interest. CRT reads could then be aligned to the known positions along the reference genome in a straightforward manner. This approach could also be used for mapping sequence reads to related genomes for comparative genomics studies. Alignment of random reads could be performed using conventional assembly algorithms, guided by the reference sequence, to produce contiguous DNA sequence information.

Although in its infancy, the potential for these emerging sequencing strategies to deliver next-generation technologies looks promising. Improvements in speed, efficiency, throughput, and sensitivity will all contribute to a reduction in cost over the next several years. The timing of these strategies coincides with an increasing demand for resequencing capacity, which will provide valuable insight into the role of specific sequence variation with common disease. Integration of multidisciplinary technologies will translate into practical and affordable sequencing devices capable of whole-genome analyses. Application of genome sequence information to health benefits could revolutionize disease prevention measures, early disease interventions, and make the possibility of personalized therapies routine.

ACKNOWLEDGMENTS

I am extremely grateful to Richard A. Gibbs, Donna M. Muzny, and Sherry Metzker for critical review of the manuscript; Steven A. Soper for technical discussion; and NHGRI for their support from grants R01 HG003573, R41 HG003072, R41 HG003265, and R21 HG002443.

REFERENCES

Alaverdian, L., Alaverdian, S., Bilenko, O., Bogdanov, I., Filippova, E., Gavrilov, D., Gorbovitski, B., Gouzman, M., Gudkov, G., Domratchev, S., et al. 2002. A family of novel DNA sequencing instruments based on single-photon detection. *Electrophoresis* **23**: 2804–2817.

Albert, T.J., Norton, J., Ott, M., Richmond, T., Nuwaysir, K., Nuwaysir, E.F., Stengele, K.-P., and Green, R.D. 2003. Light-directed 5′→3′ synthesis of complex oligonucleotide microarrays. *Nucleic Acids Res.* **31**: e35.

Backhouse, C., Caamano, M., Oaks, F., Nordman, E., Carrillo, A., Johnson, B., and Bay, S. 2000. DNA sequencing in a monolithic microchannel device. *Electrophoresis* **21**: 150–156.

Bartholomew, D.G. and Broom, A.D. 1975. One-step chemical synthesis of ribonucleosides bearing a photolabile ether protecting group. *J. Chem. Soc. Chem. Commun.* **Issue 2**: 38.

Becker, H. and Gartner, C. 2000. Polymer microfabrication methods for microfluidic analytical applications. *Electrophoresis* **21**: 12–26.

Bonham, V.L., Warshauer-Baker, E., and Collins, F.S. 2005. Race and ethnicity in the genome era: The complexity of the constructs. *Am. Psychol.* **60**: 9–15.

Boone, T., Fan, Z., Hooper, H., Ricco, A., Tan, H., and Williams, S. 2002. Plastic advances microfluidic devices. *Anal. Chem.* **74**: 78A–86A.

Braslavsky, I., Hebert, B., Kartalov, E., and Quake, S.R. 2003. Sequence information can be obtained from single DNA molecules. *Proc. Natl. Acad. Sci.* **100**: 3960–3964.

Canard, B. and Sarfati, R. 1994. DNA polymerase fluorescent substrates with reversible 3′-tags. *Gene* **148**: 1–6.

Carrilho, E. 2000. DNA sequencing by capillary array electrophoresis and microfabricated array systems. *Electrophoresis* **21**: 55–65.

Carrilho, E., Ruiz-Martinez, M.C., Berka, J., Smirnov, I., Goetzinger, W., Miller, A.W., Brady, D., and Karger, B.L. 1996. Rapid DNA sequencing of more than 1000 bases per run by capillary electrophoresis using replaceable linear polyacrylamide solutions. *Anal. Chem.* **68**: 3305–3313.

Chaisson, M., Pevzner, P., and Tang, H. 2004. Fragment assembly with short reads. *Bioinformatics* **20**: 2067–2074.

Chaulk, S. and MacMillan, A. 1998. Caged RNA: Photo-control of a ribozyme reaction. *Nucleic Acids Res.* **26**: 3173–3178.

Collins, F. and Galas, D. 1993. A new five-year plan for the U.S. Human Genome Project. *Science* **262**: 43–46.

Collins, F.S., Guyer, M.S., and Chakravarti, A. 1997. Variations on a theme: Cataloging human DNA sequence variation. *Science* **278**: 1580–1581.

Collins, F.S., Patrinos, A., Jordon, E., Chakravarti, A., Gesteland, R., Walters, L., and Members of the DOE and NIH Planning Groups. 1998. New goals for the U.S. Human Genome Project: 1998–2003. *Science* **282**: 682–689.

Collins, F.S., Green, E.D., Guttmacher, A.F., and Guyer, M.S. 2003. A vision for the future of genomics research. *Nature* **422**: 835–847.

Crawford, D.C. and Nickerson, D.A. 2005. Definition and clinical importance of haplotypes. *Annu. Rev. Med.* **56**: 303–320.

Culbertson, C.T., Jacobson, S.C., and Ramsey, J.M. 1998. Dispersion sources for compact geometries on microchips. *Anal. Chem.* **70**: 3781–3789.

Daly, M.J., Rioux, J.D., Schaffner, S.F., Hudson, T.J., and Lander, E.S. 2001. High-resolution haplotype structure in the human genome. *Nat. Genet.* **29**: 229–237.

The ENCODE Project Consortium. 2004. The ENCODE (ENCyclopedia Of DNA Elements) Project. *Science* **306**: 636–640.

Entz, P., Toliat, M.R., Hampe, J., Valentonyte, R., Jenisch, S., Nürnberg, P., and Nagy, M. 2005. New strategies for efficient typing of HLA class-II loci DQB1 and DRB1 by using pyrosequencing. *Tissue Antigens* **65**: 67–80.

Ewing, B. and Green, P. 1998. Base-calling of automated sequencer traces using Phred. II. Error probabilities. *Genome Res.* **8**: 186–194.

Ewing, B., Hillier, L., Wendl, M.C., and Green, P. 1998. Base-calling of automated sequencer traces using Phred. I. Accuracy assessment. *Genome Res.* **8:** 175–185.

Foster, M.W. and Sharp, R.R. 2002. Race, ethnicity, and genomics: Social classifications as proxies of biological heterogeneity. *Genome Res.* **12:** 844–850.

Fox, M.S., Magasanik, B., Signer, E.R., Solomon, F., Gellert, M.F., Haber, J.E., Daniel, J., Koshland, E., and Muschel, L.H. 1990. The Genome Project: Pro and con. *Science* **247:** 270.

Gabriel, S.B., Schaffner, S.F., Nguyen, H., Moore, J.M., Roy, J., Blumenstiel, B., Higgins, J., DeFelice, M., Lochner, A., Faggart, M., et al. 2002. The structure of haplotype blocks in the human genome. *Science* **296:** 2225–2229.

Gharizadeha, B., Nordströma, T., Ahmadiana, A., Ronaghi, M., and Nyrén, P. 2002. Long-read pyrosequencing using pure 2'-deoxyadenosine-5'-O'-(1-thiotriphosphate) Sp-isomer. *Anal. Biochem.* **301:** 82–90.

Guttman, A. 2002a. Capillary electrophoresis using replaceable gels. U.S. patent no. RE37,606.

———. 2002b. Capillary electrophoresis using replaceable gels. U.S. patent no. RE37,941.

Harrison, D.J., Manz, A., Fan, Z., Luedi, H., and Widmer, H.M. 1992. Capillary electrophoresis and sample injection systems integrated on a planar glass chip. *Anal. Chem.* **64:** 1926–1932.

Harrison, D.J., Fluri, K., Seiler, K., Fan, Z., Effenhauser, C.S., and Manz, A. 1993. Micromachining a miniaturized capillary electrophoresis-based chemical analysis system on a chip. *Science* **261:** 895–897.

Hayakawa, Y., Kato, H., Uchiyama, M., Kajino, H., and Noyori, R. 1986. Allyloxycarbonyl group: A versatile blocking group for nucleotide synthesis. *J. Org. Chem.* **51:** 2400–2402.

Hayakawa, Y., Hirose, M., and Noyori, R. 1993. O-Allyl protection of guanine and thymine residues in oligodeoxyribonucleotides. *J. Org. Chem.* **58:** 5551–5555.

Honda, M., Morita, H., and Nagakura, I. 1997. Deprotection of allyl groups with sulfinic acids and palladium catalyst. *J. Org. Chem.* **62:** 8932–8936.

Hyman, E.D. 1988. A new method of sequencing DNA. *Anal. Biochem.* **174:** 423–436.

The International HapMap Consortium. 2003. The International HapMap Project. *Nature* **426:** 789–796.

Jacobson, S.C., Hergenroder, R., Koutny, L.B., Warmack, R.J., and Ramsey, J.M. 1994. Effects of injection schemes and column geometry on the performance of microchip electrophoresis devices. *Anal. Chem.* **66:** 1107–1113.

Ju, J., Ruan, C., Fuller, C., Glazer, A., and Mathies, R. 1995. Fluorescence energy transfer dye-labeled primers for DNA sequencing and analysis. *Proc. Natl. Acad. Sci.* **92:** 4347–4351.

Kan, C.-W., Fredlake, C.P., Doherty, E.A.S., and Barron, A.E. 2004. DNA sequencing and genotyping in miniaturized electrophoresis systems. *Electrophoresis* **25:** 3564–3588.

Kartalov, E.P. and Quake, S.R. 2004. Microfluidic device reads up to four consecutive base pairs in DNA sequencing-by-synthesis. *Nucleic Acids Res.* **32:** 2873–2879.

Kheterpal, I., Scherer, J., Clark, S., Radhakrishnan, A., Ju, J., Ginther, C., Sensabaugh, G.F., and Mathies, R.A. 1996. DNA sequencing using a four-color confocal fluorescence capillary array scanner. *Electrophoresis* **17:** 1852–1859.

Koshland, D.E. 1989. Sequences and consequences of the human genome. *Science* **246:** 189.

Koutny, L., Schmalzing, D., Salas-Solano, O., El-Difrawy, S., Adourian, A., Buonocore, S., Abbey, K., McEwan, P., Matsudaira, P., and Ehrlich, D. 2000. Eight hundred-base sequencing in a microfabricated electrophoretic device. *Anal. Chem.* **72:** 3388–3391.

Lander, E.S. 1996. The new genomics: Global views of biology. *Science* **274:** 536–539.
Langaee, T. and Ronaghi, M. 2005. Genetic variation analyses by pyrosequencing. *Mutat. Res.* **573:** 96–102.
Lassiter, S.J., Stryjewski, W., Benjamin, J., Legendre, L., Erdmann, R., Wahl, M., Wurm, J., Peterson, R., Middendorf, L., and Soper, S.A. 2000. Time-resolved fluorescence imaging of slab gels for lifetime base-calling in DNA sequencing applications. *Anal. Chem.* **72:** 5373–5382.
Leamon, J.H., Lee, W.L., Tartaro, K.R., Lanza, J.R., Sarkis, G.J., deWinter, A.D., Berka, J., Weiner, M., Rothberg, J.M., and Lohman, K.L. 2003. A massively parallel PicoTiter-Plate™ based platform for discrete picoliter-scale polymerase chain reactions. *Electrophoresis* **24:** 3769–3777.
Lee, L., Spurgeon, S., Heiner, C., Benson, S., Rosenblum, B., Menchen, S., Graham, R., Constantinescu, A., Upadhya, K., and Cassel, J. 1997. New energy transfer dyes for DNA sequencing. *Nucleic Acids Res.* **25:** 2816–2822.
Levene, M.J., Korlach, J., Turner, S.W., Foquet, M., Craighead, H.G., and Webb, W.W. 2003. Zero-mode waveguides for single-molecule analysis at high concentrations. *Science* **299:** 682–686.
Lewis, E.K., Haaland, W.C., Nguyen, F., Heller, D.A., Allen, M.J., MacGregor, R.R., Berger, C.S., Willingham, B., Burns, L.A., Scott, G.B.I., et al. 2005. Color-blind fluorescence detection for four-color DNA sequencing. *Proc. Natl. Acad. Sci.* **102:** 5346–5351.
Li, Z., Bai, X., Ruparel, H., Kim, S., Turro, N.J., and Ju, J. 2003. A photocleavable fluorescent nucleotide for DNA sequencing and analysis. *Proc. Natl. Acad. Sci.* **100:** 414–419.
Lieberwirth, U., Arden-Jacob, J., Drexhage, K.H., Herten, D.P., Muller, R., Neumann, M., Schulz, A., Siebert, S., Sagner, G., Klingel, S., et al. 1998. Multiplex dye DNA sequencing in capillary gel electrophoresis by diode laser-based time-resolved fluorescence detection. *Anal. Chem.* **70:** 4771–4779.
Liu, S., Shi, Y., Ja, W., and Mathies, R.A. 1999. Optimization of high-speed DNA sequencing on microfabricated capillary electrophoresis channels. *Anal. Chem.* **71:** 566–573.
Liu, S., Ren, H., Gao, Q., Roach, D.J., Loder Jr., R.T., Armstrong, T.M., Mao, Q., Blaga, I., Barker, D.L., and Jovanovich, S.B. 2000. Automated parallel DNA sequencing on multiple channel microchips. *Proc. Natl. Acad. Sci.* **97:** 5369–5374.
Liu, P.-Y., Zhang, Y.-Y., Lu, Y., Long, J.-R., Shen, H., Zhao, L.-J., Xu, F.-H., Xiao, P., Xiong, D.-H., Liu, Y.-J., et al. 2005. A survey of haplotype variants at several disease candidate genes: The importance of rare variants for complex diseases. *J. Med. Genet.* **42:** 221–227.
Luria, S.E., Cooper, D.M., and Berkowitz, A. 1989. Human Genome Project. *Science* **246:** 873–874.
Madabhushi, R.S. 1998. Separation of 4-color DNA sequencing extension products in noncovalently coated capillaries using low viscosity polymer solutions. *Electrophoresis* **19:** 224–230.
Madabhushi, R.S., Menchen, S.M., Efcavitch, J.W., and Grossman, P.D. 1996. Polymers for separation of biomolecules by capillary electrophoresis. U.S. patent no. 5,567,292.
———. 1999. Polymers for separation of biomolecules by capillary electrophoresis. U.S. patent no. 5,916,426.
Margulies, M., Egholm, M., Altman, W.E., Attiya, S., Bader, J.S., Bemben, L.A., Berka, J., Braverman, M.S., Chen, Y.-J., Chen, Z., et al. 2005. Genome sequencing in microfabricated high-density picolitre reactors. *Nature* **437:** 376–380.

McDonald, J.C., Duffy, D.C., Anderson, J.R., Chiu, D.T., Wu, H., Schueller, O.J.A., and Whitesides, G.M. 2000. Fabrication of microfluidic systems in poly(dimethylsiloxane). *Electrophoresis* **21:** 27–40.

Metzker, M.L., Raghavachari, R., Richards, S., Jacutin, S.E., Civitello, A., Burgess, K., and Gibbs, R.A. 1994. Termination of DNA synthesis by novel 3′-modified deoxyribonucleoside triphosphates. *Nucleic Acids Res.* **22:** 4259–4267.

Metzker, M.L., Lu, J., and Gibbs, R.A. 1996. Electrophoretically uniform fluorescent dyes for automated DNA sequencing. *Science* **271:** 1420–1422.

Metzker, M.L., Raghavachari, R., Burgess, K., and Gibbs, R.A. 1998. Elimination of residual natural nucleotides from 3′-O-modified-dNTP syntheses by enzymatic Mop-Up. *BioTechniques* **25:** 814–817.

Mitra, R. and Church, G. 1999. In situ localized amplification and contact replication of many individual DNA molecules. *Nucleic Acids Res.* **27:** e34.

Mitra, R.D., Shendure, J., Olejnik, J., Edyta-Krzymanska-Olejnik, and Church, G.M. 2003. Fluorescent in situ sequencing on polymerase colonies. *Anal. Biochem.* **320:** 55–65.

Nunnally, B.K., He, H., Li, L.-C., Tucker, S.A., and McGown, L.B. 1997. Characterization of visible dyes for four-decay fluorescence detection in DNA sequencing. *Anal. Chem.* **69:** 2392–2397.

Ohtsuka, E., Tanaka, S., and Ikehara, M. 1974. Studies on transfer ribonucleic acids and related compounds. IX(1) Ribooligonucleotide synthesis using a photosensitive o-nitrobenzyl protection at the 2′-hydroxyl group. *Nucleic Acids Res.* **1:** 1351–1357.

Paegel, B.M., Hutt, L.D., Simpson, P.C., and Mathies, R.A. 2000. Turn geometry for minimizing band broadening in microfabricated capillary electrophoresis channels. *Anal. Chem.* **70:** 3030–3037.

Paegel, B.M., Emrich, C.A., Wedemayer, G.J., Scherer, J.R., and Mathies, R.A. 2002. High throughput DNA sequencing with a microfabricated 96-lane capillary array electrophoresis bioprocessor. *Proc. Natl. Acad. Sci.* **99:** 574–579.

Paegel, B.M., Blazej, R.G., and Mathies, R.A. 2003. Microfluidic devices for DNA sequencing: Sample preparation and electrophoretic analysis. *Curr. Opin. Biotechnol.* **14:** 42–50.

Parkhill, J., Achtman, M., James, K.D., Bentley, S.D., Churcher, C., Klee, S.R., Morelli, G., Basham, D., Brown, D., Chillingworth, T., et al. 2000a. Complete DNA sequence of a serogroup A strain of *Neisseria meningitidis* Z2491. *Nature* **404:** 502–506.

Parkhill, J., Wren, B.W., Mungall, K., Ketley, J.M., Churcher, C., Basham, D., Chillingworth, T., Davies, R.M., Feltwell, T., Holroyd, S., et al. 2000b. The genome sequence of the food-borne pathogen *Campylobacter jejuni* reveals hypervariable sequences. *Nature* **403:** 665–668.

Patil, N., Berno, A.J., Hinds, D.A., Barrett, W.A., Doshi, J.M., Hacker, C.R., Kautzer, C.R., Lee, D.H., Marjoribanks, C., McDonough, D.P., et al. 2001. Blocks of limited haplotype diversity revealed by high-resolution scanning of human chromosome 21. *Science* **294:** 1719–1723.

Pease, A.C., Solas, D., Sullivan, E.J., Cronin, M.T., Holmes, C.P., and Fodor, S.P.A. 1994. Light-generated oligonucleotide arrays for rapid DNA sequence analysis. *Proc. Natl. Acad. Sci.* **91:** 5022–5026.

Pillai, V.N.R. 1980. Photoremovable protecting groups in organic synthesis. *Synthesis* **Issue 2:** 1–26.

Prober, J., Trainor, G., Dam, R., Hobbs, F., Robertson, C., Zagursky, R., Cocuzza, A., Jensen, M., and Baumeister, K. 1987. A system for rapid DNA sequencing with fluorescent chain-terminating dideoxynucleotides. *Science* **238:** 336–341.

Quake, S. and Scherer, A. 2000. From micro- to nanofabrication with soft materials. *Science* **290:** 1536–1540.
Risch, N. and Merikangas, K. 1996. The future of genetic studies of complex human diseases. *Science* **273:** 1516–1517.
Roberts, L. 1989a. New game plan for genome mapping. *Science* **245:** 1438–1440.
———. 1989b. Watson versus Japan. *Science* **246:** 576–578.
Robertson, J.A. 2003. The $1000 genome: Ethical and legal issues in whole genome sequencing of individuals. *Am. J. Bioeth.* **3:** W-IF1.
Ronaghi, M. 2000. Improved performance of pyrosequencing using single-stranded DNA-binding protein. *Anal. Biochem.* **286:** 282–288.
———. 2001. Pyrosequencing sheds light on DNA sequencing. *Genome Res.* **11:** 3–11.
Ronaghi, M., Karamohamed, S., Pettersson, B., Uhlén, M., and Nyrén, P. 1996. Real-time DNA sequencing using detection of pyrophosphate release. *Anal. Biochem.* **242:** 84–89.
Ronaghi, M., Uhlén, M., and Nyrén, P. 1998. A sequencing method based on real-time pyrophosphate. *Science* **281:** 363, 365.
Ruiz-Martinez, M.C., Berka, J., Belenkii, A., Foret, F., Miller, A.W., and Karger, B.L. 1993. DNA sequencing by capillary electrophoresis with replaceable linear polyacrylamide and laser-induced fluorescence detection. *Anal. Chem.* **65:** 2851–2858.
Ruparel, H., Bi, L., Li, Z., Bai, X., Kim, D.H., Turro, N.J., and Ju, J. 2005. Design and synthesis of a 3′-O-allyl photocleavable fluorescent nucleotide as a reversible terminator for DNA sequencing by synthesis. *Proc. Natl. Acad. Sci.* **102:** 5932–5937.
Salas-Solano, O., Carrilho, E., Kotler, L., Miller, A.W., Goetzinger, W., Sosic, Z., and Karger, B.L. 1998. Routine DNA sequencing of 1000 bases in less than one hour by capillary electrophoresis with replaceable linear polyacrylamide solutions. *Anal. Chem.* **70:** 3996–4003.
Salas-Solano, O., Schmalzing, D., Koutny, L., Buonocore, S., Adourian, A., Matsudaira, P., and Ehrlich, D. 2000. Optimization of high-performance DNA sequencing on short microfabricated electrophoretic devices. *Anal. Chem.* **72:** 3129–3137.
Sanger, F., Nicklen, S., and Coulson, A.R. 1977. DNA sequencing with chain-terminating inhibitors. *Proc. Natl. Acad. Sci.* **74:** 5463–5467.
Schmalzing, D., Tsao, N., Koutny, L., Chisholm, D., Srivastava, A., Adourian, A., Linton, L., McEwan, P., Matsudaira, P., and Ehrlich. D. 1999. Toward real-world sequencing by microdevice electrophoresis. *Genome Res.* **9:** 853–858.
Seo, T.S., Bai, X., Ruparel, H., Li, Z., Turro, N.J., and Ju, J. 2004. Photocleavable fluorescent nucleotides for DNA sequencing on a chip constructed by site-specific coupling chemistry. *Proc. Natl. Acad. Sci.* **101:** 5488–5493.
Seo, T.S., Bai, X., Kim, D.H., Meng, Q., Shi, S., Ruparel, H., Li, Z., Turro, N.J., and Ju, J. 2005. Four-color DNA sequencing by synthesis on a chip using photocleavable fluorescent nucleotides. *Proc. Natl. Acad. Sci.* **102:** 5926–5931.
Shendure, J., Mitra, R.D., Varma, C., and Church, G.M. 2004. Advanced sequencing technologies: Methods and goals. *Nat. Rev. Genet.* **5:** 335–344.
Shi, Y. and Anderson, R.C. 2003. High-resolution single-stranded DNA analysis on 4.5 cm plastic electrophoretic microchannels. *Electrophoresis* **24:** 3371–3377.
Simpson, J.W., Ruiz-Martinez, M.C., Mulhern, G.T., Berka, J., Latimer, D.R., Ball, J.A., Rothberg, J.M., and Went, G.T. 2000. Transmission imaging spectrograph and microfabricated channel system for DNA analysis. *Electrophoresis* **21:** 135–149.

Singh-Gasson, S., Green, R.D., Yue, Y., Nelson, C., Blattner, F., Sussman, M.R., and Cerrina, F. 1999. Maskless fabrication of light-directed oligonucleotide microarrays using a digital micromirror array. *Nat. Biotechnol.* **17:** 974–978.

Smith, L., Sanders, J., Kaiser, R., Hughes, P., Dodd, C., Connell, C., Heiner, C., Kent, S., and Hood, L. 1986. Fluorescence detection in automated DNA sequence analysis. *Nature* **321:** 674–679.

Smith, L.M., Kaiser, R.J., Sanders, J.Z., and Hood, L.E. 1987. The synthesis and use of fluorescent oligonucleotides in DNA sequence analysis. *Methods Enzymol.* **155:** 260–301.

Tabor, S. and Richardson, C.C. 1989. Effect of manganese ions on the incorporation of dideoxynucleotides by bacteriophage T7 DNA polymerase and *Escherichia coli* DNA polymerase I. *Proc. Natl. Acad. Sci.* **86:** 4076–4080.

———. 1995. A single residue in DNA polymerases of the *Escherichia coli* DNA polymerase I family is critical for distinguishing between deoxy- and dideoxyribonucleotides. *Proc. Natl. Acad. Sci.* **92:** 6339–6343.

Takahashi, S., Murakami, K., Anazawa, T., and Kambara, H. 1994. Multiple sheath-flow gel capillary-array electrophoresis for multicolor fluorescent DNA detection. *Anal. Chem.* **66:** 1021–1026.

Velculescu, V.E., Zhang, L., Vogelstein, B., and Kinzler, K.W. 1995. Serial analysis of gene expression. *Science* **270:** 484–487.

Weber, J.L. and Myers, E.W. 1997. Human whole-genome shotgun sequencing. *Genome Res.* **7:** 401–409.

Woolley, A.T. and Mathies, R.A. 1995. Ultra-high-speed DNA sequencing using capillary electrophoresis chips. *Anal. Chem.* **67:** 3676–3680.

Zhang, C.-X. and Manz, A. 2001. Narrow sample channel injectors for capillary electrophoresis on microchips. *Anal. Chem.* **73:** 2656–2662.

Zhu, L., Stryjewski, W., Lassiter, S., and Soper, S.A. 2003. Fluorescence multiplexing with time-resolved and spectral discrimination using a near-IR detector. *Anal. Chem.* **75:** 2280–2291.

Zhu, L., Stryjewski, W.J., and Soper, S.A. 2004. Multiplexed fluorescence detection in microfabricated devices with both time-resolved and spectral-discrimination capabilities using near-infrared fluorescence. *Anal. Biochem.* **330:** 206–218.

20

Genome Annotation Past, Present, and Future: How to Define an Open Reading Frame at Each Locus

Michael R. Brent

Laboratory for Computational Genomics and
  Department of Computer Science
Washington University
St. Louis, Missouri 63130

THE 10 YEARS SINCE *GENOME RESEARCH* BEGAN PUBLICATION bracket a complete era of genome research—an era of stunning successes and nagging loose ends, promise exceeded and promise as yet unfulfilled. The years 1996–2005 were characterized by tremendous optimism and productivity. In 1996, the sequencing of the human genome was scheduled to be completed in 2005 (Collins and Galas 1993). Driven by competition, automation, and technology, the genomics community far exceeded its own sequencing ambitions. But there was another goal that we have not yet reached—the genome was to provide a "parts list" for the human and other major model organisms. The parts turned out to be more varied than anticipated, and we have learned wonderful things about the biology and history encoded in genome sequences (Waterston et al. 2002; Gibbs et al. 2004). But the most fundamental parts on anyone's list, then and now, must be the complete set of translated open reading frames (ORFs) and the exon–intron structures from which they are assembled. (I will use the term ORF to denote the complete exon–intron structure of the protein coding region of any mature mRNA. Thus, a primary transcript that is alternatively spliced may represent more than one ORF.) After sequencing (Lander et al. 2001; Venter et al. 2001), completing (Collins et al. 2003; The International

Human Genome Sequencing Consortium 2003), and finishing (International Human Genome Sequencing Consortium 2004) the human genome, we do not have even one complete, correct ORF for each human gene locus. In fact, we do not have a complete correct ORF for each locus in the genome of any higher eukaryote.

Among the things we have learned by analyzing mammalian genomes are the incredible abundance of alternatively spliced RNAs (Modrek and Lee 2002; Sorek et al. 2004; Kapranov et al. 2005) and retroposed pseudogenes (Torrents et al. 2003; Zhang and Gerstein 2004), as well as the importance of micro-RNAs, siRNAs, and other noncoding RNAs in gene regulation (Zamore and Haley 2005). We must ultimately work out all the functional alternative splices of mRNAs with their untranslated regions (UTRs), *cis*-regulatory sites, and basic functional categories. We must also identify all the functional noncoding transcripts, their *cis*-regulatory sites, and their basic functional categories (The ENCODE Project Consortium 2004). However, the achievement of these goals appears to be far in the future. As a concrete and achievable, if somewhat arbitrary, milestone along that path, I will focus on identifying at least one complete ORF for each protein-coding gene.

It is abundantly clear that the methods we have been using to identify ORFs for most of the last 10 years are inadequate for finishing the job. In this perspective, I argue that we cannot rely on any of the following to get us through the home stretch of ORF identification:

- obtaining EST or mRNA sequences from randomly selected cDNA clones
- aligning expressed sequences to loci other than those from which they were transcribed (e.g., to the loci of gene family members or orthologs in other species)
- sequencing more genomes
- annotating manually by using human curators

All of these things are valuable, but none of them is likely to get us to a new, higher plateau in the quest for a complete ORF at each protein-coding locus. Instead, we will have to rely on large-scale PCR amplification of specific cDNAs followed by sequencing of the amplicons. To amplify cDNAs, we need reasonably accurate, though not necessarily perfect, gene predictions to use for PCR primer design. The further a prediction is from a true gene structure, the greater the likelihood that PCR primers designed for it will fail. Each failure increases the cost per gene identified and may reduce the completeness of the resulting collection of cDNA sequences. This method is feasible and in use today (Guigó et al. 2003;

Dike et al. 2004; Wu et al. 2004; Eyras et al. 2005; Wei et al. 2005), but we must continue to drive its cost down by (1) continuing the steady and significant improvements in de novo gene prediction that have occurred over the last 10 years and (2) optimizing and automating both the informatics and wet lab components of large-scale RT-PCR.

In the end, success in translating genome to ORFeome will take the same route as success in sequencing itself—investment in technology development, process optimization, and improved automation. Of course, transcripts that are completely unexpressed except in very specific circumstances will tend to be missed, but we can use these high-throughput methods to make a qualitative leap in the completeness of our ORF annotation.

To provide historical context for the argument outlined above, I will first review the state of the major gene prediction methods roughly 10 years ago, when *Genome Research* began publishing. The second section below provides a brief summary of the progress that has been made in the last decade. The third section presents the argument that the methods used for most of the last 10 years are not suitable for the end stages of ORF identification. The final section spells out some details of the recommended path toward understanding the most basic products of a genome.

FOUNDATIONS OF THE PRESENT ERA

Over the last 10 years, we have relied on three fundamental methods for identifying ORFs in genomic sequence: (1) sequencing randomly selected cDNA clones and aligning the sequences to their genomic sources; (2) finding ORFs that could produce proteins similar to proteins that are already in databases; and (3) finding ORFs de novo, without reference to cDNA sequences or their conceptual translations. Each of these methods came of age in the middle 1990s.

Aligning cDNA and Protein Sequences

In 1996, Hillier et al. reported sequencing 280,000 human ESTs, thereby increasing the size of GenBank's human EST collection by a factor of six (Hillier et al. 1996; Wolfsberg and Landsman 1997). The ongoing scale-up of EST sequencing created a demand for tools to align these sequences to their genomic sources. While local alignment tools like BLAST can give an approximate answer, determining splice sites accurately in the presence of sequencing error requires algorithms that incorporate a stronger model of the biological processes by which pre-mRNAs are spliced. The simplest approach is to allow "intron gaps" of unbounded length with no gap extension penalty, so long as they begin with GT and end with AG—the dinucleotides that bound 99% of all known introns. EST_GENOME (Mott 1997) implemented this simple approach using an algorithm that is optimal (guaranteed to find the highest scoring alignment). Finding the optimal alignment is computationally demanding, but given the inherent difficulty of the problem, the algorithm used in EST_GENOME is quite efficient. More recent programs (Florea et al. 1998; Wheelan et al. 2001) added heuristics that speed up the computation but may not always return the absolute highest-scoring alignment. They also added some detail to the scoring system by incorporating an intron-length penalty (Florea et al. 1998) or a more accurate system of splice-site scoring (Wheelan et al. 2001). There was no clear winner in terms of accuracy. Lacking a theory about how to set the parameters appropriately for a given degree of similarity between cDNA and genome, the programs were nearly always used with the default parameters. Perhaps as a result, each system was most accurate for some species and some similarity levels. But the fact that EST_GENOME returns the best alignment according to a simple, clear scoring scheme lends it a unique appeal. Its scoring scheme turned out to be adequate, in the sense that one could not obtain alignments that were significantly more accurate for a broad range of situations, given the sequence quality and computing power available at the time.

As an example of the ambiguities that arise in cDNA-to-genome alignment, consider a short cDNA segment that can be aligned as the 3' end of a long exon with mismatches (Fig. 1A) or as a short independent exon without mismatches (Fig. 1B). Traditional spliced alignment programs, such as EST_GENOME (Mott 1997) will make this decision based on somewhat arbitrary match/mismatch scores and intron penalties that depend on whether the intron begins with GT and ends with AG. A small

```
A   ACTACCTCATGAGGTGGCgtga....atagtacgcgtaa.....ttagCACTTCTGGGGCCCA
    |||||||||||||| | ||>>>>>>>>>> 15907 >>>>>>>>>>>>|||||||||||||||
    ACTACCTCATGAGTACGC.........................CACTTCTGGGGCCCA


B   ACTACCTCATGAGgtggcgtga....atagTACGCgtaa.....ttagCACTTCTGGGGCCCA
    |||||||||||||>>>>> 12584 >>>>>|||||>>>> 3323 >>>|||||||||||||||
    ACTACCTCATGAG................TACGC............CACTTCTGGGGCCCA
```

Figure 1. (A) A fragment of an alignment of a cDNA to genomic sequence containing two mismatches and a 15,907-bp intron. (B) Another alignment of the same two sequences containing no mismatches and a 5-bp exon in the intron of alignment A. All introns are bounded by canonical GT-AG splice sites.

fraction of introns is known to be bounded by other dinucleotide pairs, such as GC-AG (~1%), AT-AC (~0.15%), and even GA-AG (1–3 known cases) (Brackenridge et al. 2003). Although some AT-AC introns are spliced via the U12 spliceosome (Tarn et al. 1995; Tarn and Steitz 1996), the initial and terminal dinucleotides do not determine whether the U2 or U12 spliceosome is used (Sharp and Burge 1997). In any case, EST_GENOME does not differentiate among GC-AG, AT-AC, or any other intron boundaries except GT-AG. Thus, it will create introns starting with TT under certain circumstances, even though there is no convincing evidence that such introns exist. Compromises like this are necessary when the TT may result from error in sequencing a GT. But when the quality of the genome sequence is high enough, the probability that an intron will start with an apparent TT approaches zero. This illustrates how the best approach to cDNA-to-genome alignment depends on the quality of the sequences involved.

A related gene prediction approach is to align protein sequences or profiles from existing databases to a genome sequence (Birney et al. 1996; Gelfand et al. 1996; Birney and Durbin 1997; Birney et al. 2004b). Because most "protein" sequences in the databases are derived by conceptual translation of cDNAs, and because the alignment algorithms for cDNA and protein sequences are similar, it is tempting to treat cDNA alignment and protein alignment as a single approach to annotation. However, there is a difference in conception and typical application. Most cDNA alignment programs are intended primarily for aligning sequences to the genomic locus from which they were transcribed, although these programs have been used for cross-species alignments (Florea et al. 1998; Wheelan et al. 2001). Protein-oriented alignment programs, on the other hand, are intended for more distant relationships, such as discovering new members of a known protein family or discovering homologs in a new species.

Philosophically, these are very different approaches. The evidence that a cDNA sequence provides about the exon–intron structure from which it is assembled is much more direct than the evidence that a protein sequence provides about the loci of putative homologs. Cross-locus protein aligners must accept a significant degree of mismatch between the protein to be aligned and the target locus, which can lead to difficulty in distinguishing between functional homologs and nontranscribed pseudogenes (Birney et al. 2004b). Systems for aligning high-quality cDNA sequences to their genomic sources, on the other hand, can require an almost perfect match, which often helps to distinguish their true loci from nontranscribed pseudogenes.

GeneWise (Birney and Durbin 1997, 2000; Birney et al. 2004b) is certainly the most important protein-to-genome alignment program, since it forms a central part of the highly influential Ensembl gene annotation pipeline. The accuracy of GeneWise is determined primarily by the degree of similarity between the protein and the locus to which it is aligned. Thus, several investigators have tried to estimate its accuracy as a function of protein-to-genome similarity. Guigó and colleagues (2000) used the P-value of the BLASTP alignment between the protein and the gene locus. The P-value reflects both the length and percent identity of the alignment, but not the fraction of the true gene covered by it. Using their Semi-Artificial Gene set, which was created by concatenating single gene sequences separated by random "intergenic" sequence, they found that GeneWise exceeded the accuracy of Genscan in exact exon prediction only when aligning highly similar proteins ($P < 10^{-50}$). When aligning proteins of moderate similarity ($10^{-50} < P < 10^{-6}$), its overall accuracy was similar to that of Genscan, but GeneWise was more conservative—it missed more of the real exons, but a higher proportion of those it predicted turned out to be exactly right.

Instead of using the P-value, Birney et al. (2004b) measured similarity using percent identity and percent of the target gene covered by the protein. Their most similar category of proteins was 85%–95% identical to the target locus and aligned to within 20 amino acids of the target's start and stop codons. On a single-gene test set, and excluding terminal exons, they found that GeneWise had 93% exact exon specificity and 75% exact exon sensitivity. Thus, while most of the exons it predicted were correct, it failed to correctly predict one fourth of the real exons. However, two caveats regarding this test are in order. First, it is impossible to *know*, when running GeneWise, whether the protein alignment is close enough to the ends of the target to meet the requirements of this accuracy level, since the genes in the target sequence are unknown. If alignments are not selected on length, GeneWise has only 40% exact exon sensitivity using proteins in the 85%–95% identity range. Second, this single gene set is likely impoverished for pseudogenes, which are a significant source of false positives for GeneWise.

De Novo Gene Prediction

De novo gene prediction, in its modern form, also appeared in the mid-1990s. Stormo and Haussler (1994) described the first Generalized

Hidden Markov model (GHMM) for gene prediction. GHMMs are mathematical models that can be used to define probabilities for all possible exon–intron annotations on a given sequence. An accurate GHMM for gene finding will assign high probabilities to correct annotations and low probabilities to incorrect annotations. GHMMs differ from ordinary HMMs in that the log probabilities, or scores, of exons and introns can depend globally on the entire sequence of the exon or intron. In ordinary HMMs, the scores of features must be the sums of the scores of individual bases within the feature. For example, the ability to compute scores from the whole feature makes it possible to create general, nonlinear models of the lengths of exons and introns. Kulp et al. (1996) were the first to use the term GHMM in the context of gene finding, the first to describe a fully general mathematical framework for all GHMM models, and the first to implement and test a GHMM-based computer program for gene finding. Burge and Karlin (1997) developed Genscan, a GHMM-based gene prediction program that could predict multiple and partial genes on both strands. This ability made Genscan suitable for annotating ORFs in anonymous DNA sequence such as that produced by sequencing random BAC clones. As the first GHMM suitable for annotating anonymous genome sequence, Genscan defined the state of the art in both technology and accuracy. Eventually, programs like Genie were enhanced to predict multiple genes on both strands (Reese et al. 2000), but Genscan remained one of the most accurate and most widely used programs for many years.

Combining Prediction Methods

The genome annotations that are most visible to the public and most widely used are created by combining predictions from all of these methods. In the mid-1990s, the results of cDNA alignments, protein alignments, and de novo predictions were integrated by human experts and were often followed by RT-PCR and sequencing experiments to test the predicted exon–intron structures (Ansari-Lari et al. 1996, 1997) (RT-PCR is PCR amplification of cDNAs made by reverse transcription from RNA). Automated "pipelines" for integrating evidence, such as Ensembl (Birney et al. 2004a; Curwen et al. 2004), OTTO (Venter et al. 2001), and the National Center for Biotechnology Information (NCBI) pipeline were not developed until shortly before the publication of the draft human genome sequence.

TRAJECTORY OF IMPROVING ACCURACY

Aligning cDNA Sequences

The accuracy of prediction systems based on aligning cDNA or protein sequences depends on the sequences that are available for alignment as well as the algorithms used to align them. There can be no doubt that both the quantity and quality of expressed sequences have improved dramatically in the last ten years. For example, the human EST database has gone from 415,000 sequences in 1997 to over 6 million in 2005. Several projects, including the Mammalian Gene Collection (MGC) (http://mgc.nci.nih.gov/) (Furey et al. 2004; The MGC Project Team 2004) have produced finished sequences from large collections of cDNA clones that appear to contain a complete ORF. Indeed, the MGC collection now contains at least one cloned transcript from about 13,000 human and 12,000 mouse gene loci. These sequences have been subjected to extremely rigorous quality control so that most produce 100% identical alignments to the reference genome, except for silent discrepancies and known polymorphisms (http://genes.cse.wustl.edu/mgc/) (Furey et al. 2004).

There have been improvements in alignment algorithms, too. Traditional cDNA-to-genome alignment programs do not explicitly model the probability of mismatches in the correct alignment (Fig. 1A) as compared with the probability of an additional intron in the correct alignment (Fig 1B). In fact, mismatches in correct alignments are either sequencing errors or differences between the reference genome and the genome from which the cDNA was transcribed. (Occasionally, they may also result from posttranscriptional events such as RNA editing.) Thus, the probabilities of these events depend on both the sequence quality and the rate of polymorphism (for within-species alignments) or divergence (for cross-species alignments). On the other hand, the probability of an additional intron depends on the frequency of introns in the species at hand.

A new generation of cDNA-to-genome alignment programs models all these things using pair hidden Markov models with parameters estimated from the specific cDNA collection and the genome sequence to be aligned (M. Arumugam and M.R. Brent, in prep.). For example, such systems can easily model the fact that sequencing errors are much less likely when aligning an MGC cDNA sequence to the finished human genome than when aligning a single-pass EST sequence to the draft dog genome. For 70%–80% of high-quality cDNA sequences, these more precise models will result in the same alignment as a program like EST_GENOME. In many of the remaining cases, however, they produce better alignments. For example, they are better able to distinguish small exons from sequencing errors. This accuracy improvement is made possible by the availability of very high quality sequences to align and the availability of sufficient computing power to run the pairHMM algorithms in a reasonable amount of time.

Single-Genome de Novo Gene Prediction

Perhaps the best indicator of how the accuracy of de novo gene prediction has changed in the last ten years is the change in how accuracy is measured. Shortly before the publication of Genscan, Burset and Guigó (1996) published the first comprehensive comparison of vertebrate prediction programs. Their test set consisted of 570 genomic sequences no longer than 50 Kb, each with a single gene on the positive strand. Burset and Guigó found that the most accurate de novo system of that time, FGENEH (Solovyev et al. 1994), predicted just 61% of the known exons correctly, GeneID (Guigó et al. 1992) got 51%, and all the rest got well below 50% of exons right. The fraction of ORFs predicted correctly was not reported, presumably because it was near zero.

Genscan (Burge and Karlin 1997) represented a breakthrough in accuracy that led to a long, slow shift in the evaluation paradigm. When tested on Burset and Guigó's single-gene set, it predicted 78% of the exons correctly, compared with just 61% for the best previous system. Furthermore, Burge and Karlin (1997) reported that Genscan predicted 43% of the ORFs in that test correctly. They also presented an analysis of Genscan's predictions on a contiguous sequence of 117 Kb containing multiple experimentally determined gene structures (Ansari-Lari et al. 1996). Although the number of genes was too small for reliable estimation of accuracy, it is interesting that only one of eight Genscan-predicted ORFs matched the annotation exactly (12%), though most were quite similar.

As test sets became more realistic, estimates of Genscan's accuracy at predicting complete human ORFs dropped. Guigó et al. (2000) published an evaluation based on a simulated human genomic sequence that they created by concatenating single-gene sequences padded by randomly generated pseudo-intergenic sequence. In this new test set, only 2.3% of the nucleotides were protein-coding, much closer to the overall average (now thought to be under 2%) than the 15% in Burset and Guigó's 1996 set. As expected, they found that prediction accuracy is much lower on contiguous sequences with typical coding density than on single-gene sequences with high coding density. However, they did not report the percentage of ORFs predicted correctly. Korf et al. (2001) tested Genscan on 7.6 Mb of mouse genome consisting of 68 contiguous sequences with an average length of 112 Kb. In this test, Genscan predicted only about 15% of annotated ORFs exactly—much lower than the 43% reported for Burset and Guigó's single-gene set. However, even this estimate turned out to be optimistic. When Genscan was finally evaluated on the entire human genome, it predicted a correct ORF at only 10% of loci containing a known ORF (9% of known ORFs, Flicek et al. 2003).

Currently, gene prediction programs are used primarily for whole genome annotation. As described above, their accuracy when evaluated on a whole genome is typically much lower than their accuracy when evaluated on isolated genes or artificially concatenated sets of single genes. Even whole chromosomes can be deceptive. For example, human chromosome 22, besides being the smallest autosome, is also unusually gene dense, with smaller than average introns and intergenic regions and above average GC content. Most gene prediction programs, including Genscan, tend to perform best on high GC, gene-dense regions. Thus, evaluation on chromosome 22 systematically overestimates the accuracy of most systems. In the current environment, the minimal standard for evaluation of gene prediction programs must be based on whole genome annotation runs. Some may argue that, since we do not know all the exon-intron structures for the human or any other genome, we cannot know the accuracy of a prediction set for the whole genome. This is true, but it should not be an impediment to evaluating whole genome annotations. Sensitivity estimates based on the subset of genes whose structures are known should be an unbiased estimate of sensitivity on all genes, to the extent that the sets of known genes and unknown genes do not differ in ways that greatly affect accuracy. While it is possible that unknown genes are radically different from known genes in this way, there is no reason to believe that they are. Specificity will be systematically underestimated when the predictions are compared to known genes rather than to all genes. Under the same assumption described above, dividing by the fraction of genes that are known (or the fraction of exons that are known, for exon-level specificity) corrects the underestimate. The exact value of that correction factor does not matter when comparing the specificities of two programs—the one with the higher raw estimate will also have the higher corrected estimate. Another approach, which seems to always give qualitatively similar results, is to use only gene predictions that overlap known genes by at least one nucleotide when computing specificity (Wei et al. 2005).

Determining gene boundaries is one of the most challenging aspects of ORF prediction—much more so than predicting the boundaries of exons with splices on both sides—and so it is also the area in which the potential for improvement is greatest. Many improvements to gene prediction algorithms have a large effect on accuracy as measured by exact ORF prediction, even though they have little effect on the accuracy of exon prediction. Thus, it is critical to include measures of exact ORF prediction in comparative evaluations of gene prediction programs.

Statistics on the exact-ORF accuracies of programs are important, but there is a legitimate argument that the value of these programs is not in predicting known genes but in predicting novel genes. Thus, the most convincing evaluation of a program or a set of programs is the extent to which its novel predictions can be verified experimentally. The trend toward publishing experimental evaluations of prediction sets (Wu et al. 2004; Brown et al. 2005; Eyras et al. 2005; Wei et al. 2005) therefore represents a significant step forward for the field of gene prediction.

Dual- and Multi-Genome de Novo Predictors

Historically, Genscan represents the apogee in the arc of improving accuracy for mammalian gene predictors using a single genomic sequence as their only input (but see Stanke and Waack 2003 for improvements on *Drosophila*). No other system robustly outperformed Genscan on large, contiguous human genomic sequences until the advent of dual-genome de novo systems, which use alignments between two genomes as a rough indicator of which nucleotides are under negative selection and hence are likely to have a function that contributes to fitness. Several such systems required that orthologous stretches of mouse and human DNA be identified in advance, or were tested only on single-gene sets that were preselected to have clearly identifiable orthologs (Bafna and Huson 2000; Batzoglou et al. 2000; Alexandersson et al. 2003). However, TWINSCAN (Korf et al. 2001; Flicek et al. 2003) and SGP2 (Parra et al. 2003) could both be run on entire human chromosomes, using alignments generated by simple, robust procedures. Neither program requires that an ortholog be present in the other genome—they can just as easily exploit similarity to paralogs, or even fragments of several genes from the same family. Flicek et al. (2003) reported that TWINSCAN was able to predict 14% of known ORFs in the human genome correctly. Although SGP2 was tested only on chromosome 22, the accuracies of the two systems were similar. Over the last few years, incremental improvements to TWINSCAN, along with better training and testing sets, have improved its accuracy to the point where it can predict a correct ORF at about 25% of human loci with known ORFs.

A new level of accuracy was achieved by N-SCAN, a version of TWINSCAN with a new, phylogenetic conservation model that is capable of considering alignments among multiple genomes (Gross and Brent 2005, 2006). N-SCAN is able to predict a correct ORF at about 35% of human loci with known ORFs. It is also notably more accurate than previous systems at the exon level, predicting 85% of known human coding exons correctly, whereas previous systems predicted fewer than 75% correctly. Furthermore, it is the first program to accurately predict the boundaries of long introns—it correctly predicts about 50% of introns in the 50- to 100-Kb length range.

Many of the challenges of de novo gene prediction that have been observed over the years remain challenges today. Even the best prediction programs tend to split and fuse genes, and they have difficulty accurately predicting stop codons and especially start codons. They only predict a single isoform at each locus, even though a large fraction of human genes are alternatively spliced. Yet there has been enormous progress. We have moved from predicting a correct ORF at one tenth of the human loci to predicting a correct ORF at one third. We can now predict long introns (Gross and Brent 2005), and we can predict spliced 5' UTRs with reasonable accuracy (Brown et al. 2005). Gene predictors are generally more accurate on more compact genomes such as those of *Caenorhabditis elegans* and *D. melanogaster* (GeneFinder: P. Green, unpubl.; Burge and Karlin 1997), but the last ten years have seen substantial progress there, too (Stanke and Waack 2003; Gross and Brent 2005; Wei et al. 2005).

Combining Prediction Methods

In the run-up to the initial publications on the human genome, it became clear that manual integration of evidence from various prediction methods would not be fast enough to provide an analysis of the entire genome in a reasonable amount of time. As an alternative, several automated "pipelines" for integrating evidence, such as OTTO (Venter et al. 2001), Ensembl (Birney et al. 2004a; Curwen et al. 2004), and the NCBI pipeline were developed. OTTO was used primarily by the team at Celera Genomics who developed it. Ensembl annotations have been used to produce the primary gene sets in many of the publications that describe the first analysis of a new vertebrate genome sequence (e.g., Lander et al. 2001; Aparicio et al. 2002; Waterston et al. 2002; Gibbs et al. 2004; Hillier et al. 2004). The NCBI annotation pipeline is also very influential because its predictions appear in GenBank as RefSeq mRNAs and proteins with "XM" and "XP" accessions, respectively. The NCBI pipeline was originally

Defining an ORF at Each Locus   451

based on GenomeScan (Yeh et al. 2001), an enhancement of Genscan that modifies the scores of potential exons depending on whether they have high-scoring alignments to proteins in the databases. More recent versions of the NCBI pipeline use an unpublished method called Gnomon (http://www.ncbi.nlm.nih.gov/genome/guide/build.html#gene). Although every pipeline works differently, both Ensembl and NCBI rely heavily on aligning protein sequences generated from one gene to the genomic loci of other genes, either within or between species. In the following discussion, I will focus on Ensembl as a representative of such annotation pipelines.

Most Ensembl gene predictions are ultimately created by GeneWise, a protein-alignment program, although Genscan is used to help identify the best proteins to align from other species (Curwen et al. 2004). Thus, GeneWise has determined the structures of a large fraction of the predicted genes used in the initial analyses of numerous vertebrate genomes. The goal in these analyses was to obtain a conservative set of predicted exons and genes—one containing few false positives. Genscan, the best available de novo gene predictor until 2003, predicts numerous false positive exons, so choosing a protein alignment method such as GeneWise made sense. Even TWINSCAN, as published in 2003, predicted only 75% of known exons correctly, as compared to 85% for Ensembl (Flicek et al. 2003).

Given the recent progress in de novo gene prediction, it is worth asking whether GeneWise is still more accurate, or even more conservative, than the best de novo predictors. A direct comparison would be most informative, but the data set that Birney et al. (2004b) used to assess the accuracy of GeneWise has been lost. However, some inferences can be drawn by comparing published results on different test sets. Gross and Brent (2005) reported that N-SCAN, when run on the whole human genome, predicts 85% of all known exons correctly; estimated specificity based on the assumption that the genome contains 200,000 exons is 86%. Considering only internal exons, as in Birney et al. (2004b), N-SCAN's estimated exon-level specificity rises to 93%. Considering that these numbers are based on whole-genome annotations, whereas the estimated 93% for GeneWise is based on a single-gene set, it is likely that predictions based on protein homology are no longer more "conservative" (i.e., specific) than de novo predictions except when the protein is very similar to the target locus. Furthermore, the sensitivity of de novo methods is much higher.

A direct comparison was recently made among integrated annotation pipelines as part of the E-GASP community evaluation (Guigó and Reese 2005; http://genome.imim.es/gencode/workshop2005.html).

Pipeline predictions were compared to manual annotation by the HAVANA group (see below) on the human ENCODE regions (The ENCODE Project Consortium 2004). Among entrants were Ensembl and a simple ad hoc pipeline with two stages. The first stage was to align full-ORF cDNA sequences from the MGC project and the RefSeq collection using a cDNA aligner called Pairagon, which is based on a strong prior model designed for highly accurate sequences (M. Arumugam and M.R. Brent, in prep.). The second stage, applied to regions not covered by cDNA alignments, was to exploit BLAT alignments of spliced human ESTs to guide de novo predictions by N-SCAN. These predictions, made by N-SCAN_EST (C. Wei and M.R. Brent, in prep.), are generally consistent with the EST alignments but may extend them with additional exons or link nonoverlapping ESTs into a single transcript. The results showed that the Pairagon+N-SCAN_EST pipeline was substantially more specific than Ensembl, and equally sensitive, in both the exact exon and exact transcript measures. Ensembl predicted more transcripts per locus than Pairagon+N-SCAN_EST, so it could predict at least one correct transcript at a higher percentage of the loci where it made predictions (gene specificity). Thus, Pairagon+N-SCAN_EST, which uses only human cDNA sequence aligned to its native locus, is at least as accurate as Ensembl, which uses cross-locus and cross-species protein alignment. However, cross-species protein homology is likely to contribute more to the annotation of species for which fewer cDNA sequences are available.

A more recent approach to integrating predictions is to score each potential exon using a weighted combination of evidence from alignment-based predictions and de novo predictions (Allen et al. 2004). The weights are derived from estimates of the accuracy of each prediction source. Thus, if several predictors that have proven accurate in the past agree on an exon, it will receive a high score. In the case of disagreement among predictors the score will generally be lower, but more weight will be given to more accurate predictors. This approach performs well in practice, especially when there are multiple evidence sources with roughly similar accuracy—empirically, it seems that different methods make different errors. Indeed, JIGSAW, a descendent of COMBINER (Allen et al. 2004), was slightly more sensitive than the Pairagon+N-SCAN_EST pipeline at the exon level in the EGASP evaluations, although Pairagon+N-SCAN_EST was more accurate in predicting exact ORFs. To achieve the accuracy it did, JIGSAW was run on 13 sources of evidence using genome coordinates provided by the UCSC browser, including Ensembl, RefSeq, Genscan, SGP, TWINSCAN, Human mRNAs, TIGR Gene Index, and UniGene (Wheeler et al. 2004).

Manual annotation has also progressed over the last 10 years. In 2000, the *Drosophila* community held an annotation "jamboree," in which fly biologists and bioinformaticians gathered at Celera Genomics for two weeks to create an initial annotation of the *Drosophila* genome (Pennisi 2000). This annotation has since been systematically revised and updated (Misra et al. 2002; Drysdale and Crosby 2005). In 2002, the Sanger Institute held two Human Annotation Workshops (known as Hawk meetings). A number of groups involved in human annotation gathered at these meetings and compared their annotations on designated sequences to "define a standard of annotation" and "draw up guidelines to help achieve the standard" (http://www.sanger.ac.uk/HGP/havana/hawk.shtml). These meetings led to the annotation standards that are used by the Sanger Institute's Human and Vertebrate Analysis project (HAVANA, http://www.sanger.ac.uk/HGP/havana/docs/guidelines.pdf). HAVANA annotators integrate information from alignments of expressed sequences and de novo predictions by Genscan and FGENESH (http://www.sanger.ac.uk/HGP/havana/). The HAVANA team has annotated human chromosomes 1, 6, 9, 10, 13, 20, 22, and X; mouse chromosomes 2, 4, 11, and X; and the entire zebrafish genome. Their annotation of the human ENCODE regions appears to have been quite complete, since RT-PCR and sequencing experiments have verified only a small handful of exons not annotated by HAVANA (R. Guigó, pers. comm.).

THE CURRENT TURNING POINT

Limits of Sequencing Random cDNA Clones

Improving the accuracy of annotations based on expressed sequences depends, to a large extent, on improving the collection of sequences that are available to align. The vast majority of ESTs and cDNA sequences currently in databases were obtained by sequencing clones selected at random from cDNA libraries. However, this method has been found to saturate well short of the full gene set (The MGC Project Team 2004). For example, the MGC project sequenced 5' ESTs from more than 110 human and 80 mouse cDNA libraries and screened them for clones that appeared likely to contain a complete ORF not already in the collection. Promising clones were then sequenced to high accuracy. This produced full-ORF clones at approximately 13,000 human and 12,000 mouse gene loci—about 50%–60% of all the genes, according to current estimates. Aligning the human ESTs to the genome produced a total of about 62,000 nonoverlapping clusters, but most of these appeared not to include the 5' end of an

ORF. EST projects for other animals have yielded qualitatively similar results (e.g., Wei et al. 2005), although the number of clones sequenced has generally been less. Thus, we cannot expect to complete the annotation of any animal genome by simply sequencing deeper into cDNA libraries.

Limits of Protein Alignment

Some genes for which we cannot obtain a full-ORF cDNA sequence can nonetheless be annotated by aligning homologous proteins to the genome. However, it appears that this approach is no more accurate than de novo prediction except when the aligned protein is nearly identical to the one encoded by the target locus. Improvements in protein alignment methods—particularly the use of alignment models that do not accept frame shifting errors—may extend this accuracy horizon. In the end, though, annotation by protein alignment will be limited by the cDNA collection from which most "proteins" are derived.

Limits of Combiners

Systems that combine predictions from many sources seem to provide at least a slight edge over the best single source. However, the accuracy of these systems is limited by the accuracy of the underlying prediction sources.

Limits of Manual Annotation

Inspection by human curators seems to be an effective method of integrating evidence from alignments of expressed sequences, alignments among genomes, and de novo predictions. Among the strengths of human curators is the ability to detect suspicious annotations, such as pseudogenes. At bottom, though, their accuracy is still limited by the accuracy of the evidence they are given to integrate. Furthermore, manual annotation is time consuming and may not be updated rapidly in response to new evidence. Most importantly, it is very expensive compared with automated integration of evidence. As a result, it seems unlikely that extensive manual annotation will be done beyond the genomes of *D. melanogaster*, human, zebrafish, and possibly mouse.

Limits of Comparative Genomics

Although the accuracy of de novo gene prediction has improved dramatically since 1996, one major source of expected improvement has not yet panned out. It has been widely anticipated that the availability of multiple genomes within a fairly narrow phylogenetic clade would lead to dramatic improvements in gene prediction accuracy. By aligning multiple mammalian genomes, we should be better able to characterize the patterns of selection operating on small sections of genomes. These patterns should be indicative of specific functions. For example, regions that contain a number of substitutions all separated by multiples of three are more likely to be coding, since the third position of a codon can often be changed without changing the amino acid it encodes. The expectation that multi-genome alignments would lead to more accurate gene prediction than alignments among two genomes is quite reasonable and such improvements may yet be achieved. However, no method for improving gene prediction accuracy by using multi-genome alignments has yet been found, despite several serious efforts. For example, EXONIPHY (Siepel and Haussler 2004), an exon prediction system based on a phylogenetic generalization of HMMs, does not exceed the accuracy of dual-genome systems like TWINSCAN or SGP2 in exact exon prediction, although it does exceed them by a few percent in nucleotide specificity (Gross and Brent 2005). N-SCAN, which is based on a different phylogenetic generalization of generalized HMMs, represents a substantial improvement in accuracy over TWINSCAN and SGP2. However, running N-SCAN on multi-genome alignments has not yet produced results that are substantially better than those obtained by running it on only two genomes (Gross and Brent 2005).

There are several possible reasons for the failure, so far, to achieve substantial accuracy improvements by using multi-genome alignments. It may be that we do not yet have the right combinations of genomes sequenced to sufficiently high quality—draft sequence may not be good enough. Or, it may be that we simply cannot align these genomes precisely enough to draw accurate inferences about selection. If these are the reasons, then finished sequences from more mammals, especially primates (Boffelli et al. 2003) may lead to the anticipated improvements. Another potential explanation is that exons and splice sites conserved throughout the mammalian lineage may be less common than originally thought. Anecdotally, it has been observed that alignments among many mammalian genomes often show that a given exon does not appear in one of the species. On the other hand, it may be that the designers of de novo gene prediction algorithms have simply not been clever enough to come up with the right methods yet. In any case, it seems that we cannot count on the availability of more genome sequences to yield substantial accuracy improvements in the immediate future.

LOOKING FORWARD

It is my conviction that a finished genome sequence should reveal the set of ORFs it encodes. Therefore, I believe we must develop a cost-effective technology for translating a genome to a set of exon–intron structures and the proteins they encode. The outlines of this technology are now becoming clear, but its cost must still be reduced through automation and optimization.

The current gold standard of evidence for gene structures is cDNA sequence aligned to the genomic locus from which it was transcribed. This leaves something to be desired, in that one must still infer the exon–intron structure by alignment and the protein product by conceptual translation of the most likely looking open reading frame. Both of these inferences are subject to error, so one might hope for confirmation by direct experimental evidence. However, there is as yet no economical, high-throughput method for obtaining such evidence. In particular, there is no analog of RT-PCR for proteins—an economical method of directly amplifying or purifying hypothesized, low-abundance proteins. Since we must rely on computational inference of protein products that are not easily picked up by high-throughput proteomics, it is possible that incorrectly processed pre-mRNAs, such as those with retained introns, would yield incorrect inferences about functional proteins. The best approach to flagging such cases may be to screen for cDNAs that are likely candidates for nonsense-mediated decay—those with splice junctions more than 50–55 nt 3' of the inferred ORF (Lejeune and Maquat 2005).

The most efficient way to obtain cDNA sequence for every protein-coding gene is to combine standard EST sequencing, gene prediction, and RT-PCR using primers designed to amplify predicted transcripts. A small to moderate collection of ESTs should be developed first by the standard method—sequencing randomly selected cDNA clones. This will produce sequence from transcripts that are relatively abundant, and will completely determine the exon-intron structures of abundant transcripts that are shorter than two read lengths (currently about 1400–1800 bp). The cost per transcript will remain relatively low as long as a fairly high proportion of sequences produced are new. By calculating the number of clones that must be sequenced to obtain a new EST and multiplying by the cost per clone one can estimate the cost per new cDNA read. When

Defining an ORF at Each Locus   457

this cost exceeds the estimated cost per new read by RT-PCR, EST sequencing should be stopped. The resulting ESTs should be aligned to the genome using cDNA-to-genome alignment tools based on strong models of gene structure, and those that do not align well should be discarded or set aside for manual inspection if time permits. High-quality EST alignments that overlap one another must then be grouped together and computational techniques used to determine which groups are likely to contain a complete ORF. Those that do form the core set of genes in the annotation.

Once the core set has been determined, the rest of the genes must be identified by a series of RT-PCR and sequencing steps, starting with the most confident predictions and progressing toward the less confident (Fig. 2). Considering the analyses described above, predictions based on cross-locus and cross-species protein alignments are more reliable than de novo predictions only when the aligned protein is highly similar to the predicted one (probably >95% identity). Such predictions should be used to design primers for the first round of RT-PCR and sequencing experiments. After each RT-PCR and sequencing step, the resulting cDNA sequences should be aligned, grouped, and sorted by

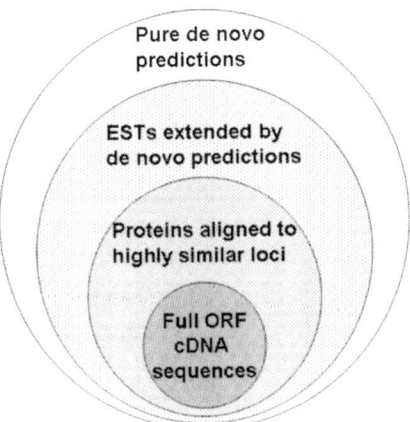

Figure 2. Recommended preference order for gene structure predictions. Alignments of full ORF cDNA sequences to the loci from which they were transcribed should take precedence. At loci where there is no full ORF cDNA alignment, alignments of highly similar proteins (probably >95% identity) should be used. Third best is a gene structure that is partially determined by an EST aligned to its native locus, possibly extended to a full ORF by de novo prediction. At loci where none of these are available, pure de novo predictions using dual- or multi-genome prediction algorithms should be used.

completeness of the predicted ORF as described above. Aligning the experimental sequences to the genome may confirm parts of the predicted gene structure, but it may also reveal errors in other parts of the predicted structure.

The updated set of full-ORF gene structures can now be used to train a de novo gene prediction algorithm. Typically, clusters of genomes within a clade are sequenced at once, so it is usually possible to use dual- or multi-genome de novo prediction methods. The EST alignments that do not cover a full ORF can be used to guide the prediction algorithms, which will predict complete structures that are consistent with the alignments, but may extend them with additional exons and/or link several ESTs together into a single predicted transcript (C. Wei and M.R. Brent, in prep.). The unconfirmed regions of predictions that extend or link EST alignments can then be tested in the next round of RT-PCR. After aligning the resulting sequences to the genome, the gene structures they define can be used as additional examples for retraining the gene predictor and as additional guidance around which the gene predictor can build models. If this process is taken to convergence, where all gene models have been tested, the result will be an annotation of exon–intron structures that is more complete than any we have now and that is fully verified by native cDNA sequences.

Several variants of this approach are also being developed. One is to use Rapid Amplification of cDNA Ends (RACE) PCR, a method in which a universal primer at one end of the cDNA is paired with a single gene-specific primer inside the predicted cDNA. Certain RACE methods selectively amplify 5′-complete mRNAs with a 7-methyl-guanine cap, allowing amplification of the 5′ end without knowing a sequence in the 5′ end. Only the sequences of one or more internal exons are needed for the design of the gene specific primer. Since only one exon needs to be predicted correctly, this method can be more sensitive than ordinary RT-PCR. Specificity is often a problem with RACE, but this can be ameliorated by a second round of PCR using a nested pair of universal and gene specific primers. McCombie and colleagues (Dike et al. 2004) have done this using mouse predictions by TWINSCAN (Flicek et al. 2003) and GenomeScan (Yeh et al. 2001), while Gingeras and colleagues have done it using genomic tiling arrays for both exon prediction and sequencing of RACE-PCR products (Cheng et al. 2005; Kapranov et al. 2005).

Of course, some transcripts are expressed transiently during development, or only under rare environmental conditions. We can increase the number of detectable transcripts by pooling RNA from many tissues. Cloning artifacts can be reduced by amplifying reverse-transcription products directly rather than using cloned cDNA libraries and by sequencing PCR products directly rather than sequencing clones. But there will still be rare transcripts that cannot be verified by a high-throughput annotation system. In the end, these will have to be identified on a case-by-case basis using traditional biochemical or genetic approaches. Nonetheless, we can use high-throughput methods to get much closer than we have so far to determining the most basic elements on the parts list of an organism.

To make this vision a reality, we must bring the cost of the RT-PCR and sequencing experiments down as far as possible. This means relying on end-to-end automation. Much of the necessary automation consists of software pipelines for selecting predictions to test, designing primer pairs to test them, and analyzing the resulting sequences to determine new gene structures. The physical processes of setting up PCR and sequencing reactions must also be optimized and automated. Finally, the accuracy of the gene predictions will be a central determinant of the cost and completeness of the resulting annotation. Prediction errors may lead to one or more PCR experiments that fail to amplify their targets and produce no useful sequence, thus raising the cost per transcript annotated. Therefore, we must continue to improve the accuracy of gene prediction by developing more complete and more realistic models of the signals in the genome sequence that guide the transcription and processing of mRNA.

The genomics community is used to rapid progress and headline-making excitement, so the temptation to "declare victory and move on" is understandable. I have heard it said numerous times that the identification of protein-coding genes is well understood, and the real challenges now are identifying transcription factor binding sites, noncoding RNA genes, and other exciting sequence elements. While these are important challenges, we must resist the temptation to leave the identification of protein-coding genes incomplete while we chase after the hottest new features. We must not forget that the defining characteristic of genomics is the all-out effort to view an organism globally by analyzing data sets that are as complete as we can possibly make them.

ACKNOWLEDGMENTS

I am grateful to Paul Flicek for help with analyzing the EGASP evaluation results and to Mark Diekhans for analysis of the MGC cDNA sequences. M.R.B. is supported in part by R01 HG02278, R01 AI051209, and U01 HG003150 from the National Institutes of Health;

in part by grant DBI-0501758 from the National Science Foundation; and in part by National Cancer Institute funds for the Mammalian Gene Collection project under Contract No. N01-CO-12400.

REFERENCES

Alexandersson, M., Cawley, S., and Pachter, L. 2003. SLAM: Cross-species gene finding and alignment with a generalized pair hidden Markov model. *Genome Res.* **13:** 496–502.

Allen, J.E., Pertea, M., and Salzberg, S.L. 2004. Computational gene prediction using multiple sources of evidence. *Genome Res.* **14:** 142–148.

Ansari-Lari, M.A., Timms, K.M., and Gibbs, R. 1996. Improved ligation-anchored PCR strategy for identification of 5' ends of transcripts. *BioTechniques* **21:** 34–38.

Ansari-Lari, M.A., Shen, Y., Muzny, D.M., Lee, W., and Gibbs, R.A. 1997. Large-scale sequencing in human chromosome 12p13: Experimental and computational gene structure determination. *Genome Res.* **7:** 268–280.

Aparicio, S., Chapman, J., Stupka, E., Putnam, N., Chia, J.M., Dehal, P., Christoffels, A., Rash, S., Hoon, S., Smit, A., et al. 2002. Whole-genome shotgun assembly and analysis of the genome of *Fugu rubripes*. *Science* **297:** 1301–1310.

Bafna, V. and Huson, D.H. 2000. The conserved exon method for gene finding. *Proc. Int. Conf. Intell. Syst. Mol. Biol.* **8:** 3–12.

Batzoglou, S., Pachter, L., Mesirov, J.P., Berger, B., and Lander, E.S. 2000. Human and mouse gene structure: Comparative analysis and application to exon prediction. *Genome Res.* **10:** 950–958.

Birney, E. and Durbin, R. 1997. Dynamite: A flexible code generating language for dynamic programming methods used in sequence comparison. *Proc. Int. Conf. Intell. Syst. Mol. Biol.* **5:** 56–64.

———. 2000. Using GeneWise in the *Drosophila* annotation experiment. *Genome Res.* **10:** 547–548.

Birney, E., Thompson, J.D., and Gibson, T.J. 1996. PairWise and SearchWise: Finding the optimal alignment in a simultaneous comparison of a protein profile against all DNA translation frames. *Nucleic Acids Res.* **24:** 2730–2739.

Birney, E., Andrews, T.D., Bevan, P., Caccamo, M., Chen, Y., Clarke, L., Coates, G., Cuff, J., Curwen, V., Cutts, T., et al. 2004a. An overview of Ensembl. *Genome Res.* **14:** 925–928.

Birney, E., Clamp, M., and Durbin, R. 2004b. GeneWise and Genomewise. *Genome Res.* **14:** 988–995.

Boffelli, D., McAuliffe, J., Ovcharenko, D., Lewis, K.D., Ovcharenko, I., Pachter, L., and Rubin, E.M. 2003. Phylogenetic shadowing of primate sequences to find functional regions of the human genome. *Science* **299:** 1391–1394.

Brackenridge, S., Wilkie, A.O., and Screaton, G.R. 2003. Efficient use of a 'dead-end' GA 5' splice site in the human fibroblast growth factor receptor genes. *EMBO J.* **22:** 1620–1631.

Brown, R.H., Gross, S.S., and Brent, M.R. 2005. Begin at the beginning: Predicting genes with 5' UTRs. *Genome Res.* **15:** 742–747.

Burge, C. and Karlin, S. 1997. Prediction of complete gene structures in human genomic DNA. *J. Mol. Biol.* **268:** 78–94.

Burset, M. and Guigó, R. 1996. Evaluation of gene structure prediction programs. *Genomics* **34:** 353–367.

Cheng, J., Kapranov, P., Drenkow, J., Dike, S., Brubaker, S., Patel, S., Long, J., Stern, D., Tammana, H., Helt, G., et al. 2005. Transcriptional maps of 10 human chromosomes at 5-nucleotide resolution. *Science* **308:** 1149–1154.

Collins, F. and Galas, D. 1993. A new five-year plan for the U.S. Human Genome Project. *Science* **262:** 43–46.

Collins, F.S., Green, E.D., Guttmacher, A.E., and Guyer, M.S. 2003. A vision for the future of genomics research. *Nature* **422:** 835.

Curwen, V., Eyras, E., Andrews, T.D., Clarke, L., Mongin, E., Searle, S.M., and Clamp, M. 2004. The Ensembl automatic gene annotation system. *Genome Res.* **14:** 942–950.

Dike, S., Balija, V.S., Nascimento, L.U., Xuan, Z., Ou, J., Zutavern, T., Palmer, L.E., Hannon, G., Zhang, M.Q., and McCombie, W.R. 2004. The mouse genome: Experimental examination of gene predictions and transcriptional start sites. *Genome Res.* **14:** 2424–2429.

Drysdale, R.A. and Crosby, M.A. 2005. FlyBase: Genes and gene models. *Nucleic Acids Res.* **33:** D390–D395.

The ENCODE Project Consortium. 2004. The ENCODE (ENCyclopedia Of DNA Elements) Project. *Science* **306:** 636–640.

Eyras, E., Reymond, A., Castelo, R., Bye, J.M., Camara, F., Flicek, P., Huckle, E.J., Parra, G., Shteynberg, D.D., Wyss, C., et al. 2005. Gene finding in the chicken genome. *BMC Bioinformatics* **6:** 131.

Flicek, P., Keibler, E., Hu, P., Korf, I., and Brent, M.R. 2003. Leveraging the mouse genome for gene prediction in human: From whole-genome shotgun reads to a global synteny map. *Genome Res.* **13:** 46–54.

Florea, L., Hartzell, G., Zhang, Z., Rubin, G.M., and Miller, W. 1998. A computer program for aligning a cDNA sequence with a genomic DNA sequence. *Genome Res.* **8:** 967–974.

Furey, T.S., Diekhans, M., Lu, Y., Graves, T.A., Oddy, L., Randall-Maher, J., Hillier, L.W., Wilson, R.K., and Haussler, D. 2004. Analysis of human mRNAs with the reference genome sequence reveals potential errors, polymorphisms, and RNA editing. *Genome Res.* **14:** 2034–2040.

Gelfand, M.S., Mironov, A.A., and Pevzner, P.A. 1996. Gene recognition via spliced sequence alignment. *Proc. Natl. Acad. Sci.* **93:** 9061–9066.

Gibbs, R.A., Weinstock, G.M., Metzker, M.L., Muzny, D.M., Sodergren, E.J., Scherer, S., Scott, G., Steffen, D., Worley, K.C., Burch, P.E., et al. 2004. Genome sequence of the Brown Norway rat yields insights into mammalian evolution. *Nature* **428:** 493–521.

Gross, S.S. and Brent, M.R. 2005. Using multiple alignments to improve gene prediction. In *9th Annual International Conference, RECOMB 2005* (eds. S. Miyano et al.), pp. 374–388. Springer, Boston.

———. 2006. Using multiple alignments to improve gene prediction. *J. Comput. Biol.* **13:** (in press).

Guigó, R. and Reese, M.G. 2005. EGASP: Collaboration through competition to find human genes. *Nat. Methods* **2:** 575–577.

Guigó, R., Knudsen, S., Drake, N., and Smith, T. 1992. Prediction of gene structure. *J. Mol. Biol.* **226:** 141–157.

Guigó, R., Agarwal, P., Abril, J.F., Burset, M., and Fickett, J.W. 2000. An assessment of gene prediction accuracy in large DNA sequences. *Genome Res.* **10:** 1631–1642.

Guigó, R., Dermitzakis, E.T., Agarwal, P., Ponting, C., Parra, G., Reymond, A., Abril, J.F., Keibler, E., Lyle, R., Ucla, C., et al. 2003. Comparison of mouse and human genomes followed by experimental verification yields an estimated 1,019 additional genes. *Proc. Natl. Acad. Sci.* **100**: 1140–1145.

Hillier, L.D., Lennon, G., Becker, M., Bonaldo, M.F., Chiapelli, B., Chissoe, S., Dietrich, N., DuBuque, T., Favello, A., Gish, W., et al. 1996. Generation and analysis of 280,000 human expressed sequence tags. *Genome Res.* **6**: 807–828.

Hillier, L.W., Miller, W., Birney, E., Warren, W., Hardison, R.C., Ponting, C.P., Bork, P., Burt, D.W., Groenen, M.A., Delany, M.E., et al. 2004. Sequence and comparative analysis of the chicken genome provide unique perspectives on vertebrate evolution. *Nature* **432**: 695–716.

International Human Genome Sequencing Consortium. 2004. Finishing the euchromatic sequence of the human genome. *Nature* **431**: 931–945.

Kapranov, P., Drenkow, J., Cheng, J., Long, J., Helt, G., Dike, S., and Gingeras, T.R. 2005. Examples of the complex architecture of the human transcriptome revealed by RACE and high-density tiling arrays. *Genome Res.* **15**: 987–997.

Korf, I., Flicek, P., Duan, D., and Brent, M.R. 2001. Integrating genomic homology into gene structure prediction. *Bioinformatics* **17 Suppl 1**: S140–S148.

Kulp, D., Haussler, D., Reese, M.G., and Eeckman, F.H. 1996. A generalized hidden Markov model for the recognition of human genes in DNA. *Proc. Int. Conf. Intell. Syst. Mol. Biol.* **4**: 134–142.

Lander, E.S., Linton, L.M., Birren, B., Nusbaum, C., Zody, M.C., Baldwin, J., Devon, K., Dewar, K., Doyle, M., FitzHugh, W., et al. 2001. Initial sequencing and analysis of the human genome. *Nature* **409**: 860–921.

Lejeune, F. and Maquat, L.E. 2005. Mechanistic links between nonsense-mediated mRNA decay and pre-mRNA splicing in mammalian cells. *Curr. Opin. Cell Biol.* **17**: 309–315.

The MGC Project Team. 2004. The status, quality, and expansion of the NIH full-length cDNA project: The Mammalian Gene Collection (MGC). *Genome Res.* **14**: 2121–2127.

Misra, S., Crosby, M.A., Mungall, C.J., Matthews, B.B., Campbell, K.S., Hradecky, P., Huang, Y., Kaminker, J.S., Millburn, G.H., Prochnik, S.E., et al. 2002. Annotation of the *Drosophila melanogaster* euchromatic genome: A systematic review. *Genome Biol.* **3**: research0083.

Modrek, B. and Lee, C. 2002. A genomic view of alternative splicing. *Nat. Genet.* **30**: 13–19.

Mott, R. 1997. EST_GENOME: A program to align spliced DNA sequences to unspliced genomic DNA. *Comput. Appl. Biosci.* **13**: 477–478.

Parra, G., Agarwal, P., Abril, J.F., Wiehe, T., Fickett, J.W., and Guigó, R. 2003. Comparative gene prediction in human and mouse. *Genome Res.* **13**: 108–117.

Pennisi, E. 2000. Ideas fly at gene-finding jamboree. *Science* **287**: 2182–2184.

Reese, M.G., Kulp, D., Tammana, H., and Haussler, D. 2000. Genie—Gene finding in *Drosophila melanogaster*. *Genome Res.* **10**: 529–538.

Sharp, P.A. and Burge, C.B. 1997. Classification of introns: U2-type or U12-type. *Cell* **91**: 875–879.

Siepel, A.C. and Haussler, D. 2004. Computational identification of evolutionarily conserved exons. In *RECOMB*. ACM, San Diego.

Solovyev, V.V., Salamov, A.A., and Lawrence, C.B. 1994. Predicting internal exons by oligonucleotide composition and discriminant analysis of spliceable open reading frames. *Nucleic Acids Res.* **22**: 5156–5163.

Sorek, R., Shamir, R., and Ast, G. 2004. How prevalent is functional alternative splicing in the human genome? *Trends Genet.* **20:** 68–71.

Stanke, M. and Waack, S. 2003. Gene prediction with a hidden Markov model and a new intron submodel. *Bioinformatics* **19 Suppl 2:** II215–II225.

Stormo, G.D. and Haussler, D. 1994. Optimally parsing a sequence into different classes based on multiple types of evidence. *Proc. Int. Conf. Intell. Syst. Mol. Biol.* **2:** 369–375.

Tarn, W.Y. and Steitz, J.A. 1996. A novel spliceosome containing U11, U12, and U5 snRNPs excises a minor class (AT-AC) intron in vitro. *Cell* **84:** 801–811.

Tarn, W.Y., Yario, T.A., and Steitz, J.A. 1995. U12 snRNA in vertebrates: Evolutionary conservation of 5′ sequences implicated in splicing of pre-mRNAs containing a minor class of introns. *RNA* **1:** 644–656.

Torrents, D., Suyama, M., Zdobnov, E., and Bork, P. 2003. A genome-wide survey of human pseudogenes. *Genome Res.* **13:** 2559–2567.

Venter, J.C., Adams, M.D., Myers, E.W., Li, P.W., Mural, R.J., Sutton, G.G., Smith, H.O., Yandell, M., Evans, C.A., Holt, R.A., et al. 2001. The sequence of the human genome. *Science* **291:** 1304–1351.

Waterston, R.H., Lindblad-Toh, K., Birney, E., Rogers, J., Abril, J.F., Agarwal, P., Agarwala, R., Ainscough, R., Alexandersson, M., An, P., et al. 2002. Initial sequencing and comparative analysis of the mouse genome. *Nature* **420:** 520–562.

Wei, C., Lamesch, P., Arumugam, M., Rosenberg, J., Hu, P., Vidal, M., and Brent, M.R. 2005. Closing in on the *C. elegans* ORFeome by cloning TWINSCAN predictions. *Genome Res.* **15:** 577–582.

Wheelan, S.J., Church, D.M., and Ostell, J.M. 2001. Spidey: A tool for mRNA-to-genomic alignments. *Genome Res.* **11:** 1952–1957.

Wheeler, D.L., Church, D.M., Edgar, R., Federhen, S., Helmberg, W., Madden, T.L., Pontius, J.U., Schuler, G.D., Schriml, L.M., Sequeira, E., et al. 2004. Database resources of the National Center for Biotechnology Information: Update. *Nucleic Acids Res.* **32:** D35–D40.

Wolfsberg, T.G. and Landsman, D. 1997. A comparison of expressed sequence tags (ESTs) to human genomic sequences. *Nucleic Acids Res.* **25:** 1626–1632.

Wu, J.Q., Shteynberg, D., Arumugam, M., Gibbs, R.A., and Brent, M.R. 2004. Identification of rat genes by TWINSCAN gene prediction, RT-PCR, and direct sequencing. *Genome Res.* **14:** 665–671.

Yeh, R.F., Lim, L.P., and Burge, C.B. 2001. Computational inference of homologous gene structures in the human genome. *Genome Res.* **11:** 803–816.

Zamore, P.D. and Haley, B. 2005. Ribo-gnome: The big world of small RNAs. *Science* **309:** 1519–1524.

Zhang, Z. and Gerstein, M. 2004. Large-scale analysis of pseudogenes in the human genome. *Curr. Opin. Genet. Dev.* **14:** 328–335.

Index

A
aMOs. *See* Antisense morpholino oligonucleotides
Antisense morpholino oligonucleotides (aMOs), *Xenopus* loss-of-function studies, 206
ApoA2, quantitative trait loci mapping, 351
Arabidopsis thaliana
  bioinformatics and modeling, 86–90
  comparative genomics and crop plant research, 84–86
  functional genomics resources, 77–79
  gene annotation, 72–74, 82
  gene expression studies, 78–82
  genome
    duplications and dynamics, 74–75
    features, 73
    polymorphisms, 75
    repeats, 76
    sequencing and study goals, 71–72
  prospects for study, 90–91
  proteomics, 83–84
AraCYC, biochemical pathways, 88
ARPKD. *See* Autosomal recessive polycystic kidney disease
Aspergillus. See Fungi
ASPM, human versus ape phenotypic differences, 368
Autosomal recessive polycystic kidney disease (ARPKD), gene discovery, 288, 292

B
BAC. *See* Bacterial artificial chromosome
Bacillus
  Bacillus anthracis pan-genome, 7
  lateral gene transfer, 6
Bacteria. *See also specific bacteria*
  genomes
    minimal genome, 2–3
    sequencing history, 1
    size variation, 2–3
  genomic prospects in microbes, 11–12
Bacterial artificial chromosome (BAC)
  bovine genome mapping, 240
  candidate gene identification in mice, 352–353
  chicken genome clones, 227
  dog genome mapping, 264–265
  pig genome mapping, 241
  rat libraries, 290
  sheep genome mapping, 241
BioMOBY, bioinformatics and modeling, 87
Bioremediation, genomics studies, 9–10
BLAST, sequence alignment, 441
Bone morphogenetic proteins, *Xenopus*
  microarray studies, 212
  synexpression groups, 209
Bordetella pertussis, genome reduction, 5
Brachyury, *Ciona* development studies, 168–169
Buffalo
  comparative genomic mapping, 244
  genomics
    health and reproduction studies, 247–248
    prospects, 248–249
    status, 238–239, 242
  livestock, 242
Burkholderia, genome rearrangement, 5

C
β-Catenin, *Ciona* development studies, 162
Caenorhabditis elegans
  advantages as model system, 117–118

465

Caenorhabditis elegans (continued)
  databases, 118
  gene annotation
    noncoding RNA genes, 123–124
    protein-coding genes, 121–123
  gene expression profiling, 122, 125–127
  gene regulation studies, 129–130
  genome sequencing
    comparison with other *Caenorhabditis* species, 120–121, 132–133
    historical perspective, 119–120
    mapping, 118
  green fluorescent protein fusion studies, 126–127
  population biology and evolution, 131–133
  prospects for study, 134–135
  proteomics, 130–131
  RNA interference studies, 128
  TILLING studies, 129
  WormBase features, 133–134
Cancer, dog models, 270
CASP12P1, human versus ape phenotypic differences, 371
Cattle
  bacterial artificial chromosome mapping, 240
  biomedical research applications, 245
  comparative genomic mapping, 243–244
  genomics
    health and reproduction studies, 247–248
    prospects, 248–249
    status, 238–240
Cblb, positional cloning in rat, 288
Cct4, positional cloning in rat, 288
Cd36, positional cloning in rat, 288
Cereals. *See* Crop plants
Chemokines, chicken studies, 232
Chicken
  bacterial artificial chromosome mapping, 227
  biomedical research impact, 221
  databases, 228
  developmental biology, 230
  food production, 222
  gene conservation in vertebrates, 227
  gene expression profiling, 227
  gene prediction, 225–227
  genome
    comparative genomics, 229–230
    organization, 223–225
    sequencing, 222–223
  immune system studies, 231–232
  prospects for study, 232–2333
  quantitative trait loci, 231
  selective breeding consequences, 222
Chimpanzee
  candidate gene approaches for human-specific differences
    comparative phenomics, 379–380
    naturally occurring human mutations, 380–381
    sequence data, 381–382
    systems approach, 382–383
  comparative genomic analysis with humans
    candidate genes and gene families contributing to phenotypic differences, 367–375
    divergence, 363–364
    focusing search
      gene duplications and retrotransposed genes, 376
      human-specific chromosomal changes, 366
      human-specific gene conversions, 378
      human-specific insertions and deletions, 366, 376
      human-specific rapid gene evolution, 376–377
      human-specific repetitive element insertion, 377
      noncoding region changes, 378
    human-specific gene expression differences, 378–379
    narrowing search to important differences
      intraspecies polymorphism exclusion, 365–366
      outgroup defining of human-specific changes, 364
    prospects for study, 383–386
  genome sequencing, 363
  human phenotypic trait comparison with great apes, 358–361

medical condition severity comparison with humans, 361–362
ChIP. *See* Chromatin immunoprecipitation
Chromatin immunoprecipitation (ChIP), transcription factor binding site identification, 28–29
Ciona intestinalis
  advantages as model system, 159–160
  approaches in postgenomic research, 160, 162
  databases, 170
  developmental networks
    neural specification, 163–168
    notochord morphogenesis, 168
    overview, 159–161
  electroporation of constructs in fertilized eggs, 165–166
  gene expression and function studies, 162–163
  prospects for study, 171–172
  resources for study, 169–172
CMAH, human versus ape phenotypic differences, 367, 382
Collaborative Cross, goals, 335
Complex Trait Consortium (CTC), goals, 335
Congenic strains, rats, 293–294
Consomic strains, rats, 294–296
Corn. *See* Maize
COX, human versus ape phenotypic differences, 372
CpG islands, chicken genome, 223–224
Crop plants. *See also* Maize; Rice; Sorghum
  Arabidopsis thaliana comparative genomics, 84, 86
  comparative analysis
    conserved noncoding sequences, 108–109
    gene fates, 107–108
    heterochromatin, 109–110
  genome sequencing projects, 85
  genome size variation in cereals, 97
  phylogeny, 98–99
  prospects for study, 110–112
  rationale for study, 98
CRT. *See* Cyclic reversible termination
CST, human versus ape phenotypic differences, 374

CTC. *See* Complex Trait Consortium
Cyclic reversible termination (CRT)
  principles of DNA sequencing, 416, 428–429
  terminators, 429–431
  cycle efficiency, 431

D
Dishevelled, Ciona development studies, 169
DNA microarray
  Arabidopsis thaliana studies, 78–82
  Caenorhabditis elegans, 125
  Ciona intestinalis, 171
  Xenopus, 211–213
  chicken, 227
DNA sequencing
  costs, 415, 418
  cyclic reversible termination
    cycle efficiency, 431
    principles, 416, 428–429
    terminators, 429–431
  Human Genome Project approach, 413
  prospects, 431–432
  Sanger sequencing
    automation, 417
    fluorescence detection, 420–424
    high-throughput sequencing, 417–418
    microfluidic separation platforms, 418–421
    principles, 415–417
  single-nucleotide addition fluorescence in situ sequencing, 427–428
  nucleobase mixture strategy, 428
  principles, 416
  pyrosequencing, 424–427
  single-pair fluorescence resonance energy transfer, 427
  single-nucleotide polymorphisms, 414
  commercial resources, 415, 417
Dog
  behavioral genetics, 272
  breeds
    diversity and genetic structure, 260
    linkage disequilibrium, 261, 266–268, 273

Dog (*continued*)
  origins and relationships, 261–264
  disease gene mapping, 268–270
  domestication, 258–260
  genome mapping and sequencing, 255, 264–265
  morphology genetics, 270–272
  phylogeny and evolutionary framework, 256–258
  prospects for study, 273
Double Muscling, cattle, 245–246
Drosophila melanogaster
  gene annotation, 145–146, 149–150
  gene cloning, 146–147
  gene expression neighborhoods, 147–148, 151
  gene regulatory network, 150
  genome sequencing, 144–145
  history of study, 143–144
  prospects for study, 148–152
Dsg4, positional cloning in rat, 288

E
ELN, human versus ape phenotypic differences, 370
Embryonic stem (ES) cell
  mouse
    chromosomal rearrangement engineering, 332
    mutagenesis, 328–331
  rat, 298
EMPReSS, applications, 334
EMR4, human versus ape phenotypic differences, 369
ENCODE project, goals, 398–399, 401, 414, 440
Ensembl, exon prediction, 451
ENU mutagenesis. *See* *N*-Ethyl-*N*-nitrosourea mutagenesis
Environmental genomics, fish studies, 189–190
Epilepsy, dogs, 269–270
ES cell. *See* Embryonic stem cell
Escherichia coli
  metabolic engineering, 10
  strain O157:H7 genome sequencing, 1
EST_GENOME, sequence alignment, 442–443, 451

N-Ethyl-*N*-nitrosourea (ENU) mutagenesis
  mouse, 327–328
  rat, 296–297
EUCOMM, goals, 329–330
EUMORPHIA, 333–334
Exofish, exon identification, 181
EXONIPHY, exon prediction, 455

F
FANTOM, features, 322
FCGR1, human versus ape phenotypic differences, 373
FGI. *See* Fungal Genome Initiative
Fibroblast growth factors
  Ciona development studies, 162, 165–167
  Xenopus
    microarray studies, 212
    synexpression groups, 209
Fish. *See also* *Takifugu rubripes*; Medaka; *Tetraodon nigroviridis*; Zebrafish
  genome duplications, 183–188
  phylogeny, 177–179
FISSEQ. *See* Fluorescence in situ sequencing
Flcn, positional cloning in rat, 288
Fluorescence detection, Sanger sequencing, 420–424
Fluorescence in situ sequencing (FISSEQ), principles, 427–428
FoxD, *Ciona* development studies, 162
Foxn1, positional cloning in rat, 288
FOXP2
  human versus ape phenotypic differences, 367
  mutation and developmental delay, 380
Fungal Genome Initiative (FGI), overview, 44
Fungi. *See also* Yeast
  gene annotation
    alternative splicing, 54–55
    gene prediction, 53–54
  genome evolution
    Aspergillus species comparison, 55
    intron evolution, 56–58
    repeat induced point mutation, 59
    RNA interference, 59–60

genome sequencing projects
  approaches, 51–53
  completed projects, 45–47
  projects in progress, 48–49
  history and resources, 43–44
  human infection, 62–63
  phylogeny, 41–43, 50–51
  plant pathogen studies, 60–62
  prospects for study, 63–64

G

Gdf8, knockout and muscular hypertrophy, 245–247
Gene annotation
  Arabidopsis thaliana, 72–74, 82
  Caenorhabditis elegans
    noncoding RNA genes, 123–124
    protein-coding genes, 121–123
  combining prediction methods, 445, 450–453
  de novo prediction
    dual- and multi-genome de novo predictors, 449–450
    overview, 444–445
    single-genome de novo gene prediction, 447–449
    TWINSCAN, 54, 123, 225, 449–450
  Drosophila melanogaster, 145–146, 149–150
  fungi
    alternative splicing, 54–55
    gene prediction, 53–54
  gene structure prediction process, 457–458
  human, 398, 407
  limitations
    combiners, 454
    comparative genomics, 454–456
    manual annotation, 454
    protein alignments, 454
    sequencing random cDNA clones, 453–454
  manual annotation, 453
  open reading frame identification, 440–441
  overview, 4–6
  prospects, 456–459
  sequence alignment
    cDNA sequences, 446–447

  programs, 441–444
Gene Ontology (GO)
  Drosophila melanogaster, 149–150
  inference validation from expression data, 27–28
  limitations in fish–mammal comparisons, 182
  overview, 25, 27
GeneChip. *See* DNA microarray
Generalized Hidden Markov model (GHMM), gene prediction, 445
Generic Model Organism Database (GMOD)
  bioinformatics and modeling, 87
  overview, 31, 229
GeneWise, sequence alignment, 444
Genome duplication
  diploidization and gene removal, 183–184
  double-conserved synteny, 185–187
  history of study, 183
Genome rearrangement, overview, 4–5
Genome reduction, overview, 4–5
GENEVESTIGATOR, microarray data analysis, 80
Geobacter sulfurreducens, bioremediation, 9–10
GHMM. *See* Generalized Hidden Markov model
Gimap5, positional cloning in rat, 288
GMOD. *See* Generic Model Organism Database
GO. *See* Gene Ontology
Goat
  comparative genomic mapping, 244
  genomics
    health and reproduction studies, 247–248
    prospects, 248–249
    status, 238–239, 243
GYP, human versus ape phenotypic differences, 373

H

H19, imprinting control regions, 230
Haemophilus influenzae, genome sequencing, 1

HAPPY mapping, fungal genome
    sequencing, 52
Hc, quantitative trait loci mapping, 349
Heterochromatin
    crop studies, 109–110
    Drosophila studies, 149
Horse
    comparative genomic mapping, 244
    genomics
        health and reproduction studies,
            247–248
        prospects, 248–249
        status, 238–239, 241–242
HOX clusters
    fish versus mammals, 183, 185
    humans, 400
Human genome
    ancestry studies, 404–406
    comparative genome analysis
        chimpanzee. *See* Chimpanzee
        mouse
            differences, 319–323
            similarities, 317–319
        Tetraodon nigroviridis, 181
    genes
        annotation, 398, 407
        distribution within DNA, 400
        elements controlling expression,
            400–403
        non-protein-coding transcripts,
            398–400
        number, 398
    genome browsers, 407–408
    Human Genome Project, 395–396, 405,
        407–408, 413
    International HapMap project, 401,
        403–404
    large-scale structures, 403
    phenotypic trait comparison with great
        apes, 358–361
    prospects for study, 408–409
    sequence assembly, 396–397
    ultra-conserved elements, 401–403

I
IGKV, human versus ape phenotypic
    differences, 373
IL8, mouse versus human genes, 320, 322

IL9R, human versus ape phenotypic
    differences, 369
INGENIUM, mouse mutagenesis, 328

J
JIGSAW, de novo gene prediction, 452

K
KIR, human versus ape phenotypic
    differences, 373
KRT, human versus ape phenotypic
    differences, 374
KRTHAP1, human versus ape phenotypic
    differences, 370

L
Lafora disease, dog model, 270
Lateral gene transfer (LGT)
    overview, 4, 6
    phylogenetic analysis, 6–7
LCE, human versus ape phenotypic
    differences, 374
LD. *See* Linkage disequilibrium
Lep, positional cloning in rat, 289
Lepr, positional cloning in rat, 289
LGT. *See* Lateral gene transfer
LILR, human versus ape phenotypic
    differences, 373
Limbin, mutation and dwarfism, 245
LINEs
    chicken, 225
    chimpanzee versus human, 377
    mice versus humans, 322–323
    puffer fish, 182
Linkage disequilibrium (LD)
    dog studies, 261, 266–268, 273
    limitations of mouse studies,
        347–348
Livestock. *See* Buffalo; Cattle; Goat; Horse;
    Pig; Sheep

M
Macho-1. Ciona development studies, 163
Maize. *See also* Crop plants
    advantages as model system, 105–106
    databases, 112
    gene functional assignments and
        resources, 106–107

genome organization and sequence, 85, 106
origins, 105
Major histocompatibility complex (MHC)
chicken studies, 231
dog domestication studies, 260
HLA allele sequencing, 425
MAOA, human versus ape phenotypic differences, 368
Massively parallel signature sequencing (MPSS), noncoding RNA identification in *Arabidopsis*, 79
MCPH1, human versus ape phenotypic differences, 368
MCR4, mutation and obesity, 245
mDNA. *See* Mitochondrial DNA
MDR1, dog breed studies, 264
Medaka
advantages of study, 188
genomic applications, 188–189
Meiotic silencing by unpaired DNA (MSUD), fungi, 59–60
Mertk, positional cloning in rat, 288
Metabolomics, overview, 33
Metagenomics, overview, 10–11
MHC. *See* Major histocompatibility complex
MICER, 352
Microarray. *See* DNA microarray
Microfluidic separation platforms, Sanger sequencing, 418–421
MicroRNA, human genes, 399
Minimal genome, bacteria studies, 2–3, 8
Mitochondrial DNA (mDNA), dog domestication studies, 259
Morpheus, human versus ape phenotypic differences, 372
Mouse
advantages as model, 313
chromosomal rearrangement engineering, 332
comparative genomic analysis with humans
differences, 319–323
similarities, 317–319
complex trait genetic analysis, 334–337
databases, 314–316
genome sequencing, 314–317, 337

history of genetics studies, 313–314
mutagenesis
embryonic stem cells, 328–331
N-ethyl-*N*-nitrosourea-induced, 327–328
screening, 328
spontaneous mutations, 326–327
trappable genes, 331
phenotyping programs, 333–334
phylogeny, 324–325
prospects for study, 325–326, 337–338
quantitative trait loci
candidate gene identification, 349–351
causative gene identification, 351–354
detection and localization, 346–348
expression quantitative trait loci, 350
identification strategies, 346
polygenic disease studies, 345
positional cloning, 345
strain distribution pattern comparisons, 348
RNA interference, 337
single-nucleotide polymorphisms
candidate gene testing, 353–354
overview, 323–324
strain development, 337–338
MPSS. *See* Massively parallel signature sequencing
MSUD. *See* Meiotic silencing by unpaired DNA
MSX2, dog morphology role, 270
MURR1, copper toxicosis role, 269
Mycobacterium tuberculosis, genome reduction, 5
Mycoplasma, minimal genome studies, 2
MYH16, human versus ape phenotypic differences, 367, 381–382
Myostatin, loss-of-function effects, 245, 247

N

Narcolepsy, dogs, 269
NCBI pipeline, de novo gene prediction, 450–451
Ncf1, positional cloning in rat, 288
Neisseria meningitidis, transcriptomics and drug discovery, 9

Neurospora. See Fungi
NHLRC1, epilepsy role, 270
N-SCAN, de novo gene prediction, 451–452

O
OAS genes, mouse versus human genes, 319–320
OBO. See Open Biological Ontologies
Obsessive compulsive disease (OCD), dogs, 272
OCD. See Obsessive compulsive disease
Olfactory receptors, human versus ape phenotypic differences, 371
Open Biological Ontologies (OBO), overview, 31
Optical mapping, fungal genome sequencing, 52
OTTO, de novo gene prediction, 450
Otx, *Ciona* development studies, 163, 165–167

P
PCDH11Y, human versus ape phenotypic differences, 369
PicoTiter Plate (PTP), DNA sequencing, 426
Pig
bacteria artificial chromosome mapping, 241
biomedical research applications, 245
comparative genomic mapping, 244
genomics
health and reproduction studies, 247–248
prospects, 248–249
status, 238–241
Pkhd1, positional cloning in rat, 288
PME. See Pulsed multiline excitation
Prickle, *Ciona* development studies, 168–169
PSG, human versus ape phenotypic differences, 374
PTP. See PicoTiter Plate
Pufferfish. See *Takifugu rubripes*, *Tetraodon nigroviridis*
Pulsed multiline excitation (PME), Sanger sequencing, 422–424
P-value, sequence alignment, 444
Pyrosequencing, 424–427

Q
QTL. See Quantitative trait loci
Quantitative trait loci (QTL)
chicken, 231
dog, 271
livestock species, 247–248
mouse
candidate gene identification, 349–351
causative gene identification, 351–354
detection and localization, 346–348
expression quantitative trait loci, 350
identification strategies, 346
polygenic disease studies, 345
positional cloning, 345
strain distribution pattern comparisons, 348
rat, mapping, 285–288, 292, 301–303

R
Rab38, positional cloning in rat, 288
RACE. See Rapid amplification of cDNA ends
Radiation hybrid (RH) mapping
bovine genome, 240
dog genome, 255, 264
horse genome, 242
Rapid amplification of cDNA ends (RACE), gene structure prediction, 458–458
Rat
biomedical research importance, 282–282
comparative genomic mapping, 291–292
congenic strains, 293–294
consomic strains, 294–296
embryonic stem cell generation, 298
N-ethyl-*N*-nitrosourea mutagenesis, 296–297
functional cloning, 290–291
gene therapy experiments, 292–293
genome project, 282, 284
genomics resources
ACP Haplotyper, 300
databases, 283, 298–299
Gene Annotation Tool, 300
Genome Browser, 299
Genome Scanner, 300

Index   473

GViewer Tool, 299
VCMap, 300
heterogeneous stocks, 297–298
positional cloning, 288–289
prospects for study, 301–303
quantitative trait loci mapping, 285–288, 292, 301–303
recombinant inbred strains, 296
single nucleotide polymorphism project, 302–303
strain characterization, 284–285
target validation in gene discovery, 291
transgenic rat, 289–290
Recombinant inbred strains, rats, 296
Reln, positional cloning in rat, 288
Renin, mouse versus human genes, 319
Repeat induced point mutation (RIP), fungal genome evolution studies, 59
Reverse transcriptase-polymerase chain reaction (RT-PCR)
  Caenorhabditis elegans transcripts, 123
  gene structure prediction, 458–459
RH mapping. *See* Radiation hybrid mapping
Rice. *See also* Crop plants
  databases, 111–112
  gene functional assignments and resources, 102–103
  genome organization and sequence, 85, 101–102
Rice blast, fungal genomics studies, 61–61
RIP. *See* Repeat induced point mutation
RLN, human versus ape phenotypic differences, 370
RNA interference
  Arabidopsis thaliana studies, 78
  Caenorhabditis elegans studies, 128
  chicken studies, 227
  fungal genome evolution studies, 59, 59–60
  mouse, 337
RT-PCR. *See* Reverse transcriptase-polymerase chain reaction
Runx-2, dog morphology role, 271

S

Saccharomyces cerevisiae. *See* Yeast
SAGE. *See* Serial analysis of gene expression

Sanger sequencing
  automation, 417
  fluorescence detection, 420–424
  high-throughput sequencing, 417–418
  microfluidic separation platforms, 418–421
  principles, 415–417
SELEX, binding motif identification, 130
Self-organizing maps (SOMs), genome-scale data analysis, 25
Sequencing. *See* DNA sequencing
Serial analysis of gene expression (SAGE)
  Caenorhabditis elegans, 122, 125–126
  human sequence tag identification, 424
  yeast systems biology, 23
SGA. *See* Synthetic genetic array
SGP2, exon prediction, 455
Sheep
  comparative genomic mapping, 244
  genomics
    health and reproduction studies, 247–248
    prospects, 248–249
    status, 238–239, 241
Shigella, genome reduction, 5
Short interspersed elements (SINEs)
  chimpanzee versus human, 377
  mice versus humans, 322–323
  puffer fish, 182
SIGLEC, human versus ape phenotypic differences, 370–372, 378
SINEs. *See* Short interspersed elements
Single-nucleotide addition (SNA)
  fluorescence in situ sequencing, 427–428
  mice, 323–324, 353–354
  nucleobase mixture strategy, 428
  principles of DNA sequencing, 416
  pyrosequencing, 424–427
  single-pair fluorescence resonance energy transfer, 427
Single-nucleotide polymorphisms (SNPs)
  dbSNP database, 414
  dog genome, 267–268
  International HapMap project, 401, 403–404
  mouse genome

Single-nucleotide polymorphisms (SNPs) (*continued*)
  candidate gene testing, 353–354
  overview, 323–342
  rat genotyping, 302–303
Single-pair fluorescence resonance energy transfer, principles of DNA sequencing, 427
SNA. *See* Single nucleotide addition
SNPs. *See* Single-nucleotide polymorphisms
SOMs. *See* Self-organizing maps
Sorghum. *See also* Crop plants
  databases, 112
  gene functional assignments and resources, 104–105
  genome organization and sequence, 85, 104
  uses, 103–104
SOX2, chicken development studies, 230
SPANX, human versus ape phenotypic differences, 372, 376
Splice variants, fungi, 54–55
SPRY3, human versus ape phenotypic differences, 369–370
ST6GAL1, human versus ape phenotypic differences, 368
Staphylococcus, lateral gene transfer, 6
Streptococcus agalactiae, pan-genome, 7
SYBL1, human versus ape phenotypic differences, 370
Synthetic genetic array (SGA), genetic interaction analysis, 29

T

Takifugu rubripes
  advantages of study, 179
  genome features in draft sequences, 182
  genome sequencing, 180–182
  prospects for study, 189–192
Tbce, mutation and disease, 331–332
TCF1, human gene sequencing, 424–425
TCOF1, dog morphology role, 270
Tetraodon nigroviridis
  genome duplication studies, 185, 187
  genome features in draft sequences, 182
  genome sequencing, 181, 183
  human comparative genome analysis, 181

TILLING
  Arabidopsis thaliana studies, 78
  Caenorhabditis elegans studies, 129
  zebrafish studies, 190
Tinman, Ciona homolog, 160
Tnfsf4, quantitative trait loci mapping, 353
TRG, human versus ape phenotypic differences, 373
Tsc2, positional cloning in rat, 288
TTR, human versus ape phenotypic differences, 368
TWINSCAN, gene prediction, 54, 123, 225, 449–450

U

UCEs. *See* Ultra-conserved elements
UCRs. *See* Ultra-conserved regions
Ultra-conserved elements (UCEs), human genome, 401–403
Ultra-conserved regions (UCRs), chicken genome, 229
Unc5h3, positional cloning in rat, 288

V

Vaccines, genomics in development, 9
Vkorc1, positional cloning in rat, 288

W

WFDC, human versus ape phenotypic differences, 375
Whole genome shotgun sequencing, fungal genomes, 51
Wilson's disease, dog model, 269
WormBase, features, 133–134

X

Xenopus
  functional screens
    gain-of-function
      embryos, 201–203
      genes identified, 202
      in vitro systems, 203–205
      oocyte expression screens, 200–201
    loss-of-function, 205–207
    gene expression profiling, 210–213
    genome sequencing of *Xenopus tropicalis*, 199

whole-mount in situ hybridization
   screens
   cell fate mapping, 207
   prospects, 210
   synexpression groups, 208–209
   tissue relatedness concept, 208

Y

YAC. *See* Yeast artificial chromosome
Yeast. *See also* Fungi
   comparative genomics, 19–21
   genes
      abundance, 22
      annotation, 22–23
      databases, 22, 49
      genome sequencing history, 19–20
      interaction networks, 29
      prospects for study, 31, 33
      public resources for genomics, 31–32
   systems biology

   Gene Ontology, 25, 27
   gene expression studies, 23, 27
   genome-scale data analysis and
      display, 25
   genomics research goal, 21–22
   module identification, 23–25
   transcriptomics, 28–29
Yeast artificial chromosome (YAC),
   Caenorhabditis elegans genome
   sequencing, 118–120

Z

Zebrafish
   advantages of study, 179
   genome features in draft sequences,
      182, 199
   green fluorescence protein reporter
      constructs, 190–191
   ZF-MODEL consortium, 190
Zoo-FISH, livestock species, 243–244